# Adaptive Learning Methods for Nonlinear System Modeling

# Adaptive Learning Methods for Nonlinear System Modeling

Edited by

**Danilo Comminiello**

Department of Information Engineering, Electronics and Telecommunications,
Sapienza University of Rome, Italy
danilo.comminiello@uniroma1.it

**José C. Príncipe**

Department of Electrical and Computer Engineering,
University of Florida, Gainesville, FL, USA
principe@cnel.ufl.edu

ELSEVIER

Butterworth-Heinemann
An imprint of Elsevier

Butterworth-Heinemann is an imprint of Elsevier
The Boulevard, Langford Lane, Kidlington, Oxford OX5 1GB, United Kingdom
50 Hampshire Street, 5th Floor, Cambridge, MA 02139, United States



**Library of Congress Cataloging-in-Publication Data**
A catalog record for this book is available from the Library of Congress

**British Library Cataloguing-in-Publication Data**
A catalogue record for this book is available from the British Library

ISBN: 978-0-12-812976-0

For information on all Butterworth-Heinemann publications
visit our website at https://www.elsevier.com/books-and-journals



*Publisher:* Mara Conner
*Acquisition Editor:* Sonnini Ruiz Yura
*Editorial Project Manager:* Katie Chan
*Production Project Manager:* Anitha Sivaraj
*Designer:* Mark Rogers

Typeset by VTeX

# Contents

## PART 3 NONLINEAR MODELING WITH MULTIPLE LEARNING MACHINES

# Contributors

**Jerónimo Arenas-García**
Universidad Carlos III de Madrid, Madrid, Spain

**Luis A. Azpicueta-Ruiz**
Universidad Carlos III de Madrid, Madrid, Spain

**Theodore W. Berger**
University of Southern California, Los Angeles, CA, United States

**Pantelis Bouboulis**
National and Kapodistrian University of Athens, Athens, Greece

**Renato Candido**
Universidade de São Paulo, São Paulo, SP, Brasil

**Alberto Carini**
University of Urbino Carlo Bo, Urbino, Italy

**Stefania Cecchi**
Università Politecnica delle Marche, Ancona, Italy

**Badong Chen**
Xi'an Jiaotong University, Xi'an, China

**Jie Chen**
Northwestern Polytechnical University, Xi'an, China

**Symeon Chouvardas**
Capital Fund Management, Paris, France

**Danilo Comminiello**
Sapienza University of Rome, Rome, Italy

**Konstantinos Diamantaras**
TEI of Thessaloniki, Thessaloniki, Greece

**Georgios B. Giannakis**
University of Minnesota, Minneapolis, MN, United States

**Christian Hofmann**
Friedrich-Alexander-Universität Erlangen-Nürnberg, Erlangen, Germany

**Vassilis N. Ioannidis**
University of Minnesota, Minneapolis, MN, United States

**Walter Kellermann**
Friedrich-Alexander-Universität Erlangen-Nürnberg, Erlangen, Germany

**Bahare Kiumarsi**
University of Texas at Arlington Research Institute, Fort Worth, TX, United States

**Suleyman Serdar Kozat**
Bilkent University, Ankara, Turkey

**S.Y. Kung**
Princeton University, Princeton, NJ, United States

**Frank L. Lewis**
University of Texas at Arlington Research Institute, Fort Worth, TX, United States
Northeastern University, Shenyang, China

**Zhe Li**
Southeast University, Nanjing, China

**Derong Liu**
Guangdong University of Technology, Guangzhou, China

**Biao Luo**
Institute of Automation, Chinese Academy of Sciences, Beijing, China

**Meng Ma**
University of Minnesota, Minneapolis, MN, United States

**Danilo P. Mandic**
Imperial College London, London, United Kingdom

**Tomas McKelvey**
Chalmers University of Technology, Electrical Engineering, Gothenburg, Sweden

**Hamidreza Modares**
Missouri University of Science and Technology, Rolla, MO, United States

**Athanasios N. Nikolakopoulos**
University of Minnesota, Minneapolis, MN, United States

**Simone Orcioni**
Università Politecnica delle Marche, Ancona, Italy

**Raffaele Parisi**
Sapienza University of Rome, Rome, Italy

**José C. Príncipe**
University of Florida, Gainesville, FL, United States
Xi'an Jiaotong University, Xi'an, China

**Cédric Richard**
Université Côte d'Azur, Nice, France

**Brian S. Robinson**
University of Southern California, Los Angeles, CA, United States

**Daniel Romero**
University of Agder, Grimstad, Norway

**Simone Scardapane**
Sapienza University of Rome, Rome, Italy

**Michele Scarpiniti**
Sapienza University of Rome, Rome, Italy

**Magno T.M. Silva**
Universidade de São Paulo, São Paulo, SP, Brasil

**Dong Song**
University of Southern California, Los Angeles, CA, United States

**Sergios Theodoridis**
National and Kapodistrian University of Athens, Athens, Greece

**Aurelio Uncini**
Sapienza University of Rome, Rome, Italy

**Kyriakos G. Vamvoudakis**
Virginia Tech, Blacksburg, VA, United States

**Nuri Denizcan Vanli**
Massachusetts Institute of Technology, Cambridge, MA, United States

**Xin Wang**
Xi'an Jiaotong University, Xi'an, China

**Huai-Ning Wu**
Beihang University, Beijing, China

**Yili Xia**
Southeast University, Nanjing, China

**Min Xiang**
Imperial College London, London, United Kingdom

**Yinan Yu**
Chalmers University of Technology, Electrical Engineering, Gothenburg, Sweden

# Preface

This book introduces recent advances on adaptive algorithms and learning methods designed for nonlinear system modeling and identification.

With the recent technology progress, the role of online learning, and in particular of adaptive filtering, has given the possibility to develop effective nonlinear methods that are also computationally efficient and suited for real-life signal processing applications. This is the reason why the goal of this book is to provide a view on the recent methodologies that involve advanced theories aimed at the identification and modeling of several nonlinear systems with higher complexity. This book also aims to provide a view on recent examples of applications where such methods can be applied in order to understand how nonlinear modeling methodologies are evolving and/or changing, in order to imagine what the future perspective in this field are.

The book is intended as a research book and it is specially addressed to graduate students, PhD students and researchers both in science and engineering. The new trends of statistical learning applied to streaming data is becoming very important, and the preprocessing stage design can be accomplished with some of these models directly integrated with decision stages as classifiers. Engineers and practitioners shall also benefit from the reading of selected chapters, since they can find good ideas and subjects to deepen their knowledge with the goal of designing future projects. A possible pool of users for this book may be represented by those people who are interested in nonlinear problems, such as channel identification, wind prediction, vibration control, high-power amplifier modeling, loudspeaker modeling and data processing. The topics of this book are not covered in undergraduate courses, but can be found in postgraduate courses. The book material addresses topics in the cutting edge of the field rather than an extensive review on the subject. However, selected chapters can be used, where appropriate, as supplementary class material in degree courses.

Overall, this book benefits the reader who wants to know what are the most promising and recent adaptive methods in the field of nonlinear signal processing, who wants to understand what are the most significant challenges in modeling nonlinear systems, who wants to know what are the nonlinear systems that need to be addressed in recent applications, who wants to learn how to address a nonlinear modeling problem and how to develop an effective method for it and who wants to know what are the key trends and the future perspectives in the field of nonlinear signal processing.

Danilo Comminiello
José C. Príncipe

# Acknowledgments

We would like to sincerely thank all the authors who have contributed to this book. They quickly understood the scope of the book and its message and embodied and enhanced these goals in their chapters. Their interest and enthusiasm for this project are critical for the success of this initiative. We took in consideration their advice and suggestions and we appreciated a lot of their efforts in providing high-quality contributions. Without them, this book would never have been possible. We would like also to deeply thank Sonnini Ruiz Yura and Katie Chan from Elsevier for their work and assistance. We also thank all those people who are not directly involved in this book but provided us with precious and invaluable advice.

Danilo Comminiello
José C. Príncipe

CHAPTER 1

# INTRODUCTION

**Danilo Comminiello***, **José C. Príncipe**†

**Sapienza University of Rome, Rome, Italy †University of Florida, Gainesville, FL, United States*

## CONTENTS



## 1.1 NONLINEAR SYSTEM MODELING: BACKGROUND, MOTIVATION AND OPPORTUNITIES

### 1.1.1 MODELING AN UNKNOWN SYSTEM

In an engineering context, system modeling refers to the study and analysis of a particular real-world plant (or simply referred to hereinafter as "system"), whose characteristics are partially or completely unknown, and to the estimation of some of its behavioral properties, allowing to build a "model" capable of reproducing, predicting or simulating the same behavior of the analyzed plant that can then be utilized in a control framework (system identification) or simply as a predictor of future outputs (decision making).

In many real-world applications, the modeling of an unknown plant represents a key step to have a full understanding of the capabilities and behavior properties of such a plant or even of an entire industrial installation that includes the plant. Often, the knowledge of the system behavior is crucial for the prediction or the simulation of a result, depending on certain input data, on any external factor or on a change in the environment for decision making. Through the analysis of an unknown plant it may be possible to have full control over it and possibly over the full installation. This point is important, for example, for the quality monitoring of a certain working flow, rather than detecting anomalies, faults or criticalities. Analyzing an unknown plant can also be very important for the development of new processes. In fact, it is possible to exploit a system estimate to design a new model that has the task of properly correcting any abnormal or critical system behaviors when they occur.

In the current applications of statistical science to streaming data, the problem is very similar, except now the identification of the system that produces the data stream can be directly connected to decision devices such as classifiers that produce online, and sometimes even in real time, depending on the strings of decisions about the content of the streaming data flow.

**Adaptive Learning Methods for Nonlinear System Modeling. DOI: 10.1016/B978-0-12-812976-0.00001-4**

These properties imply a great demand for modeling techniques, which can even be very different, depending on the application area. For this reason, system modeling has always been the object of constant research, with a constant need to update modeling methods for technological advancements and for new and emerging issues. In that sense, this book aims to highlight the new requirements and the resulting methodological novelties in system modeling.

### 1.1.2 SUITABLE MODELS FOR REAL-WORLD PROBLEMS

In this section, we want to describe the problems that will be addressed in this book and the models that can be adopted to provide a feasible solution. To this end, we want to introduce a brief description of the classes of models that will be taken into account.

Unknown systems can be categorized according to their natural mapping function, *i.e.*, how they transform their inputs into outputs, based on which they can be defined as linear or nonlinear. *Linear systems* represent a complete and well-defined class where input–output relationships and derivation of parameters are uniquely determined by a theoretical derivation [1–5]. Alternatively, systems that do not satisfy the superposition principle of linear theory are classified as *nonlinear systems*, which complicates the theoretical foundations, but they still need to be further characterized, using techniques based on their characteristics. This classification of a system affects its modeling and, in particular, the choice and selection of a suitable model.

Very often, especially in the past, modeling of an unknown plant was performed by using methods derived from linear system theory, often regardless of the plant nature. The reasons for this choice can be found mainly in three important features:

- **Ease of design.** Linear system theory has been widely studied and it offers a wide range of system modeling solutions. Such solutions are often based on theoretic optimization and robustness that often allow a model estimate to describe, albeit with a certain degree of approximation, the unknown plant.
- **Ease of implementation.** Linear system theory has always been the subject of studies not only for performance improvement but also for computational efficiency. The computational constraint has always represented a constant issue in application studies that require real-time implementation. For this reason, a wide variety of alternative linear methods can be found that, besides offering a more or less approximate solution to the plant, also allow the required computational load to be reduced as much as possible.
- **Ease of handling.** Over the years, linear system theory has developed robust and reliable solutions. Very often, once properly initialized, a well-designed linear model cannot undergo major deterioration, thus the control of such systems is rather straightforward. In addition, the behavior of a linear model is often quite easy to predict during the testing phase, which further simplifies any corrective action on the model, when needed.

However, real-life problems always entail a certain degree of nonlinearity, which makes linear models a nonoptimal choice. The issue is that if the plant is nonlinear and the model adopted is linear, there will be always an error which is outside the control of the designer, so although the solution found is theoretically optimal, it may be practically useless. In order to evaluate the behavior of such systems in these conditions, it is necessary to adopt nonlinear models, whose choice is more complex than a linear one and requires further analysis of the system class to be modeled.

To this end, another distinctive feature of nonlinear systems to be modeled is represented by the availability of any *a priori* knowledge or physical detail that can be exploited to select a suitable model for the identification [2,3,6,5]. In that sense, when prior information is completely available it is possible to choose a model that perfectly fits the unknown system. In this case, the adopted model is also called a *white box*. When only some *a priori* information or physical insights are known, but several parameters remain to be estimated from input data, a *gray box* model can be used, which can exploit a well-defined structure. Finally, if no *a priori* information on the system is available a *black box* model must be chosen, which needs to show high capabilities in fitting an unknown solution. Modeling in this case is often based on behavioral approaches.

Real-world problems often require this last class of models, whose choice is made more complex by the nonlinear nature of a system. The complexity of selecting such models lies in the wide range of possible choices, which in many cases will be limited by a number of conditions, *e.g.*, concerning flexibility, approximation capabilities to the solution, computational complexity and handling capacity.

Recent advances on this kind of models are mainly introduced in this book for the modeling of nonlinear systems.

### 1.1.3 CHALLENGES AND OPPORTUNITIES IN INVESTIGATING NONLINEAR MODELS

As stated, the choice of employing nonlinear models has the drawback of losing simplicity and universality due to linear theory outlined above. However, from the point of view of the research and development such a choice entails countless opportunities, among which the main ones are discussed below:

1. Designing a nonlinear method can be much more complicated and risky than a linear method. While the linear system theory benefits from an extensive literature (see [1,4,5], among others), nonlinear systems are not characterized by a unified theory [7–13]. This is certainly a reason for encouraging the development of theories that can fit well-defined scenarios, and it is also the reason why the modeling of nonlinear systems has always been a very active and investigated field of research. Designing nonlinear models may show some difficulty and theoretical challenges, but it may lead to significant improvement in performance results, depending on the degree of nonlinearity introduced by the system to be estimated.
2. The implementation stage of a nonlinear model is another point that has always discouraged the use of nonlinear models, since these ones are almost always more complex than linear models and very often they require much higher computational load than the linear models. This is the case, for example, for those applications that require real-time processing, like acoustic echo cancellation, in which, over the years, linear methods have been preferred to nonlinear solutions due to much lower computational cost, often at the expense of performance and, therefore, of quality. However, recent years have seen a significant technological development that has led to greater availability of computational resources. This has allowed to reconsider many theories developed for nonlinear models, thus creating many advantageous opportunities for development.
3. In addition to the points above, linear models have often been preferred in the past due to their ease of use and reliability. We cannot say the same about nonlinear models that require further stability-driven efforts as well as a careful choice of an exact model. In fact, an inappropriate choice of the model, or even just of any configuration of it, can cause a loss of performance, even

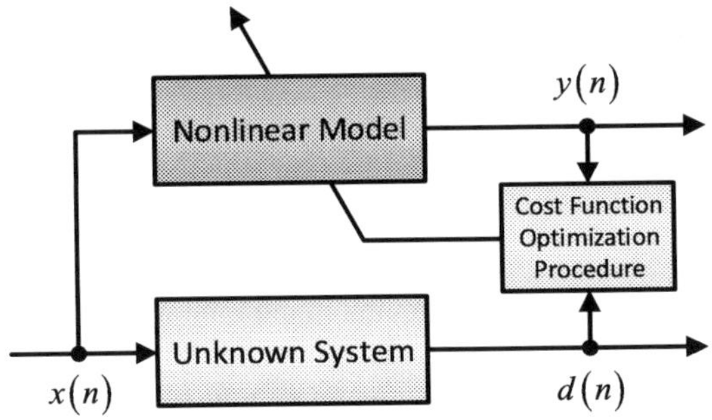


**FIGURE 1.1**

Adaptive nonlinear modeling of an unknown system.

worse than using approximate (*i.e.*, not optimally configured) linear systems. However, this point too represents a huge reason to investigate nonlinear models.

4. Emerging fields of application, including big data analysis, sensor data processing, internet of things and the 4.0 paradigm [14], exploit several concepts from nonlinear system identification to infer or define laws, relationships, dependencies or causal effects from large sets of data [12]. This inevitably leads to defining new problems and scenarios to be addressed and to creating new opportunities for studying and developing novel nonlinear modeling methodologies.

## 1.2 KEY FACTORS IN DEFINING ADAPTIVE LEARNING METHODS FOR NONLINEAR SYSTEM MODELING

Nonlinear system modeling is a very wide field of research. This book mainly focuses on those methodologies for nonlinear modeling that involve any adaptive learning approach to process data coming from an unknown nonlinear system. Such adaptive learning methods aim at estimating the nonlinearity introduced by the unknown system on the available data. This allows one to identify and model the unknown nonlinear system.

A typical scheme of the adaptive modeling of an unknown nonlinear system is shown in Fig. 1.1, where the system to be modeled receives an input $x(n)$ and generates a nonlinear output $d(n)$, possibly including any additive noise. A nonlinear model, receiving the same input $x(n)$ of the unknown system, is employed to "learn" the behavior of a system from its response and try to perform exactly like it. In order to estimate at best the behavior of the unknown system, a suitable choice of the model must be made. Moreover, the parameters of the chosen nonlinear model need to be optimized by an adaptive procedure, usually based on a cost function involving the response $d(n)$ of the unknown system and the nonlinear output $y(n)$ produced by the nonlinear model.

Adapting a model can depend on many factors, which can strongly affect its strategy. The knowledge of such information allows one to treat an unknown system as a gray box and therefore to provide a reliable modeling. Here we want to focus on key aspects that can affect the design of the learning strategy.

One of the first aspects that are evaluated in a model design is the *formulation of the problem* to be addressed, according to which adaptation models can be different. For example, different choices can be made for a problem of classification rather than a regression one, for identification rather than prediction, etc.

However, one of the most significant factors that determine the learning of a model are *input data*. Most of the modern theory in machine learning and signal processing is based on the *learning from data* paradigm (see [15,16], among others). According to data received by a model, a learning strategy is able to identify the desired solution within a set of hypotheses by minimizing a chosen cost function. Input data can be received from a single source or from multiple sources, even asynchronously and heterogeneously, as is often the case in the era of big data. In this sense, the amount of data to process is definitely crucial information for model design.

In addition to the amount of data, it is important to know the *rate* with which these data arrive at the model. Based on this information, input data can be analyzed altogether (*offline learning*) or progressively (*online learning*) (see for example [17,18]). This book will focus in particular on online learning methods for nonlinear modeling. Online learning can be performed in the following ways:

- *example-by-example learning* (or often simply *online learning*): for each iteration, only one new example is available at the model input;
- *batch learning*: a whole set of examples is available at each iteration;
- *mini-batch learning* (or *block-based learning*): at each iteration a subset of the possible examples is available.

A special case of online learning is *real-time learning*, which occurs when the iteration index coincides with the time index.

Another aspect that cannot be neglected concerns the *architecture* of the proposed method. Depending on the problem, the nonlinear modeling method can be implemented by a simple filter, or by a neural network, rather than by more complex structures of distributed and cooperating agents on networks or even on graphs. In this book, the recent proposals in this respect will be presented.

Finally, modeling a nonlinearity with an online approach allows us to consider a wide range of application situations in which *nonlinearity can vary* during iterations, showing nonstationary and dynamic behaviors. This is definitely one of the key aspects for choosing an online approach to modeling nonlinear systems. Furthermore, due to the recent technology progress and, in particular, to the always growing availability of computational resources and calculation tools, online learning strategies have given the possibility to develop effective nonlinear methods that are also computationally efficient and even suited for real-time data analysis and processing. This is the reason why this book aims to provide a view on the recent methodologies that show an adaptive learning theory aimed at the identification and modeling of several and complex nonlinear systems.

## 1.3 BOOK ORGANIZATION

This book contains 15 chapters, including this one. We aim at covering the most significant topics of the recent years among the field of nonlinear system modeling in the next 14 chapters, divided in four parts.

## Part 1

Part 1 is composed of three chapters (2–4) and it deals with linear-in-the-parameter (LIP) nonlinear filters. This class of nonlinear filters won popularity in the past with adaptive Volterra filters, but it is again receiving increasing interest in recent years due to the possibility of combining the strength of complex nonlinear models with the flexibility of linear adaptive filters.

**Chapter 2**, by Alberto Carini *et al.*, provides an exhaustive overview on LIP nonlinear filters, including the theoretical aspects that contributed to making these methods effective and reliable. This class of filters includes several kinds of models and it is presented under a unified framework. In particular, the chapter introduces the orthogonal LIP nonlinear filters, which leverage the orthogonality of basis functions to improve the robustness of the model. In order to guarantee the orthogonality of the basis functions, the development and use of perfect periodic sequences, together with the multiple-variance identification method, is also introduced. Finally, the chapter proves by experimental results that this kind of filters is capable of providing fast convergence of gradient descent adaptation algorithms and efficient identification of the nonlinear systems.

**Chapter 3**, by Michele Scarpiniti *et al.*, presents an overview of a family of LIP nonlinear filters, known as Spline Adaptive Filters (SAFs), consisting of a cascade of a linear adaptive filter and a flexible memoryless nonlinear spline function. The peculiarity of SAFs is that the shape of the spline function can be modified during learning, by including the spline control points in the overall adaptive optimization procedure. SAF mainly uses B-splines and Catmull–Rom splines, since they allow one to impose simple constraints on control parameters. Due to the flexibility and ease of implementation, SAFs can be applied to several real-world nonlinear problems and some examples can be found in this chapter.

**Chapter 4**, by Christian Hofmann *et al.*, introduces and reviews the Significance-Aware (SA) filtering concept, which aims at decomposing the model for an unknown system and its adaptation mechanism into synergetic subsystems to achieve high computational efficiency. The decomposition can be performed according to serial or parallel strategies. The SA approach perfectly fits with LIP nonlinear filters, as proved in this chapter. Among LIP filters, cascade models, cascade group models and bilinear cascade models have been taken into account using both time-domain and frequency-domain adaptive algorithms. The derived models have been applied to the challenging problem of nonlinear acoustic echo cancellation.

## Part 2

Part 2 is composed of four chapters (5–8) and it presents some of the most significant recent advances on nonlinear modeling by using adaptive methods based on kernel functions. This family of methods is closely related to the concept of LIP nonlinear filters, since it involves online kernel learning methods, which are implemented as linear learning methods with input data being transformed into a high-dimensional feature space. The difference with respect to the methods presented in Part 1 mainly lies in the feature space, which is induced by applying Mercer's theorem [19]. This property entails a strong and consolidated theory with several desirable features for nonlinear modeling, which have given rise, in the recent years, to an extensive literature on these models. This is the reason why we devote an entire part of the book to these methods.

One of most popular kernel-based methods is represented by the kernel adaptive filter (KAF). Most of the existing KAFs are obtained by minimizing the well-known mean square error (MSE), which is computationally simple and very easy to implement. However, the MSE may show a lack of robustness in the presence of non-Gaussian noises. **Chapter 5**, by Badong Chen *et al.*, addresses this problem by leveraging the correntropy, which is a similarity measure involving all even-order moments, thus being robust to the presence of heavy-tailed non-Gaussian noises. The chapter shows that KAFs based on the maximum correntropy criterion (MCC) provide an effective solution for the modeling of those nonlinearities that bare nonstationarity and non-Gaussianity properties.

Kernel-based methods are also widely used in machine learning applications. One example is described in **Chapter 6**, by Yinan Yu *et al.*, where kernel approximation is represented as a subspace learning technique for feature construction for pattern recognition. Kernel approximation aims at reducing the computational complexity, which belongs to the most challenging issue for kernels, since complexity does not scale well with respect to the training size. Considering the kernel approximation as an approximated subspace in a high-dimensional space, it is possible to extract feature vectors onto the approximated subspace, thus subsequently applying a linear learning technique. The chapter also introduces possible learning criteria for adaptive kernel approximation and suitable infrastructures for an efficient implementation.

The problem of the growing size of the kernels is addressed for a different field of applications in **Chapter 7**, by Pantelis Bouboulis *et al.* In particular, the chapter introduces an approach based on random Fourier features to approximate kernels without compromising performance. The approximation procedure of random Fourier features is applied to several online kernel-based schemes. More importantly, this rationale has also been applied to learning over distributed networks, in which several nodes collect the observations in a sequential fashion and each node communicates with a subset of nodes to reach a consensus for the overall solution.

The problem of inferring a function defined over the nodes of a network is also addressed in **Chapter 8**, by Vassilis N. Ioannidis *et al.* Nodal attributes can be represented as signals that take values over the vertices of the underlying graph, thus allowing the associated inference tasks to leverage node dependencies captured by a graph structure. This chapter presents a unifying framework for tackling signal reconstruction problems over graphs by using kernel-based learning both in the traditional time-invariant, as well as in the more challenging time-varying setting. Furthermore, a flexible model is also introduced based on online multikernel learning (MKL) that selects an optimal combination of kernels "on-the-fly".

## Part 3

Part 3 contains three chapters (9–11) and deals with algorithms and architectures for the modeling of complex nonlinear systems. Very often in real-world problems, one model is not sufficient to estimate an unknown nonlinear environment, due to the complexity of the nonlinearity or often to its variability during the modeling process, which may require a change of the parameter setting of the chosen model or even the change of the adopted nonlinear model. This leads to the use of complex architectures composed of different learning machines, which can even partially contribute to the overall nonlinear modeling. Complex architectures provide performance improvement, but also some issues like overfitting, gradient noise introduction and weak stability, among others. In this part, some recent advances on nonlinear modeling with multiple learning machines are presented, with the goal of achieving accurate nonlinear modeling while avoiding unwanted issues.

**Chapter 9**, by Nuri Denizcan Vanli *et al.*, introduces a solution for overcoming issues like overfitting, stability and convergence, which can affect nonlinear methods. To this end, self-organizing trees (SOTs) are proposed to be used for regression and classification algorithms. SOTs adaptively partition the feature space into small regions and combine simple local learners defined in such regions. The cumulative loss is minimized by learning both the partitioning of the feature space and the local learners. The overall output of this architecture is achieved by a particular combination of the partition outputs. The proposed method can be applied to different constructions of the tree in order to solve a wide range of complex nonlinear problems, *e.g.*, characterized by nonstationary environments, big data inputs, chaotic signals and very complex nonlinear functions.

A particular cause of complexity in nonlinear modeling may be represented by problems in distributed environments, *e.g.*, wireless sensor networks and peer-to-peer networks. In such problems, a number of agents are required to estimate a common nonlinear model, based on received input data, by combining local computations and information coming from neighboring agents. **Chapter 10**, by Simone Scardapane *et al.*, addresses this kind of problems, which can be distinguished in single-task problems, characterized by a common model for all the agents, and multitask problems, where each agent might converge to a slightly different model due to some specific factors. Several nonlinear adaptive methods can be used for this kind of architectures, and in this chapter an example is presented involving a kernel-based algorithm for the multitask case.

Nonlinear methods are often difficult to be tuned over an unknown nonlinearity. Moreover, complex nonlinearities, *e.g.*, varying in time, can create even more uncertainty in the nonlinear model to be adopted. In order to tackle this problem, **Chapter 11**, by Luis A. Azpicueta-Ruiz *et al.*, introduces combined filtering architectures for nonlinear modeling. One of the most useful approaches involves the combination of filters from the same family but having different parameter setting, which aims at simplifying the parameter selection and/or at improving the overall performance. A second combination approach is also presented in which, given a nonlinear architecture with different constituents (*e.g.*, filters, filter partitions or general learning machines), only relevant nonlinear elements are adaptively chosen by optimizing the modeling performance. Several nonlinear methods can be adopted to be used in the presented combined schemes and some examples are shown in this chapter.

## Part 4

Part 4 presents four chapters (12–15) dealing with nonlinear modeling by neural processing systems. Neural methods are powerful tools to model several kinds of nonlinearities that can hardly be modeled by other nonlinear methods. One example is represented by dynamical nonlinearities, which often occur in real-world applications. In this book, we show that neural-based models can be applied and used in different systems and for very different applications. In particular, neural systems are here used for nonlinear modeling problems for multidimensional data processing, for the behavioral analysis of spiking activities and, in the last two chapters, for adaptive control systems.

**Chapter 12**, by Yili Xia *et al.*, deals with the processing of real-world multidimensional data, typically 3D or 4D, such as measurements from body sensor networks or from ultrasonic anemometers, which often entail nonlinear and nonstationary behavior. Since the quaternion domain offers a convenient means to process this kind of data, this chapter proposes quaternion-valued learning methods for the nonlinear modeling of multidimensional data. In particular, quaternion-valued echo state networks (QESNs) are introduced based on quaternion nonlinear activation functions with local analytic properties. QESNs provide efficient computational load for training and favorable analytical conditions in developing full quaternion-valued nonlinearities. Moreover, the proposed method is proved to be second-order optimal for the generality of quaternion signals. The method is evaluated with both simulated and real-world scenarios, including 3D body motion tracking and wind forecasting.

**Chapter 13**, by Dong Song *et al.*, addresses the problem of identifying short-term and long-term synaptic plasticity from spiking activities. In particular, the chapter proposes a nonstationary dynamical modeling approach in which the synaptic strength is represented as input–output dynamics between neurons, the short-term synaptic plasticity is defined as input–output nonlinear dynamics, the long-term synaptic plasticity is formulated as the nonstationarity of such nonlinear dynamics, and the synaptic learning rule is the function governing the characteristics of the long-term synaptic plasticity based on the input–output spiking patterns.

In **Chapter 14**, by Hamidreza Modares *et al.*, neural networks are used for the implementation of reinforcement learning algorithms for the adaptive $H_{\text{inf}}$ control of nonlinear systems. In particular, this chapter introduces an adaptive optimal controller for nonlinear continuous-time systems without any knowledge about the system dynamics or the command generator dynamics. First, an augmented system is constructed from the tracking error dynamics and the command generator dynamics to introduce a new performance function. Then a reinforcement learning approach is used to solve the optimal tracking control problem. The proposed method offers several possibilities to be employed in real-world applications, including mobile robots, unmanned aerial vehicles and power systems.

The optimal adaptive control problem is also addressed in **Chapter 15**, by Biao Luo *et al.*, but with reference to distributed parameter systems (DPSs). The resulting problem is described by highly dissipative nonlinear partial differential equations. The proposed method exploits empirical eigenfunctions of the DPS and the singular perturbation technique to obtain a finite-dimensional slow subsystem of ordinary differential equations that accurately describes the dominant dynamics of the DPS. Then the optimal control problem is reformulated on the basis of the slow subsystem that leads to solve a Hamilton–Jacobi–Bellman equation by using adaptive dynamic programming methods based on neural networks. An adaptive optimal control method based on policy iteration is developed for partially unknown DPSs.

## 1.4 FURTHER READINGS

Topics included in this book, described in the previous section, aim to provide an overview on the recent advances about adaptive learning methods for nonlinear system modeling. However, several noteworthy works on this topic have been populating the literature of the very last years. Among these works, we want to mention some related to the following aspects of nonlinear modeling: LIP nonlinear filters [20–24], time series modeling [25–28], Bayesian learning methods [29,30], kernel-based methods [31–36], nonlinear filtering architectures [37–41], nonlinear dynamic systems [42–44] and nonlinear system control [45–49].

## REFERENCES

[1] T. Kailath, Linear Systems, Information and System Science, Prentice-Hall, 1980.
[2] T. Söderström, P. Stoica, System Identification, Prentice-Hall, London, 1989.
[3] L. Ljung, T. Glad, Modeling of Dynamic Systems, Prentice-Hall, Englewood Cliffs, NJ, 1994.
[4] W.J. Rugh, Linear System Theory, 2nd edition, Information and System Sciences, Prentice-Hall, 1996.
[5] L. Ljung, System Identification: Theory for the Users, Information and System Sciences, Prentice-Hall, 1999.
[6] A. Juditsky, H. Hjalmarsson, A. Benveniste, B. Delyon, L. Ljung, J. Sjöberg, Q. Zhang, Nonlinear black-box models in system identification: mathematical foundations, Automatica 31 (12) (1995) 1725–1750.

[7] V.J. Mathews, G.L. Sicuranza, Polynomial Signal Processing, John Wiley & Sons, Inc., New York, NY, 2000.

[8] O. Nelles, Nonlinear System Identification: From Classical Approaches to Neural Networks and Fuzzy Models, Springer-Verlag, Berlin–Heidelberg, 2001.

[9] S. Marshall, G.L. Sicuranza (Eds.), Advances in Nonlinear Signal and Image Processing, Signal Processing and Communications, Hindawi Publishing Corporation, 2006.

[10] T. Ogunfunmi, Adaptive Nonlinear System Identification: The Volterra and Wiener Model Approaches, Signals and Communication Technology, Springer-Varlag, Boston, MA, 2007.

[11] W.H. Haddad, V.S. Chellaboina, Nonlinear Dynamical Systems and Control: A Lyapunov-Based Approach, Princeton University Press, February 2008.

[12] S.A. Billings, Nonlinear System Identification: NARMAX Methods in the Time, Frequency, and Spatio-Temporal Domains, John Wiley & Sons, Inc., Chichester, West Sussex, United Kingdom, September 2013.

[13] H.K. Khalil, Nonlinear Systems, Pearson, 2014.

[14] T. Hey, S. Tansley, K. Tolle (Eds.), The Fourth Paradigm: Data-Intensive Scientific Discovery, Microsoft Corporation, Redmond, WA, October 2009.

[15] T. Hastie, R. Tibshirani, J. Friedman, The Elements of Statistical Learning: Data Mining, Inference, and Prediction, 2nd edition, Statistics, Springer-Verlag, New York, NY, 2009.

[16] S. Theodoridis, Machine Learning: A Bayesian and Optimization Perspective, Elsevier, 2015.

[17] A.H. Sayed, Adaptive Filters, John Wiley & Sons, Hoboken, NJ, 2008.

[18] A. Uncini, Fundamentals of Adaptive Signal Processing, Signal and Communication Technology, Springer International Publishing AG, Cham, Switzerland, ISBN 978-3-319-02806-4, 2015.

[19] W. Liu, J.C. Príncipe, S. Haykin, Kernel Adaptive Filtering: A Comprehensive Introduction, John Wiley & Sons, Inc., 2010.

[20] A. Carini, G.L. Sicuranza, J.V. Mathews, Efficient adaptive identification of linear-in-the-parameters nonlinear filters using periodic input sequences, Signal Processing 93 (5) (May 2013) 1210–1220.

[21] D. Comminiello, M. Scarpiniti, L.A. Azpicueta-Ruiz, J. Arenas-García, A. Uncini, Functional link adaptive filters for nonlinear acoustic echo cancellation, IEEE Transactions on Audio, Speech, and Language Processing 21 (7) (July 2013) 1502–1512.

[22] P. Shah, I. Lewis, S. Grant, S. Angrignon, Nonlinear acoustic echo cancellation using voltage and current feedback, IEEE/ACM Transactions on Audio, Speech and Language Processing 23 (10) (2015) 1589–1599.

[23] Y. Mao, F. Ding, A. Alsaedi, T. Hayat, Adaptive filtering parameter estimation algorithms for Hammerstein nonlinear systems, Signal Processing 128 (2016) 417–425.

[24] V. Patel, V. Gandhi, S. Heda, N.V. George, Design of adaptive exponential functional link network-based nonlinear filters, IEEE Transactions on Circuits and Systems I: Regular Papers 63 (9) (September 2016) 1434–3442.

[25] C. Sheng, J. Zhao, W. Wang, H. Leung, Prediction intervals for a noisy nonlinear time series based on a bootstrapping reservoir computing network ensemble, IEEE Transactions on Neural Networks and Learning Systems 24 (7) (July 2013) 1036–1048.

[26] P. Zhu, B. Chen, J.C. Príncipe, Learning nonlinear generative models of time series with a Kalman filter in RKHS, IEEE transactions on Signal Processing 62 (1) (January 2014) 141–155.

[27] M. Gan, C.L. Philip Chen, H.-X. Li, L. Chen, Gradient radial basis function based varying-coefficient autoregressive model for nonlinear and nonstationary time series, IEEE Signal Processing Letters 22 (7) (July 2015) 809–812.

[28] Q. Han, J. Ding, E.M. Airoldi, V. Tarokh, SLANTS: sequential adaptive nonlinear modeling of time series, IEEE Transactions on Signal Processing 65 (19) (October 2017) 4994–5005.

[29] S. Malik, G. Enzner, A variational Bayesian learning approach for nonlinear acoustic echo control, IEEE Transactions on Signal Processing 61 (23) (December 2013) 5853–5867.

[30] F.R. Ávila, H.T. Carvalho, L.W.P. Biscainho, Bayesian blind identification of nonlinear distortion with memory for audio applications, IEEE Signal Processing Letters 23 (4) (February 2016) 414–418.

[31] D. Tuia, J. Muñoz Marí, J.L. Rojo-Álvarez, M. Martínez-Ramón, G. Camps-Valls, Explicit recursive and adaptive filtering in reproducing kernel Hilbert spaces, IEEE Transactions on Neural Networks and Learning Systems 25 (7) (July 2014) 1413–1419.

[32] M. Takizawa, M. Yukawa, Adaptive nonlinear estimation based on parallel projection along affine subspaces in reproducing kernel Hilbert space, IEEE Transactions on Signal Processing 63 (16) (August 2015) 4257–4269.

[33] S. Scardapane, D. Comminiello, M. Scarpiniti, A. Uncini, Online sequential extreme learning machine with kernels, IEEE Transactions on Neural Networks and Learning Systems 26 (9) (September 2015) 2214–2220.

[34] F. Tobar, P.M. Djurić, D.P. Mandic, Unsupervised state-space modeling using reproducing kernels, IEEE Transactions on Signal Processing 63 (19) (October 2015) 5210–5221.

[35] B. Chen, L. Xing, B. Xu, H. Zhao, N. Zheng, J.C. Príncipe, Kernel risk-sensitive loss: definition, properties and application to robust adaptive filtering, IEEE Transactions on Signal Processing 65 (11) (June 2017) 2888–2901.

[36] K. Li, J.C. Príncipe, Transfer learning in adaptive filters: the nearest instance centroid-estimation kernel least-mean-square algorithm, IEEE Transactions on Signal Processing 65 (24) (December 2017) 6520–6535.

[37] D. Comminiello, M. Scarpiniti, S. Scardapane, R. Parisi, A. Uncini, Improving nonlinear modeling capabilities of functional link adaptive filters, Neural Networks 69 (2015) 51–59.

[38] M. Scarpiniti, D. Comminiello, R. Parisi, A. Uncini, Novel cascade spline architectures for the identification of nonlinear systems, IEEE Transactions on Circuits and Systems I: Regular Papers 62 (7) (July 2015) 1825–1835.

[39] G. Battistelli, L. Chisci, C. Fantacci, Parallel consensus on likelihoods and priors for networked nonlinear filtering, IEEE Signal Processing Letters 21 (7) (July 2014) 787–791.

[40] A.K. Sahu, S. Kar, J.M.F. Moura, V.H. Poor, Distributed constrained recursive nonlinear least-squares estimation: algorithms and asymptotics, IEEE Transactions on Signal and Information Processing over Networks 2 (4) (October 2016) 426–441.

[41] D. Comminiello, M. Scarpiniti, L.A. Azpicueta-Ruiz, J. Arenas-García, A. Uncini, Combined nonlinear filtering architectures involving sparse functional link adaptive filters, Signal Processing 135 (2017) 168–178.

[42] B.-S. Chen, C.-F. Wu, Robust scheduling filter design for a class of nonlinear stochastic Poisson signal systems, IEEE Transactions on Signal Processing 63 (23) (December 2015) 6245–6257.

[43] K. Geng, V.Z. Marmarelis, Methodology of recurrent Laguerre–Volterra network for modeling nonlinear dynamic systems, IEEE Transactions on Neural Networks and Learning Systems 28 (9) (September 2017) 2196–2208.

[44] Y. Zhang, T. Chai, D. Wang, An alternating identification algorithm for a class of nonlinear dynamical systems, IEEE Transactions on Neural Networks and Learning Systems 28 (7) (July 2017) 1606–1617.

[45] M. Chen, G. Tao, B. Jiang, Dynamic surface control using neural networks for a class of uncertain nonlinear systems with input saturation, IEEE Transactions on Neural Networks and Learning Systems 26 (9) (September 2015) 2086–2097.

[46] Y.-J. Liu, L. Tang, S. Tong, C.L. Philip Chen, D.-J. Li, Reinforcement learning design-based adaptive tracking control with less learning parameters for nonlinear discrete-time MIMO systems, IEEE Transactions on Neural Networks and Learning Systems 26 (1) (January 2015) 165–176.

[47] Y. Li, S. Tong, T. Li, Hybrid fuzzy adaptive output feedback control design for uncertain MIMO nonlinear systems with time-varying delays and input saturation, IEEE Transactions on Fuzzy Systems 24 (4) (August 2016) 841–853.

[48] X. Yang, H. He, Self-learning robust optimal control for continuous-time nonlinear systems with mismatched disturbances, Neural Networks 99 (March 2018) 19–30.

[49] H. Gao, Y. Song, C. Wen, Backstepping design of adaptive neural fault-tolerant control for mimo nonlinear systems, IEEE Transactions on Neural Networks and Learning Systems 28 (11) (November 2017) 2605–2613.

PART

# 1

# LINEAR-IN-THE-PARAMETERS NONLINEAR FILTERS

CHAPTER

# ORTHOGONAL LIP NONLINEAR FILTERS

# 2

**Alberto Carini***, **Stefania Cecchi**†, **Simone Orcioni**†

*University of Urbino Carlo Bo, DiSPeA, Urbino, Italy †Università Politecnica delle Marche, DII, Ancona, Italy*

## CONTENTS



Adaptive Learning Methods for Nonlinear System Modeling. DOI: 10.1016/B978-0-12-812976-0.00003-8

**CHAPTER POINTS**

- The chapter provides an overview of orthogonal linear-in-the-parameters (LIP) nonlinear filters.
- It presents different classes of orthogonal LIP nonlinear filters, under the unified framework of functional link polynomial (FLiP) filters.
- It discusses how perfect periodic sequences (PPSs) for FLiP filters can be developed and how they can be used for system identification using the cross-correlation method.
- Multiple-variance identification techniques, which allow one to contrast the problem of locality of the solution, are also discussed.

## 2.1 INTRODUCTION

The history of linear-in-the-parameters (LIP) nonlinear filters starts at the end of the 19th century, when the Italian mathematician Vito Volterra developed his *functional power series*, later called Volterra series, as an extension to functionals of the Taylor series expansion [1,2]. In 1910 Fréchet showed that a sum of *homogeneous functionals of degree n* can approximate, arbitrarily well, any continuous real-valued functional, defined on compact sets as, for example, the set of functions defined over intervals [3]. Fréchet obtained his results as a generalization of the Weierstrass theorem, which states that the set of polynomial functions is complete. Fréchet showed in a similar manner that the set of Volterra functionals is complete.

The restriction to a compact domain is an important limitation when the input signals are defined over an infinite (or semiinfinite) time interval (*i.e.*, periodic stimuli). Years later, this gave rise to many works on the so-called fading memory systems [4,5]. However, this is not a limitation in practical cases when most systems are implemented in the form of discrete-time, finite-memory filters.

A key factor for spreading the use of polynomial series was the work of Wiener on Brownian motion linear transformation, on higher degree functionals of Brownian motion and on their orthogonalization [6], published during his years at the MIT Research Laboratory of Electronics. Along with the work of his colleagues [7], Wiener continued to develop his theories by combining Volterra series and his previous works, finally defining the so called Wiener series.

In real applications, the implementation of Wiener's method was difficult due to the use of Brownian motion. A further step in the usability of Volterra and Wiener series was due to the work of Lee and Schetzen [8], who developed a method that uses white Gaussian noise (WGN) as inputs and cross-correlation for parameter estimation of Wiener series. This was possible because, differently from what happens in the Volterra series, the Wiener functionals of any order are orthogonal to each other when the input is a WGN. Exploiting the relationship between Wiener and Volterra coefficients, it is also possible to obtain an equivalent Volterra representation of Wiener series. A rigorous proof on the range of validity of this method for continuous-time time-invariant nonlinear systems can be found in [9]. The same paper showed unavoidable difficulties in the continuous-time identification of terms with at least two equal indices, called diagonal elements, of kernels of order greater than or equal to two. Fortunately, these theoretical difficulties disappear in most real applications concerning discrete-time systems [10], but a great inaccuracy in the identification of diagonal elements still remains [11]. Effective solutions that overcome this problem have been proposed in the literature, for series up to the third order in [12] and for a generic order in [13], where a comparison between the two methods is also provided.

While in analytical power series (infinite sum of elements) the identification with cross-correlation is independent of the input variance, this is no more true with truncated power series, where the approximation error depends on the variance used in the identification: a Volterra series is optimal only for inputs with variances in a neighborhood of that used for identification [14]. An improved cross-correlation method for nonlinear system identification based on multiple variances has been proposed for Wiener–Volterra series and Gaussian noise in [14]: low input variances are used to identify lower-order Wiener kernels, while the input variance is gradually increased for higher-order kernels.

The truncated Volterra filter, which derives from a double truncation with respect to the order and memory of the Volterra series, has been one of the first LIP nonlinear filters to be used in theory and applications. It is still today one of the most popular LIP filters. Its success derives from (i) the simplicity of the LIP input–output relationship and (ii) the property of being a universal approximator according to the Stone–Weierstrass theorem. On the other hand, the main limitations of Volterra filters are (i) the very large number of coefficients involved, which increases exponentially with the filter order and geometrically with the memory length, and (ii) the difficulties in coefficient estimation. For example, the gradient descent adaptation algorithms provide for any input signal slow convergence speed, mainly due to the poor conditioning of the input autocorrelation matrix. The algorithms used to overcome these convergence problems have often a high computational complexity.

It is possible to facilitate the nonlinear system identification with the use of Wiener nonlinear (WN) filters, which derive from the double truncation of the Wiener series. The WN filter is a first example of nonlinear filters with orthogonal basis functions, in particular, orthogonal for a white Gaussian input signal. The orthogonality of the basis functions allows the efficient identification of the filter coefficients with the cross-correlation method, as in [8]. Moreover, the orthogonality guarantees fast convergence of gradient descent algorithms. Deterministic periodic signals, called perfect periodic sequence (PPS), that guarantee the orthogonality of the basis functions on a finite period can also be developed. Expressing the WN filter as a linear combination of basis functions and using a PPS input signal, problems in the estimation of the kernel diagonal points can be avoided [15].

To reduce the model complexity, different block structured models have been proposed in the literature. They consist of a nonlinear memoryless block preceded or followed by a linear system and can often be considered subclasses of the Volterra or Wiener model. To this class belong the Hammerstein filters [16–20], the memory polynomial filters [21,22] and functional link artificial neural networks (FLANNs) based on polynomial expansions [23–30]. All these filters involve nonlinearities of the input samples at the same time instant, without the so-called cross-terms, *i.e.*, products of samples at different instants.

Block structured models based on nonpolynomial functions have also been proposed, such as FLANNs based on orthogonal trigonometric expansions of the input samples [23] and the radial basis function networks [31].

Among the block structured models involving also cross-terms, the generalized memory polynomial filters [22], the Wiener systems, the Hammerstein–Wiener model and generalized FLANN filters [32] can be mentioned.

Many of the block structured models previously considered are LIP filters. Very few of them, *e.g.*, the FLANN filters based on orthogonal expansions, have orthogonal basis functions. In most cases, the reduction in the computational complexity obtained from the use of a block structure is paid with the loss of the property of universal approximation. Thus, the block structure models cannot arbitrarily well approximate any nonlinear system.

Recently, different families of LIP nonlinear filters have been proposed considering a set of basis functions that can arbitrarily well approximate the $N$-dimensional nonlinear function defining a nonlinear system of memory $N$. These LIP filters are universal approximators for discrete-time, causal, time-invariant, finite-memory, continuous nonlinear systems and have orthogonal basis functions for appropriate distributions of the input signal. To this class of LIP nonlinear filters belong the Fourier nonlinear (FN) filters [33], the even mirror FN (EMFN) filters [33,34], the Legendre nonlinear (LN) filters [35,36] and the Chebyshev nonlinear (CN) filters [37].

It is also possible to develop perfect periodic sequences (PPSs) that guarantee the orthogonality of the basis functions of EMFN, LN and CN filters on a finite period [38–40,36,37]. Using PPSs as input signals, the filter coefficients can be estimated with the cross-correlation method. It is worth noting that also adaptive identification algorithms can benefit from the use of PPS inputs since the orthogonality of the basis functions guarantees a fast convergence of gradient descent algorithms [32].

Volterra, WN, EMFN, LN and CN filters belong to the general class of functional link polynomial (FLiP) filters [41], which include also many of the block structured models previously mentioned.

In what follows, this chapter aims to:

- review the theory of the principal LIP nonlinear filters that are universal approximators according to the Stone–Weierstrass theorem;
- present Volterra, WN, EMFN, LN and CN filters under the unified framework of FLiP filters;
- provide an overview of orthogonal LIP nonlinear filters and in particular of orthogonal FLiP filters;
- discuss some recent identification techniques for orthogonal LIP nonlinear filters.

The rest of the chapter is organized as follows. Section 2.2 provides a review of LIP nonlinear filters, of the Stone–Weierstrass theorem and of the Volterra and Wiener theory, discusses FLiP filters and orthogonal FLiP filters and provides simplified structures that can be used to reduce the computational complexity of the filters. Classical identification methods, like minimum mean square estimation and the cross-correlation method, are also introduced in this section. Section 2.3 presents an overview of new identification techniques that can be applied to orthogonal LIP nonlinear filters. Specifically, identification based on PPSs and the multiple-variance method is discussed. Section 2.4 provides experimental results that illustrate the ability of the discussed nonlinear filters and identification techniques to model the nonlinearities encountered in real nonlinear systems. Concluding remarks are provided in Section 2.5.

The following notation is used throughout the chapter. Intervals are represented with square brackets, $\mathbb{N}$ is the set of natural numbers, $\mathbb{R}$ is the set of real numbers, $\mathbb{R}^+$ is the set of positive real numbers, $\mathbb{R}_1$ is the unit interval $[-1,+1]$, $< x(n) >_L$ indicates time average over $L$ successive samples of $x(n)$, $\binom{n}{m}$ denotes the number of combinations of $n$ things taken $m$ at a time, $\delta_{ij}$ is equal to 1 when $i = j$ and is equal to 0 otherwise, $E(\cdot)$ indicates mathematical expectation, $\mathcal{N}(m_x, \sigma_x^2)$ is the normal distribution with mean $m_x$ and variance $\sigma_x^2$ and $(Hx)(n)$ is an operator acting on the input sequence. Sets are represented with curly brackets and intervals with square brackets, while the following convention for brackets $\{[(\ldots)]\}$ is used elsewhere.

## 2.2 LIP NONLINEAR FILTERS

In this section, we first discuss the approximation of nonlinear filters using LIP nonlinear filters and we review the Stone–Weierstrass theorem reporting the classical Volterra and Wiener theories. Then we introduce the class of FLiPs, which includes many families of nonlinear filters currently used in common practice. Some families of orthogonal FLiP filters are also discussed. Simplified structures for an efficient implementation of FLiP filters are eventually presented.

### 2.2.1 NONLINEAR FILTERS AND STONE–WEIERSTRASS THEOREM

We are interested in developing families of LIP nonlinear filters capable of arbitrarily well approximating any discrete-time, time-invariant, finite-memory, causal, continuous nonlinear system. The input–output relationship of the nonlinear system can be expressed in the following form:

$$y(n) = f[x(n), x(n-1), \ldots, x(n-N+1)], \tag{2.1}$$

where $x(n)$ is the input signal, which is assumed to belong to the unit interval, *i.e.*, $x(n) \in \mathbb{R}_1$, $y(n) \in \mathbb{R}$ is the output signal and $N$ is the system memory length. The function $f$ in Eq. (2.1) can be interpreted as a multidimensional function over the space $\mathbb{R}_1^N$. Each delayed input sample corresponds to a different dimension. Taking into account this notation, it is possible to expand the function $f$ in Eq. (2.1) with a series of basis functions $f_i$,

$$f[x(n), x(n-1), \ldots, x(n-N+1)] = \sum_{i=1}^{+\infty} c_i f_i[x(n), x(n-1), \ldots, x(n-N+1)], \tag{2.2}$$

where $c_i \in \mathbb{R}$ and each $f_i$ is a continuous function from $\mathbb{R}_1^N$ to $\mathbb{R}$. Every choice of the set of basis functions $f_i$ leads to a different family of LIP nonlinear filters, which can be used to identify or to approximate the nonlinear systems in Eq. (2.1). For compactness, in what follows $f_i[x(n), x(n-1), \ldots, x(n-N+1)]$ will be indicated as $f_i(n)$.

We are particularly interested in sets of basis functions that satisfy the conditions of the Stone–Weierstrass theorem [42].

**Theorem 1** (Stone–Weierstrass Theorem). *Let $\mathcal{A}$ be an algebra of real continuous functions on a compact set $K$. If (i) $\mathcal{A}$ separates points on $K$ and if (ii) $\mathcal{A}$ vanishes at no point of $K$, then the uniform closure $\mathcal{B}$ of $\mathcal{A}$ consists of all real continuous functions on $K$.*

According to this theorem, every algebra of real continuous functions on the compact $\mathbb{R}_1^N$ that separates points and vanishes at no point is able to arbitrarily well approximate the function $f$ in Eq. (2.1).

**Definition 1.** A family $\mathcal{A}$ of real functions is said to be an *algebra* if $\mathcal{A}$ is closed under addition, multiplication and scalar multiplication, *i.e.*, if (i) $f + g \in \mathcal{A}$, (ii) $f \cdot g \in \mathcal{A}$ and (iii) $cf \in \mathcal{A}$, for all $f \in \mathcal{A}$, $g \in \mathcal{A}$ and for any real constant $c$.

For example, the set of basis functions formed by all possible products of the samples $x(n)$, $x(n-1)$, ..., $x(n-N+1)$ form an algebra that satisfies all conditions of the Stone–Weierstrass theorem. The linear combination of these basis functions defines the class of truncated Volterra filters.

### 2.2.2 THE VOLTERRA AND WIENER THEORY

Before introducing the class of FLiP filters, which encompasses many of the LIP filters currently used in practice, we first review the classical theory of discrete-time Volterra and Wiener filters.

The Volterra series can be defined as a summation of nonlinear Volterra operators of order $p$, as

$$y(n) = (Vx)(n) = \sum_{p=0}^{P} (H_p x)(n), \tag{2.3}$$

where the Volterra operator of order $p$ is defined as

$$(H_p x)(n) = \sum_{\tau_1=0}^{M} \sum_{\tau_2=0}^{M} \cdots \sum_{\tau_p=0}^{M} h_{p,\tau_1,\tau_2\ldots,\tau_p} x(n-\tau_1)x(n-\tau_2)\cdots x(n-\tau_p), \tag{2.4}$$

with $P = M = +\infty$ in the most general formulation and $(H_0 x)(n) = h_0$ being a constant term. In Eq. (2.4), the summations start from zero since only causal systems are considered. The set of coefficients $h_{p,\tau_1,\tau_2,\ldots,\tau_p}$ is the $p$th-order Volterra kernel.

Truncated Volterra filters derive from the double truncation of the Volterra series in Eq. (2.3). Specifically, they are obtained by truncating the order of the series, setting $P = K < +\infty$, and by limiting the memory of the Volterra operators in (2.4) to $N$ samples, *i.e.*, considering $M = N - 1 < +\infty$.

Without loss of generality, the kernel $h_{p,\tau_1,\ldots,\tau_p}$ can always be considered symmetrical. In fact, for the commutativity of the multiplication it is always possible to symmetrize the kernel without changing $(H_p x)$. It is also possible to reduce the summation in Eq. (2.4) to an upper triangular form, considering

$$(H_p x)(n) = \sum_{\tau_1=0}^{M} \sum_{\tau_2=\tau_1}^{M} \cdots \sum_{\tau_p=\tau_{p-1}}^{M} h_{p,\tau_1,\tau_2\ldots,\tau_p} x(n-\tau_1)x(n-\tau_2)\cdots x(n-\tau_p), \tag{2.5}$$

for a proper choice of the coefficients $h_{p,\tau_1,\tau_2\ldots,\tau_p}$, obtaining the so called *triangular representation* of Volterra filters.

Once established with the Stone–Weierstrass theorem that any discrete-time, causal, time-invariant, continuous system can be arbitrarily well approximated with a truncated Volterra series, the problem of estimating its parameters comes up. To this end we will present the Wiener's orthogonal classic representation and its estimation by means of the cross-correlation method by Lee and Schetzen [8]. The WN filters, which derive from the double truncation of the Wiener series, represent one of the first examples of nonlinear filters with orthogonal basis functions proposed in the literature.

#### *2.2.2.1 Orthogonalization*

In order to allow identification by a cross-correlation method, the Volterra series must be rearranged in terms of orthogonal operators:

$$y(n) = \sum_{p=0}^{P} (H_p x)(n) \equiv \sum_{p=0}^{P} (G_p x)(n). \tag{2.6}$$

These operators, commonly known as Wiener G–functionals, are orthogonal to each other and, also, to Volterra operators of lower orders if the input $x(n)$ is a stationary white-noise signal with zero mean and variance $\sigma_x^2$, *i.e.*,

$$\begin{aligned} E[(H_p x)(n)(G_q x)(n)] &= 0, \quad p < q, \\ E[(G_p x)(n)(G_q x)(n)] &= 0, \quad p \neq q, \end{aligned} \tag{2.7}$$

whenever $(H_p x)(n)$ is an arbitrary homogeneous Volterra operator and $x(n)$ is a stationary white noise with zero mean and variance $\sigma_x^2$, *i.e.*, $x(n) \in \mathcal{N}(0, \sigma_x^2)$. By means of the Gram–Schmidt orthogonalization method, the following expressions for the first Wiener G–functionals can be obtained:

$$\begin{aligned} (G_0 x)(n) &= k_0, \\ (G_1 x)(n) &= \sum_{\tau_1=0}^{M} k_{1,\tau_1} x(n-\tau_1), \\ (G_2 x)(n) &= \sum_{\tau_1=0}^{M} \sum_{\tau_2=0}^{M} k_{2,\tau_1,\tau_2} x(n-\tau_1) x(n-\tau_2) - \sigma_x^2 \sum_{\tau_1=0}^{M} k_{2,\tau_1,\tau_1}, \\ (G_3 x)(n) &= \sum_{\tau_1=0}^{M} \sum_{\tau_2=0}^{M} \sum_{\tau_3=0}^{M} k_{3,\tau_1,\tau_2,\tau_3} x(n-\tau_1) x(n-\tau_2) x(n-\tau_3) + \\ &\quad - 3\sigma_x^2 \sum_{\tau_1=0}^{M} \sum_{\tau_2=0}^{M} k_{3,\tau_1,\tau_2,\tau_2} x(n-\tau_1). \end{aligned} \tag{2.8}$$

As can be seen, the Wiener G–functionals are nonhomogeneous operators. The order $P$ and the memory $M$ are infinite in the case of the Wiener series, while they are truncated to finite values, $P = K < +\infty$ and $M = N - 1 < +\infty$, in the case of a WN filter. For simplicity, the Wiener kernels $k_p$ have been assumed symmetric, but the triangular representation is also possible.

From the expression of the Wiener G–functionals, it is not immediate to determine the basis functions of WN filters. In Section 2.2.3, we will derive an alternative representation of the WN filters under the framework of FLiP filters. Moreover, it will be shown that the basis functions are orthogonal for $x(n) \in \mathcal{N}(0, \sigma_x^2)$.

The Volterra kernels of a truncated series can be derived from the Wiener kernels by means of simple equivalence formulas. In the following, the Wiener–Volterra formulas valid in the case of a third-order Volterra system are reported:

$$\begin{aligned} h_3 &= k_3, \\ h_2 &= k_2, \\ h_1 &= k_1 - 3\sigma_x^2 \sum_{\tau_2=0}^{M} k_{3,\tau_1,\tau_2,\tau_2}, \\ h_0 &= k_0 - \sigma_x^2 \sum_{\tau_1=0}^{M} k_{2,\tau_1,\tau_1}. \end{aligned} \tag{2.9}$$

The reader is referred to [14] for higher-order formulas.

#### 2.2.2.2 Cross-Correlation

The Wiener kernels can be estimated with the cross-correlation method. Recalling that every Volterra functional is orthogonal to all Wiener functionals of greater order and considering the following Volterra functional:

$$(H_p^* x)(n) = \prod_{j=1}^{p} x(n-\tau_j), \tag{2.10}$$

it is possible to write

$$E\left[y(n)(H_p^* x)(n)\right] = E\left[\sum_{q=0}^{P} (G_q x)(n)(H_p^* x)(n)\right]. \tag{2.11}$$

If $x(n) \in \mathcal{N}(0, \sigma_x^2)$, $\tau_1 \neq \tau_2 \neq \ldots \neq \tau_p$, *i.e.*, excluding the so called diagonal terms,

$$E\left[y(n)\prod_{j=1}^{p} x(n-\tau_j)\right] = E\left[(G_p x)(n)\prod_{j=1}^{p} x(n-\tau_j)\right] = p!\sigma_x^{2p} k_{p,\tau_1,\ldots,\tau_p}$$

and thus

$$k_{p,\tau_1,\ldots,\tau_p} = \frac{E\left\{y(n)x(n-\tau_1)\cdots x(n-\tau_p)\right\}}{p!\sigma_x^{2p}}.$$

To consider also the diagonal elements, Lee and Schetzen proposed to subtract from the output $y(n)$ the Wiener operators of lower orders that, in correlation with the operator of Eq. (2.10), give raise to impulses. However, if in Eq. (2.11) we write explicitly the expectation of the product between the $G$ functionals and products of delay input samples, we obtain

$$\begin{aligned} E[y(n)\,x(n-\tau_1)x(n-\tau_2)] &= 2!\sigma_x^4 k_{2,\tau_1,\tau_2} - \sigma_x^2 k_0 \delta_{\tau_1\tau_2}, \\ E[y(n)\,x(n-\tau_1)x(n-\tau_2)x(n-\tau_3)] &= 3!\sigma_x^6 k_{3,\tau_1,\tau_2,\tau_3} \\ &\quad -\sigma_x^4\left(k_{1,\tau_1}\delta_{\tau_2\tau_3} + k_{1,\tau_2}\delta_{\tau_1\tau_3} + k_{1,\tau_3}\delta_{\tau_1\tau_2}\right), \end{aligned} \tag{2.12}$$

where $\delta_{\tau_i\tau_j} \triangleq \delta(\tau_i - \tau_j)$ is the unitary impulse delayed by $\tau_i - \tau_j$.

By properly considering these impulses, in [13] the authors proposed a formula for an efficient implementation of the Lee–Schetzen method for the estimation of Wiener series of any order. The formula simplifies to the classic Lee–Schetzen one if nondiagonal terms of the Wiener kernel are identified. The reader is referred to [13] for the implementation details.

In the cross-correlation method described above, the nonlinear kernels are estimated computing the cross-correlation between the unknown system output and products of input samples and solving a system of linear equations as in Eq. (2.12). In the next section, we introduce a large class of nonlinear systems, the FLiP filters, which includes many subclasses of filters with orthogonal basis functions. In the orthogonal FLiP filters, the nonlinear kernels can be directly estimated with the aforementioned cross-correlation method.

### 2.2.3 FLIP FILTERS AND ORTHOGONAL FLIP FILTERS

FLiP filters are a subclass of LIP filters whose basis functions are polynomials of nonlinear expansions of delayed input samples that follow the constructive rule of the triangular representation of Volterra filters. All families of FLiP nonlinear filters are universal approximators. As mentioned in the introduction, Volterra, WN, EMFN, LN and CN filters belong to the general class of FLiP filters. EMFN filters, like FN, are based on trigonometric function expansions of the input signal samples and do not include a linear term among the basis functions. EMFN filters should be preferred to FN filters, because they often provide a much more compact representation of nonlinear systems. Nevertheless, also FN filters have been found useful for modeling nonlinear systems [43]. LN, CN and WN filters are based on Legendre, Chebyshev and Hermite polynomial expansions, respectively, and have a linear term formed by the first-order basis functions. The basis functions of FN, EMFN, LN, CN and WN filters form algebras that satisfy all the requirements of the Stone–Weierstrass approximation theorem.

Let us consider an ordered set of one-dimensional basis functions that satisfies all requirements of the Stone–Weierstrass theorem on $\mathbb{R}_1$,

$$\{g_0[x(n)], g_1[x(n)], g_2[x(n)], g_3[x(n)], \ldots\}, \tag{2.13}$$

where $g_0[x(n)]$ is the basis function of order 0, usually equal to 1, $g_{2k}[x(n)]$ is an even basis function of order $2k$, $g_{2k+1}[x(n)]$ is an odd basis function of order $2k+1$, for any $k$ in $\mathbb{N}$.

This set of basis functions can arbitrarily well approximate the system in Eq. (2.1) for $N = 1$. To develop a set of basis functions for the more general case of $N > 1$, we first write the basis functions $g_0, g_1, g_2, \ldots$, for $x(n), x(n-1), x(n-2), \ldots$,

$$\begin{gathered} g_0[x(n)], g_0[x(n-1)], g_0[x(n-2)], \ldots \\ g_1[x(n)], g_1[x(n-1)], g_1[x(n-2)], \ldots \\ g_2[x(n)], g_2[x(n-1)], g_2[x(n-2)], \ldots \\ \vdots \end{gathered}$$

Then the functions $g_j$ of different variables are multiplied with each other in any possible manner, avoiding repetitions, as in the triangular representation of Volterra filters. The resulting basis functions $f_i$ are formed by cross-products of the functions $g_j$. The basis functions and their linear combinations satisfy all requirements of the Stone–Weierstrass theorem on $\mathbb{R}_1^N$ and thus FLiP filters are universal approximators. Indeed, by construction the set is an algebra since it is closed under addition, multiplication and scalar multiplication, *i.e.*, it satisfies properties (i), (ii) and (iii) of Definition 1. It is closed under multiplication because any product of the functions $g_j[x(n)]$ can be expanded in a linear combination of the functions $g_j[x(n)]$ and, consequently, any product of the functions $f_i$ can be expanded in a linear combination of the functions $f_i$. The set of basis functions vanishes at no point, thanks to $g_0[x(n)] = 1$, satisfying property (ii) of Theorem 1. As required in property (i) of Theorem 1, the set separates points, because two different points must have at least one different coordinate $x(n-k)$ and the functions $g_j[x(n-k)]$ separates it. Thus, the algebra of FLiP basis functions can arbitrarily well approximate the nonlinear system in Eq. (2.1).

The order of each $N$-dimensional basis function $f_i(n)$ is defined as the sum of the orders of the constituent one-dimensional basis functions. Since any one-dimensional basis function $g_{2k}[x(n)]$ is

**Table 2.1 Basis functions $f_i(n)$ of FLiP filters**

| | |
|---|---|
| Order 0: | $g_0[x(n)] = 1.$ |
| Order 1: | $g_1[x(n)], \ldots, g_1[x(n-N+1)].$ |
| Order 2: | $g_2[x(n)], \ldots, g_2[x(n-N+1)],$<br>$g_1[x(n)]g_1[x(n-1)], \ldots, g_1[x(n-N+2)]g_1[x(n-N+1)],$<br>$g_1[x(n)]g_1[x(n-2)], \ldots, g_1[x(n-N+3)]g_1[x(n-N+1)],$<br>$\vdots$<br>$g_1[x(n)]g_1[x(n-N+1)].$ |
| Order 3: | $g_3[x(n)], \ldots, g_3[x(n-N+1)],$<br>$g_2[x(n)]g_1[x(n-1)], \ldots, g_2[x(x-N+2)]g_1[x(n-N+1)],$<br>$\vdots$<br>$g_2[x(n)]g_1[x(n-N+1)],$<br>$g_1[x(n)]g_2[x(n-1)], \ldots, g_1[x(n-N+2)]g_2[x(n-N+1)],$<br>$\vdots$<br>$g_1[x(n)]g_2[x(n-N+1)],$<br>$g_1[x(n)]g_1[x(n-1)]g_1[x(n-2)], \ldots$<br>$g_1[x(n-N+3)]g_1[x(n-N+2)]g_1[x(n-N+1)],$<br>$\vdots$<br>$g_1[x(n)]g_1[x(n-N+2)]g_1[x(n-N+1)].$ |

even and any basis function $g_{2k+1}[x(n)]$ is odd, for any $k \in \mathbb{N}$, by construction all $N$-dimensional basis functions $f_i(n)$ of even order are even and all $N$-dimensional basis functions $f_i(n)$ of odd order are odd.

A FLiP filter of order $K$, memory $N$, is originated by the linear combination of all basis functions $f_i$ having orders from 0 to $K$, which depends on $x(n), x(n-1), \ldots, x(n-N+1)$.

For example, Table 2.1 summarizes the basis functions $f_i$ of a FLiP filter of order $K = 3$ and memory $N$. In particular, in Table 2.1 $f_0(n) = g_0[x(n)]$, $f_1(n) = g_1[x(n)]$, $f_2(n) = g_1[x(n-1)]$, ..., $f_N(n) = g_1[x(n-N+1)]$, $f_{N+1}(n) = g_2[x(n)]$, and so on for the other basis functions of order 2 and 3, till $f_{N_T-1}(n) = g_1[x(n)]g_1[x(n-N+2)]g_1[x(n-N+1)]$, where $N_T$ is the total number of basis functions.

Any choice of the set of one-dimensional basis functions $g_j[x(n)]$ defines a different family of FLiP filters. Many families of LIP nonlinear filters studied in the literature or used in applications belong to the class of FLiP filters. For example, the truncated Volterra filter is a FLiP filter obtained using as $g_j[x(n)]$ the monomials

$$\left\{1, x(n), x^2(n), x^3(n), \ldots\right\}. \tag{2.14}$$

In what follows, we discuss four families of nonlinear filters with orthogonal basis functions. Specifically, we consider:

1. the EMFN filters [33],
2. the LN filters [36],
3. the CN filters [37] and
4. the WN filters [15].

For each of these filters, the expression of the one-dimensional basis functions $g_j$ is provided and their orthogonality properties are discussed. At the end of the section, it is also explained how to develop a family of polynomial FLiP filters whose basis functions are orthogonal for any desired distribution of the input samples.

#### 2.2.3.1 Even-Mirror Fourier Nonlinear Filters

EMFN filters are defined using as $g_j[x(n)]$ the following trigonometric functions:

$$\left\{1, \sin\left[\frac{1}{2}\pi x(n)\right], \cos[\pi x(n)], \sin\left[\frac{3}{2}\pi x(n)\right], \ldots, \cos[j\pi x(n)], \sin\left[\frac{2j+1}{2}\pi x(n)\right], \ldots\right\}, \tag{2.15}$$

where 1 is the basis function of order 0, $\sin[\frac{2j+1}{2}\pi x(n)]$ is a basis function of order $2j+1$, and $\cos[j\pi x(n)]$ is a basis function of order $2j$, with $j \in \mathbb{N}$. Note that all even-basis functions are even and all odd-basis functions are odd. The basis functions $g_j[x(n)]$ and, consequently, the corresponding FLiP basis functions $f_i$ are even mirror-symmetric, which explains the name of the filters.

The basis functions $g_j$ and their linear combinations form an algebra that satisfies all the requirements of the Stone–Weierstrass theorem on the compact $\mathbb{R}_1$. Indeed, the set is complete under addition, scalar multiplication and multiplication, since

$$\sin\left(\frac{2i+1}{2}\pi x\right)\cos(j\pi x) = \frac{1}{2}\sin\left[\frac{2(i-j)+1}{2}\pi x\right] + \frac{1}{2}\sin\left[\frac{2(i+j)+1}{2}\pi x\right], \tag{2.16}$$

$$\cos(i\pi x)\cos(j\pi x) = \frac{1}{2}\cos\left[(i-j)\pi x\right] + \frac{1}{2}\cos\left[(i+j)\pi x\right], \tag{2.17}$$

$$\sin\left(\frac{2i+1}{2}\pi x\right)\sin\left(\frac{2j+1}{2}\pi x\right) = \frac{1}{2}\cos\left[(i-j)\pi x\right] + \frac{1}{2}\cos\left[(i+j+1)\pi x\right], \tag{2.18}$$

it separates points (*e.g.*, with $\sin\left[\frac{1}{2}\pi x(n)\right]$) and it vanishes at no point (*e.g.*, for $g_0 = 1$).

The basis functions are orthogonal for a white uniform distribution in $\mathbb{R}_1$, *i.e.*,

$$\begin{aligned}\int_{-1}^{+1}\cdots\int_{-1}^{+1} & f_i[x(n), \ldots, x(n-N+1)]\, f_l[x(n), \ldots, x(n-N+1)] \cdot \\ & \cdot p[x(n), \ldots, x(n-N+1)]\mathrm{d}x(n)\cdots \mathrm{d}x(n-N+1) = 0,\end{aligned} \tag{2.19}$$

where $i \neq l$ and $p[x(n), \ldots, x(n-N+1)]$ is the probability density function of the $N$-tuple $[x(n), \ldots, x(n-N+1)]$, equal to the constant $\dfrac{1}{2^N}$ for a white uniform distribution of the input signal

| Table 2.2 Legendre polynomials |
|---|
| $\text{leg}_0(x) = 1$ |
| $\text{leg}_1(x) = x$ |
| $\text{leg}_2(x) = \frac{1}{2}(3x^2 - 1)$ |
| $\text{leg}_3(x) = \frac{1}{2}x(5x^2 - 3)$ |
| $\text{leg}_4(x) = \frac{1}{8}(35x^4 - 30x^2 + 3)$ |
| $\text{leg}_5(x) = \frac{1}{8}x(63x^4 - 70x^2 + 15)$ |

$x(n)$ in $\mathbb{R}_1$. Indeed, by constructions, the basis functions $f_i$ and $f_l$ are products of one-dimensional basis functions $g_j$. For any $j \in \mathbb{N}$, it is

$$\int_{-1}^{+1} \sin\left(\frac{2j+1}{2}\pi x\right) \mathrm{d}x = 0 \tag{2.20}$$

and for any $j \in \mathbb{N}^+$ it is

$$\int_{-1}^{+1} \cos(j\pi x)\, \mathrm{d}x = 0. \tag{2.21}$$

Since the product of two different one-dimensional basis functions $g_j$, according to Eqs. (2.16), (2.17) and (2.18), is a linear combination of basis functions $g_j$ different from $g_0$, it immediately follows Eq. (2.19).

#### 2.2.3.2 Legendre Nonlinear Filters

LN filters are defined by using as $g_j[x(n)]$ the Legendre polynomials, so

$$\text{leg}_0[x(n)], \text{leg}_1[x(n)], \text{leg}_2[x(n)], \text{leg}_3[x(n)], \ldots. \tag{2.22}$$

The Legendre polynomials are obtained from the recursive relation

$$\text{leg}_{j+1}[x(n)] = \frac{2j+1}{j+1} x(n)\text{leg}_j[x(n)] - \frac{j}{j+1}\text{leg}_{j-1}[x(n)], \tag{2.23}$$

where $\text{leg}_0[x(n)] = 1$, $\text{leg}_1[x(n)] = x(n)$ and $j \in \mathbb{N}$ is the order of the basis function.

Table 2.2 summarizes the first Legendre polynomials.

The product of two Legendre polynomials of order $i$ and $j$, respectively, can be expressed as a linear combination of Legendre polynomials up to the order $i + j$ [44]:

$$\text{leg}_i(x)\text{leg}_j(x) = \sum_{m=0}^{\min(i,j)} \frac{A(j-m)A(i-m)A(m)}{A(i+j-m)} \frac{2i+2j-4m+1}{2i+2j-2m+1} \text{leg}_{i+j-2m}(x), \tag{2.24}$$

where

$$A(t) = \frac{1 \cdot 3 \cdot 5 \cdot \ldots \cdot (2t-1)}{1 \cdot 2 \cdot 3 \cdot \ldots \cdot t} = \frac{(2t)!}{2^t (t!)^2} = \frac{1}{2^t}\binom{2t}{t}. \tag{2.25}$$

Exploiting this property it can be proved that the set of Legendre polynomials and their linear combinations satisfy all the requirements of the Stone–Weierstrass theorem on the compact $\mathbb{R}_1$.

The Legendre polynomials are orthogonal in $\mathbb{R}_1$, since

$$\int_{-1}^{1} \mathrm{leg}_i(x)\mathrm{leg}_j(x)\mathrm{d}x = \frac{2}{2i+1}\delta_{ij} \tag{2.26}$$

and, thus, the LN basis functions satisfy Eq. (2.19) and are orthogonal for a white uniform distribution of the input signal $x(n)$ in $\mathbb{R}_1$.

#### 2.2.3.3 Chebyshev Nonlinear Filters

In CN filters, the one-dimensional basis functions $g_j[x(n)]$ are Chebyshev polynomials of the first kind [45,46], *i.e.*, $g_j[x(n)] = T_j[x(n)]$, and $T_j$ are the Chebyshev polynomials of order $j$. Chebyshev polynomials of the first kind are orthogonal polynomials generated by the following recursive relation:

$$T_{j+1}(x) = 2xT_j(x) - T_{j-1}(x), \tag{2.27}$$

with $T_0(x) = 1$, $T_1(x) = x$. The polynomials are orthogonal in $\mathbb{R}_1$ with respect to the weighting function $\dfrac{1}{\pi\sqrt{1-x^2}}$, since

$$\int_{-1}^{+1} T_j(x)T_k(x)\frac{1}{\pi\sqrt{1-x^2}}\mathrm{d}x = \begin{cases} 0, & n \neq m, \\ 1, & n = m = 0, \\ 1/2, & n = m \neq 0. \end{cases} \tag{2.28}$$

For any $x \in \mathbb{R}_1$ also $T_j(x) \in \mathbb{R}_1$ and it has multiple maxima and minima equal to $+1$ and $-1$, respectively. Thus, the $T_j(x)$ are equiripple functions in $\mathbb{R}_1$.

The Chebyshev polynomials and their linear combinations form an algebra on the compact $\mathbb{R}_1$ that satisfies all the requirements of the Stone–Weierstrass theorem, as can be proved since the product of two polynomials of order $i$ and $j$ is

$$2T_i(x)T_j(x) = T_{i+j}(x) + T_{|i-j|}(x). \tag{2.29}$$

It is interesting to note that the approximation of any continuous function $f(x) = \sum_{n=1}^{+\infty} c_n T_n(x)$ with a linear combination $h(x)$ of Chebyshev polynomials up to a degree $N$ is very close to a min–max approximation [46]. Indeed, the approximation error can be expressed as

$$\epsilon(x) = f(x) - h(x) = \sum_{N+1}^{+\infty} c_n T_n(x). \tag{2.30}$$

If the function is continuous and differentiable, the sequence of coefficients $c_n$ converges rapidly to 0, such that $\epsilon(x) \simeq c_{N+1}T_{N+1}(x)$, which is an equiripple function for the properties of Chebyshev polynomials.

Table 2.3 summarizes the first Chebyshev polynomials of the first kind.

By exploiting the orthogonality property of the Chebyshev polynomials in Eq. (2.28), it can be verified that the CN basis functions $f_i[x(n), x(n-1), \ldots, x(n-N+1)]$ are orthogonal in $\mathbb{R}_1^N$ with

**Table 2.3 Chebyshev polynomials of the first kind**

| |
|---|
| $T_0(x) = 1$ |
| $T_1(x) = x$ |
| $T_2(x) = 2x^2 - 1$ |
| $T_3(x) = 4x^3 - 3x$ |
| $T_4(x) = 8x^4 - 8x^2 + 1$ |
| $T_5(x) = 16x^5 - 20x^3 + 5x$ |

weighting function $\frac{1}{\pi\sqrt{1-x^2(n)}} \cdots \frac{1}{\pi\sqrt{1-x^2(n-N+1)}}$. Taking two different basis functions $f_i$ and $f_j$, according to Eq. (2.28), we have

$$\int_{-1}^{+1} \cdots \int_{-1}^{+1} f_i[x(n), \ldots, x(n-N+1)] \cdot f_j[x(n), \ldots, x(n-N+1)] \cdot \frac{dx(n) \cdots dx(n-N+1)}{\pi\sqrt{1-x^2(n)} \cdots \pi\sqrt{1-x^2(n-N+1)}} = 0. \tag{2.31}$$

Thus, the basis functions are orthogonal for a white distribution of the input signal in $\mathbb{R}_1$ having the probability density function

$$p_x(x) = \frac{1}{\pi\sqrt{1-x^2}}. \tag{2.32}$$

Indeed, in Eq. (2.31) the factor

$$\frac{1}{\pi\sqrt{1-x^2(n)} \cdots \pi\sqrt{1-x^2(n-N+1)}}$$

can be interpreted as the joint probability density function of the $N$-tuple $[x(n), \ldots, x(n-N+1)]$.

An input signal $x(n)$ with probability density function (2.32) can be obtained by transforming a signal $u(n)$, white and uniformly distributed in $\mathbb{R}_1$, with the following mapping:

$$x(n) = \sin[\frac{\pi}{2}u(n)]. \tag{2.33}$$

This can be derived considering that the probability density of $x(n)$, $p_x(x)$, is related to the probability density of $u(n)$, $p_u(u)$, as follows:

$$p_x(x) = \left|\frac{du}{dx}\right| p_u(u). \tag{2.34}$$

By inserting in Eq. (2.34) the inverse of Eq. (2.33),

$$u(n) = \frac{2}{\pi} \arcsin[x(n)], \tag{2.35}$$

and considering that $p_u(u) = \frac{1}{2}$, it is immediate to obtain Eq. (2.32).

**Table 2.4 Hermite polynomials of variance $\sigma_x^2$**

| |
|---|
| $W_0(x) = 1$ |
| $W_1(x) = x$ |
| $W_2(x) = x^2 - \sigma_x^2$ |
| $W_3(x) = x^3 - 3\sigma_x^2 x$ |
| $W_4(x) = x^4 - 6\sigma_x^2 x^2 + 3\sigma_x^4$ |
| $W_5(x) = x^5 - 10\sigma_x^2 x^3 + 15\sigma_x^4 x$ |

#### 2.2.3.4 Wiener Nonlinear Filters

The WN nonlinear filters defined in Section 2.2.2 are polynomial filters with orthogonal basis functions for any white Gaussian input signal $x(n) \in \mathcal{N}(0, \sigma_x^2)$. The corresponding one-dimensional basis functions $g_j$ can be determined by applying the Gram–Schmidt orthogonalization to the set of polynomials

$$\{1,\ x,\ x^2,\ x^3,\ \ldots\}, \tag{2.36}$$

taking into account that, for $x \in \mathcal{N}(0, \sigma_x^2)$, the $k$th moment of $x$ is

$$\mu_k = E[x^k] = \begin{cases} 0 & \text{for } k \text{ odd}, \\ \sigma_x^k (k-1)!! & \text{for } k \text{ even}, \end{cases} \tag{2.37}$$

with $q!! = q \cdot (q-2) \cdot (q-4) \cdot \ldots \cdot 1$. The resulting polynomials, here labeled as $W_j(x)$ for order $j$, are reported in Table 2.4 and are related to the Hermite polynomials [47]. In particular, they are "Hermite polynomials of variance $\sigma_x^2$" according to the definition reported in [48]. They can be generated with the following recursive relation:

$$W_{j+1}(x) = x W_j(x) - j\sigma_x^2 W_{j-1}(x), \tag{2.38}$$

with $W_0(x) = 1$ and $W_1(x) = x$. The set of polynomials $W_j(x)$ and their linear combinations satisfy the conditions of the Stone–Weierstrass theorem on any compact interval $[-A, +A]$, with $A \in \mathbb{R}^+$. Thus, also the resulting $N$-dimensional WN basis functions $f_i$ satisfy the condition of the Stone–Weierstrass theorem on the compact $[-A, +A]^N$.

By constructions, the polynomials $W_j(x)$ are orthogonal for $x \in \mathcal{N}(0, \sigma_x^2)$, *i.e.*, for any $i, j$ with $i \neq j$ we have

$$\int_{-\infty}^{+\infty} W_i(x) W_j(x) \frac{e^{\frac{x^2}{2\sigma_x^2}}}{\sigma_x \sqrt{2\pi}} \mathrm{d}x = 0. \tag{2.39}$$

From Eq. (2.39), the orthogonality of the WN basis function $f_i$ for any white Gaussian input signal $x(n) \in \mathcal{N}(0, \sigma_x^2)$ also follows.

#### 2.2.3.5 Orthogonal FLiP Filters for a Generic Probability Distribution $\mathcal{P}$

The procedure followed to derive the polynomials $W_j(x)$ can easily be applied to derive a family of orthogonal polynomials for any probability distribution $\mathcal{P}$. Once the moments $\mu_k = E[x^k]$ are computed, the Gram–Schmidt orthogonalization procedure can easily be applied to the polynomial set (2.36), deriving a family of polynomials $g_j(x)$ orthogonal for $x \in \mathcal{P}$. From this, it is possible to define a family of FLiP filters orthogonal for any white input signal $x(n) \in \mathcal{P}$.

#### 2.2.3.6 Least Square Error Estimation

The input–output relationship of any FLiP filter of order $K$, with memory of $N$ samples, can be written as

$$y(n) = \sum_{i=0}^{N_{\mathrm{T}}-1} c_i f_i(n), \tag{2.40}$$

where $f_i(n), i = 0, \ldots, N_{\mathrm{T}} - 1$, are $N$-dimensional basis functions taken from Table 2.1. Using vector notation, Eq. (2.40) becomes

$$y(n) = \mathbf{c}^{\mathrm{T}} \mathbf{f}_n, \tag{2.41}$$

where

$$\mathbf{c} = [c_0, c_1, \ldots, c_{N_{\mathrm{T}}-1}]^{\mathrm{T}}$$

and

$$\mathbf{f}_n = [f_0(n), f_1(n), \ldots, f_{N_{\mathrm{T}}-1}(n)]^{\mathrm{T}}.$$

Then the minimum mean square error (MSE) estimation gives as optimal solution of (2.41)

$$\mathbf{c}_o = \mathbf{R}_{ff}^{-1} \mathbf{p}_{yf}, \tag{2.42}$$

since the filter output is linear with respect to the coefficients [49]. In Eq. (2.42), $\mathbf{R}_{ff}$ is the $N_{\mathrm{T}} \times N_{\mathrm{T}}$ auto-correlation matrix of the basis functions $f_i(n)$ and $\mathbf{p}_{yf}$ is the $N_{\mathrm{T}} \times 1$ cross-correlation vector between the output $y(n)$ of the unknown system and the basis functions $f_i$, evaluated using the given input sequence $x(n)$. When the elements of $\mathbf{R}_{ff}$ and $\mathbf{p}_{yf}$ are computed by time averages over $L$ samples of the input $x(n)$, $\mathbf{c}_o$ is called the optimal least square solution of (2.41). When the basis functions are orthogonal, the autocorrelation matrix is diagonal and Eq. (2.42) reduces to the formulas of the cross-correlation method.

#### 2.2.3.7 Cross-Correlation Method

The orthogonality of the basis functions for a stochastic signal makes it possible to estimate the coefficients of the FLiP filter with the cross-correlation method. The method is here applied computing the cross-correlation between the basis functions and the system output. Exploiting orthogonality of the basis functions $f_i(n)$, the $i$th entry of Eq. (2.42) reduces to

$$c_i = \frac{E[y(n) f_i(n)]}{E[f_i^2(n)]}. \tag{2.43}$$

The expectations can be estimated using time averages, providing a feasible identification method.

The main problem of the cross-correlation method for stochastic inputs is the huge number of samples (in the order of millions) often necessary to obtain an accurate estimate of the coefficients $c_i$.

Moreover, it can be difficult to generate white input signals with the desired distribution. For example, for white Gaussian inputs the maximum amplitude and the maximum bandwidth have to be limited, altering the orthogonality of the basis functions of the Wiener nonlinear filter. The problem will be solved in Section 2.3, where some deterministic signals, the PPSs, capable of guaranteeing the orthogonality of the basis functions on a finite period, will be developed.

### 2.2.4 SIMPLIFIED FLIP FILTERS

All FLiP filters discussed in the previous subsection have been obtained following the constructive rule of the triangular representation of Volterra filters and have the same number of basis functions and coefficients. For a FLiP filter of order $K$, memory $N$, this number is equal to

$$N_{\mathrm{T}} = N_T(K, N) = \binom{N+K}{N}, \tag{2.44}$$

as in the triangular representation of truncated Volterra filters.

According to Eq. (2.44), the number of coefficients of a FLiP filter increases geometrically with the memory of the filter $N$ and exponentially with the order $K$ of the filter. Even for small values of $K$ (*e.g.*, 2 or 3) and moderate values of $N$, the number of coefficients can be prohibitively large. This problem is considered the main drawback of all FLiP filters. To alleviate this difficulty, it is possible to resort to the same strategy of simplified Volterra filters [50]. The FLiP filter can be first expressed according to its diagonal representation, which was first introduced for finite-memory symmetric Volterra filters in [51]. In the diagonal representation rather than using Cartesian coordinates, the filter is expressed considering coordinates aligned with the diagonals of the hypercubes representing the nonlinear kernels. For comparison, the second-order kernel of a FLiP filter of memory $N$ in the Cartesian representation is

$$\sum_{i=0}^{N-1} c_{i,i} g_2[x(n-i)] + \sum_{i=0}^{N-1} \sum_{j=i+1}^{N-1} c_{i,j} g_1[x(n-i)] g_1[x(n-j)], \tag{2.45}$$

while in the diagonal representation it is

$$\sum_{i=0}^{N-1} c_{i,i} g_2[x(n-i)] + \sum_{i=1}^{N-1} \sum_{j=0}^{N-1-i} c_{j,i+j} g_1[x(n-j)] g_1[x(n-i-j)]. \tag{2.46}$$

The diagonal number of a basis function is defined as the maximum time difference between the samples involved in its expression. For example, $g_1[x(n)]g_1[x(n-2)]$ has diagonal number 2; $g_2[x(n-r)]g_1[x(n-r-3)]$ has diagonal number 3 for any $r \in \mathbb{Z}$. Simplified FLiP filters are obtained by considering only the basis functions that have a *diagonal number* lower than or equal to a certain value $D$. The diagonal number is always comprised in the range $0 < D \leq N-1$. When $D = 0$, only the basis functions of the principal diagonals of hypercubes representing the nonlinear kernels are selected. When $D = N-1$, all basis functions of nonlinear kernels are selected. The second-order

**Table 2.5 Number of coefficients $N_D$ of a FLiP filter of order $K$, memory $N$ and diagonal number $D$ and number of coefficients $N_T$ of a full FLiP filter with the same order and memory**

| $K$ | $N$ | $N_D$ | | | | | $N_T$ |
|---|---|---|---|---|---|---|---|
| | | $D$ | | | | | |
| | | 0 | 2 | 4 | 6 | 8 | |
| | 10 | 21 | 38 | 51 | 60 | 65 | 66 |
| 2 | 50 | 101 | 198 | 291 | 380 | 465 | 1326 |
| | 100 | 201 | 398 | 591 | 780 | 965 | 5151 |
| | 10 | 31 | 90 | 161 | 228 | 275 | 286 |
| 3 | 50 | 151 | 490 | 1001 | 1668 | 2475 | 23426 |
| | 100 | 301 | 990 | 2051 | 3468 | 5225 | 176851 |

kernel of a FLiP filter of memory $N$ and diagonal number $D$ reduces to

$$\sum_{i=0}^{N-1} c_{i,i} g_2[x(n-i)] + \sum_{i=1}^{D} \sum_{j=0}^{N-1-i} c_{j,i+j} g_1[x(n-j)] g_1[x(n-i-j)]. \tag{2.47}$$

The motivation for limiting the diagonal number of the basis functions used to define the FLiP filter derives from the observation that the "energy" of the kernels, used to represent real-world nonlinear systems, is often concentrated around the principal diagonals. This situation has been observed in the fields of nonlinear acoustic echo cancellation [50,52,53], nonlinear active noise control [54] and identification of nonlinear systems [55] such as saturated amplifiers or guitar pedals [38,40].

A FLiP filter of memory $N$, order $K$, diagonal number $D$ has a number of coefficients $N_D$ equal to

$$N_D = \binom{D+K+1}{D+1} + \binom{D+K}{D+1} \cdot (N-1-D). \tag{2.48}$$

In the particular case of $D = N-1$, $N_D = N_T$.

Table 2.5 provides a comparison of the numbers of coefficients $N_D$ and $N_T$ of FLiP filters of order 2 and 3, for different memory lengths and diagonal numbers. Reducing the number of diagonals, it is possible to obtain a significant reduction in the computational complexity of the FLiP filter.

## 2.3 RECENT IDENTIFICATION METHODS FOR ORTHOGONAL FILTERS

Most algorithms suitable for the identification of linear filters can also be used for the identification of LIP nonlinear filters. For example, the adaptive algorithms suitable for the identification of linear FIR filters can easily be adapted to LIP filters. The Least Mean Square (LMS), the Recursive Least Square (RLS), the Affine Projection (AP), the Filtered-X Least Mean Square (FX-LMS) and the Filtered-X Affine Projection (FX-AP) algorithms have derivations and expressions formally identical to the linear case. The reader is referred to [32] for a review of these algorithms in the context of LIP nonlinear filters.

In what follows, we review some recent identification techniques suitable for orthogonal LIP nonlinear filters. In particular, we consider the identification of FLiP filters using PPSs, *i.e.*, deterministic periodic sequences that guarantee the orthogonality of the FLiP basis functions on a finite period. Then we consider the multiple-variance identification method that can be used for the identification of Volterra filters avoiding the problem of locality of the solution, *i.e.*, the fact that the estimated model well approximates the unknown system only at the same input variance of the measurement.

### 2.3.1 PERFECT PERIODIC SEQUENCES

A recent solution for the identification of linear and nonlinear models exploits deterministic input signals able to guarantee the orthogonality of the basis functions on a finite interval. As a matter of fact, PPSs have been proposed for the identification of linear filters, as shown in [56–58]. A periodic sequence, applied as input to a linear or nonlinear system, is called perfect if all cross-correlations between any two different basis functions of the modeling filter, estimated over a period, are zero. In the case of linear filters, using a PPS input of period $L$ the basis functions $x(n-i)$, with $i$ ranging from 0 to $L-1$, form an orthogonal set. It has been shown in the literature that the PPSs optimize the convergence speed of the NLMS algorithm [59,60]. Moreover, in the absence of output noise, a PPS can identify a linear filter in just one period [61,62]. Exploiting periodic input signals, it is also possible to identify the unknown system using an adaptive algorithm that requires just a multiplication, an addition and a subtraction per time sample [63].

#### 2.3.1.1 Development of Perfect Periodic Sequences

In what follows, we discuss how we can develop PPSs suitable for the identification of an orthogonal FLiP filter. Different strategies have been devised for this purpose. Let us indicate with $f_i(n)$ the $i$th FLiP basis function with $i$ ranging from 1 to the cardinality of the set it belongs to, with $S_f(K,N)$ being the set of basis functions of order less than or equal to $K$ and memory $N$ and $S_{f.n}(K,N)$ being the subset of $S_f(K,N)$ formed by all basis functions having $x(n)$ as one of their arguments and with $L$ as the PPS period. $S_f(K,N)$ has cardinality $N_{\mathrm{T}}(K,N)$, defined in (2.44), and $S_{f.n}(K,N)$ has cardinality $N_{\mathrm{T}}(K-1,N)$. The $L$ samples of the fundamental period of the PPS are indicated as

$$x_0, x_1, \ldots, x_{L-1}.$$

PPSs for orthogonal FLiP filters can be developed by directly imposing the orthogonality of any couple of basis functions over a period, *i.e.*,

$$< f_i(n) f_j(n) >_L = 0, \tag{2.49}$$

for all $f_i(n) \in S_{f.n}(K,N)$, $f_j(n) \in S_f(K,N)$ with $f_i(n) \neq f_j(n)$. It should be noted that $f_i(n) \in S_{f.n}(K,N)$ is assumed because if Eq. (2.49) is satisfied, then also

$$< f_i(n-l) f_j(n-l) >_L = 0,$$

for all $l \in \mathbb{Z}$.

Together with the conditions in (2.49), it is also convenient to impose the condition that

$$< f_i^2(n) >_L = P_i, \tag{2.50}$$

for all $f_i(n) \in S_{f,n}(K, N)$, with $P_i$ being the power of the basis functions $f_i(n)$ for the white stochastic input signal that guarantees the orthogonality of the basis functions (*i.e.*, for a white uniform input signal in case of EMFN and LN filters, for a white input signal with the probability density function in (2.32) for CN filters and for a white Gaussian input signal with variance $\sigma_x^2$ for WN filters). With (2.50), the power of each basis function is *a priori* known. Moreover, the condition in (2.50) allows also to easily form an orthonormal set of basis functions by dividing each basis function $f_i(n)$ by $\sqrt{P_i}$.

The system of nonlinear equations defined in (2.49) and (2.50) can be proved to be equivalent to the following simpler system [39,36]:

$$< f_i(n) >_L = 0, \tag{2.51}$$

for all $f_i(n) \in S_{f,n}(2K, N)$.

For sufficiently large $L$, Eq. (2.51) provides an under-determined system of nonlinear equations in the variables $x_0, x_1, \ldots, x_{L-1}$, which is expected to have infinite solutions. The system in (2.51) has $Q = N_T(2K-1, N)$ equations and, for $L$ ranging between $1.5Q$ and $6Q$, depending on the particular FLiP filter, a solution for this system has always been found. The nonlinear equation system in (2.51) has been solved using the Newton–Raphson method, implemented as described in [64, Ch. 9.7], with the Jacobian matrix computed analytically. In the case of EMFN, LN and CN filters, where the input samples belong to $\mathbb{R}_1$, the variables $x_0, x_1, \ldots, x_{L-1}$ have been reflected in $\mathbb{R}_1$ any time they exceeded the range at some iteration.

Another strategy for deriving PPSs for orthogonal polynomial FLiP filters is also possible. We can impose the PPS to have the same joint moments of $\mathcal{P}$, which guarantees the orthogonality of the basis functions, for all variables involved in the system in (2.51), *i.e.*, we can impose

$$< x_n^{i_0} x_{n-1}^{i_1} \cdots\cdots\cdots x_{n-N+1}^{i_{N-1}} >_L = E[x^{i_0}(n) x^{i_1}(n-1) \cdots\cdots\cdots x^{i_{N-1}}(n-N+1)], \tag{2.52}$$

where $x_n$ is the PPS sample at time $n$, $x(n) \in \mathcal{P}$, $i_0 \in [1, 2K]$, $i_1, i_2, \ldots, i_{N-1} \in [0, 2K]$ and $i_0 + i_1 + \ldots i_{N-1} \le 2K$. For example, this strategy has been applied to develop PPSs for WN filters. The system in (2.52) has the same number of equations as that in (2.51) and can again be solved with the Newton–Raphson method.

The number of equations $Q$ of the nonlinear systems in (2.51) and (2.52) increases exponentially with the order $K$ and geometrically with the memory $N$. Even for low orders and memory lengths, $Q$ can be unacceptably large. Periodic sequences with specific structures can be adopted to reduce the number of equations and variables in the nonlinear system. For example, the following conditions can almost halve the number of equations and variables [39]:

1. Symmetry: if in the PPS for any $N$-tuple of samples $a_1, a_2, \ldots, a_N$ there is also the reversed one $a_N, a_{N-1}, \ldots, a_1$, then, for any couple of symmetric basis functions, only one needs to be considered.
2. Oddness: if in the PPS for any $N$-tuple of samples $a_1, a_2, \ldots, a_N$ there is also the negated one $-a_1, -a_2, \ldots, -a_N$, all odd basis functions have *a priori* zero average and can be discarded from the system.
3. Oddness-1: if in the PPS for any $N$-tuple of samples $a_1, a_2, \ldots, a_N$ there is also that obtained by alternatively negating one every two terms, $a_1, -a_2, a_3, -a_4, \ldots, -a_N$, all Odd-1 functions have *a priori* zero average and can be discarded from the system.

By definition, Odd-1 are all those basis functions that change their sign by alternatively negating one every two samples, *e.g.*, $f_1[x(n)]f_1[x(n-1)]$. Similarly Odd-2, Odd-4, ... functions can be defined and exploited to reduce the complexity of the nonlinear system in (2.51). Moreover, two or more conditions can be considered together.

The Newton–Raphson algorithm has memory and processing time requirements that grow with $Q^3$. Thus, the reduction in the number of equations obtained with symmetry and oddness conditions is often fundamental to be able to solve the system in (2.51), but it is paid for with a longer period of the resulting PPS. Another strategy to reduce the computational complexity of the system in (2.51) for large orders and memory lengths resorts to the use of simplified models, as those introduced in Section 2.2.4.

#### 2.3.1.2 System Identification using Perfect Periodic Sequences

Let us assume we want to identify a time-invariant, finite-memory, causal, continuous, nonlinear system. When the input–output relationship of the nonlinear system can be expressed as a linear combination of FLiP basis functions up to order $K$ and memory $N$ as in Eq. (2.40), the coefficients $c_i$ can be estimated using as input signal a PPS suitable for the specific FLiP filter with at least order $K$ and memory $N$.

The coefficients of the FLiP filter can be estimated with the cross-correlation method, *i.e.*, computing the cross-correlation between the basis functions and the system output over a period of the PPS,

$$\widehat{c_i} = \frac{< f_i(n)y(n) >_L}{< f_i^2(n) >_L}. \tag{2.53}$$

Since the autocorrelations $< f_i^2(n) >_L = P_i$, with $P_i$ *a priori* known, the computational cost of (2.53) is simply a multiplication and an addition per basis function and per input sample.

In what follows, we discuss the errors caused by an under-determination of the unknown system order or memory.

#### 2.3.1.3 Under-Determined Order

Assume that a PPS for FLiP filters of order $K$, memory $N$, is used to identify a system formed by a linear combination of FLiP basis functions with memory $N$ but maximum order greater than $K$, *i.e.*,

$$y(n) = \sum_{i=0}^{N_T} c_i f_i(n) + O_{K+1}(n), \tag{2.54}$$

where $f_i(n)$ has maximum order $K$ and $O_{K+1}(n)$ is a linear combination of basis functions of order greater than $K$. Assume also $O_{K+1}(n)$ is formed by basis functions of order lower than $2K$ and that the power of the nonlinear kernels reduces with the kernel order. In these conditions, the estimated coefficient $\widehat{c_i}$ is affected by an error generated by $O_{K+1}(n)$. The error influences mainly the coefficients of the higher-order basis functions and, in general, does not influence or influences only mildly the coefficients of the lower-order basis functions. This can be justified by first considering $O_{K+1}(n)$ a linear combination of basis functions of order $K+1$. The estimated coefficients $\widehat{c_i}$ of all basis functions of order less than $K$ are not affected by the error, because for the PPS construction rules the cross-correlation of these basis functions with those of $O_{K+1}(n)$ is zero. In fact, the product of a basis function of an order less than $K$ with a basis function of order $K+1$ is a linear combination of basis functions of order lower than or equal to $2K$, which satisfy (2.51) for a PPS input. Only the coefficients of the basis functions of order $K$ are affected, because their cross-correlation with $O_{K+1}(n)$ is in general different from zero. If $O_{K+1}(n)$ is a linear combination of basis functions of order $K+2$, the estimated coefficients $\widehat{c}_i$ of all basis functions of order less than $K-1$ are not affected by the error, while the coefficients of the basis functions of order $K-1$ and $K$ are affected by an error. Eventually, consider $O_{K+1}(n)$ a linear combination of basis functions of order ranging from $K+1$ to $K+A$, with $A < K$, whose kernels have power decreasing with the order. Then the estimated coefficients $\widehat{c}_i$ of all basis functions of order lower than $K-A+1$ are not affected by the error, while the coefficients of the basis functions of order greater than $K-A$ are affected. The error in the coefficients $\widehat{c}_i$ of the basis functions of order $K-A+1$ depends only on the kernel of order $K+A$ of $O_{K+1}(n)$, which has the lowest power. The error in the coefficients $\widehat{c}_i$ of the basis functions of order $K-A+2$ depends only on the kernels of order $K+A-1$ and $K+A$ of $O_{K+1}(n)$. The error in the coefficients $\widehat{c}_i$ of the basis functions of order $K$ depends on all the kernels of $O_{K+1}(n)$ and can be expected to be the most relevant.

#### 2.3.1.4 Under-Determined Memory

Assume next that a PPS for FLiP filters of order $K$, memory $N$, is used to identify a system that is a linear combination of FLiP basis functions with order $K$ but memory greater than $N$, *i.e.*,

$$y(n) = \sum_{i=0}^{N_T} c_i f_i(n) + M_{N+1}(n), \tag{2.55}$$

where $f_i(n)$ has order $K$, memory $N$, and $M_{N+1}(n)$ is a linear combination of basis functions of memory greater than $N$ but lower than $2N$. Also in this case, the estimated coefficients $\widehat{c}_i$ are affected by an aliasing error generated by $M_{N+1}(n)$. The error affects mainly the coefficients of the basis functions associated with the most recent samples, $x(n), x(n-1), \ldots$, and, in general, affects only mildly or does not affect the coefficients of the basis functions associated with the less recent samples, $x(n-N+1), x(n-N+2), \ldots$. To illustrate this property, let us first consider the identification of the system of memory $\widetilde{N} = N+1$,

$$y(n) = \sum_{i=0}^{\widetilde{N}-1} c_i g_1[x(n-i)], \tag{2.56}$$

where $g_1$ is the one-dimensional basis function of order 1 defining the FLiP filter. The system is identified using a PPS for a FLiP filter of order 1, memory $N$. In this case, the estimate of $c_0$ in (2.53) is affected by an error, since we have not imposed $< g_1[x(n)]g_1[x(n-N)] >_L$ to be zero in the construction of the PPS. In contrast, $c_1, \ldots, c_{N-1}$ are not affected by any error since $< g_1[x(n-l)]g_1[x(n-N)] >_L = 0$ for all $1 \le l \le N-1$ according to the construction rules of the PPS. If the system in (2.56) has memory $\widetilde{N} = N+2$, the estimate of $c_0$ and $c_1$ in (2.53) is affected by an error, since we have not imposed $< g_1[x(n)]g_1[x(n-N)] >_L = < g_1[x(n-1)]g_1[x(n-N-1)] >_L$ and $< g_1[x(n)]g_1[x(n-N-1)] >_L$ to be zero, and in (2.53)

$$< y(n)g_1[x(n)] >_L = c_0 < g_1^2[x(n)] >_L + c_N < g_1[x(n)]g_1[x(n-N)] >_L$$
$$+ c_{N+1} < g_1[x(n)]g_1[x(n-N-1)] >_L, \tag{2.57}$$
$$< y(n)g_1[x(n-1)] >_L = c_1 < g_1^2[x(n-1)] >_L$$
$$+ c_{N+1} < g_1[x(n-1)]g_1[x(n-N-1)] >_L . \tag{2.58}$$

In contrast, $c_2, \ldots, c_{N-1}$ are not affected by any error. When the system in (2.56) has memory $\widetilde{N}$, with $N < \widetilde{N} < 2N$, the coefficients $c_0, c_1, \ldots, c_{\widetilde{N}-N-1}$ are affected by an error, while the coefficients $c_{\widetilde{N}-N}, \ldots, c_{N-1}$ are not affected. Moreover, the error in the estimate of $c_0$ depends on $c_N, c_{N+1}, \ldots, c_{\widetilde{N}}$; the error in the estimate of $c_1$ depends on $c_{N+1}, \ldots, c_{\widetilde{N}}$; and the error in the estimate of $c_{\widetilde{N}-N-1}$ depends only on $c_{\widetilde{N}}$. Also considering that in real systems the coefficients $c_i$ tends to fade for increasing $i$, we can expect the error to affect mainly the coefficients of the basis functions associated with the most recent samples, $x(n), x(n-1), \ldots$ and only mildly those of the basis functions associated with the less recent samples. The same considerations can be applied to more general forms of nonlinear systems than (2.56).

#### 2.3.1.5 Most Relevant Basis Function and Information Criteria

When a FLiP filter is identified with a PPS and the cross-correlation method, its basis functions can easily be ranked according to the MSE reduction they produce. In fact, for the orthogonality property over a period induced by the PPS, the MSE reduction produced by the $i$th basis function is

$$\delta \text{MSE}_i = \frac{< f_i(n)y(n) >_L^2}{< f_i^2(n) >_L}. \tag{2.59}$$

Eqs. (2.53) and (2.59) can be combined with the minimization of an information criterion to determine the number of most relevant basis functions and thus obtain a compact representation for the nonlinear system. The most popular information criteria available in the literature are:

- the Akaike information criterion (AIC) [65]:

$$\text{AIC}(\alpha) = L \log_e[\sigma_\epsilon^2(n_p)] + \alpha n_p, \tag{2.60}$$

  with $\alpha > 0$,
- the Final Prediction Error (FPE) [65]:

$$\text{FPE} = L \log_e[\sigma_\epsilon^2(n_p)] + L \log_e \frac{L + n_p}{L - n_p}, \tag{2.61}$$

- the Khundrin law of iterated logarithm criterion (LILC) [66]:

$$\text{LILC} = L \log_e[\sigma_\epsilon^2(n_p)] + 2n_p \log_e[\log_e(L)], \tag{2.62}$$

- the Bayesian information criterion (BIC) or Schwarz criterion [67]:

$$\text{BIC} = L \log_e[\sigma_\epsilon^2(n_p)] + n_p \log_e[L], \tag{2.63}$$

where $\sigma_\epsilon^2(n_p)$ is the variance of the residual estimation error for a model formed by $n_p$ basis functions, *i.e.*, in our case

$$\sigma_\epsilon^2(n_p) = < y^2(n) >_L - \sum_{i=0}^{n_p-1} \delta\text{MSE}_i, \tag{2.64}$$

and $L$ is the number of data used for the model estimation.

## 2.3.2 MULTIPLE-VARIANCE METHODS

Let us consider again the Volterra and Wiener theory. It has been shown in [11] how the input non-idealities affect kernel estimation. In particular, an explicit expression of residuals that occur when there is a difference between the moments of an ideal input and those produced by a pseudorandom finite length input is given. The residual causes an error in system identification even if the target system itself is a Volterra series having the same order as the model. Another important source of error are the truncation errors that arise if the order of the Volterra filter is lower than that of the system to be identified. These factors cause limitations in Volterra filter approximation capabilities, which are particularly evident in the phenomenon of locality of the solution. In most cases, the model well represents the nonlinear system only in a limited range of input signal powers, close to the variance $\sigma_x^2$ used for the model estimate.

The aim of the multiple-variance methods is to lower the MSE of the Volterra filter when signals with variances different from that used in the identification are input of the system. To do this, multiple-variance inputs are used for Wiener kernel identification.

In the traditional orthogonal algorithm, using high $\sigma_x^2$ inputs has the advantage of stimulating high-order nonlinearities, so as to achieve more accurate high-order kernel identification. As a drawback, the use of high $\sigma_x^2$ values causes higher identification errors in lower-order kernels, as shown in [14].

The rationale should be to use a low $\sigma_x^2$ value for the lower-order kernels and to gradually increase it for higher-order kernels. For the peculiar expression of Wiener functionals, see Eq. (2.8) and the resulting Wiener-to-Volterra formulas in Eq. (2.9); an appropriate normalization is needed in Wiener kernel expressions and in Wiener-to-Volterra conversion formulas, for taking into account the use of multiple variances.

New formulas for Wiener kernel identification were proposed in [14], and they are explicitly reported in the following up to the order 3:

$$k_0^{(0)} = E[y^{(0)}(n)], \tag{2.65}$$

$$k_{1,\tau_1}^{(1)} = \frac{1}{\sigma_{x,(1)}^2} E\left[y^{(1)}(n)\,x^{(1)}(n-\tau_1)\right], \tag{2.66}$$

$$k_{2,\tau_1,\tau_2}^{(2)} = \frac{1}{2!\sigma_{x,(2)}^4}\left\{E\left[y^{(2)}(n)\,x^{(2)}(n-\tau_1)x^{(2)}(n-\tau_2)\right] - \sigma_{x,(2)}^2 k_0^{(2)}\delta_{\tau_1\tau_2}\right\}, \tag{2.67}$$

$$\begin{aligned} k_{3,\tau_1,\tau_2,\tau_3}^{(3)} &= \frac{1}{3!\sigma_{x,(3)}^6}\left\{E\left[y^{(3)}(n)\,x^{(3)}(n-\tau_1)x^{(3)}(n-\tau_2)x^{(3)}(n-\tau_3)\right]\right. \\ &\left. - \sigma_{x,(3)}^4\left[k_{1,\tau_1}^{(3)}\delta_{\tau_2\tau_3} + k_{1,\tau_2}^{(3)}\delta_{\tau_1\tau_3} + k_{1,\tau_3}^{(3)}\delta_{\tau_1\tau_2}\right]\right\}, \end{aligned} \tag{2.68}$$

where $k_p^{(j)}$ and $y^{(j)}(n)$ are obtained with the input $x^{(j)}(n)$ of variance equal to $\sigma_{x,(j)}^2$.

As can be clearly seen in the explicit formulas, the drawback with respect to the classic formula is that, for the identification of a $p$th-order kernel, all lower kernels must be identified again with the higher variance. However, an outstanding improvement in the output MSE is obtained if the Volterra kernels are obtained from the previous estimated Wiener kernels, using the following Wiener to Volterra formulas:

$$h_3 = k_3^{(3)}, \tag{2.69}$$

$$h_2 = k_2^{(2)}, \tag{2.70}$$

$$h_1 = k_1^{(1)} - 3\sigma_{x,(1)}^2 \sum_{\tau_2=1}^{M} k_{3,\tau_1,\tau_2,\tau_2}^{(3)}, \tag{2.71}$$

$$h_0 = k_0^{(0)} - \sigma_{x,(0)}^2 \sum_{\tau_1=1}^{M} k_{2,\tau_1,\tau_1}^{(2)}. \tag{2.72}$$

These formulas are valid for a third-order Volterra filter. By comparing them with Eq. (2.9), we note that in each Volterra kernel the diagonal Wiener coefficients of higher order are multiplied by the variance corresponding to the order of the Volterra kernel instead of the corresponding Wiener kernel.

The multiple-variance method is straightforward when used with the Wiener basis functions of Section 2.2.3.4, since each Wiener kernel can be determined computing only the cross-correlation between the system output and the basis functions of the same order. Using the Wiener basis functions, recomputing the kernels of lower orders as in (2.67) and (2.68) can be avoided [68].

The multiple-variance method can also be used with PPSs as inputs. In this case the residual due to input nonidealities are greatly reduced, if not completely canceled, and the estimation of the kernel diagonals is not a problem anymore. However, the approximation of the estimated model is still affected by truncation errors in the memory or the order of the model, so multiple-variance methods can still be useful. PPSs with multiple variances suitable for WN filters can easily be developed. Indeed, if a PPS for a WN filter of order $P$, memory $N$ and input variance $\sigma_x^2$ is amplified by a factor $A$, by construction the resulting sequence is a PPS for a WN filter of the same order and memory and variance $A^2\sigma_x^2$. Each Wiener kernel $k_i^{(i)}$ of order $i$ can be separately estimated using a PPS of variance $\sigma_{x,(i)}^2$ using (2.53). Eventually, Eqs. (2.69)–(2.72) can be used to estimate the corresponding Volterra kernels.

## 2.4 EXPERIMENTAL RESULTS

In this section, we provide a couple of experimental results illustrating the identification and emulation of real nonlinear devices using orthogonal nonlinear filters.

### 2.4.1 IDENTIFICATION OF NONLINEAR DEVICES USING PERFECT PERIODIC SEQUENCES

In the first experiment, we consider the identification of a real device using different FLiP filters. The experiment aims to illustrate the advantages offered by PPSs for the identification of nonlinear devices and the very compact models that can be obtained by selecting the most significant basis functions

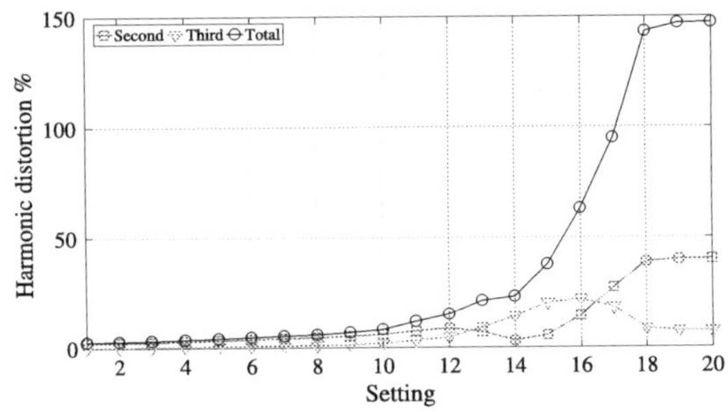


**FIGURE 2.1**

Second, third and total harmonic distortion.

according to some information criterion. In particular, the identification of an audiophile vacuum tube preamplifier, *i.e.*, Behringer Tube Ultragain Mic 100, is reported. The experiment takes advantage of the preamplifier gain setting that is used for introducing different levels of nonlinear distortion. The results of twenty different distortion conditions are reported. Fig. 2.1 shows the second, third and total harmonic distortions on a 200-Hz signal at the maximum used amplitude at the different settings. Clearly, the last settings that have total harmonic distortion close to or greater than 100% provide extreme distortion conditions. The preamplifier has been identified with EMFN, LN and CN filters using three PPSs, suitable for the three filters with order 3, memory 20 and with a period of 655 408 samples. The coefficients of the filters were estimated with the cross-correlation method and the most relevant basis functions were selected according to the Bayesian information criterion, minimizing Eq. (2.63). The preamplifier has also been identified with a Volterra filter on the same data used for the CN filter identification. The cross-correlation method cannot be applied to the Volterra filter estimation since its basis functions are not orthogonal. Thus, the Volterra filter has been identified with the method of [69], which belongs to the most computationally efficient identification methods for LIP nonlinear systems available in the literature. In all conditions, the signal-to-noise ratio was greater than 65 dB.

Fig. 2.2 shows the number of selected terms and the percentage of unexplained power (*i.e.*, the ratio in percent between the residual MSE and the power of the output signal). The percentage of unexplained power is very low, close to 1%, till settings 16. Then it tends to increase, indicating that a third-order model is inadequate to represent the nonlinear system at those high nonlinear distortions. CN, LN and Volterra filters are all polynomial filters and each filter can be converted into one of the other representations. Thus, for the same input signal the filters should provide very similar results, with just a possible little change in the number of selected basis functions. This is confirmed in Fig. 2.2 for Volterra and CN filters, which are estimated for the same input signal. Also LN filters, which are estimated for a different PPS input, provide very similar results in Volterra and CN filters for small and medium nonlinear distortions. On the contrary, because of the different input signal, they produce larger errors for higher distortions. When the Volterra, LN and CN filters are estimated on the same signal, they provide very similar results. EMFN filters for low and medium distortions originate slightly worse results than the other filters, because they lack a linear term, but for higher distortions they are able to provide better results than the other filters, also when the estimate is performed on the same

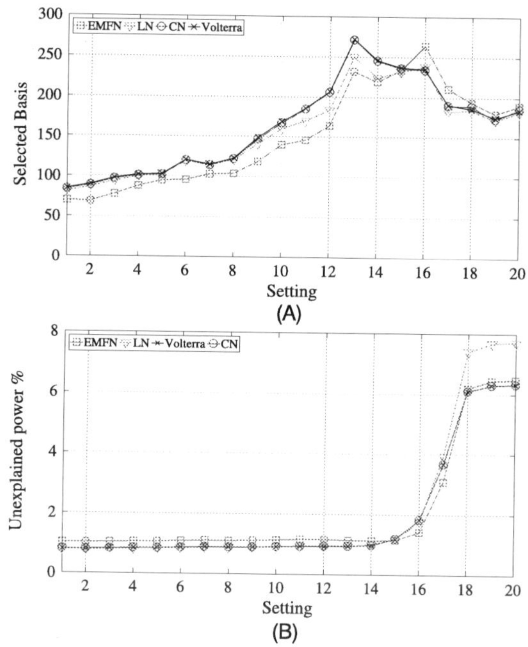


**FIGURE 2.2**

Number of selected bases (A) and unexplained power (B) for EMFN, LN, CN and Volterra filters.

signal. Note that the number of selected basis functions is very low compared with the total number of basis functions $N_T$ of a FLiP filter of order 3, memory 20, which is $N_T = 1771$.

There is here a significant difference between the effort necessary to estimate the CN, LN and EMFN filters and that for estimating the Volterra filters. Obtaining CN, LN and EMFN filters using PPSs and the cross-correlation method required only a few hours of computer time. On the contrary, computing the results for Volterra filters with the method in [69] required days of simulations on the same computer. As a matter of fact, if $T$ indicates the number of samples used for the identification, $B$ the number of candidate basis functions and $S$ the number of selected basis functions, the computational cost of the method in [69] is of the order of $TBS^2$ operations, while the cross-correlation method requires an order of $TB$ operations.

### 2.4.2 MULTIPLE-VARIANCE SYSTEM IDENTIFICATION AND EMULATION

In the second experiment, we illustrate the advantage offered by the multiple-variance methods in reducing the problem of the solution locality and in improving the emulation of nonlinear devices. For this purpose, several measurements have been carried out to test the exploitation of multiple-variance PPSs in the cross-correlation method and making comparisons with multiple-variance WGN [14]. Different test sessions have been accomplished, considering a real-world nonlinear device, modeled using a third-order Volterra series with memory 10 and 25, respectively. The adopted PPS sequence has order $P = 3$, memory $N = 25$, period $L = 1\,393\,024$ and variance $\sigma_{\text{pps}}^2 = 1/12$, and the WGN sequence has length $L$. A sampling frequency $f_s = 44.1$ kHz has been adopted. For the multiple-variance cross-correlation method, since in the considered examples the second-order kernel is dominant at high input variances, $\sigma_{x,0}^2 = \sigma_{x,1}^2 = \frac{\sigma_{\text{pps}}^2}{16}$ and $\sigma_{x,2}^2 = \sigma_{x,3}^2 = \sigma_{\text{pps}}^2$ have been assumed. With reference to the traditional cross-correlation method, the same two values for variance have been considered. Therefore, the identification of the model has been performed considering the following settings: (1) WGN with single variance $\sigma_{x,0}^2$, (2) WGN with single variance $\sigma_{x,3}^2$, (3) WGN with multiple variances, (4) PPS with single variance $\sigma_{x,0}^2$, (5) PPS with single variance $\sigma_{x,3}^2$, (6) PPS with multiple variances.

Then a performance evaluation has been performed applying WGN and music of length 44 100 samples to the system under investigation and to the model, assuming several input variances. Results are reported in terms of the normalized MSE (NMSE) in the frequency domain between the output of the system under investigation $y(n)$ and the output of the identified Volterra series $\hat{y}(n)$, according to the following formula:

$$\text{NMSE} = 10 \log_{10} \frac{\sum_{n=1}^{N} \left[ |Y(f_n)| - \left|\hat{Y}(f_n)\right| \right]^2}{\sum_{n=1}^{N} |Y(f_n)|^2}. \tag{2.73}$$

The real-world device used is the Presonus TubePRE microphone/instrument tube preamplifier. It provides a drive potentiometer controlling the amount of tube saturation, *i.e.*, the amount of applied distortion. The device has been set to provide a second harmonic distortion of 4.2% and third harmonic distortion of 0.5% on a 1-kHz tone signal. Once the model has been obtained, inputs with variance $\sigma_x^2$ in the interval $\left[(1/4096, \ldots, 1/4, 1/2, 1)\,\sigma_{\text{pps}}^2\right]$ have been adopted and the output of the real device has been compared with the output of the model. The results reported in Fig. 2.3 show that the exploitation of multiple-variance PPSs provides two advantages at the same time, *i.e.*, the input region in which the error has acceptable values both for noise and music is widened, thus overcoming the locality problem, and accuracy is improved with respect to stochastic input. In particular, the improvements provided by multiple-variance PPSs can be noted especially for music input as reported in Fig. 2.3 B.

## 2.5 CONCLUDING REMARKS

The chapter has discussed different families of orthogonal LIP nonlinear filters. FLiP filters have provided the common framework for introducing the orthogonal LIP filters. FLiP filters are able to arbitrarily well approximate any causal, time-invariant, finite-memory, continuous, nonlinear system. In orthogonal FLiP filters the basis functions are orthogonal for some stochastic distribution of the input signal. For example, EMFN and LN filters have orthogonal basis functions for a white uniform input signal in $\mathbb{R}_1$, CN filters for a white signal with the distribution in (2.32) in $\mathbb{R}_1$, WN filters for a white Gaussian input signal. The orthogonality of the basis functions guarantees a fast convergence of gradient descent adaptive identification algorithms and allows the identification of nonlinear filters with the cross-correlation method.

The cross-correlation method applied to stochastic inputs requires a huge amount of input samples for an accurate estimation of the coefficients. To solve this problem, PPSs have been proposed since

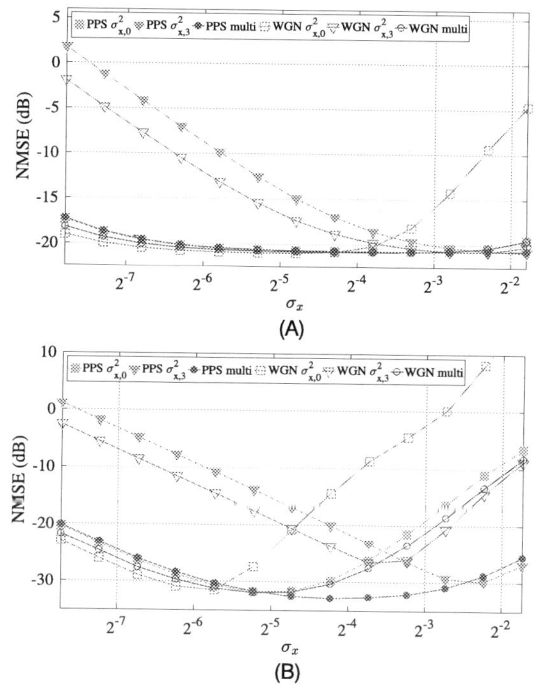


**FIGURE 2.3**

NMSE obtained for (A) noise and (B) music signals.

they are deterministic sequences that guarantee orthogonality of basis functions on a finite period. The PPSs allow also to avoid problems in the estimation of kernel diagonal points, which afflict the classical cross-correlation methods.

The nonlinear filter estimation is often affected by the problem of locality of the solution: the estimated filter well models the unknown system only at the same input power used for the identification. Multiple-variance methods can be used in conjunction with PPSs to improve the accuracy of the estimation and contrast the solution locality.

Experimental results have illustrated nonlinear system identification with FLiP filters using PPSs and the multiple-variance technique, highlighting the benefits provided by these methodologies.

## REFERENCES

[1] V. Volterra, Sopra le funzioni che dipendono da altre funzioni, in: Rend. R. Accademia dei Lincei, vol. III of IV, 1887, pp. 97–105.

[2] V. Volterra, Leçons sur les Fonctions de Lignes, Gauthier Villars, Paris, 1913.

[3] M. Frechét, Sur les fonctionnelles continues, in: Annales scientifiques de l'École Normale Supérieure, vol. 27 of 3, 1910, pp. 193–216.

[4] S. Boyd, L.O. Chua, Fading memory and the problem of approximating nonlinear operators with Volterra series, IEEE Transactions on Circuits and Systems CAS-32 (11) (1985) 1150–1161.
[5] I.W. Sandberg, $\mathbb{R}_+$ fading memory and extensions of input–output maps, IEEE Transactions on Circuits and Systems I 49 (11) (2002) 1586–1592.
[6] N. Wiener, Nonlinear Problems in Random Theory, John Wiley, New York, 1958.
[7] M.B. Brilliant, Theory of the Analysis of Nonlinear Systems, RLE Technical Report 345, MIT, Cambridge, MA, 1958.
[8] Y.W. Lee, M. Schetzen, Measurement of the Wiener kernels of a nonlinear system by crosscorrelation, International Journal of Control 2 (3) (1965) 237–254.
[9] G. Palm, T. Poggio, The Volterra representation and the Wiener expansion: validity and pitfalls, SIAM Journal of Applied Mathematics 33 (2) (1977) 195–216.
[10] G. Palm, T. Poggio, Stochastic identification methods for nonlinear systems: an extension of the Wiener theory, SIAM Journal of Applied Mathematics 34 (3) (1978) 524–534.
[11] S. Orcioni, M. Pirani, C. Turchetti, Advances in Lee–Schetzen method for Volterra filter identification, Multidimensional Systems and Signal Processing 16 (3) (2005) 265–284.
[12] Y. Goussard, W.C. Krenz, L. Stark, An improvement of the Lee and Schetzen cross-correlation method, IEEE Transactions on Automatic Control AC-30 (9) (1985) 895–898.
[13] M. Pirani, S. Orcioni, C. Turchetti, Diagonal kernel point estimation of n-th order discrete Volterra–Wiener systems, EURASIP Journal on Applied Signal Processing 2004 (12) (2004) 1807–1816.
[14] S. Orcioni, Improving the approximation ability of Volterra series identified with a cross-correlation method, Nonlinear Dynamics 78 (4) (2014) 2861–2869.
[15] A. Carini, S. Cecchi, L. Romoli, S. Orcioni, Perfect periodic sequences for nonlinear Wiener filters, in: Proc. of EUSIPCO 2016, European Signal Processing Conference, Budapest, Hungary, 2016.
[16] K. Narendra, P. Gallman, An iterative method for the identification of nonlinear systems using a Hammerstein model, IEEE Transactions on Automatic Control 11 (3) (1966) 546–550.
[17] E.L.O. Batista, R. Seara, A new perspective on the convergence and stability of NLMS Hammerstein filters, in: Proc. of ISPA 2013, 8th International Symposium on Image and Signal Processing and Analysis, Trieste, Italy, 2013, pp. 336–341.
[18] M. Gasparini, L. Romoli, S. Cecchi, F. Piazza, Identification of Hammerstein model using cubic splines and FIR filtering, in: Proc. of ISPA 2013, 8th International Symposium on Image and Signal Processing and Analysis, Trieste, Italy, 2013, pp. 347–352.
[19] J.M. Gil-Cacho, T. van Waterschoot, M. Moonen, S.H. Jensen, Linear-in-the-Parameters Nonlinear Adaptive Filters for Loudspeaker Modeling in Acoustic Echo Cancellation, Internal Report 13-110, ESAT-SISTA, K.U. Leuven, Leuven, Belgium.
[20] L. Romoli, M. Gasparini, S. Cecchi, A. Primavera, F. Piazza, Adaptive identification of nonlinear models using orthogonal nonlinear functions, in: Proc. AES 48th Int. Conf., Munich, Germany, Sep. 21–23, 2012, pp. 1597–1601.
[21] J. Kim, K. Konstantinou, Digital predistortion of wideband signals based on power amplifier model with memory, Electronics Letters 37 (2001) 1417–1418.
[22] D.R. Morgan, Z. Ma, J. Kim, M.G. Zierdt, J. Pastalan, A generalized memory polynomial model for digital predistortion of RF power amplifiers, IEEE Transactions on Signal Processing 54 (2006) 3852–3860.
[23] Y.H. Pao, Adaptive Pattern Recognition and Neural Networks, Addison-Wesley Longman Publishing Co., Inc., Boston, MA, USA, 1989.
[24] J.C. Patra, R.N. Pal, A functional link artificial neural network for adaptive channel equalization, Signal Processing 43 (1995) 181–195.
[25] J. Patra, A. Kot, Y.Q. Chen, Chebyshev functional link artificial neural networks for nonlinear dynamic system identification, in: Proc. of IEEE International Conference on Systems, Man, and Cybernetics, 2000, vol. 4, 2000, pp. 2655–2660.
[26] J. Patra, W. Chin, P. Meher, G. Chakraborty, Legendre-FLANN-based nonlinear channel equalization in wireless communication system, in: Proc. of SMC 2008, IEEE International Conference on Systems, Man and Cybernetics, 2008, pp. 1826–1831.
[27] H. Zhao, J. Zhang, Functional link neural network cascaded with Chebyshev orthogonal polynomial for nonlinear channel equalization, Signal Processing 88 (2008) 1946–1957.
[28] H. Zhao, J. Zhang, Adaptively combined FIR and functional link artificial neural network equalizer for nonlinear communication channel, IEEE Transactions on Neural Networks 20 (4) (2009) 665–674.
[29] K. Das, J. Satapathy, Legendre neural network for nonlinear active noise cancellation with nonlinear secondary path, in: Proc. of IMPACT 2011, International Conference on Multimedia, Signal Processing and Communication Technologies, 2011, pp. 40–43.

[30] N. George, G. Panda, A reduced complexity adaptive Legendre neural network for nonlinear active noise control, in: Proc. of IWSSIP 2012, International Conference on Systems, Signals and Image Processing, 2012, pp. 560–563.
[31] M.D. Buhmann, Radial Basis Functions: Theory and Implementations, University Press, Cambridge, 2003.
[32] G.L. Sicuranza, A. Carini, On a class of nonlinear filters, in: M.G.I. Tabus, K. Egiazarian (Eds.), Festschrift in Honor of Jaakko Astola on the Occasion of His 60th Birthday, vol. TICSP Report #47, 2009, pp. 115–144.
[33] A. Carini, G.L. Sicuranza, Fourier nonlinear filters, Signal Processing 94 (2014) 183–194.
[34] A. Carini, G.L. Sicuranza, Even mirror Fourier nonlinear filters, in: Proc. of ICASSP 2013, International Conference on Acoustic, Speech, Signal Processing, Vancouver, Canada, 2013, pp. 5608–5612.
[35] A. Carini, S. Cecchi, M. Gasparini, G.L. Sicuranza, Introducing Legendre nonlinear filters, in: Proc. of ICASSP 2014, International Conference on Acoustic, Speech, Signal Processing, Florence, Italy, 2014, pp. 7989–7993.
[36] A. Carini, S. Cecchi, L. Romoli, G.L. Sicuranza, Legendre nonlinear filters, Signal Processing 109 (2015) 84–94.
[37] A. Carini, G.L. Sicuranza, A study about Chebyshev nonlinear filters, Signal Processing 122 (2016) 24–32.
[38] A. Carini, G.L. Sicuranza, Perfect periodic sequences for identification of even mirror Fourier nonlinear filters, in: Proc. of ICASSP 2014, International Conference on Acoustic, Speech, Signal Processing, Florence, Italy, 2014, pp. 8009–8013.
[39] A. Carini, G.L. Sicuranza, Perfect periodic sequences for even mirror Fourier nonlinear filters, Signal Processing 104 (2014) 80–93.
[40] A. Carini, S. Cecchi, L. Romoli, G.L. Sicuranza, Perfect periodic sequences for Legendre nonlinear filters, in: Proc. of EUSIPCO 2014, European Signal Processing Conference, Lisbon, Portugal, 2014.
[41] G.L. Sicuranza, A. Carini, Nonlinear system identification using quasi-perfect periodic sequences, Signal Processing 120 (2016) 174–184.
[42] W. Rudin, Principles of Mathematical Analysis, McGraw-Hill, New York, 1976.
[43] N.D. Vanli, S.S. Kozat, A comprehensive approach to universal piecewise nonlinear regression based on trees, IEEE Transactions on Signal Processing 62 (20) (2014) 5471–5485.
[44] J.C. Adams, On the expression of the product of any two Legendre's coefficients by means of a series of Legendre's coefficients, Proceedings of the Royal Society of London 27 (1878) 63–71.
[45] T.J. Rivlin, Chebyshev Polynomials: From Approximation Theory to Algebra and Number Theory, John Wiley & Sons, New York, USA, 1990.
[46] A. Gil, J. Segura, N. Temme, Numerical Methods for Special Functions, Society for Industrial and Applied Mathematics, 2007.
[47] M. Abramowitz, I.A. Stegun, Handbook of Mathematical Functions: With Formulas, Graphs, and Mathematical Tables, Dover Publications, New York, 1972.
[48] S. Roman, The Umbral Calculus, Academic Press, San Diego, 1984.
[49] V.J. Mathews, G.L. Sicuranza, Polynomial Signal Processing, Wiley, New York, 2000.
[50] A. Fermo, A. Carini, G.L. Sicuranza, Simplified Volterra filters for acoustic echo cancellation in GSM receivers, in: Proc. of EUSIPCO 2000, European Signal Processing Conference, Tampere, Finland, Sept. 5–8, 2000.
[51] G.M. Raz, B.D.V. Veen, Baseband Volterra filters for implementing carrier based nonlinearities, IEEE Transactions on Signal Processing 46 (1) (1998) 103–114.
[52] L.A. Azpicueta-Ruiz, M. Zeller, A.R. Figueiras-Vidal, J. Arenas-Garcia, W. Kellermann, Adaptive combination of Volterra kernels and its application to nonlinear acoustic echo cancellation, IEEE Transactions on Audio, Speech and Language Processing 19 (11) (2011) 97–110.
[53] A. Fermo, A. Carini, G.L. Sicuranza, Low-complexity nonlinear adaptive filters for acoustic echo cancellation in GSM handset receivers, European Transactions on Telecommunications 14 (2) (2003) 161–169.
[54] G.L. Sicuranza, A. Carini, A new recursive controller for nonlinear active noise control, in: Proc. ISPA 2013, 8th International Symposium on Image and Signal Processing and Analysis, Trieste, Italy, 2013, pp. 626–631.
[55] A. Carini, G.L. Sicuranza, Recursive even mirror Fourier nonlinear filters and simplified structures, IEEE Transactions on Signal Processing 62 (24) (2014) 6534–6544.
[56] V. Ipatov, Ternary sequences with ideal periodic autocorrelation properties, Radio Engineering and Electronic Physics 24 (1979) 75–79.
[57] A. Milewski, Periodic sequences with optimal properties for channel estimation and fast start-up equalization, IBM Journal of Research and Development 27 (5) (1983) 426–431.
[58] R.H. Kwong, E.W. Johnston, A variable step size LMS algorithm, IEEE Transactions on Signal Processing 40 (1992) 1633–1642.

[59] C. Antweiler, M. Dörbecker, Perfect sequence excitation of the NLMS algorithm and its application to acoustic echo control, Annales des Telecommunications 49 (7–8) (1994) 386–397.

[60] C. Antweiler, M. Antweiler, System identification with perfect sequences based on the NLMS algorithm, International Journal of Electronics and Communications (AEU) 49 (3) (1995) 129–134.

[61] C. Antweiler, Multi-channel system identification with perfect sequences, in: R. Martin, U. Heute, C. Antweiler (Eds.), Advances in Digital Speech Transmission, SPIE, John Wiley & Sons, 2008, pp. 171–198.

[62] C. Antweiler, A. Telle, P. Vary, NLMS-type system identification of MISO systems with shifted perfect sequences, in: Proc. IWAENC 2008, International Workshop on Acoustic Echo and Noise Control, Seattle, Washington, USA, 2008.

[63] A. Carini, Efficient NLMS and RLS algorithms for perfect and imperfect periodic sequences, IEEE Transactions on Signal Processing 58 (4) (2010) 2048–2059.

[64] W.H. Press, S.A. Teukolsky, W.T. Vetterling, B.P. Flannery, Numerical Recipes in C: The Art of Scientific Computing, Cambridge University Press, New York, NY, USA, 1995.

[65] H. Akaike, A new look at the statistical model identification, IEEE Transactions on Signal Processing 19 (6) (1974) 716–723.

[66] E.J. Hannan, B.G. Quinn, The determination of the order of an autoregression, Journal of the Royal Statistical Society. Series B 41 (2) (1979) 190–195.

[67] G. Schwartz, Estimating the dimension of a model, The Annals of Statistics 6 (1978) 416–464.

[68] S. Orcioni, L. Romoli, S. Cecchi, A. Carini, Multivariance nonlinear system identification using Wiener basis functions and perfect sequences, in: Proc. of EUSIPCO 2017, European Signal Processing Conference, 2017.

[69] K. Li, J.-X. Peng, G.W. Irwin, A fast nonlinear model identification method, IEEE Transactions on Automatic Control 50 (8) (2005) 1211–1216.

CHAPTER

# SPLINE ADAPTIVE FILTERS

## THEORY AND APPLICATIONS

3

**Michele Scarpiniti*, Danilo Comminiello*, Raffaele Parisi*, Aurelio Uncini***
**Sapienza University of Rome, Rome, Italy*

## CONTENTS



## CHAPTER POINTS

- We describe a nonlinear filtering approach based on spline nonlinear functions.
- The proposed approach can be successfully used for the identification of nonlinear systems.
- The proposed approach outperforms other approaches based on Volterra filters, both in speed of convergence and in performance.
- We provide upper bounds on the choice of learning rates and the theoretical EMSE estimates.

**Adaptive Learning Methods for Nonlinear System Modeling. DOI: 10.1016/B978-0-12-812976-0.00004-X**

## 3.1 INTRODUCTION

The theory of linear adaptive filtering (LAF) is well established in the literature [1–3]. In addition, the use of LAF methods for the solution of real-world problems is extensive and represents a central paradigm for many strategic applications [4]. The basic discrete-time linear adaptive filter model is made of a simple linear combiner whose length is determined by following some proper criterion and whose weights are adapted by choosing the most convenient optimization strategy [1].

In contrast, in many practical situations, the nature of the process of interest is often affected by any nonlinearity and thus requires the use of nonlinear adaptive filter (NAF) models. In the ideal case, if the nature of the nonlinearities is known *a priori*, it is possible to determine the NAF structure, in analogy with the mathematical model of nonlinearities. In this case the NAF is referred to as *white-box* (or *structural*) *model* and the process of adaptation consists of the estimation of the free model parameters (*e.g.*, the coefficients of the nonlinear difference equation) [5].

On the contrary, if any *a priori* knowledge about the nature of nonlinearity is not available, it is preferred to use a *black-box* approach (*functional model*) [6,7]. In these cases, a unique framework for the solution does not exist and the existence of the solution cannot be guaranteed.

One of the best-known functional models that provide a good and general solution is based on the Volterra series [8]. Unfortunately, due to the large number of free parameters required, the Volterra adaptive filter (VAF) is generally employed only in situations of mild nonlinearity [8,9]. Neural networks (NNs) [10,11] represent another flexible tool to realize black-box NAFs, but they generally require a high computational cost and are not always feasible in online applications.

In the literature on nonlinear filtering, one of the most used structures is the so-called *block-oriented* representation, in which linear time-invariant (LTI) models are connected with memoryless nonlinear functions. The basic classes of block-oriented nonlinear systems are represented by the Wiener and Hammerstein models [12–14] and all those architectures originated by the connection of these two classes according to different topologies (*i.e.*, parallel, cascade, feedback, etc. [15]). More specifically, the Wiener model consists of a cascade of an LTI filter followed by a static nonlinear function and sometimes it is known as linear–nonlinear (LN) model [16–19]. Conversely, the Hammerstein model consists of a cascade connection of a static nonlinear function followed by an LTI filter [20–22] and is sometimes indicated as nonlinear–linear (NL) model.

In a general nonlinear problem, with no *a priori* information available, the best adaptive model (Hammerstein or Wiener) is that most similar to the one to be identified. This fact implies that Wiener and Hammerstein models could be too restrictive in practical applications. Hence, other cascade models, such as the linear–nonlinear–linear (LNL) or the nonlinear–linear–nonlinear (NLN) models, were also introduced [12,23,24]. These cascade models, here called *sandwich models*, constitute a relatively simple but important class of nonlinear models that can approximate a wide class of nonlinear systems [25].

These models assume a key role in practical applications, since they are often related to specialized classes of nonlinear systems. As a matter of fact, these architectures can be successfully used in several applications in all fields of engineering [26], from circuit and system theory to modeling nonlinear characteristics of electronic devices and resistive networks [27–29], to signal processing [19, 30], biomedical data analysis [31], digital communications [32,33] and even to hydraulics [34] and chemistry [35].

A crucial point in adopting a block-oriented model is the choice of the specific nonlinear function to be adopted. Obviously, several choices for this nonlinearity are possible, namely some invertible squashing functions, piece-wise linear functions [36,37,27], polynomial nonlinearities [38], splines [39] and more.

Recently a novel approach for the identification of Wiener, Hammerstein and sandwich models based on Spline Adaptive Filters (SAFs) has been proposed [30,40–44]. Among the advantages of these solutions, in order to yield great representative potential, high performance, low computational cost and easy implementability in dedicated hardware, all the nonlinearities involved in such systems are flexible and adapted by means of a fixed-length look-up table (LUT) that collects a suitable number of samples of the nonlinearity. Moreover, in order to have a continuous and derivable smooth function, the LUT samples are interpolated by a low-order uniform spline, like done in [40] for the Wiener SAF (WSAF), in [41] for the Hammerstein SAF (HSAF) and in [43] for the two sandwich SAFs (S1SAF and S2SAF).

SAFs have been successfully adopted in the literature in different scenarios, such as nonlinear echo cancellation [30] and active noise control [45–47], where signals are distorted by the low-cost hardware equipments. The standard LMS adaptation scheme has also been extended to the normalized version in [48] and to the maximum correntropy criterion [49]. The latter approach is particularly efficient in the case when data are corrupted by impulsive noise or contain large outliers. Finally, the SAF approach has been recently extended to the diffusion case [50], involving a network of distributed sensors.

In this chapter we aim at providing a deep and unifying overview of SAF models. After the description of the basic architectures and related learning algorithms, the main theoretical results on the choice of the step-size and the steady-state mean square error (MSE) are also provided. Finally some experimental results show the effectiveness of this family of nonlinear adaptive filters to solve real-world applications.

The rest of the chapter is organized as follows. Section 3.2 gives some background of spline interpolation. Section 3.3 provides a complete description of SAF models including the computational cost. Section 3.4 is devoted to the introduction of convergence results, while Section 3.5 provides some applicative scenarios and numerical results. Finally, Section 3.6 concludes the chapter.

### 3.1.1 NOTATION

In this chapter matrices are represented by boldface capital letters, *i.e.*, $\mathbf{A} \in \mathbb{R}^{M \times N}$, while $\mathrm{Tr}(\cdot)$ denotes the trace of a matrix. All vectors are column vectors, denoted by boldface lowercase letters, like $\mathbf{w} \in \mathbb{R}^{M\times 1} = [w[0], w[1], \ldots, w[M-1]]^T$, where $w[i]$ denotes the $i$th individual entry of $\mathbf{w}$. In the recursive algorithm definition, a discrete-time subscript index $n$ is added. For example, the weight vector, calculated according to some law, is written as $\mathbf{w}_{n+1} = \mathbf{w}_n + \Delta\mathbf{w}_n$. In the case of signal regression, vectors are indicated as $\mathbf{x}_n \in \mathbb{R}^{M\times 1} = [x[n], x[n-1], \ldots, x[n-M+1]]^T$. Moreover, regarding the interpolated nonlinearity, the LUT control points are denoted $\mathbf{q}_n \in \mathbb{R}^{Q\times 1} = \left[q_1, q_2, \ldots, q_Q\right]^T$, where $q_j$ is the $j$th control point, while the $i$th span is denoted by $\mathbf{q}_{i,n} \in \mathbb{R}^{4\times 1} = \left[q_i, q_{i+1}, q_{i+2}, q_{i+3}\right]^T$. Finally, the Euclidean norm and the transpose of a vector are denoted by $\|\cdot\|$ and $(\cdot)^T$, respectively.

## 3.2 FOUNDATION OF SPLINE INTERPOLATION

Historically derived from elastic rulers that cross a certain number of points, called *knots*, and used in the past for technical drawing, *splines* are differentiable, up to a given order, polynomial curves for interpolation or approximation of a succession of $Q+1$ knots inside a convex hull. The knots $\mathbf{Q}_i = [q_{x,i}, q_{y,i}]^T$, for $i = 0, 1, \ldots, Q$, represent a tabulated function on the plane $x$-$y$ [51,39,52] and, in order to construct an appropriate function, they are also constrained as

$$q_{x,0} < q_{x,1} < \cdots < q_{x,Q}. \tag{3.1}$$

Literature on splines is vast. Without loss of generality, in this brief review we consider the univariate basis or blending functions developed by Schoenberg, who introduced the name B-splines, deriving them from Bézier splines [51].

Let $u \in [q_{x,i}, q_{x,i+1}]$ be the abscissa value between two consecutive knots. The spline curve is defined as an affine combination of some knots of a suitable $P$ degree spline basis function $N_i^P(u)$ with minimal support and certain continuity properties [52]:

$$\varphi(u) = \sum_{i=0}^{Q-P-1} \mathbf{Q}_i N_i^P(u), \quad u \in [q_{x,n}, q_{x,Q-P}]. \tag{3.2}$$

Given $Q+1$ knots, the univariate spline basis functions $N_i^P(u)$ can be defined by the Cox–de Boor recursion [39]

$$\begin{aligned} N_i^0(u) &= \begin{cases} 1, & q_{x,i} \le u < q_{x,i+1}, \\ 0, & \text{otherwise}, \end{cases} \\ N_i^P(u) &= \frac{u - q_{x,i}}{q_{x,i+P} - q_{x,i}} N_i^{P-1}(u) + \frac{q_{x,i+P+1} - u}{q_{x,i+P+1} - q_{x,i+1}} N_{i+1}^{P-1}(u), \end{aligned} \tag{3.3}$$

where $i = 0, 1, \ldots, Q-P-1$. The blending functions $N_i^P(u)$ are polynomials of degree $P$ with continuity of the first $P-1$ derivatives. For example, for $P=1$, we have

$$N_i^1(u) = \begin{cases} \frac{u - q_{x,i}}{q_{x,i+1} - q_{x,i}}, & q_{x,i} \le u < q_{x,i+1}, \\ \frac{q_{x,i+2} - u}{q_{x,i+2} - q_{x,i+1}}, & q_{x,i+1} \le u < q_{x,i+2}, \\ 0, & \text{otherwise}, \end{cases}$$

which results in a classical linear interpolation scheme, so $N_i^0(u)$ is a rectangular function, $N_i^1(u)$ is a linear double-ramp function, $N_i^2(u)$ is a quadratic double-ramp function. The generic $N_i^P(u)$ can be evaluated as a sequential convolution of $P$ rectangular pulse functions $N_i^P = N_i^0 * \ldots * N_i^0$.

Hence, the curve (3.2) can be written as

$$\varphi(u) = \varphi_i(u), \quad \forall i = 0, \ldots, Q \text{ and } u \in [0, 1), \tag{3.4}$$

where $\varphi_i(u)$ is a $P$ degree local polynomial called curve *span*, entirely defined from $P+1$ knots, that are sometimes called control points. Moreover, scaling or translating the knot vector does not modify the basis functions, that can be fixed *a priori* for all control points.

From (3.4), there must be a unique mapping that allows us to calculate the local parameter $u$, as well as the proper curve span index $i$, from the global abscissa parameter (that in our case is the input sequence to the nonlinearity). In this way, we can represent any point lying on the curve $\varphi(u)$ as a point belonging to the single $\varphi_i(u)$ curve span.

### 3.2.1 UNIFORM CUBIC SPLINE

The recursive Eq. (3.3) can be entirely pre-calculated with considerable computational saving, by imposing a uniform distribution knot-vector, *i.e.*, constant sampling step of the abscissa $\Delta x = q_{x,i} - q_{x,i+1}$ [39]. This implies that the local polynomial $\varphi_i(u)$ in (3.4) can be written in a very simple form. A similar approach was successfully used in multilayer feed-forward neural networks [53].

Regarding the choice of the optimal degree $P$ of a spline basis, a good compromise between flexibility and approximation behavior is represented by uniform cubic spline nonlinearities [54], as they provide good approximation capabilities avoiding the problem of over-fitting.

In the case of our interest where $P = 3$, the precalculation of the recursion (3.3) provides [39,54, 53,40]

$$N_i^3(u) = \begin{cases} \frac{1}{6}\left[-(u-q_i)^3 + 3(u-q_i)^2 - 3(u-q_i) + 1\right], & q_{x,i} \le u < q_{x,i+1}, \\ \frac{1}{6}\left[3(u-q_{i+1})^3 - 6(u-q_{i+1})^2 + 4, \right], & q_{x,i+1} \le u < q_{x,i+2}, \\ \frac{1}{6}\left[-3(u-q_{i+2})^3 + 3(u-q_{i+2})^2 + 3(u-q_{i+2}) + 1\right], & q_{x,i+2} \le u < q_{x,i+3}, \\ \frac{1}{6}(u-q_{i+3})^3, & q_{x,i+3} \le u < q_{x,i+4}, \end{cases}$$

which can be reformulated in a matrix form as

$$\varphi_i(u) = \begin{bmatrix} u^3 & u^2 & u & 1 \end{bmatrix} \mathbf{C} \begin{bmatrix} q_i \\ q_{i+1} \\ q_{i+2} \\ q_{i+3} \end{bmatrix}, \tag{3.5}$$

for $i = 0, 1, \ldots, Q$ and $u \in [0, 1)$ and where the basis matrix $\mathbf{C}$ is defined as

$$\mathbf{C} = \frac{1}{6}\begin{bmatrix} -1 & 3 & -3 & 1 \\ 3 & -6 & 3 & 0 \\ -3 & 0 & 3 & 0 \\ 1 & 4 & 1 & 0 \end{bmatrix}. \tag{3.6}$$

Imposing different constraints to the approximation relationship, several spline bases with different properties can be evaluated in a similar manner. An example of such a basis is the *Catmull–Rom* (CR) spline [55], very important in many applications. The interpolation scheme is the same of (3.5), while the CR-spline basis matrix $\mathbf{C}$ has the form

$$\mathbf{C} = \frac{1}{2}\begin{bmatrix} -1 & 3 & -3 & 1 \\ 2 & -5 & 4 & -1 \\ -1 & 0 & 1 & 0 \\ 0 & 2 & 0 & 0 \end{bmatrix}. \tag{3.7}$$

Note that CR-splines, initially developed for computer graphics purposes, represent a local interpolating scheme. The CR spline, in fact, specifies a curve that passes through all of the control points, a feature which is not necessarily true for other spline methodologies. Overall, the CR-spline results in a more local approximation with respect to the B-spline.

For both the CR- and B-basis the curve has a continuous first derivative. For the B-spline the second derivative is also a continuous function, while for the CR spline the second derivative is not continuous at the knots. However, there is a regularization property common to most of the polynomial spline basis set (CR- and B-spline included) called *variation diminishing property* [39], which ensures the absence of unwanted oscillations of the curve between two consecutive control points, as well as the exact representation of linear segments.

Note also that the entries of the matrix **C** satisfy an important general property: the sum of all the entries is equal to one. This property is a direct consequence of the fact that the spline basis functions, derived from Bézier curves, are a generalization of the Bernstein polynomials and hence they represent a partition of unit (*i.e.*, the sum of all basis function is one) [39,51].

In the next section, we present the main ideas on the SAFs and the derivation of the basic SAF models.

## 3.3 SPLINE ADAPTIVE FILTERS

A Spline Adaptive Filter (SAF) consists of the cascade of a suitable number of linear filters and memoryless nonlinear functions obtained by the spline interpolation scheme described in Section 3.2.

In order to compute the nonlinear output $y[n] = \varphi(x[n])$, which is a spline interpolation of some adaptive control points contained in a LUT, we have to determine the explicit dependence between the input signal $x[n]$ and $y[n]$. This can easily be done by considering that $y[n]$ is a function of two local parameters $u_n$ and $i$ that depend on $x[n]$. From this onward we denote the local abscissa $u_n$ with the subscript $n$ in order to highlight the dependence on the input sample $x[n]$.

In the simple case of a uniformly spaced spline and a third-order ($P = 3$) local polynomial interpolator (adopted in this work), each spline segment, denoted *span*, is controlled by four ($P + 1$) control points, also known as *knots*, addressed by the span index $i$ that points at the first element of the related four points contained in the LUT. The computation procedure for the determination of the span index $i$ and the local parameters $u_n$ can be expressed by the following equations [53]:

$$\begin{aligned} u_n &= \frac{x[n]}{\Delta x} - \left\lfloor \frac{x[n]}{\Delta x} \right\rfloor, \\ i &= \left\lfloor \frac{x[n]}{\Delta x} \right\rfloor + \frac{Q-1}{2}, \end{aligned} \tag{3.8}$$

where $\Delta x = q_{i+1} - q_i$ is the uniform space between knots, $\lfloor \bullet \rfloor$ is the floor operator and $Q$ is the fixed LUT length, *i.e.*, the total number of control points. The second term on the right side of the second equation is an offset value needed to force $i$ to be always nonnegative. Note that the index $i$ depends on the time index $n$, *i.e.*, $i_n$; for simplicity of notation in the following we will write simply $i_n \equiv i$.

Referring to [40] and to [41] and with reference to (3.5), the output of the nonlinearity can be evaluated as

$$y[n] = \varphi_i(u_n) = \mathbf{u}_n^T \mathbf{C} \mathbf{q}_{i,n}, \tag{3.9}$$

where the vector $\mathbf{u}_n$ is defined as $\mathbf{u}_n \in \mathbb{R}^{4\times 1} = \left[u_n^3,\ u_n^2,\ u_n,\ 1\right]^T$, $\mathbf{C} \in \mathbb{R}^{4\times 4}$ is the precomputed matrix in (3.6) or (3.7), called *spline basis matrix*, and the vector $\mathbf{q}_{i,n}$ contains the control points at the

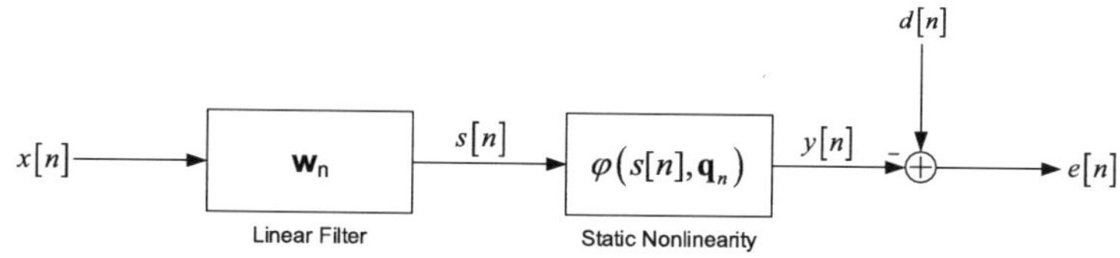


**FIGURE 3.1**

Block diagram of a WSAF.

instant $n$. More specifically, $\mathbf{q}_{i,n}$ is defined by $\mathbf{q}_{i,n} \in \mathbb{R}^{4\times 1} = \left[q_i,\ q_{i+1},\ q_{i+2},\ q_{i+3}\right]^T$, where $q_k$ is the $k$th entry in the LUT.

It could be very important to compute the derivative of (3.9) with respect to its input. It is easily evaluated as

$$\varphi_i'(u_n) = \dot{\mathbf{u}}_n^T \mathbf{C}\mathbf{q}_{i,n}, \tag{3.10}$$

where $\dot{\mathbf{u}}_n \in \mathbb{R}^{4\times 1} = \left[3u_n^2,\ 2u_n,\ 1,\ 0\right]^T$.

The online learning algorithm can be derived by considering the cost function[1] (CF) $\hat{J}(\mathbf{w}_n, \mathbf{q}_{i,n}) = E\{e^2[n]\}$. In order to derive the LMS algorithm, this CF is usually approximated by considering only the instantaneous error

$$J(\mathbf{w}_n, \mathbf{q}_{i,n}) = e^2[n]. \tag{3.11}$$

The expression of the error signal $e[n]$ in (3.11) depends on the particular architecture used.

In the following subsections we provide the description of the four basic SAF architectures.

### 3.3.1 WIENER SAF

The Wiener SAF (WSAF) architecture consists of the cascade of a linear filter, represented by the filter coefficient weight vector $\mathbf{w}_n$, and a memoryless nonlinear function implemented by a spline interpolation scheme whose knots are denoted by the vector $\mathbf{q}_n$. The Wiener architecture is shown in Fig. 3.1, where $s[n]$ denotes the output of the linear filter and it is the input to the spline function.

With reference to Fig. 3.1, let us define the output error $e[n]$ as

$$e[n] = d[n] - y[n] = d[n] - \varphi_i(u_n), \tag{3.12}$$

where $d[n]$ is the reference signal, *i.e.*, the output of the model to be identified. The online learning algorithm can be derived by the minimization of (3.11). We proceed by applying the stochastic gradient adaptation [40], obtaining the LMS iterative algorithm

$$\mathbf{w}_{n+1} = \mathbf{w}_n + \mu_w[n]e[n]\varphi_i'(u)\mathbf{x}_n, \tag{3.13}$$

[1] In this work we consider only real-valued variables.

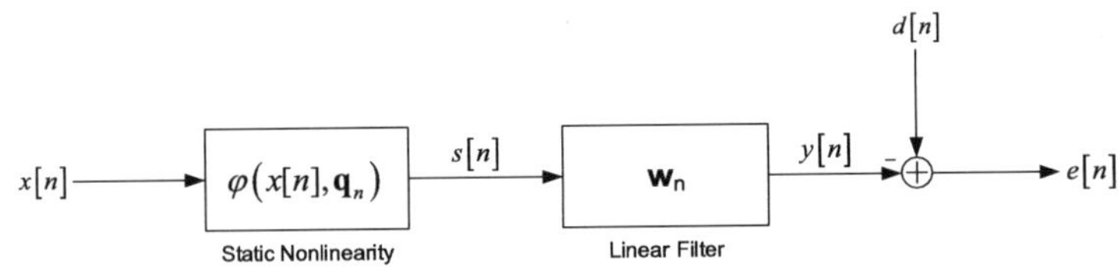


**FIGURE 3.2**

Block diagram of an HSAF.

for the weights of the linear filter and

$$\mathbf{q}_{i,n+1} = \mathbf{q}_{i,n} + \mu_q[n]e[n]\mathbf{C}^T\mathbf{u}_n, \tag{3.14}$$

for the spline control points. The parameters $\mu_w[n]$ and $\mu_q[n]$ represent the learning rates for the weights and for the control points, respectively, and, for simplicity, incorporate the other constant values.

### 3.3.2 HAMMERSTEIN SAF

The Hammerstein SAF (HSAF) architecture consists of the cascade of a memoryless nonlinear function implemented by a spline interpolation scheme whose knots are denoted by the vector $\mathbf{q}_n$ and a linear filter, represented by the filter coefficient weight vector $\mathbf{w}_n$. The Hammerstein architecture is shown in Fig. 3.2, where $s[n]$ denotes the output of the nonlinear function.

With reference to Fig. 3.2, we can define the output error $e[n]$ as

$$e[n] = d[n] - y[n] = d[n] - \mathbf{w}_n^T \cdot \mathbf{s}_n, \tag{3.15}$$

where $d[n]$ is the reference signal, $\mathbf{s}_n = [s[n], s[n-1], \ldots, s[n-M+1]]^T$, and each sample $s[n-k]$ is evaluated by using (3.9).

As in the Wiener case, the learning algorithm is derived by minimizing the CF (3.11). We proceed by applying the stochastic gradient adaptation as done in [41], obtaining the LMS iterative algorithm

$$\mathbf{w}_{n+1} = \mathbf{w}_n + \mu_w[n]e[n]\mathbf{s}_n \tag{3.16}$$

for the weights of the linear filter and

$$\mathbf{q}_{i,n+1} = \mathbf{q}_{i,n} + \mu_q[n]e[n]\mathbf{C}^T\mathbf{U}_{i,n}\mathbf{w}_n \tag{3.17}$$

for the spline control points. In (3.17), $\mathbf{U}_{i,n} \in \mathbb{R}^{4\times M} = \left[\mathbf{u}_{i,n}, \mathbf{u}_{i,n-1}, \ldots, \mathbf{u}_{i,n-M+1}\right]$ is a matrix which collects $M$ past vectors $\mathbf{u}_{i,n-k}$, where each vector assumes the value of $\mathbf{u}_{n-k}$ if evaluated in the same span $i$ of the current sample input $x[n]$ or in an overlapped span; otherwise it is a zero vector.

As in the Wiener case, the learning rates $\mu_w[n]$ and $\mu_q[n]$ incorporate the other constant values.

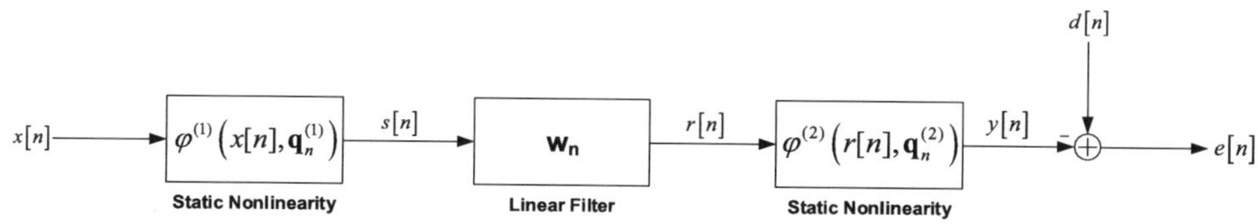


**FIGURE 3.3**

Block diagram of an S1SAF.

### 3.3.3 SANDWICH 1 SAF

The Sandwich 1 (S1SAF) architecture [43] consists of the cascade of a memoryless nonlinear function implemented by a spline interpolation scheme whose knots are denoted by the vector $\mathbf{q}_n^{(1)}$, a linear filter, represented by the filter coefficient weight vector $\mathbf{w}_n$, and a second memoryless nonlinear function implemented by a spline interpolation scheme whose knots are denoted by the vector $\mathbf{q}_n^{(2)}$. The S1 architecture is shown in Fig. 3.3, where $s[n]$ denotes the output of the first nonlinear function and $r[n]$ is the output of the linear filter.

With reference to Fig. 3.3, we can define the output error $e[n]$ as

$$e[n] = d[n] - y[n] = d[n] - \varphi^{(2)}\left(\mathbf{w}_n^T \cdot \mathbf{s}_n\right), \tag{3.18}$$

where $d[n]$ is the reference signal. As in the previous cases, the learning algorithm is derived by minimizing the CF (3.11). We proceed by applying the stochastic gradient adaptation as done in [43], obtaining the LMS iterative algorithm

$$\mathbf{q}_{i,n+1}^{(2)} = \mathbf{q}_{i,n}^{(2)} + \mu_q^{(2)}[n]e[n]\mathbf{C}^T\mathbf{u}_n^{(2)} \tag{3.19}$$

for the control points of the second nonlinearity,

$$\mathbf{w}_{n+1} = \mathbf{w}_n + \mu_w[n]e[n]\varphi_i'(u_n^{(2)})\mathbf{s}_n \tag{3.20}$$

for the weights of the linear filter and

$$\mathbf{q}_{i,n+1}^{(1)} = \mathbf{q}_{i,n}^{(1)} + \mu_q^{(1)}[n]e[n]\varphi_i'(u_n^{(2)})\mathbf{C}^T\mathbf{U}_{i,n}^{(1)}\mathbf{w}_n \tag{3.21}$$

for the control points of the first nonlinearity.

As in the previous cases, the learning rates $\mu_w[n]$, $\mu_q^{(1)}[n]$ and $\mu_q^{(2)}[n]$ incorporate the other constant values.

### 3.3.4 SANDWICH 2 SAF

The Sandwich 2 (S2SAF) architecture [43] consists of the cascade of a linear filter, represented by the filter coefficient weight vector $\mathbf{w}_n^{(1)}$, a memoryless nonlinear function implemented by a spline interpolation scheme whose knots are denoted by the vector $\mathbf{q}_n$ and a second linear filter, represented

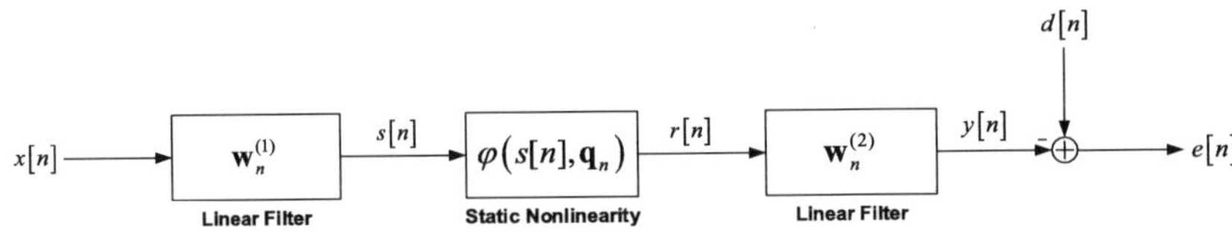


**FIGURE 3.4**

Block diagram of an S2SAF.

by the filter coefficient weight vector $\mathbf{w}_n^{(2)}$. The S2 architecture is shown in Fig. 3.4, where $s[n]$ denotes the output of the first linear filter and $r[n]$ is the output of the nonlinear function.

With reference to Fig. 3.4, we can define the output error $e[n]$ as

$$e[n] = d[n] - y[n] = d[n] - \mathbf{w}_n^{(2)T} \cdot \varphi\left(\mathbf{w}_n^{(1)T} \cdot \mathbf{x}_n\right), \tag{3.22}$$

where again $d[n]$ is the reference signal. As in the previous cases, the learning algorithm is derived by minimizing the CF (3.11). We proceed by applying the stochastic gradient adaptation as done in [43], obtaining the LMS iterative algorithm

$$\mathbf{w}_{n+1}^{(2)} = \mathbf{w}_n^{(2)} + \mu_w^{(2)}[n]e[n]\mathbf{r}_n \tag{3.23}$$

for the weights of the second linear filter,

$$\mathbf{q}_{i,n+1} = \mathbf{q}_{i,n} + \mu_q[n]e[n]\mathbf{C}^T\mathbf{U}_{i,n}\mathbf{w}_n^{(2)} \tag{3.24}$$

for the control points of the nonlinearity and

$$\mathbf{w}_{n+1}^{(1)} = \mathbf{w}_n^{(1)} + \mu_w^{(1)}[n]e[n]\varphi_i'(u_n)\mathbf{X}_n\mathbf{w}_n^{(2)} \tag{3.25}$$

for the weights of the first linear filter.

As in the previous cases, the learning rates $\mu_w^{(1)}[n]$, $\mu_w^{(2)}[n]$ and $\mu_q[n]$ incorporate the other constant values.

### 3.3.5 OTHER ARCHITECTURES

It is possible to derive other kinds of block-oriented architectures by suitable combinations of the basic architectures introduced in the previous subsections, thus obtaining particular cascade and/or parallel systems that can be successfully used to identify specific types of nonlinear systems or by using an IIR linear filter instead of a FIR one, like done in [42]. A more general situation of particular interest is the case of a convex (and affine) combination of the former architectures, like the ones proposed in [56] for another important class of nonlinear adaptive filters. In this scenario, the whole combined filter is able to automatically detect the subsystem that can provide better results. The extension of SAF to a combined scenario will be postponed to future works.

In a similar fashion, the work [46] introduces a hybrid nonlinear spline filter, which is designed as a cascade of an adaptive spline function and a single-layer adaptive nonlinear network, devoted to the problem of removing noise in the acoustic domain. Another work worthy to consider is [57], where the HSAF architecture is generalized to a Functional Link network. The proposed work shows very interesting results also in terms of classification ability.

Another type of generalization of SAFs is based on a suitable choice of the learning schemes. In most works, SAFs are adapted by using a general LMS strategy. However, this situation cannot be the best one. An extension to the more general normalized LMS (NLMS) is provided in [48], while the extension to the correntropy criterion can be found in [49]. The latter strategy is particularly useful when data are corrupted by impulsive noise or contain large outliers.

Finally, [50] derives a diffused version of the WSAF (D-SAF) by applying ideas from the Diffusion Adaptation (DA) framework. DA algorithms allow a network of agents to collectively estimate a parameter vector, by jointly minimizing the sum of their local cost functions. Experimental evaluations show that the D-SAF is able to robustly learn the underlying nonlinear model, with a significant gain compared with a noncooperative solution [50]. A detailed description of D-SAFs can be found in Chapter 9.

### 3.3.6 COMPUTATIONAL COST

From a computational point of view each SAF architecture requires the evaluation of two or three processing blocks. For each block the computational cost of the forward phase has to be evaluated (*i.e.*, the evaluation of the system output) and also that of the backward phase (*i.e.*, the adaptation of the adaptive system). It is well known from the literature that the computational cost of the LMS algorithm is $2M + 1$ multiplications plus $2M$ additions for each input sample [1], if $M$ denotes the filter length.

In addition the cost of the spline evaluation must be considered. For each iteration only the $i$th span of the curve is modified by calculating the quantities $u_n$, $i$ and the expressions $\mathbf{u}_n^T \mathbf{C} \mathbf{q}_{i,n}$, $\dot{\mathbf{u}}_n^T \mathbf{C} \mathbf{q}_{i,n}$ and $\mathbf{C}^T \mathbf{u}_n$. Note that the calculation of the quantity $\mathbf{C} \mathbf{q}_{i,n}$ is executed during the output computation, as well as in the backward phase (the spline derivatives). However, most of the computations can be done through the reuse of past calculations (*e.g.*, by explicating the polynomial and by use of the well-known Horner rule [58]). The cost for the spline output computation and its adaptation is $4K_M$ multiplications plus $4K_A$ additions, where $K_M$ and $K_A$ are constants (less than 16), depending on the implementation structure (see for example [53]). In any case, for high deep memory SAF, where the filter length is $M >> 4$, the computational overhead, for the nonlinear function computation and its adaptation, can be neglected with respect to a simple linear filter. It is straightforward to note that the computational cost can be further reduced by using some smart trick, like choosing the sampling step $\Delta x$ or the learning rates $\mu_j[n]$ as a power of two. In fact, these constraints transform the multiplications into shift operations.

The total computational cost, inclusive of the forward and backward computation, for the presented four SAF architectures is summarized in Table 3.1. In this table $M$ denotes the linear filter length. In the case of S2SAF architecture, $M_1$ and $M_2$ are the length of the first and second linear filter, respectively.

## 3.4 CONVERGENCE PROPERTIES

The convergence properties of the proposed learning algorithms could be evaluated in terms of their stability behavior, convergence speed, accuracy of the steady state and transient results and tracking

**Table 3.1 Computational cost of the SAF architectures in terms of additions and multiplications**

| | Additions | Multiplications |
|---|---|---|
| WSAF | $2M+48$ | $2M+67$ |
| HSAF | $23M+20$ | $30M+14$ |
| S1SAF | $23M+68$ | $30M+81$ |
| S2SAF | $2M_1M_2+M_1+22M_2+35$ | $2M_1M_2+30M_2+38$ |

ability as a function of their MSEs. For simplicity, in this chapter the attention is focused only on the optimal choice of the learning rates and the theoretical formulation of the excess MSE (EMSE) for the Wiener and Hammerstein architectures. Additional details and other useful results can be found in [40,41,43,44].

### 3.4.1 BOUNDS ON THE LEARNING RATES

A necessary but not sufficient condition for the algorithm convergence is the identification of a bound on the learning rate choice. These bounds on the learning rates can be derived by the Taylor series expansion of the error $e[n+1]$ around the time index $n$, stopped at the first order. This series expansion involves the derivatives of the error with respect to the free parameters of the considered architecture.

By performing these evaluations for the four SAF architectures introduced in Section 3.3 and by using the previous notation, we can easily obtain

$$0 < \mu_w[n]\varphi_i'^2(u)\,\|\mathbf{x}_n\|^2 + \mu_q[n]\left\|\mathbf{C}^T\mathbf{u}\right\|^2 < 2, \tag{3.26}$$

$$0 < \mu_w[n]\,\|\mathbf{s}_n\|^2 + \mu_q[n]\left\|\mathbf{C}^T\mathbf{U}_{i,n}\mathbf{w}_n\right\|^2 < 2, \tag{3.27}$$

$$0 < \mu_q^{(1)}[n]\frac{\varphi_i'^2(u_n^{(2)})}{\Delta x^{(2)}}\left\|\mathbf{C}^T\mathbf{U}_{i,n}^{(1)}\mathbf{w}_n\right\|^2 + \mu_q^{(2)}[n]\left\|\mathbf{C}^T\mathbf{u}_n^{(2)}\right\|^2 + \mu_w[n]\frac{\varphi_i'^2(u_n^{(2)})}{\Delta x^{(2)}}\|\mathbf{s}_n\|^2 < 2, \tag{3.28}$$

$$0 < \mu_w^{(1)}[n]\frac{\varphi_i'^2(u_n)}{\Delta x}\left\|\mathbf{X}_n\mathbf{w}_n^{(2)}\right\|^2 + \mu_w^{(2)}[n]\,\|\mathbf{r}_n\|^2 + \mu_q[n]\left\|\mathbf{C}^T\mathbf{U}_{i,n}\mathbf{w}_n^{(2)}\right\|^2 < 2. \tag{3.29}$$

The full derivation of these bounds can be found in [40,41,43].

### 3.4.2 STEADY-STATE MSE

The aim of this paragraph is to evaluate the mean square performance of some of the proposed architectures at steady state. In particular, we are interested in the derivation of the EMSE of the nonlinear adaptive filter [2]. For simplicity we limit our analysis to the case of the Wiener and Hammerstein systems. The steady-state analysis is performed in a separate manner, by considering first the case of the linear filter alone and then the nonlinear spline function alone. Note that, since we are interested only in the steady state, the fixed subsystem in each substep can be considered adapted. The analysis is conducted by using a model of the same type of the used one, as shown in Fig. 3.5 for the particular case of the WSAF. Similar schemes are valid for the other kinds of architectures.

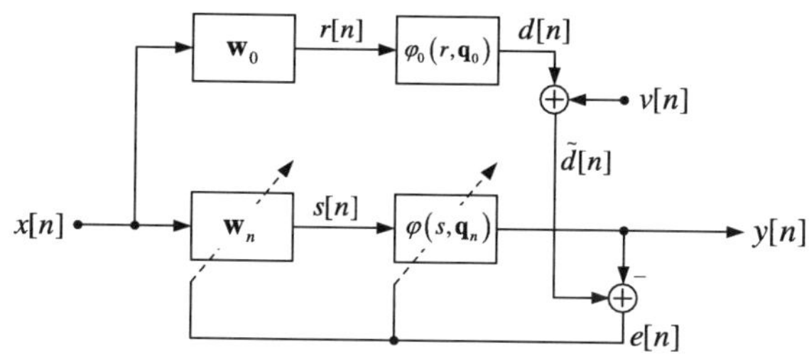


**FIGURE 3.5**

Model used for statistical performance evaluation, in the particular case of the WSAF.

In the following we denote with $\varepsilon[n]$ the *a priori* error of the whole system, with $\varepsilon_w[n]$ the *a priori* error when only the linear filter is adapted while the spline control points are fixed and similarly with $\varepsilon_q[n]$ the *a priori* error when only the control points are adapted while the linear filter is fixed. In order to make tractable the mathematical derivation, some assumption must be introduced:

**A1.** The noise sequence $v[n]$ is independent and identically distributed with variance $\sigma_v^2$ and zero mean.

**A2.** The noise sequence $v[n]$ is independent of $\mathbf{x}_n$, $\mathbf{s}_n$, $\varepsilon[n]$, $\varepsilon_w[n]$ and $\varepsilon_q[n]$.

**A3.** The input sequence $x[n]$ is zero mean with variance $\sigma_x^2$ and at steady state it is independent of $\mathbf{w}_n$.

**A4.** At steady state, the error sequence $e[n]$ and the *a priori* error sequence $\varepsilon[n]$ are independent of $\varphi_i'(u_n)$, $\varphi_i'^2(u_n)$, $\|\mathbf{x}_n\|$, $\|\mathbf{s}_n\|$, $\left\|\mathbf{C}^T\mathbf{u}_n\right\|$ and $\left\|\mathbf{C}^T\mathbf{U}_{i,n}\mathbf{w}_n\right\|$.

The EMSE $\zeta$ is defined at steady state as

$$\zeta \equiv \lim_{n\to\infty} E\left\{\varepsilon^2[n]\right\}. \tag{3.30}$$

Similar definitions hold for the EMSE $\zeta_w$ due to the linear filter component $\mathbf{w}_n$ only and the EMSE $\zeta_q$ due to the spline nonlinearity $\mathbf{q}_n$ only.

The MSE analysis has been performed by the energy conservation method [2]. Following the derivation presented in [41] and [44] we obtain an interesting property.

**Proposition 1.** *The EMSE $\zeta$ of a Wiener and a Hammerstein spline adaptive filter satisfies the condition*

$$\zeta \approx \zeta_w + \zeta_q. \tag{3.31}$$

The values of $\zeta_w$ and $\zeta_q$ are dependent on the particular SAF architecture considered. Specifically, for the Wiener case, we have [44]

$$\zeta_w = \frac{\mu_w[n]\sigma_v^2 E\left\{\varphi_i'^2(u_n)\left\|\mathbf{x}_n\right\|^2\right\}}{2\Delta x^2 E\left\{\frac{\varphi_i'(u_n)}{\mathbf{c}_3\mathbf{q}_{i,n}}\right\} - \mu_w[n]E\left\{\varphi_i'^2(u_n)\left\|\mathbf{x}_n\right\|^2\right\}}, \tag{3.32}$$

where $\mathbf{c}_3$ is the third row of the $\mathbf{C}$ matrix and

$$\zeta_q = \frac{\mu_q[n]\sigma_v^2 E\left\{\left\|\mathbf{C}^{\mathrm{T}}\mathbf{u}_n\right\|^2\right\}}{2-\mu_q[n]E\left\{\left\|\mathbf{C}^{\mathrm{T}}\mathbf{u}_n\right\|^2\right\}}. \tag{3.33}$$

Other useful properties on the steady-state performances of the WSAF can be found in [44]. For the Hammerstein case, we have [41]

$$\zeta_w = \frac{\mu_w[n]\sigma_v^2 E\left\{\|\mathbf{s}_n\|^2\right\}}{2-\mu_w[n]E\left\{\|\mathbf{s}_n\|^2\right\}} \tag{3.34}$$

and

$$\zeta_q = \frac{\mu_q[n]\sigma_v^2 E\left\{\left\|\mathbf{C}^T\mathbf{U}_{i,n}\mathbf{w}_n\right\|^2\right\}}{2-\mu_q[n]E\left\{\left\|\mathbf{C}^T\mathbf{U}_{i,n}\mathbf{w}_n\right\|^2\right\}}. \tag{3.35}$$

## 3.5 EXPERIMENTAL RESULTS

In order to validate the proposed nonlinear adaptive filtering solutions, several experiments were performed. In our previous works [40–44], several convergence tests showed the effectiveness of the proposed architectures in converging to the optimal solution. Moreover, these works also reported some experimental tests in system identification scenarios.[2]

In this section, we provide two kinds of experimental results on the identification of nonlinear systems: (i) tests taken from the literature on the field and (ii) tests on data measured from real-world applications.

### 3.5.1 RESULTS FROM SIMULATED SCENARIOS

Regarding the experiments on nonlinear systems taken from the literature, we compare results obtained by SAFs with those of a full third-order Volterra adaptive filter [59] and the approach proposed by Jeraj and Mathews in [60] where the nonlinearity is implemented as a third-order polynomial. In addition, Sandwich architectures are also compared with the LNL approach proposed by Hegde *et al.* in [61, 62]. In all experiments the input signal $x[n]$ consists of 50 000 samples of the signal generated by the following relationship:

$$x[n] = ax[n-1] + \sqrt{1-a^2}\xi[n], \tag{3.36}$$

[2]Some Matlab source codes, implementing the SAF architectures and experimental results, are available at the following web page: https://github.com/ispamm/SAF-Toolbox.

where $\xi[n]$ is a zero mean white Gaussian noise with unitary variance and $a$ $(0 \le a < 1)$ is a parameter that determines the level of correlation between adjacent samples. Experiments were conducted with $a$ set to 0.1 and 0.95. In addition an additive white noise $v[n]$ with variance $\sigma_v^2$ is considered.

In the following we propose four tests in order to verify the Wiener, Hammerstein, Sandwich 1 and Sandwich 2 architectures described in Section 3.3. All the SAF models use a CR-spline basis in (3.7) with $\Delta x = 0.2$.

*WSAF*—The Wiener architecture to be identified is composed of the Back and Tsoi NARMA model reported in [63]. This model consists of a cascade of the third-order IIR filter

$$H(z) = \frac{0.0154 + 0.0462z^{-1} + 0.0462z^{-2} + 0.0154z^{-3}}{1 - 1.99z^{-1} + 1.572z^{-2} - 0.4583z^{-3}} \tag{3.37}$$

and the following nonlinearity:

$$y[n] = \sin(x[n]). \tag{3.38}$$

The input signal $x[n]$ is the colored signal obtained with (3.36), choosing $a = 0.95$. The learning rates are set to $\mu_w = \mu_q = 0.1$ and the filter length is $M = 15$ taps. The third-order Volterra filter has $M_v = 10$ filter taps and the learning rate is set to $\mu_v = 0.1$, while the Jeraj and Mathews architecture is characterized by $M_a = 1$, $M_b = 15$, a nonlinearity of order $N = 3$ and learning rate $\mu_L = \mu_{NL} = 10^{-5}$. The MSE, averaged over 100 trials, is reported in Fig. 3.6A.

*HSAF*—The Hammerstein architecture to be identified, drawn from [64], is composed of the following nonlinearity:

$$s[n] = \frac{1}{8} + \lfloor 2x[n] - 1 \rfloor, \tag{3.39}$$

where $\lfloor \bullet \rfloor$ is the floor operator, and the FIR filter $\mathbf{h} = \frac{1}{2}[1,\ 0.75,\ 0.5,\ 0.25,\ 0,\ -0.25]$. The learning rates are set to $\mu_w = 0.01$ and $\mu_q = 0.02$, while the parameters of Volterra and Jeraj and Mathews architectures are the same as in the Wiener case. The MSE, averaged over 100 trials, is reported in Fig. 3.6B.

*S1SAF*—The identification of a Sandwich 1 system is drawn from [65]. The first and last blocks are the following two static nonlinear functions:

$$s[n] = \frac{x[n]}{\sqrt{0.1 + 0.9x^2[n]}} \tag{3.40}$$

and

$$y[n] = r[n] - 0.5r^2[n] + 0.2r^3[n], \tag{3.41}$$

while the second block is the following FIR filter:

$$H(z) = 1 + 0.5z - 0.25z^{-1} + 0.05z^{-2} + 0.001z^{-3}. \tag{3.42}$$

The input signal $x[n]$ is taken from (3.36) with $a = 0.1$. Also SNR = 30 dB is considered. The learning rates are set to $\mu_w = \mu_q^{(1)} = \mu_q^{(2)} = 0.05$ and the length is set to $M = 15$. The proposed methods were compared with a standard full third-order Volterra filter, the approach of Jeraj and Mathews

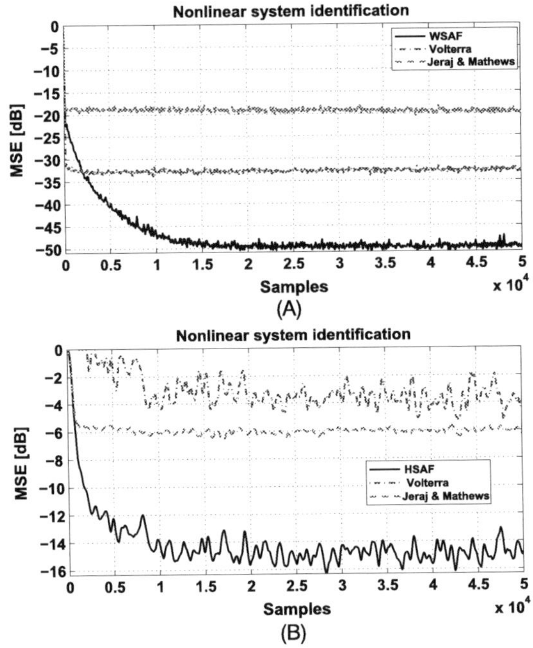


**FIGURE 3.6**

Results of the proposed system identification tests for WSAF and HSAF. (A) MSE comparisons in the Wiener case. (B) MSE comparisons in the Hammerstein case.

with the same parameters of the Wiener case and the LNL approach proposed by Hegde *et al.* with $M_1 = M_2 = 5$, $\mu_A = \mu_B = 10^{-4}$ and $\mu_C = 10^{-3}$. The MSE, averaged over 100 trials, is reported in Fig. 3.7A.

*S2SAF*—Finally, a Sandwich 2 to be identified is drawn from [66]. The first and last blocks are two fourth-order IIR filters, Butterworth and Chebyshev, respectively, with transfer functions

$$H_B(z) = \frac{(0.2851 + 0.5704z^{-1} + 0.2851z^{-2})}{(1 - 0.1024z^{-1} + 0.4475z^{-2})} \cdot \frac{(0.2851 + 0.5701z^{-1} + 0.2851z^{-2})}{(1 - 0.0736z^{-1} + 0.0408z^{-2})} \tag{3.43}$$

and

$$H_C(z) = \frac{(0.2025 + 0.2880z^{-1} + 0.2025z^{-2})}{(1 - 1.01z^{-1} + 0.5861z^{-2})} \cdot \frac{(0.2025 + 0.0034z^{-1} + 0.2025z^{-2})}{(1 - 0.6591z^{-1} + 0.1498z^{-2})}, \tag{3.44}$$

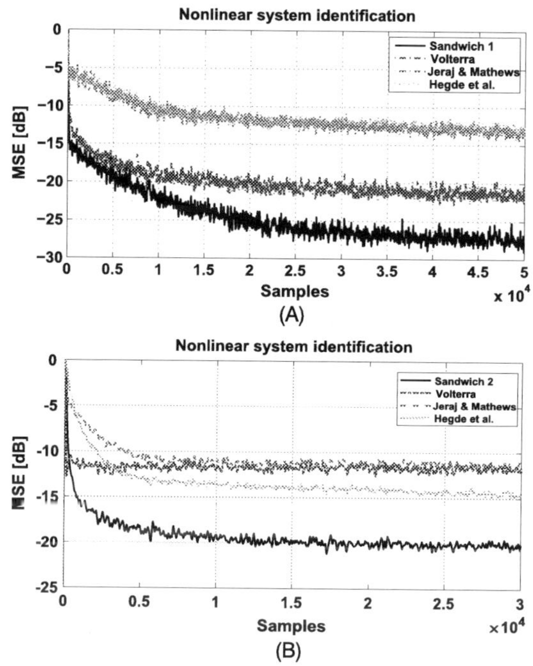


**FIGURE 3.7**

Results of the proposed system identification tests for S1SAF and S2SAF. (A) MSE comparisons in the Sandwich 1 case. (B) MSE comparisons in the Sandwich 2 case.

while the second block is the following nonlinearity:

$$y[n] = \frac{2x[n]}{1 + |x[n]|^2}. \tag{3.45}$$

This system is similar to radio frequency amplifiers used in satellite communications (high-power amplifiers), in which the linear filters model the dispersive transmission paths, while the nonlinearity models the amplifier saturation. The input signal $x[n]$ is as in (3.36) with $a = 0.1$. The learning rates are set to $\mu_w^{(1)} = 0.005$, $\mu_w^{(2)} = 0.009$ and $\mu_q = 0.04$ and the lengths are set to $M_1 = M_2 = 5$. Comparisons with Volterra, Jeraj and Mathews and Hegde *et al.* have been performed with the previous parameters. The MSE, averaged over 100 trials, is reported in Fig. 3.7B.

### 3.5.2 RESULTS FROM REAL-WORLD SCENARIOS

In evaluating nonlinear architectures, it is important to test the performance of the proposed algorithm on completely unknown systems. For this purpose, we test our architecture from data sets collected

**Table 3.2 Comparison of the asymptotic computational complexity of the algorithms used in experimental comparisons**

| Algorithm | Asymptotic complexity |
|---|---|
| WSAF [40] | $\mathcal{O}(M)$ |
| HSAF [41] | $\mathcal{O}(M)$ |
| S1 SAF [43] | $\mathcal{O}(M)$ |
| S2 SAF [43] | $\mathcal{O}(M_1 M_2)$ |
| Third-order Volterra [59] | $\mathcal{O}(M_v^3)$ |
| Hegde *et al.* [61] | $\mathcal{O}(M_1 M_2)$ |
| Jeraj and Mathews [60] | $\mathcal{O}((M_b M_a N)^2)$ |

from laboratory processes that can be described by block-oriented dynamic models. In particular, a number of interesting databases for system identification is represented by DaISy.[3]

From the DaISy database we have collected the following interesting and challenge data sets:

- Exchanger [97-002]: the process is a liquid-saturated steam heat exchanger, where water is heated by pressurized saturated steam through a copper tube. The output variable is the outlet liquid temperature while the input variable is the liquid flow rate.
- CD player arm [96-007]: data measured from the mechanical construction of a CD player arm. The input is the force of the mechanical actuators while the output is related to the tracking accuracy of the arm.
- Thermic wall [96-011]: Heat flow density through a two-layer wall (brick and insulation layer). The input is the external temperature of the wall while the output is the heat flow density through the wall.
- Heating system [99-001]: The experiment is a simple SISO heating system; the input drives a 300-Watt Halogen lamp, suspended several inches above a thin steel plate while the output is a thermocouple measurement taken from the back of the plate.
- Continuous stirred tank reactor (CSTR) [98-002]: The process is a model of a CSTR, where the reaction is exothermic and the concentration is controlled by regulating the coolant flow. The input is the coolant flow while the output is the temperature.

All results have been compared with other state-of-the-art approaches used in previous experiments, namely a full third-order Volterra architecture [59] (with $M_v = 15$), the LNL approach proposed by Hegde *et al.* in [61,62] (with $M_1 = M_2 = 5$), where the nonlinearity is implemented as a third-order polynomial and the approach proposed by Jeraj and Mathews in [60] (with $M_a = 1$, $M_b = 5$ and $N = 3$). In order to provide a fair comparison of the performance of such algorithms, Table 3.2 shows their asymptotic computational complexity.

Moreover, we have chosen two additional measured data from coupled electric drives that consist of two electric motors that drive a pulley using a flexible belt and two measured data from a fluid level

[3] Available at http://homes.esat.kuleuven.be/~smc/daisy/daisydata.html.

**Table 3.3 MSE in dB of the SAF architectures, compared with other referenced approaches, using real-world data**

| | WSAF | HSAF | S1SAF | S2SAF | Volterra | Jeraj & Mathews | Hedge *et al.* |
|---|---|---|---|---|---|---|---|
| Exchanger | −34.18 | −25.12 | −39.14 | −32.38 | −16.66 | −7.77 | −7.87 |
| CD player arm | −13.89 | −10.68 | −19.94 | −12.50 | −19.23 | −5.53 | −5.58 |
| Thermic wall | −16.29 | −15.66 | −22.71 | −16.27 | −24.26 | −7.58 | −7.66 |
| Heating system | −17.99 | −17.94 | −22.09 | −17.87 | −8.79 | −16.04 | −16.02 |
| CSTR | −39.53 | −42.03 | −53.85 | −41.42 | −21.51 | −26.16 | −26.29 |
| Drive 1 | −15.81 | −22.52 | −12.51 | −13.29 | −13.51 | −2.33 | −2.55 |
| Drive 2 | −14.09 | −2.65 | −19.17 | −15.25 | −6.54 | −9.59 | −8.34 |
| Tank 1 | −1.08 | −17.33 | −1.13 | −21.60 | −10.81 | −14.62 | −18.54 |
| Tank 2 | −2.65 | −14.59 | −5.70 | −20.49 | −2.50 | −11.84 | −15.39 |
| Diode | −15.88 | −11.69 | −11.98 | −19.79 | −7.02 | −14.02 | −17.25 |

control system consisting of two cascaded tanks with free outlets fed by a pump. Additional details as regards these systems and data sets can be found in [67].

A last data set is taken from [68,69] and consist of the identification of a nonlinear electric circuit composed of the cascade of a third-order Chebyshev linear filter, a nonlinearity built by using a diode circuit and a second linear filter designed as a third-order inverse Chebyshev filter with a transmission of zero in the frequency band of interest.

Results in terms of MSEs for all data sets are summarized in Table 3.3. It can be seen that the proposed SAF approaches provide the best performance in almost all cases.

### 3.5.3 APPLICATIVE SCENARIOS

In the previous subsections, we provided some numerical results on the general problem of nonlinear system identification. However, SAFs have been used in applicative scenarios providing interesting and very good results.

A first applicative scenario is the nonlinear acoustic echo cancellation (NAEC) problem. The SAF approach has been successfully applied to NAEC in [30], where it is shown that the proposed solutions outperform other linear and nonlinear approaches.

Another challenging scenario is the nonlinear adaptive noise cancellation (NANC) problem. In particular, in [45] authors show the effectiveness of SAF approach to other state-of-the-art algorithms. Moreover, the proposed solution has been extended with additional refinements in [46] and [47].

Finally, as underlined in Section 3.3.5, the SAF approach has been recently extended to the diffusion case [50], involving a distributed (wireless) sensors network. The authors of this work derive a D-SAF by applying ideas from the DA framework, as explained in Chapter 9. Experimental evaluations on the D-SAF approach show that it is able to robustly learn the underlying nonlinear model, with a significant gain compared with a noncooperative solution.

## 3.6 CONCLUSION

In this chapter, a complete overview of a recent class of adaptive nonlinear filters, SAFs, has been provided. Specifically, we have described the basic block-oriented architecture implemented by SAFs, namely the Wiener, Hammerstein, NLN and LNL ones, as well as the related learning rules. Additionally, we have referred to the main theoretical results found in the literature on SAFs; the bounds on the choice of the learning rates and the theoretical evaluation of the EMSE for the Wiener and Hammerstein architectures.

Finally, after the introduction of some possible real-world applicative scenarios, some numerical results have shown the effectiveness of the proposed architectures.

## ACKNOWLEDGMENTS

This chapter has been partially supported by the PRIN2015 project no. 2015YPXH4W_004: "A green adaptive Fog computing and networking architecture (GAUChO)", funded by the Italian MIUR, and by the project: "Vehicular Fog for energy-efficient QoS mining and dissemination of multimedia Big Data streams (V-FOG)," funded by Sapienza University of Rome.

## REFERENCES

[1] A. Uncini, Fundamentals of Adaptive Signal Processing, Signals and Communication Technology, Springer International Publishing, Switzerland, 2015.
[2] A.H. Sayed, Fundamentals of Adaptive Filtering, John Wiley & Sons, 2003.
[3] S. Haykin, Adaptive Filter Theory, 4th edition, Prentice-Hall, 2002.
[4] L. Garcia (Ed.), Adaptive Filtering Applications, InTech, 2011.
[5] N. Wiener, Nonlinear Problems in Random Theory, John Wiley & Sons, 1958.
[6] O. Montiel, O. Castillo, P. Melin, R. Sepúlveda, Black box evolutionary mathematical modeling applied to linear systems, International Journal of Intelligent Systems 20 (2) (2005) 293–311.
[7] R.K. Pearson, M. Pottmann, Gray-box identification of block-oriented nonlinear models, Journal of Process Control 10 (4) (2000) 301–3115.
[8] V.J. Mathews, G.L. Sicuranza, Polynomial Signal Processing, John Wiley & Sons, 2000.
[9] D. Mirri, G. Iuculano, F. Filicori, G. Pasini, G. Vannini, G. Pellegrini Gualtieri, A modified Volterra series approach for nonlinear dynamic systems modeling, IEEE Transactions on Circuits and Systems—I: Fundamental Theory and Applications 49 (8) (2002) 1118–1128.
[10] S. Haykin, Neural Networks and Learning Machines, 2nd edition, Pearson Publishing, 2009.
[11] J. de Jesus Rubio, W. Yu, Stability analysis of nonlinear system identification via delayed neural networks, IEEE Transactions on Circuits and Systems—II: Express Briefs 54 (2) (2007) 161–165.
[12] L. Ljung, System Identification—Theory for the User, 2nd edition, Prentice Hall, Upper Saddle River, N.J., 1999.
[13] A.E. Nordsjo, L.H. Zetterberg, Identification of certain time-varying Wiener and Hammerstein systems, IEEE Transactions on Signal Processing 49 (3) (2001) 577–592.
[14] S. Boyd, L.O. Chua, Uniqueness of a basic nonlinear structure, IEEE Transactions on Circuits and Systems CAS-30 (9) (1983) 648–651.
[15] R. Haber, H. Unbehauen, Structure identification of nonlinear dynamic systems—a survey on input/output approaches, Automatica 26 (4) (1990) 651–677.
[16] E.W. Bai, Frequency domain identification of Wiener models, Automatica 39 (2003) 1521–1530.
[17] M. Korenberg, I. Hunter, Two methods for identifying Wiener cascades having noninvertible static nonlinearities, Annals of Biomedical Engineering 27 (6) (1999) 793–804.

[18] D. Westwick, M. Verhaegen, Identifying MIMO Wiener systems using subspace model identification methods, Signal Processing 52 (2) (1996) 235–258.

[19] G. Mzyk, A censored sample mean approach to nonparametric identification of nonlinearities in Wiener systems, IEEE Transactions on Circuits and Systems—II: Express Briefs 54 (10) (2007) 897–901.

[20] E.W. Bai, Frequency domain identification of Hammerstein models, IEEE Transactions on Automatic Control 48 (2003) 530–542.

[21] I.J. Umoh, T. Ogunfunmi, An affine projection-based algorithm for identification of nonlinear Hammerstein systems, Signal Processing 90 (6) (2010) 2020–2030.

[22] J. Vörös, Identification of Hammerstein systems with time-varying piecewise-linear characteristics, IEEE Transactions on Circuits and Systems—II: Express Briefs 52 (12) (2005) 865–869.

[23] L. Ljung, Identification of nonlinear systems, in: 9th International Conference on Control, Automation, Robotics and Vision, ICARCV 2006, Singapore, 2006, pp. 1–9.

[24] H.-E. Liao, W.A. Sethares, Cross-term analysis of LNL models, IEEE Transactions on Circuits and Systems—I: Fundamentals Theory and Applications 43 (4) (1996) 333–335.

[25] S. Billings, S.Y. Fakhouri, Identification of a class of nonlinear systems using correlation analysis, Proceedings of IEEE 125 (7) (1978) 691–697, https://doi.org/10.1049/piee.1978.0161.

[26] O. De Feo, M. Storace, PWL identification of dynamical systems: some examples, in: Proceedings of the 2004 International Symposium on Circuits and Systems, ISCAS2004, vol. 4, 2004, pp. 665–668.

[27] C. Wen, X. Ma, A basis-function canonical piecewise-linear approximation, IEEE Transactions on Circuits and Systems—I. Regular Papers 55 (5) (2008) 1328–1331, https://doi.org/10.1109/TCSI.2008.916552.

[28] B.N. Bond, Z. Mahmood, Y. Li, R. Sredojević, A. Megretski, V. Stojanović, Y. Avniel, L. Daniel, Compact modeling of nonlinear analog circuits using system identification via semidefinite programming and incremental stability certification, IEEE Transactions on Computer-Aided Design of Integrated Circuits and Systems 29 (8) (2010) 1149–1162, https://doi.org/10.1109/TCAD.2010.2049155.

[29] S. Orcioni, Approximation and analog implementation of nonlinear systems represented by Volterra series, in: 6th IEEE International Conference on Electronics, Circuits and Systems, 1999, pp. 1109–1113.

[30] M. Scarpiniti, D. Comminiello, R. Parisi, A. Uncini, Comparison of Hammerstein and Wiener systems for nonlinear acoustic echo cancelers in reverberant environments, in: Proc. of 17th International Conference on Digital Signal Processing, DSP2011, Corfu, Greece, 2011, pp. 1–6.

[31] K.J. Hunt, M. Munih, N.N. Donaldson, F.M.D. Barr, Investigation of the Hammerstein hypothesis in the modeling of electrically stimulated muscle, IEEE Transactions on Biomedical Engineering 45 (8) (1998) 99–1009.

[32] Z. Zhu, H. Leung, Adaptive identification of nonlinear systems with application to chaotic communications, IEEE Transactions on Circuits and Systems—I: Fundamental Theory and Applications 47 (7) (2000) 1072–1080.

[33] L. Peng, H. Ma, Design and implementation of software-defined radio receiver based on blind nonlinear system identification and compensation, IEEE Transactions on Circuits and Systems—I: Regular Papers 58 (11) (2011) 2776–2789, https://doi.org/10.1109/TCSI.2011.2151050.

[34] B. Kwak, A.E. Yagle, J.A. Levitt, Nonlinear system identification of hydraulic actuator friction dynamics using a Hammerstein model, in: Proceedings of the IEEE ASSP'98, vol. 4, Seattle, WA, 1998, pp. 1933–1936.

[35] J.-C. Jeng, M.-W. Lee, H.-P. Huang, Identification of block-oriented nonlinear processes using designed relay feedback tests, Industrial & Engineering Chemistry Research 44 (7) (2005) 2145–2155.

[36] O. De Feo, M. Storace, Piecewise-linear identification of nonlinear dynamical systems in view of their circuit implementations, IEEE Transactions on Circuits and Systems—I: Regular Papers 54 (7) (2007) 1542–1554, https://doi.org/10.1109/TCSI.2007.899613.

[37] C. Wen, H. Ma, A canonical piecewise-linear representation theorem: geometrical structures determine representation capability, IEEE Transactions on Circuits and Systems—II: Express Briefs 58 (12) (2011) 936–940, https://doi.org/10.1109/TCSII.2011.2172715.

[38] A. Stenger, W. Kellerman, Adaptation of a memoryless preprocessor for nonlinear acoustic echo cancelling, Signal Processing 80 (9) (2000) 1747–1760.

[39] C. de Boor, A Practical Guide to Splines, revised edition, Springer-Verlag, 2001.

[40] M. Scarpiniti, D. Comminiello, R. Parisi, A. Uncini, Nonlinear spline adaptive filtering, Signal Processing 93 (4) (2013) 772–783, https://doi.org/10.1016/j.sigpro.2012.09.021.

[41] M. Scarpiniti, D. Comminiello, R. Parisi, A. Uncini, Hammerstein uniform cubic spline adaptive filters: learning and convergence properties, Signal Processing 100 (2014) 112–123, https://doi.org/10.1016/j.sigpro.2014.01.019.

[42] M. Scarpiniti, D. Comminiello, R. Parisi, A. Uncini, Nonlinear system identification using IIR spline adaptive filters, Signal Processing 108 (2015) 30–35, https://doi.org/10.1016/j.sigpro.2014.08.045.

[43] M. Scarpiniti, D. Comminiello, R. Parisi, A. Uncini, Novel cascade spline architectures for the identification of nonlinear systems, IEEE Transactions on Circuits and Systems—I: Regular Papers 62 (7) (2015) 1825–1835, https://doi.org/10.1109/TCSI.2015.2423791.

[44] M. Scarpiniti, D. Comminiello, G. Scarano, R. Parisi, A. Uncini, Steady-state performance of spline adaptive filters, IEEE Transactions on Signal Processing 64 (4) (2016) 816–828, https://doi.org/10.1109/TSP.2015.2493986.

[45] V. Patel, N.V. George, Nonlinear active noise control using spline adaptive filters, Applied Accoustics 93 (2015) 38–43, https://doi.org/10.1016/j.apacoust.2015.01.009.

[46] V. Patel, D. Comminiello, M. Scarpiniti, N.V. George, A. Uncini, Design of hybrid nonlinear spline adaptive filters for active noise control, in: Proc. of 2016 International Joint Conference on Neural Networks, IJCNN2016, Vancouver, Canada, 2016, pp. 3420–3425.

[47] V. Patel, N.V. George, Compensating acoustic feedback in feed-forward active noise control systems using spline adaptive filters, Signal Processing 120 (2016) 448–455, https://doi.org/10.1016/j.sigpro.2015.10.003.

[48] S. Guan, Z. Li, Normalised spline adaptive filtering algorithm for nonlinear system identification, Neural Processing Letters (2017) 1–13, https://doi.org/10.1007/s11063-017-9606-6.

[49] S. Peng, Z. Wu, X. Zhang, B. Chen, Nonlinear spline adaptive filtering under maximum correntropy criterion, in: 2015 IEEE Region 10 Conference, TENCON 2015, 2015, pp. 1–5.

[50] S. Scardapane, M. Scarpiniti, D. Comminiello, A. Uncini, Diffusion spline adaptive filtering, in: Proc. of the 24-th European Signal Processing Conference, EUSIPCO2016, Budapest, Hungary, 2016, pp. 1498–1502.

[51] L. Schumaker, Spline Functions: Basic Theory, 3rd edition, Krieger Publishing Company, 1993.

[52] H. Prautzsch, W. Boehm, M. Paluszny, Bézier and B-Spline Techniques, 1st edition, Springer-Verlag, 2002.

[53] S. Guarnieri, F. Piazza, A. Uncini, Multilayer feedforward networks with adaptive spline activation function, IEEE Transaction on Neural Network 10 (3) (1999) 672–683.

[54] R.H. Bartels, J.C. Beatty, B.A. Barsky, An Introduction to Splines for Use in Computer Graphics & Geometric Modeling, Morgan Kaufmann, 1987.

[55] E. Catmull, R. Rom, A class of local interpolating splines, in: R.E. Barnhill, R.F. Riesenfeld (Eds.), Computer-Aided Geometric Design, New York Academic, 1974, pp. 317–326.

[56] D. Comminiello, M. Scarpiniti, L.A. Azpicueta-Ruiz, J. Arenas-Garcìa, A. Uncini, Combined nonlinear filtering architectures involving sparse functional link adaptive filters, Signal Processing 135 (2017) 168–178, https://doi.org/10.1016/j.sigpro.2017.01.009.

[57] M. Rathod, V. Patel, N.V. George, Generalized spline nonlinear adaptive filters, Expert Systems with Applications 48 (2017) 122–130, https://doi.org/10.1016/j.eswa.2017.04.043.

[58] P. Borwein, T. Erdelyi, Polynomials and Polynomial Inequalities, Springer-Verlag, 1995.

[59] M. Schetzen, Nonlinear system modeling based on the Wiener theory, Proceedings of the IEEE 69 (12) (1981) 1557–1573.

[60] J. Jeraj, V.J. Mathews, A stable adaptive Hammerstein filter employing partial orthogonalization of the input signals, IEEE Transactions on Signal Processing 54 (4) (2006) 1412–1420.

[61] V. Hegde, C. Radhakrishnan, D.J. Krusienski, W.K. Jenkins, Architectures and algorithms for nonlinear adaptive filters, in: Conference Record of the 36th Asilomar Conference on Signals, Systems and Computers, vol. 2, Pacific Grove, CA, USA, 2002, pp. 1015–1018.

[62] V. Hegde, C. Radhakrishnan, D.J. Krusienski, W.K. Jenkins, Series-cascade nonlinear adaptive filters, in: Proceedings of the 45th Midwest Symposium on Circuits and Systems, MWSCAS2002, vol. 3, Tulsa, Oklahoma, 2002, pp. 219–222.

[63] A.D. Back, A.C. Tsoi, A simplified gradient algorithm for IIR synapse multilayer perceptrons, Neural Computation 5 (3) (1993) 456–462.

[64] Z. Hasiewicz, M. Pawlak, P. Sliwinski, Nonparametric identification of nonlinearities in block-oriented systems by orthogonal wavelets with compact support, IEEE Transactions on Circuits and Systems—I: Regular Papers 52 (2) (2005) 427–442.

[65] Y. Zhu, Estimation of an N–L–N Hammerstein–Wiener model, Automatica 38 (9) (2002) 1607–1614.

[66] T.M. Panicker, V.J. Mathews, G.L. Sicuranza, Adaptive parallel-cascade truncated Volterra filters, IEEE Transactions on Signal Processing 46 (10) (1998) 2664–2673.

[67] T. Wigren, J. Schoukens, Three free data sets for development and benchmarking in nonlinear system identification, in: Proc. of 2013 European Control Conference, ECC2013, Zürich, Switzerland, 2013, pp. 2933–2938.
[68] Y. Han, R.A. de Callafon, Identification of Wiener–Hammerstein benchmark model via rank minimization, Control Engineering Practice 20 (11) (2012) 1149–1155, https://doi.org/10.1016/j.conengprac.2012.05.004.
[69] J. Schoukens, J. Suykens, L. Ljung, Wiener–Hammerstein benchmark, in: 15th IFAC Symposium on System Identification, SYSID 2009, St. Malo, France, 2009, pp. 1–4.

CHAPTER

# 4

# RECENT ADVANCES ON LIP NONLINEAR FILTERS AND THEIR APPLICATIONS

## EFFICIENT SOLUTIONS AND SIGNIFICANCE-AWARE FILTERING

**Christian Hofmann***, **Walter Kellermann***

*Friedrich-Alexander-Universität Erlangen-Nürnberg, Lehrstuhl für Multimediakommunikation und Signalverarbeitung, Erlangen, Germany

## CONTENTS



Adaptive Learning Methods for Nonlinear System Modeling. DOI: 10.1016/B978-0-12-812976-0.00005-1

## CHAPTER POINTS

- LIP nonlinear filters with simple linear dependence on the parameters can be identified in a unified way by classical partitioned-block frequency-domain normalized least mean square algorithms.
- Algorithms to adapt LIP nonlinear filters with bilinear dependence on parameter subsets can be treated in a unified way as filtered-X algorithms.
- Significance-aware filtering can be applied in both cases for the efficient and effective identification of LIP nonlinear filters.

## 4.1 INTRODUCTION

Linear-In-the-Parameters (LIP) nonlinear filters constitute a broad class of nonlinear systems and are used for modeling in a broad range of scientific areas, such as in biology (*e.g.*, [1,2]), control engineering (*e.g.*, [3]), communications (*e.g.*, [4–6]) and in acoustic signal processing, where they are, for instance, employed for loudspeaker linearization (*e.g.*, [7–9]) and Acoustic Echo Cancellation (AEC) (*e.g.*, [10–16]). A nonlinear system with input $x(n)$ and output $y(n)$, where $n$ is the discrete-time sample index, will be described by the input/output relation

$$y(n) = f_{\mathrm{I/O}}\{x(n)\}, \tag{4.1}$$

and will be called a memoryless nonlinearity if and only if $y(n)\delta(n) = f_{\mathrm{I/O}}\{x(n)\delta(n)\}$ for any $n$ and $x(n)$, where $\delta(n)$ is the unit impulse. Otherwise, $f_{\mathrm{I/O}}\{x(n)\}$ will be called a nonlinearity with memory. In both cases, a nonlinearity may either be considered as time-invariant and known, or as parametrized by a set of unknown parameters. The following treatment will focus on LIP nonlinear filters, where the output depends linearly on the parameters, *e.g.*, on a vector $\mathbf{a}$, or bilinearly on two parameter subsets, *e.g.*, on parameter vectors $\mathbf{a}_1$ and $\mathbf{a}_2$.

The adaptive identification of a parametric model for an unknown nonlinear system $f_{\mathrm{I/O}}\{\cdot\}$ is illustrated in Fig. 4.1. Therein, an adaptive nonlinear filter (the parametric model) produces an estimate $\hat{y}(n)$ for an observation $y(n)$ and the filter's parameters are iteratively refined (adapted) to minimize a cost function derived from the error signal $e(n) = y(n) - \hat{y}(n)$, based on noisy observations $y(n) = f_{\mathrm{I/O}}\{x(n)\} + s(n)$. When obtaining the observations by measuring a physical signal, $s(n)$ may be the superposition of sensor self-noise and other processes coupling into the sensors as well.

As will be explained in the following, AEC is a particularly challenging example for system identification. For AEC, $y(n)$ in Fig. 4.1 corresponds to a microphone signal composed of local sound sources $s(n)$ (*e.g.*, the local speaker in a telephone communication) and acoustic echoes $f_{\mathrm{I/O}}\{x(n)\}$ from a played-back loudspeaker signal $x(n)$ (*e.g.*, the far-end speaker in a telephone communication). The AEC system aims at removing the acoustic echoes $f_{\mathrm{I/O}}\{x(n)\}$ from the microphone signal $y(n) = f_{\mathrm{I/O}}\{x(n)\} + s(n)$, using an echo signal estimate $\hat{y}(n)$. In periods of local sound source activity, the error signal $e(n)$ constitutes an estimate for the local sound sources $s(n)$. In single-talk periods where only the echo signal $f_{\mathrm{I/O}}\{x(n)\}$ contributes to the microphone signal, the error signal can be employed to refine the adaptive filter. The detection of local sound sources (double-talk detection) is out of the scope of this chapter and solutions for this problem exist (*e.g.*, [17,18]). Therefore, we will assume that no local sources are active when considering AEC, which means $s(n) = 0$.

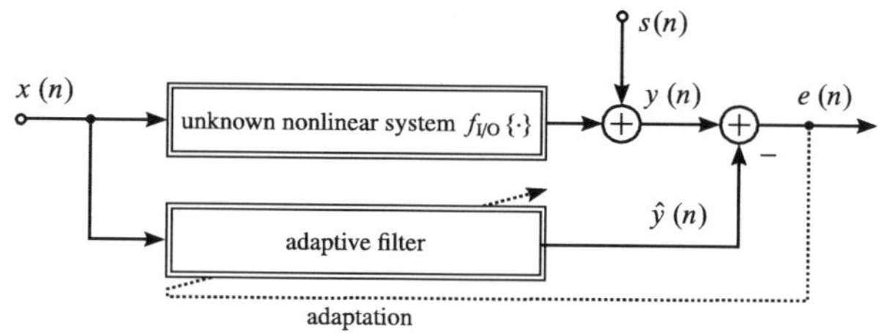


**FIGURE 4.1**

The generic task of adaptive system identification.

AEC is a particularly challenging example for system identification due to the complex acoustic echo paths to be modeled, which typically requires nonlinear adaptive filters with large memory. Furthermore, adaptive filters have to identify a system excited by speech or audio signals (super-Gaussian, non-white and highly instationary signals) instead of signals tailored to system identification. Thus, constituting an especially challenging application, AEC will be used as a reference for evaluating the methods in this chapter. However, the models described and introduced in this chapter are not limited to AEC. Potential other applications will be highlighted at the beginning of Section 4.5, where the required terminology on nonlinear systems and their adaptation will have been established.

The organization of this chapter will be summarized in the following. After introducing the fundamental notations and terminology in the remainder of this section, state-of-the-art LIP nonlinear filters will be categorized in Section 4.2 and fundamental algorithms for parameter estimation for such filters will be presented in Section 4.3. Resulting from a unified treatment of time-domain Normalized Least Mean Square (NLMS), Partitioned-Block Frequency-domain Normalized Least Mean Square (PB-FNLMS), as well as Filtered-X (FX) algorithms for LIP nonlinear filters, a novel and efficient FX PB-FNLMS algorithm will be introduced. In Section 4.4, the recently proposed, computationally very efficient concept of Significance-Aware (SA) filtering will be summarized and a novel SA FX PB-FNLMS algorithm with even further increased efficiency will be proposed. The adaptive filtering techniques described in this chapter will be evaluated and compared in Section 4.5. Finally, similarities and differences of SA filters and nonlinear adaptive filtering techniques based on convex filter combinations for model structure estimation will be reviewed in Section 4.6, before summarizing this chapter in Section 4.7.

### Notations and Fundamentals

As common in system identification literature, the terms filter and system are used interchangeably. The input/output relationship of a linear, causal Finite Impulse Response (FIR) system of length $L$ with input signal $x(n)$ and output $y(n)$ will be written as

$$y(n) = h(n) * x(n) = \sum_{k=0}^{L-1} h(k) \cdot x(n-k) = \langle \mathbf{h}, \check{\mathbf{x}}_n \rangle, \tag{4.2}$$

where $h(k)$ denotes the $k^{\text{th}}$ tap of the system's Impulse Response (IR), $*$ denotes convolution, and $\langle \mathbf{h}, \check{\mathbf{x}}_n \rangle$ represents the scalar product between the vectors $\mathbf{h}$ and $\check{\mathbf{x}}_n$, where

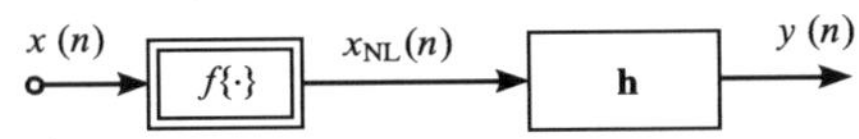


**FIGURE 4.2**

A CM, which is commonly referred to as HM for the special case of a memoryless nonlinearity $f\{\cdot\}$.

$$\mathbf{h} = [h(0), h(1), \ldots, h(L-1)]^{\mathrm{T}}, \tag{4.3}$$

$$\check{\mathbf{x}}_n = [x(n), x(n-1), \ldots, x(n-L+1)]^{\mathrm{T}} \tag{4.4}$$

denote the IR vector and the time-reversed input signal vector, respectively, and the time reversal is indicated by the caron accent. Furthermore, $\mathbf{0}_M$ and $\mathbf{I}_M$ will denote the $M \times M$ matrix of zeros and the $M \times M$ identity matrix, respectively, and are used as components of the windowing matrices

$$\mathbf{W}_{01} = \begin{bmatrix} \mathbf{0}_M & \mathbf{0}_M \\ \mathbf{0}_M & \mathbf{I}_M \end{bmatrix} \quad \text{and} \quad \mathbf{W}_{10} = \begin{bmatrix} \mathbf{I}_M & \mathbf{0}_M \\ \mathbf{0}_M & \mathbf{0}_M \end{bmatrix}, \tag{4.5}$$

which set the first and second half of a vector of length $N = 2M$ to zero, respectively. In addition, $\mathbf{F}$ will refer to the $N \times N$ unitary Discrete Fourier Transform (DFT) matrix and $\mathbf{a}_1 \odot \mathbf{a}_2$ and $\mathbf{a}_1 \oslash \mathbf{a}_2$ will denote element-wise multiplication and division of the vectors $\mathbf{a}_1$ and $\mathbf{a}_2$, respectively. The element-wise magnitude square computation for the entries of a vector $\mathbf{a}$ will be written as $|\mathbf{a}|^2$.

## 4.2 A CONCISE CATEGORIZATION OF STATE-OF-THE-ART LIP NONLINEAR FILTERS

In this section, generic state-of-the-art LIP nonlinear filter structures are introduced as building blocks for the recently developed and newly proposed adaptive filtering methods in Sections 4.3 and 4.4.

### 4.2.1 HAMMERSTEIN MODELS AND CASCADE MODELS

A very simple class of LIP nonlinear filters can be described with the cascade structure depicted in Fig. 4.2, where a nonlinearity $f\{\cdot\}$ is followed by a linear system with IR vector $\mathbf{h}$. The output depends linearly on the parameters $\mathbf{h}$ of the linear system and can be expressed in the notation of Eq. (4.2) as

$$y(n) = f\{x(n)\} * h(n) = x_{\mathrm{NL}}(n) * h(n). \tag{4.6}$$

If $f\{\cdot\}$ in Fig. 4.2 is a memoryless nonlinearity, the structure is widely known as HM [19], whereas the more general case, where $f\{\cdot\}$ is a nonlinearity with memory, will simply be referred to as CM in the following (the attribute nonlinear will be implied). HMs can, for instance, be employed for loudspeaker linearization (*e.g.*, [20,21]) and AEC (*e.g.*, [22–32]). HMs are simple and powerful models for acoustic echo paths if the major source of nonlinearity is the playback equipment (amplifiers and loudspeakers operating at the limits of their dynamic range) and can be mapped to $f\{\cdot\}$, while the acoustic IR (describing sound propagation through the acoustic environment) can be described by $\mathbf{h}$ [22].

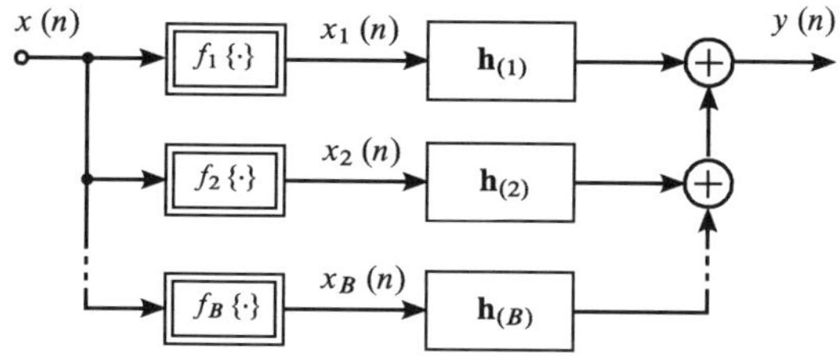


**FIGURE 4.3**

The general structure of a CGM, which is also referred to as HGM for the special case of memoryless nonlinearities $f_b\{\cdot\}, b \in \{1, \ldots, B\}$.

### 4.2.2 HAMMERSTEIN GROUP MODELS AND CASCADE GROUP MODELS

The term Cascade Group Model (CGM) will refer to models with the structure depicted in Fig. 4.3. Such CGMs are composed of $B$ parallel branches, each of which is a nonlinear CM (see Section 4.2.1) with a branch nonlinearity $f_b\{\cdot\}$ and IR $\mathbf{h}_{(b)}$. For memoryless branch nonlinearities $f_b\{\cdot\}$, each branch becomes an HM and the entire CGM will be referred to as HGM in the following. HGMs have successfully been employed to describe both acoustic systems, *e.g.*, in [33–37], and nonacoustic systems, *e.g.*, [38,39]. In their treatment in the literature, HGMs have been addressed by various names, such as the Uryson model [3,39], or are simply treated as an approximation of a particular CGM. The most prominent examples of such HGM/CGM pairs are the so-called power filters [37,40,41] and Volterra filters [19]. Power filters are HGMs with monomes as branch nonlinearities $f_b\{\cdot\}$, *e.g.*,

$$f_1\{\cdot\} = (\cdot), f_2\{\cdot\} = (\cdot)^2, f_3\{\cdot\} = (\cdot)^3, \ldots, f_B\{\cdot\} = (\cdot)^B.$$

Volterra filters result from power filters by augmenting the latter with additional branches where the branch nonlinearities $f_b\{\cdot\}$ perform time-lagged products of powers of the input (time-lagged products of the power filter branch nonlinearities), *e.g.*, the set

$$f_1\{x(n)\} = x(n), \quad f_2\{x(n)\} = x^2(n), \quad f_3\{x(n)\} = x(n) \cdot x(n-1),$$
$$\ldots, f_{2+R}\{x(n)\} = x(n) \cdot x(n-R), \quad f_{3+R}\{x(n)\} = x^3(n), \ldots$$

leads to a Volterra filter in Diagonal Coordinate Representation (DCR) [6]. The set of IRs $\mathbf{h}_{(b)}$ of branches where $f_b\{\cdot\}$ performs (potentially time-lagged) products of $o$ input signal values is called $o$th-order Volterra kernel and the branches within a kernel are also termed diagonals of the respective kernel. In the aforementioned example, the second-order kernel is modeled by $R+1$ diagonals ($b \in \{2, \ldots, 2+R\}$). Volterra filters are universal approximators for a large class of nonlinear systems and have therefore been applied in biology (*e.g.* [1]), control engineering (*e.g.*, [3]) and communications (*e.g.*, [4–6]), as well as in acoustic signal processing (*e.g.*, [7–16,42,43]). Recently, HGMs and CGMs have also been realized with other sets of branch nonlinearities, such as Fourier basis functions [44], Legendre polynomials [45,46] or Chebyshev polynomials [47]. Also, Functional-Link Adaptive Filters (FLAFs) [34,35,48–50] can be viewed as HGM or CGM structures. All these models can be described by the input/output relationship

$$y(n) = \sum_{b=1}^{B}\sum_{k=0}^{L-1} \underbrace{f_b\{x(n-k)\}}_{x_b(n-k)} h_{(b)}(k) = \sum_{b=1}^{B} x_b(n) * h_{(b)}(n), \tag{4.7}$$

where $h_{(b)}(k)$ denotes the $k$th tap of the linear subsystem in branch $b$ and $x_b(n)$ is the $b$th branch signal. Thus, algorithms developed for one particular realization (*i.e.*, set of basis functions $f_b\{\cdot\}$) of CGMs can often immediately be applied to other realizations. This holds for the adaptation algorithms for coefficient estimation described in Sections 4.3 and 4.4, as well as for the structure estimation methods outlined in Section 4.6.

### 4.2.3 BILINEAR CASCADE MODELS

For applications where CGMs with a large number of branches are computationally too demanding, *e.g.*, for AEC in mobile devices with very limited processing or battery power, CM-structured adaptive filters are particularly attractive. Yet, the nonlinearity itself is usually unknown. Therefore, the nonlinearity $f\{\cdot\}$ of a CM is typically modeled by a parametric preprocessor according to

$$f\{x(n)\} = f_{\mathrm{pp}}\{x(n)\} = \sum_{b=1}^{B} w_b f_b\{x(n)\} = x_{\mathrm{pp}}(n), \tag{4.8}$$

where $B$ nonlinearities $f_b\{\cdot\}$ are combined with weights $w_b$, which are parameters to be identified in addition to the linear subsystem. The subindex "pp" of the nonlinear preprocessor and its output signal are used to emphasize the parametrical structure of the preprocessor. The general structure of a CM with a preprocessor according to Eq. (4.8) is depicted in Fig. 4.4A. Most prominently, $f_b\{\cdot\}$ are chosen as monomes [20,21,24,51–54], *e.g.*, $f_b\{\cdot\} = (\cdot)^b$, such that the nonlinearity $f_{\mathrm{pp}}\{\cdot\}$ can be seen as a truncated Taylor series expansion. Fourier series- [55] and Legendre series-based approximations [28,30,31,56] have also recently been employed in AEC. Expanding the preprocessor of Fig. 4.4A in the block diagram results in Fig. 4.4B. Obviously, such a system has the input/output relationship

$$y(n) = x_{\mathrm{pp}}(n) * h(n) \tag{4.9}$$

$$= h(n) * \sum_{b=1}^{B} \overbrace{w_b \underbrace{f_b\{x(n)\}}_{x_b(n)}}^{y_b(n)}, \tag{4.10}$$

and the paths from $x_b(n)$ to $y(n)$ consist not only of a single linear system, but of cascades of $w_b$ (a single-tap linear system) with the linear system $\mathbf{h}$. Therefore, the output of $y(n)$ linearly depends on both the parameter vector $\mathbf{w} = [w_1, w_2, \ldots, w_B]^{\mathrm{T}}$ and the parameter vector $\mathbf{h}$, which is why such systems will be referred to as bilinear[1] CMs in the following. Bilinearity holds regardless of $f_b\{\cdot\}$

[1] This should not be confused with the bilinear filters in [19], which are recursive filters where the difference equation contains linear and bilinear terms in $x(n)$ and $y(n)$.

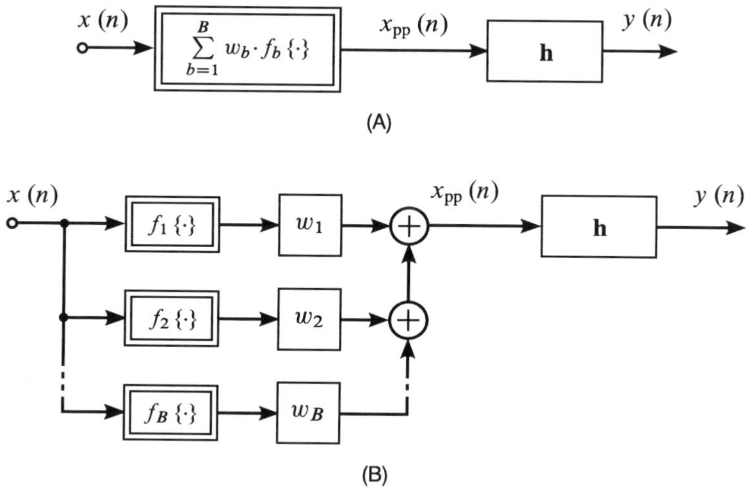


**FIGURE 4.4**

Alternative representations of a CM with parametric preprocessor. (A) Compact representation of a CM with parametric preprocessor as nonlinear–linear cascade. To emphasize the parametric structure of the preprocessor, the nonlinearly distorted input signal is denoted $x_{pp}(n)$ instead of $x_{NL}(n)$. (B) Expansion of the memoryless nonlinearity in (A), revealing a bilinear dependency of the model output $y(n)$ on the parameter vectors $\mathbf{w} = [w_1, w_2, \ldots, w_B]^T$ and $\mathbf{h}$.

being memoryless or with memory and also for general linear filters with IRs $\mathbf{h}_{(b)}$ replacing the single-tap filters $w_b$, resulting in CGM preprocessors.[2] Thus, in the following, all considerations on simple preprocessor systems according to Fig. 4.4A translate to bilinear CMs with CGM preprocessors and *vice versa*, unless noted otherwise.

## 4.3 FUNDAMENTAL METHODS FOR COEFFICIENT ADAPTATION

The most common methods for parameter estimation for linear and LIP nonlinear filters are stochastic gradient-type algorithms, such as the Least Mean Square (LMS) algorithm or the NLMS algorithm [57]. These can be derived as an approximation of a gradient descent w.r.t. the mean squared error cost function $J = \mathcal{E}\left\{|e(n)|^2\right\}$, where $\mathcal{E}\{\cdot\}$ denotes mathematical expectation and $e(n) = y(n) - \hat{y}(n)$ is the error signal, computed as difference between the unknown system's output $y(n)$ and its estimate $\hat{y}(n)$, obtained with the adaptive filter (*cf.* Fig. 4.1). For a comprehensive treatment of different types of LMS-type algorithms, we refer to [57].

In the following, classical system identification will be employed in Section 4.3.1 to adapt $\mathbf{h}_{(b)}$ of CGMs from Section 4.2.2 (*cf.* Fig. 4.3). In Section 4.3.2, the filtered-X method is employed to convert

[2]Note that each filter tap in $\mathbf{h}_{(b)}$ can alternatively be expressed as an additional preprocessor branch with an appropriately delayed version of $f_b\{\cdot\}$ as branch nonlinearity.

the adaptation of a preprocessor system according to Section 4.2.3 (*cf.* Fig. 4.4) into two ordinary system identification tasks, one for $\mathbf{h}_{(b)}$ and one for $w_b$, by reordering $w_b$ and $\mathbf{h}_{(b)}$.

### 4.3.1 NLMS FOR CASCADE GROUP MODELS

In this section, the adaptation of CGMs according to Section 4.2.2 by a time-domain and a partitioned-block frequency-domain NLMS will be described in Sections 4.3.1.1 and 4.3.1.2, respectively. Note that the branch nonlinearities will be considered as time-invariant and known, such that only the linear subsystems of $\mathbf{h}_{(b)}$ of the CGMs from Section 4.2.2 are adaptive.

#### 4.3.1.1 Time-Domain NLMS

To implement and adapt a CGM, the input signal $x(n)$ has to be mapped nonlinearly to a set of branch signals

$$x_b(n) = f_b\{x(n)\},\ b \in \{1, \ldots, B\}, \tag{4.11}$$

as inputs for the linear $B \times 1$ Multiple-Input/Single-Output (MISO) subsystem depicted in the right half of Fig. 4.3. This allows one to compute output samples

$$\hat{y}(n) = \sum_{b=1}^{B} \left\langle \hat{\mathbf{h}}_{(b),n-1}, \check{\mathbf{x}}_{(b),n} \right\rangle, \tag{4.12}$$

where $\hat{\mathbf{h}}_{(b),n-1}$ is the length-$L$ IR estimate in branch $b$ and has been obtained at time index $n-1$ and where $\check{\mathbf{x}}_{(b),n}$ is the time-reversed branch signal vector, structured analogously to Eq. (4.4). This allows one to compute the error signal

$$e(n) = y(n) - \hat{y}(n) \tag{4.13}$$

before performing for each branch $b$ a filter update according to

$$\hat{\mathbf{h}}_{(b),n} = \hat{\mathbf{h}}_{(b),n-1} + \mu_b \cdot \frac{e(n)}{E_{x_b}(n) + \delta} \check{\mathbf{x}}_{(b),n}, \tag{4.14}$$

with the branch-specific adaptation step sizes $\mu_b \in (0, 2)$, a nonnegative regularization constant $\delta$ for numerical stability and the branch signal energies

$$E_{x_b}(n) = \left\langle \check{\mathbf{x}}_{(b),n}, \check{\mathbf{x}}_{(b),n} \right\rangle. \tag{4.15}$$

The computational effort of such a time-domain NLMS-adaptive CGM grows linearly with the number of branches $B$. Note that the CGM adaptation includes the CM adaptation for $B = 1$ and the adaptation of linear systems for $B = 1$ and $f_1\{x(n)\} = x(n)$.

#### 4.3.1.2 Partitioned-Block Frequency-Domain NLMS

To efficiently adapt LIP nonlinear filters with a low input/output delay, Partitioned Block (PB) frequency-domain algorithms can be applied. For this purpose, a PB-FNLMS will be summarized here for CGMs.

### Partitioned-Block Convolution

For partitioned-block convolution, the summation in Eq. (4.2) is split into partial sums and each partial sum (the convolution of an IR partition with an appropriately delayed segment of the input signal) is computed for $M$ consecutive samples via fast DFT-domain convolution, like in [58–62]. This leads to block processing of the input signal $x(n)$ with a frameshift of $M$ samples and a frame size of $N = 2M$. In this notation, let the input signal vector $\mathbf{x}_\kappa$ at frame $\kappa$, the IR partition vector $\mathbf{h}^{(p)}$ (containing the $p$th partition of $\mathbf{h}$) and the output signal vector $\mathbf{y}_\kappa$ at frame $\kappa$ be defined as

$$\mathbf{x}_\kappa = \left[x(\kappa M - M), \ldots, x(\kappa M + M - 1)\right]^{\mathrm{T}}, \tag{4.16}$$

$$\mathbf{h}^{(p)} = \left[h(pM), \ldots, h(pM + M - 1), 0, \ldots, 0\right]^{\mathrm{T}} \text{ and} \tag{4.17}$$

$$\mathbf{y}_\kappa = \left[0, \ldots, 0,\ y(\kappa M), \ldots, y(\kappa M + M - 1)\right]^{\mathrm{T}}, \tag{4.18}$$

respectively. Employing the length-$N$ DFT matrix $\mathbf{F}$, DFT-domain representations

$$\underline{\mathbf{x}}_\kappa = \mathbf{F}\mathbf{x}_\kappa \quad \text{and} \tag{4.19}$$

$$\underline{\mathbf{h}}^{(p)} = \mathbf{F}\mathbf{h}^{(p)} \tag{4.20}$$

can be obtained, and the output signal vector can be computed as

$$\mathbf{y}_\kappa = \mathbf{W}_{01}\, \overbrace{\mathbf{F}^{\mathrm{H}} \underbrace{\sum_{p=0}^{P-1} \underline{\mathbf{x}}_{\kappa-p} \odot \underline{\mathbf{h}}^{(p)}}_{\underline{\mathbf{y}}^{\circ}_{\kappa}}}^{\mathbf{y}^{\circ}_{\kappa}}, \tag{4.21}$$

where $\underline{\mathbf{y}}^{\circ}_{\kappa}$ results from the accumulation over all $P$ partitions' DFT-domain products and contains cyclic convolution artifacts in its time-domain representation $\mathbf{y}^{\circ}_{\kappa}$. Thus, the time-domain output signal vector $\mathbf{y}_\kappa$ emerges from $\mathbf{y}^{\circ}_{\kappa}$ by premultiplication with the windowing matrix $\mathbf{W}_{01}$, defined in Eq. (4.5).

### Application of Partitioned-Block Convolution for Adaptive Filtering

As in the previous paragraph's partitioned convolution scheme, an adaptive CGM estimate for the linear subsystem in branch $b$ at frame $\kappa$ can be split into partitions $\hat{\mathbf{h}}^{(p)}_{(b),\kappa}$. Their DFT-domain representations $\hat{\underline{\mathbf{h}}}^{(p)}_{(b),\kappa}$ can be adapted efficiently on a frame-wise basis by an NLMS-type algorithm. To this end, branch signal vectors

$$\mathbf{x}_{(b),\kappa} = \begin{bmatrix} [\mathbf{0}_M, \mathbf{I}_M]\mathbf{x}_{(b),\kappa-1} \\ \left[f_b\{x(\kappa M)\}, \ldots, f_b\{x(\kappa M + M - 1)\}\right]^{\mathrm{T}} \end{bmatrix} \tag{4.22}$$

are determined, which contain $M$ samples of the previous frame (upper half) and $M$ newly computed branch signal samples (lower half), and they are transformed into the DFT domain according to

$$\underline{\mathbf{x}}_{(b),\kappa} = \mathbf{F}\mathbf{x}_{(b),\kappa}. \tag{4.23}$$

Applying the fast partitioned convolution scheme of Eq. (4.21) in all $B$ branches and summing up all branches in the DFT domain yields the output estimate with cyclic convolution artifacts

$$\hat{\underline{\mathbf{y}}}^{\circ}_{\kappa} = \sum_{b=1}^{B} \sum_{p=0}^{P-1} \underline{\mathbf{x}}_{(b),\kappa-p} \odot \hat{\underline{\mathbf{h}}}^{(p)}_{(b),\kappa-1}. \tag{4.24}$$

This result allows one to compute the DFT-domain error signal vector

$$\underline{\mathbf{e}}_{\kappa} = \mathbf{F}\left(\mathbf{y}_{\kappa} - \underbrace{\mathbf{W}_{01}\mathbf{F}^{\mathrm{H}}\hat{\underline{\mathbf{y}}}^{\circ}_{\kappa}}_{\hat{\mathbf{y}}_{\kappa}}\right), \tag{4.25}$$

where $\hat{\mathbf{y}}_{\kappa}$ represents the output signal vector estimate, obtained from $\hat{\underline{\mathbf{y}}}^{\circ}_{\kappa}$ after an Inverse DFT (IDFT) and windowing. For NLMS adaptation, the DFT-domain signal energy can be estimated using a recursive averaging

$$\mathbf{s}_{x_b,\kappa} = \gamma_{\mathrm{E}}\mathbf{s}_{x_b,\kappa-1} + (1-\gamma_{\mathrm{E}})\left|\underline{\mathbf{x}}_{(b),\kappa}\right|^2, \tag{4.26}$$

where $\gamma_{\mathrm{E}}$ is a smoothing factor ($\gamma_{\mathrm{E}} = 0.9$ will be assumed as default) and will be used to compute normalized branch signals

$$\underline{\mathbf{x}}_{\mathrm{norm},(b),\kappa} = \mu_b \left(\underline{\mathbf{x}}_{(b),\kappa}\right)^{*} \oslash \left(\mathbf{s}_{x_b,\kappa} + \delta\right), \tag{4.27}$$

where $\mu_b \in (0,2)$ is the adaptation step size and $\delta$ is a small positive constant for numerical stability, which is added to each element of $\mathbf{s}_{x_b,\kappa}$. Based on $\underline{\mathbf{x}}_{\mathrm{norm},(b),\kappa}$, the update of the filter partitions can be expressed as

$$\hat{\underline{\mathbf{h}}}^{\circ(p)}_{(b),\kappa} = \hat{\underline{\mathbf{h}}}^{(p)}_{(b),\kappa-1} + \underline{\mathbf{x}}_{\mathrm{norm},(b),\kappa-p} \odot \underline{\mathbf{e}}_{\kappa}, \tag{4.28}$$

which is also known as unconstrained update in the literature [58] and where cyclic convolution artifacts lead to nonzero filter taps in the second half of $\hat{\mathbf{h}}^{\circ(p)}_{(b),\kappa} = \mathbf{F}^{\mathrm{H}}\hat{\underline{\mathbf{h}}}^{\circ(p)}_{(b),\kappa}$. Limiting the temporal support of the partitions to $M$ samples is possible by

$$\hat{\underline{\mathbf{h}}}^{(p)}_{(b),\kappa} = \mathbf{F}\underbrace{\mathbf{W}_{10}\mathbf{F}^{\mathrm{H}}\left(\hat{\underline{\mathbf{h}}}^{\circ(p)}_{(b),\kappa}\right)}_{\hat{\mathbf{h}}^{(p)}_{(b),\kappa}} \tag{4.29}$$

and corresponds to the so-called constrained update in the literature, which is typically formulated by introducing the constraint on the update instead of the partitions (see [58]). Alternatively, a soft-partitioned update is also possible [62,63], which shapes the temporal support of $\hat{\mathbf{h}}^{(p)}_{(b),\kappa}$ by convolution in the DFT domain with a very short DFT-domain sequence in order to save the DFTs in Eq. (4.29).

Corresponding to the time-domain adaptation in Section 4.3.1.1, this section's PB-FNLMS CGM adaptation includes the CM adaptation for $B = 1$ and the adaptation of linear systems for $B = 1$ and $f_1\{x(n)\} = x(n)$. Furthermore, a nonpartitioned Frequency-domain Normalized Least Mean Square

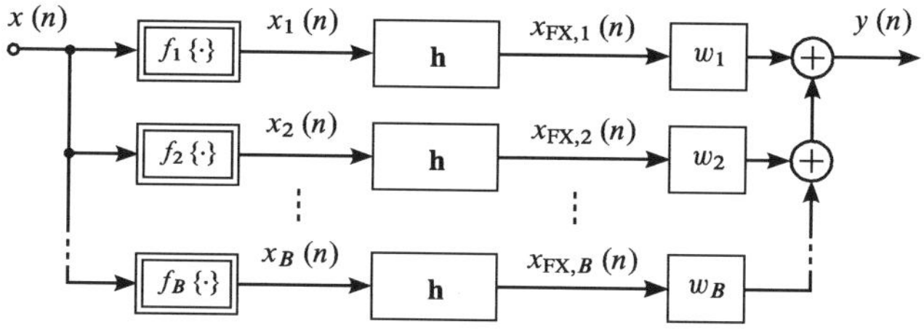


**FIGURE 4.5**

A classical FX structure for an HM with parametric preprocessor according to Eq. (4.8) as a result of exchanging the filtering order of $\mathbf{h}$ and $w_b$ in Fig. 4.4B (exploiting linearity and assuming time invariance), which forms the common core of a class of adaptive algorithms for nonlinear system identification [24].

(FNLMS) algorithm results from $P = 1$. Taking the number of DFTs as measure of computational complexity, a CGM with $B = 5$ branches and $P = 4$ partitions is about 4.3 times as complex as a linear model with $P = 4$ partitions ($B = 1$).

### 4.3.2 FILTERED-X ADAPTATION FOR BILINEAR CASCADE MODELS

In this section, the FX algorithm will first be introduced and discussed on a conceptual level for bilinear CMs, before specifying it in detail and proposing an efficient realization for CGM preprocessors with single-tap branch filters (*i.e.*, parametric preprocessors according to Eq. (4.8)).

#### The Generic Filtered-X Structure

FX algorithms are frequently employed in Active Noise Control (ANC) [64] for prefilter adaptation. Thereby, the FX algorithm exploits the fact that the ordering of linear time-invariant systems in a cascade can be reversed without altering the cascade's output: for the preprocessor-based CM in Fig. 4.4B, joint linearity and time invariance allow incorporating $\mathbf{h}$ into each of the branches, as depicted in Fig. 4.5. Therein, the prefiltered inputs to the former prefilters $w_b$ are often termed FX signals. The representation of the CM in Fig. 4.5 allows one to directly apply an NLMS-type adaptation algorithm for linear MISO systems from Section 4.3.1 to adapt $w_b$, as their inputs are known and their combined output is directly matched to the target signal (in system identification: an unknown system's output). Mathematically, an FX LMS algorithm can also be derived directly like an LMS algorithm as stochastic gradient descent algorithm w.r.t. the partial derivative w.r.t. the prefilters. The exchange of the filtering orders results simply from an advantageous summation order in the gradient calculation. Note that adaptive filters for $w_b$ and $\mathbf{h}$ actually violate the time invariance required for exchanging the filter order. However, the time variance appears to be uncritical in practice, where FX algorithms have been employed for ANC for more than three decades [64,65].

#### Review of Known Algorithms for Filtered-X Adaptation of Bilinear Cascade Models

Many FX-like algorithms have been derived for the adaptation of bilinear LIP nonlinear systems but do not establish the link to the filtered-X algorithms and the exchange of filtering orders. The following description of these methods will highlight the fact that these algorithms can be viewed and implemented as FX algorithms.

In [24], the FX preprocessor adaptation was derived and applied to the adaptation of a polynomial preprocessor with $f_b\{\cdot\} = (\cdot)^b$—a nonadaptive offline version of this mechanism was proposed in [23], which resembles the iterative version of [51] for AutoRegressive Moving Average (ARMA) models as linear submodels. More recent work on such cascade models considered a generalization of [24] by allowing longer IRs than a single tap [37,66] or considered piece-wise linear functions for $f_b\{\cdot\}$ in conjunction with a modified joint normalization of the linear and nonlinear filter coefficients [27]. In particular, [37] employs a power filter as preprocessor to the linear subsystem. The resulting algorithm corresponds to a time-domain FX algorithm. In [66], a particular CGM preprocessor is discussed, which corresponds to a Volterra filter in triangular representation (see, *e.g.*, [19] for the triangular representation) with memory lengths of $N_1$ and $N_2$ for the kernels of orders 1 and 2, respectively, and no memory at all for higher-order Volterra kernels (becoming a simple memoryless preprocessor[3] for these orders). Structures similar to this preprocessor also result from the so-called EVOLutionary Volterra Estimation (EVOLVE), which will be outlined later on in Section 4.6. While evaluating their algorithm for a memoryless preprocessor ($N_1 = N_2 = 0$), [66] also evaluates an extension of [24] by adapting the linear subsystem with a frequency-domain NLMS with unconstrained update. The preprocessor is adapted, as in [24], by a Recursive Least Squares (RLS)-like algorithm based on the FX signals. Still, the two RLS descriptions differ as [66] employs the direct inversion of the involved correlation matrix, whereas [24] makes use of the alternative variant with a recursively computed inverse (see [57] for RLS-adaptive filters). Another extension of [24] is treated in [25] (in German), where the signal flow and benefit of a subband-AEC variant of [24] is discussed on a theoretical level.

Note that none of these approaches describes the preprocessor adaptation explicitly as an FX algorithm.[4] As opposed to this, the remainder of this section introduces an FX PB-FNLMS algorithm for CMs with CGM preprocessors and tailors this algorithm to the special case of memoryless preprocessors according to Eq. (4.8).

### Filtered-X Adaptation of Preprocessors of Partitioned-Block CMs

In the following, consider the structure of Fig. 4.4B as a single-tap CGM, for which DFT-domain branch signal vectors can be computed according to Eq. (4.23). Based on that, preprocessed DFT-domain input signal vectors can be determined by

$$\underline{\mathbf{x}}_{\mathrm{pp},\kappa} = \sum_{b=1}^{B} \hat{w}_b(\kappa-1)\,\underline{\mathbf{x}}_{(b),\kappa} \tag{4.30}$$

and the DFT-domain FX signal vectors $\underline{\mathbf{x}}_{\mathrm{FX},(b),\kappa}$ (see Fig. 4.5 illustrating the FX signals) can be computed by partitioned convolution in a two-step procedure as

$$\underline{\mathbf{x}}^{\circ}_{\mathrm{FX},(b),\kappa} = \sum_{p=0}^{P-1} \underline{\mathbf{x}}_{(b),\kappa-p} \odot \hat{\underline{\mathbf{h}}}^{(p)}_{\kappa-1} \tag{4.31}$$

[3]Note that [66] uses a different notation than this chapter and refers to an HM with a monome-based preprocessor as power filter.
[4]Also note that not every bilinear cascade calls for FX adaptation, *e.g.*, in the so-called sparse-interpolated Volterra filters of [67], linear interpolation filters are prefilters to subsampled Volterra kernels to interpolate the latter. As the interpolation filters are time-invariant, they do not need to be adapted and consequently do not require an FX adaptation.

$$\underline{\mathbf{x}}_{\mathrm{FX},(b),\kappa} = \mathbf{F} \underbrace{\begin{bmatrix} [\mathbf{0}_M, \mathbf{I}_M]\mathbf{x}_{\mathrm{FX},(b),\kappa-1} \\ [\mathbf{0}_M, \mathbf{I}_M]\mathbf{F}^{\mathrm{H}}\underline{\mathbf{x}}^{\circ}_{\mathrm{FX},(b),\kappa} \end{bmatrix}}_{\mathbf{x}_{\mathrm{FX},(b),\kappa}}. \tag{4.32}$$

With the introduced signals, the adaptive filter's output can alternatively be determined by one of the two equations

$$\hat{\mathbf{y}}_\kappa = \mathbf{W}_{01}\mathbf{F}^{\mathrm{H}} \sum_{p=0}^{P-1} \underline{\mathbf{x}}_{\mathrm{pp},\kappa-p} \underline{\hat{\mathbf{h}}}_p^{(\kappa)} \tag{4.33}$$

$$\approx \mathbf{W}_{01}\mathbf{F}^{\mathrm{H}} \underbrace{\sum_{b=1}^{B} \hat{w}_b(\kappa-1)\, \underline{\mathbf{x}}_{\mathrm{FX},(b),\kappa}}_{\underline{\mathbf{y}}^{\circ}_{\kappa}}, \tag{4.34}$$

where Eq. (4.33) corresponds to block processing with the signal flow in Fig. 4.4B and Eq. (4.34) corresponds to the signal flow in Fig. 4.5. For time-invariant systems, Eqs. (4.33) and (4.34) are identical. This allows one to compute the error signal vector $\mathbf{e}_\kappa = \mathbf{y}_\kappa - \hat{\mathbf{y}}_\kappa$ and its DFT

$$\underline{\mathbf{e}}_\kappa = \mathbf{F}\mathbf{e}_\kappa, \tag{4.35}$$

both of which will be used for NLMS-type adaptation of the parameters. Analogously to Section 4.3.1.2, a PB-FNLMS adaptation of $\underline{\hat{\mathbf{h}}}_\kappa^{(p)}$ can be expressed as

$$\underline{\hat{\mathbf{h}}}_\kappa^{\circ(p)} = \underline{\hat{\mathbf{h}}}_{\kappa-1}^{(p)} + \underline{\mathbf{x}}_{\mathrm{norm,pp},\kappa-p} \odot \underline{\mathbf{e}}_\kappa, \tag{4.36}$$

$$\underline{\hat{\mathbf{h}}}_\kappa^{(p)} = \mathbf{F} \underbrace{\mathbf{W}_{10}\mathbf{F}^{\mathrm{H}} \left(\underline{\hat{\mathbf{h}}}_\kappa^{\circ(p)}\right)}_{\hat{\mathbf{h}}_\kappa^{(p)}}, \tag{4.37}$$

where $\underline{\mathbf{x}}_{\mathrm{norm,pp},\kappa}$ is the normalized preprocessed input signal vector computed analogously to Eqs. (4.26) and (4.27) (containing the adaptation step size) and where Eq. (4.36) and Eq. (4.37) correspond to the unconstrained and the constrained update, respectively.

Similar to [24], the weights $\hat{w}_b(\kappa)$ are estimated by an NLMS algorithm employing the FX signals as

$$\hat{w}_b(\kappa) = \hat{w}_b(\kappa-1) + \frac{\mu_{\mathrm{FX}}}{E_b + \delta} \left\langle \mathbf{x}_{\mathrm{FX},(b),\kappa}, \mathbf{e}_\kappa \right\rangle \tag{4.38}$$

with the adaptation step size $\mu_{\mathrm{FX}}$ and

$$E_b = \left\langle \mathbf{x}_{\mathrm{FX},(b),\kappa}, \mathbf{x}_{\mathrm{FX},(b),\kappa} \right\rangle. \tag{4.39}$$

For the more general case where a partitioned-block CGM with $P_{\mathrm{pp}}$ partitions $\underline{\dot{\mathbf{h}}}_{(b),\kappa}^{(p)}$ replaces the preprocessor with weights $\hat{w}_b(\kappa)$, all linear combinations with the weights, like in Eq. (4.30), have to be replaced by actual partitioned convolutions and PB-FNLMS updates can be computed for the preprocessor according to

$$\underline{\hat{\mathbf{h}}}^{\circ(p)}_{(b),\kappa} = \underline{\hat{\mathbf{h}}}^{(p)}_{(b),\kappa-1} + \underline{\mathbf{x}}_{\mathrm{FX,norm},(b),\kappa-p} \odot \underline{\mathbf{e}}_{\kappa} \tag{4.40}$$

and

$$\underline{\hat{\mathbf{h}}}^{(p)}_{(b),\kappa} = \underbrace{\mathbf{F}\,\mathbf{W}_{10}\mathbf{F}^{\mathrm{H}}\left(\underline{\hat{\mathbf{h}}}^{\circ(p)}_{(b),\kappa}\right)}_{\hat{\mathbf{h}}^{(p)}_{(b),\kappa}} \tag{4.41}$$

with the normalized DFT-domain FX branch signal vectors $\underline{\mathbf{x}}_{\mathrm{FX,norm},(b),\kappa-p}$ determined from the FX branch signals $\mathbf{x}_{\mathrm{FX},(b),\kappa}$ analogously to Eqs. (4.26) and (4.27). This corresponds to a generalization of [37] from time-domain power filter preprocessor adaptation to PB-FNLMS adaptation of general CGM preprocessors.

Thereby, the adaptation of bilinear CMs has been decomposed into two ordinary PB-FNLMS system identification tasks, plus the additional prefiltering operations generating the FX branch signals. All these components can be implemented efficiently using partitioned-block convolution techniques.

### *An Adaptation Tailored to Single-Tap CGM Preprocessors*

In the following, the implications of a parametric preprocessor according to Eq. (4.30) (a single-tap CGM preprocessor, *cf.* Fig. 4.4B) will be considered in detail. Interestingly, Eq. (4.34) does not introduce cyclic convolution artifacts into $\underline{\mathbf{y}}^{\circ}_{\kappa}$ for single-tap filters $\hat{w}_b(\kappa)$. Consequently, the cyclic convolution artifacts present in $\mathbf{x}^{\circ}_{\mathrm{FX},(b),\kappa}$ do not spread by filtering with the weights, allowing to compute the output estimate

$$\hat{\mathbf{y}}_{\kappa} = \mathbf{W}_{01}\mathbf{F}^{\mathrm{H}} \sum_{b=1}^{B} \hat{w}_b(\kappa-1) \cdot \underline{\mathbf{x}}^{\circ}_{\mathrm{FX},(b),\kappa} \tag{4.42}$$

without Eq. (4.32). Furthermore, the zeros in the vector $\mathbf{e}_{\kappa}$ and the unitarity of the DFT imply

$$\begin{aligned}\left\langle \mathbf{x}_{\mathrm{FX},(b),\kappa}, \mathbf{e}_{\kappa}\right\rangle &= \left\langle \mathbf{x}^{\circ}_{\mathrm{FX},(b),\kappa}, \mathbf{e}_{\kappa}\right\rangle &(4.43)\\ &= \left\langle \underline{\mathbf{x}}^{\circ}_{\mathrm{FX},(b),\kappa}, \underline{\mathbf{e}}_{\kappa}\right\rangle. &(4.44)\end{aligned}$$

Further identifying

$$E_b \approx E_b^{\circ} = \left\langle \underline{\mathbf{x}}^{\circ}_{\mathrm{FX},(b),\kappa}, \underline{\mathbf{x}}^{\circ}_{\mathrm{FX},(b),\kappa}\right\rangle \tag{4.45}$$

as a reasonable approximation of the branch signal energy within a frame allows one to compute the weights according to

$$\hat{w}_b(\kappa) = \hat{w}_b(\kappa-1) + \frac{\mu_{\mathrm{FX}}}{E_b^{\circ}+\delta}\left\langle \underline{\mathbf{x}}^{\circ}_{\mathrm{FX},(b),\kappa}, \underline{\mathbf{e}}_{\kappa}\right\rangle \tag{4.46}$$

without Eq. (4.32) as well. Thus, Eq. (4.32) does not need to be evaluated at all, which allows saving $2B$ DFTs of length $N$.

**Table 4.1 Summary of operations of a time-domain NLMS algorithm for CGMs as executed for each sampling instant $n$**

| computed quantity | computation formula | | equation no. |
|---|---|---|---|
| branch signals | $x_b(n) = f_b\{x(n)\}$ | $\forall b$ | Eq. (4.11) |
| time-reversed branch signal vectors | $\check{\mathbf{x}}_{(b),n} = [x_b(n), \ldots, x_b(n-L+1)]^{\mathrm{T}}$ | $\forall b$ | – |
| output signal | $\hat{y}(n) = \sum_{b=1}^{B} \left\langle \hat{\mathbf{h}}_{(b),n-1}, \check{\mathbf{x}}_{(b),n} \right\rangle$ | | Eq. (4.12) |
| error signal | $e(n) = y(n) - \hat{y}(n)$ | | Eq. (4.13) |
| frame energy | $E_{x_b}(n) = \langle \check{\mathbf{x}}_{(b),n}, \check{\mathbf{x}}_{(b),n} \rangle$ | $\forall b$ | Eq. (4.15) |
| filter coefficients | $\hat{\mathbf{h}}_{(b),n} = \hat{\mathbf{h}}_{(b),n-1} + \mu_b \cdot \frac{e(n)}{E_{x_b}(n)+\delta} \check{\mathbf{x}}_{(b),n}$ | $\forall b$ | Eq. (4.14) |

The remaining DFTs and IDFTs are $B+2$ transforms for computing DFT-domain branch signals prior to Eq. (4.31), the output estimate in Eq. (4.34), and the DFT-domain error signal in Eq. (4.35), as well as $2P$ DFTs for a constrained update of the partitioned linear subsystem in Eq. (4.37). The computational effort for non-DFT operations has a component growing proportional to $B \cdot P$ due to Eq. (4.31). However, as the effort for a length-$N$ DFT is typically much higher than the effort for a vector multiplication in Eq. (4.31), the number of DFTs may serve as a first indicator of the computational complexity. Thus, the relative computational complexity $R_{\text{CGM}}^{\text{FX-CM}}$ of an FX CM and a full-length CGM can be approximated by

$$\tilde{R}_{\text{CGM}}^{\text{FX-CM}} = \frac{B+2+2P}{B+2+2BP} + \varepsilon, \tag{4.47}$$

which is the ratio of the numbers of involved DFTs plus an overhead $\varepsilon$ representing the number of operations which are not DFTs and grow with $B \cdot P$ in both the FX-CM and the CGM (filtering and filter coefficient updates). To give an example, $P = 4$ and $B = 5$ leads to $\tilde{R}_{\text{CGM}}^{\text{FX-CM}} \approx 0.32 + \varepsilon$ and suggests that the CM can be identified with about one-third of the effort of a CGM of the same length.

### 4.3.3 SUMMARY OF ALGORITHMS

The main algorithms presented in Section 4.3 are summarized in tabular form to support an implementation in practice. The most basic algorithm is the time-domain NLMS algorithm summarized in Table 4.1. The operations listed therein have to be evaluated at each sample index $n$. For processing an entire block of $M$ subsequent samples, as in block- and frequency-domain processing, all equations in Table 4.1 have to be evaluated sample by sample for each of the block's samples—$M$ times in total. For PB-FNLMS adaptation, the operations for processing a block of $M$ samples are summarized in Table 4.2. The operations listed therein have to be executed for each signal frame $\kappa$ (once every $M$ samples). In the same way, the operations required for FX adaptation of a bilinear CM with a single-tap CGM as nonlinear preprocessor are listed in Table 4.3.

These algorithms will be employed as building blocks of the computationally very efficient SA filters in the following section, Section 4.4. Therein, the aforementioned basic algorithms will only be referenced without repeating all equations.

**Table 4.2 Summary of operations of a PB-FNLMS algorithm for CGMs as executed for each frame $\kappa$**

| computed quantity | computation formula | | equation no. |
|---|---|---|---|
| branch signals | $\mathbf{x}_{(b),\kappa} = \begin{bmatrix} [\mathbf{0}_M, \mathbf{I}_M]\mathbf{x}_{(b),\kappa-1} \\ \left[ f_b\{x(\kappa M)\}, \ldots, f_b\{x(\kappa M + M - 1)\}\right]^{\mathrm{T}} \end{bmatrix}$ | $\forall b$ | Eq. (4.22) |
| $\sim$ in DFT domain | $\underline{\mathbf{x}}_{(b),\kappa} = \mathbf{F}\mathbf{x}_{(b),\kappa}$ | $\forall b$ | Eq. (4.23) |
| DFT-domain output estimate with cyclic convolution artifacts | $\hat{\underline{\mathbf{y}}}^{\circ}_{\kappa} = \sum_{b=1}^{B}\sum_{p=0}^{P-1} \underline{\mathbf{x}}_{(b),\kappa-p} \odot \hat{\underline{\mathbf{h}}}^{(p)}_{(b),\kappa-1}$ | | Eq. (4.24) |
| DFT-domain error signal | $\underline{\mathbf{e}}_{\kappa} = \mathbf{F}\left(\mathbf{y}_{\kappa} - \mathbf{W}_{01}\mathbf{F}^{\mathrm{H}}\hat{\underline{\mathbf{y}}}^{\circ}_{\kappa}\right)$ | | Eq. (4.25) |
| branch PSDs | $\mathbf{s}_{x_b,\kappa} = \gamma_{\mathrm{E}}\mathbf{s}_{x_b,\kappa-1} + (1-\gamma_{\mathrm{E}})\left\lvert\underline{\mathbf{x}}_{(b),\kappa}\right\rvert^2$ | $\forall b$ | Eq. (4.26) |
| normalized branch signals | $\underline{\mathbf{x}}_{\mathrm{norm},(b),\kappa} = \mu_b\left(\underline{\mathbf{x}}_{(b),\kappa}\right)^* \oslash \left(\mathbf{s}_{x_b,\kappa} + \delta\right)$ | $\forall b$ | Eq. (4.27) |
| updated filters | $\hat{\underline{\mathbf{h}}}^{\circ(p)}_{(b),\kappa} = \hat{\underline{\mathbf{h}}}^{(p)}_{(b),\kappa-1} + \underline{\mathbf{x}}_{\mathrm{norm},(b),\kappa-p} \odot \underline{\mathbf{e}}_{\kappa}$ | $\forall b, p$ | Eq. (4.28) |
| constrained update | $\hat{\underline{\mathbf{h}}}^{(p)}_{(b),\kappa} = \mathbf{F}\mathbf{W}_{10}\mathbf{F}^{\mathrm{H}}\left(\hat{\underline{\mathbf{h}}}^{\circ(p)}_{(b),\kappa}\right)$ | $\forall b, p$ | Eq. (4.29) |

## 4.4 SIGNIFICANCE-AWARE FILTERING

In [28], the concept of SA filtering has been introduced, in order to estimate the parameters of nonlinear CMs by decomposing the adaptation process in a divide-and-conquer manner into synergetic adaptive subsystems [28,30–32] which are LIP. Thereby, the estimation of the nonlinearity is separated from the estimation of a possibly long linear subsystem, as it would be characteristic for acoustic echo paths. As a consequence, the nonlinearity can be estimated computationally efficiently by nonlinear models with a very low number of parameters.

In the following, two different SA decompositions will be introduced in Section 4.4.1 and employed for the Serial SA CMs (SSA-CMs) in Section 4.4.2 (a generalization of the Equalization-based SA (ESA)-HM from [32]), the Parallel SA CGMs (PSA-CGMs) in Section 4.4.3 (a generalization of the SA-HGM from [32]) and a novel Parallel SA Filtered-X (PSAFX) algorithm in Section 4.4.4.

### 4.4.1 SIGNIFICANCE-AWARE DECOMPOSITIONS OF LIP NONLINEAR SYSTEMS

In [32], two different SA decompositions have been employed, which will be denoted Serial SA (SSA) decomposition (employed for ESA filtering in [32]) and Parallel SA (PSA) decomposition (employed for so-called "classical" SA filtering in [32]). Both decompositions allow for the estimation of the nonlinearity with a low-complexity nonlinear adaptive filter, as will be explained in the following.

#### *Serial Significance-Aware Decomposition*

The SSA decomposition is depicted schematically in Fig. 4.6, where the unknown system is cascaded with an equalizer filter $\hat{\mathbf{h}}_{\mathrm{eq}}$. The overall response of this serial connection will ideally be a (delayed) version of the nonlinear preprocessor $f\{\cdot\}$. Thus, an estimate for the physically inaccessible inter-

**Table 4.3 Summary of operations of an FX algorithm tailored to the adaptation of bilinear CMs with single-tap CGM preprocessors as executed for each frame $\kappa$**

| computed quantity | computation formula | | equation no. |
|---|---|---|---|
| branch signals | $\mathbf{x}_{(b),\kappa} = \begin{bmatrix} [\mathbf{0}_M, \mathbf{I}_M]\mathbf{x}_{(b),\kappa-1} \\ [f_b\{x(\kappa M)\}, \ldots, f_b\{x(\kappa M + M - 1)\}]^{\mathrm{T}} \end{bmatrix}$ | $\forall b$ | Eq. (4.22) |
| $\sim$ to DFT domain | $\underline{\mathbf{x}}_{(b),\kappa} = \mathbf{F}\mathbf{x}_{(b),\kappa}$ | $\forall b$ | Eq. (4.23) |
| DFT-domain preprocessed signal | $\underline{\mathbf{x}}_{\mathrm{pp},\kappa} = \sum_{b=1}^{B} \hat{w}_b(\kappa-1)\,\underline{\mathbf{x}}_{(b),\kappa}$ | | Eq. (4.30) |
| DFT-domain FX branch signals with cyclic convolution artifacts | $\underline{\mathbf{x}}^{\circ}_{\mathrm{FX},(b),\kappa} = \sum_{p=0}^{P-1} \underline{\mathbf{x}}_{(b),\kappa-p} \odot \hat{\underline{\mathbf{h}}}^{(p)}_{\kappa-1}$ | $\forall b, p$ | Eq. (4.31) |
| output signal | $\hat{\mathbf{y}}_\kappa = \mathbf{W}_{01}\mathbf{F}^{\mathrm{H}} \sum_{b=1}^{B} \hat{w}_b(\kappa-1)\cdot \underline{\mathbf{x}}^{\circ}_{\mathrm{FX},(b),\kappa}$ | | Eq. (4.33) |
| error signal | $\mathbf{e}_\kappa = \mathbf{y}_\kappa - \hat{\mathbf{y}}_\kappa$ | | – |
| $\sim$ to DFT domain | $\underline{\mathbf{e}}_\kappa = \mathbf{F}\mathbf{e}_\kappa$ | | Eq. (4.35) |
| preprocessed signal PSD | $\mathbf{s}_{x_{\mathrm{pp}},\kappa} = \gamma_{\mathrm{E}}\mathbf{s}_{x_{\mathrm{pp}},\kappa-1} + (1-\gamma_{\mathrm{E}})\left\lvert\underline{\mathbf{x}}_{\mathrm{pp},\kappa}\right\rvert^{2}$ | | – |
| normalized input | $\underline{\mathbf{x}}_{\mathrm{norm,pp},\kappa} = \left(\underline{\mathbf{x}}_{\mathrm{pp},\kappa}\right)^{*} \odot \left(\mu_b \oslash \left(\mathbf{s}_{x_{\mathrm{pp}},\kappa} + \delta\right)\right)$ | | – |
| updated filters | $\hat{\underline{\mathbf{h}}}^{\circ(p)}_{\kappa} = \hat{\underline{\mathbf{h}}}^{(p)}_{\kappa-1} + \underline{\mathbf{x}}_{\mathrm{norm,pp},\kappa-p} \odot \underline{\mathbf{e}}_\kappa,$ | $\forall p$ | Eq. (4.40) |
| constrained update | $\hat{\underline{\mathbf{h}}}^{(p)}_{\kappa} = \mathbf{F}\underbrace{\mathbf{W}_{10}\mathbf{F}^{\mathrm{H}}\left(\hat{\underline{\mathbf{h}}}^{\circ(p)}_{\kappa}\right)}_{\hat{\mathbf{h}}^{(p)}_{\kappa}},$ | $\forall p$ | Eq. (4.41) |
| FX energies | $E_b \approx E^{\circ}_b = \left\langle \underline{\mathbf{x}}^{\circ}_{\mathrm{FX},(b),\kappa}, \underline{\mathbf{x}}^{\circ}_{\mathrm{FX},(b),\kappa} \right\rangle$ | $\forall b > 1$ | Eq. (4.45) |
| weight update | $\hat{w}_b(\kappa) = \hat{w}_b(\kappa-1) + \frac{\mu_{\mathrm{FX}}}{E^{\circ}_b+\delta}\left\langle \underline{\mathbf{x}}^{\circ}_{\mathrm{FX},(b),\kappa}, \underline{\mathbf{e}}_\kappa \right\rangle$ | $\forall b > 1$ | Eq. (4.46) |

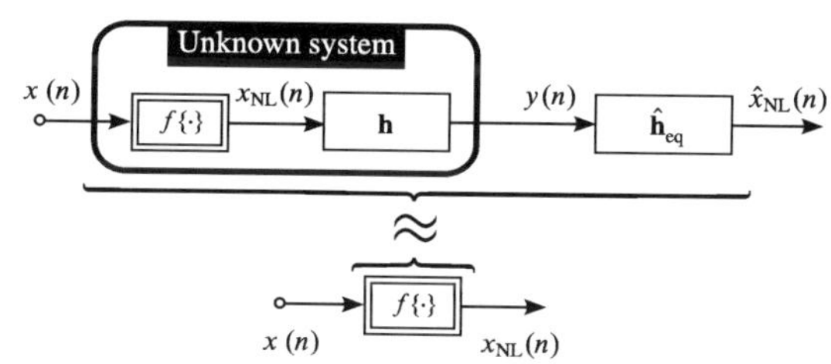


**FIGURE 4.6**

Serial decomposition of the unknown system exploited for SSA filtering. Cascading the unknown system with a linear equalizer filter allows one to approximate the cascade as the nonlinear component of the unknown system (disregarding the processing delay due to causal filtering).

mediate signal $x_{\mathrm{NL}}(n)$ is obtained, which allows one to estimate the nonlinearity directly from this decomposition efficiently by a nonlinear model with a short temporal support. An adaptive system using this decomposition will be specified in Section 4.4.2.

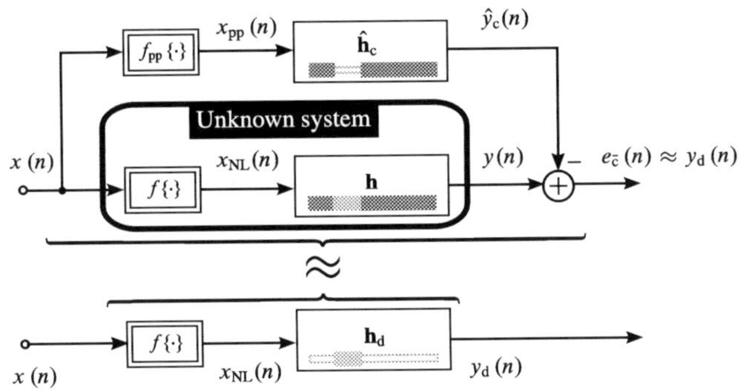


**FIGURE 4.7**

Parallel decomposition of the unknown system exploited for PSA filtering. The temporal support modeled by IR vectors is illustrated below the respective vectors' symbols as follows. The entire vector **h** is represented by a flat white box with a gray frame. Nonzero entries are marked by light gray boxes if they correspond to the dominant region and by dark gray boxes if they describe the complementary region.

### *Parallel Significance-Aware Decomposition*

The PSA decomposition is illustrated in Fig. 4.7. The LIP property of a nonlinear CM is employed to decompose its IR vector into a dominant region $\mathbf{h}_d$ (light gray), describing, *e.g.*, direct-path propagation, and a complementary region $\mathbf{h}_c$ (dark gray). Augmenting the unknown system with a parallel system modeling the complementary-region output signal component $y_c(n)$ allows one to compute the dominant-region output component $y_d(n) = y(n) - y_c(n)$. As $f\{\cdot\}$, $x_{NL}(n)$, $\mathbf{h}_c$, and thereby $y_c(n)$ are not known during system identification, they have to be replaced by estimates $f_{pp}\{\cdot\}$, $x_{pp}(n)$, $\hat{\mathbf{h}}_c$ and $\hat{y}_c(n)$ in practice, as depicted in the upper part of Fig. 4.7. Still, the output of the parallel arrangement yields an estimate $e_{\bar{c}}(n) = y(n) - \hat{y}_c(n) \approx y_d(n)$ of the dominant-region component of the unknown system (bottom of Fig. 4.7). As the dominant-region component of the system describes the transmission of a significant amount of energy, a nonlinear dominant-region model can be estimated at a high Signal to Noise Ratio (SNR) and will suffer much less from gradient noise than a reverberation tail model. Adaptive systems using this decomposition will be specified in Sections 4.4.3 and 4.4.4.

### *Estimating a Preprocessor System From a CGM*

Another key component for the SA models described in Sections 4.4.2 and 4.4.3 below is the possibility to extract a CM's preprocessor from an identified CGM-structured estimate of the CM system. To this end, assume that the CM's nonlinearity can be expressed as weighted sum of the CGM's branch nonlinearities $f_b\{\cdot\}$, resulting in an expression identical to Eq. (4.8). The CM's linear subsystem will be denoted $\mathbf{h}$. Then, the CM can be expressed as CGM with linear subsystems $\mathbf{h}_{(b)} = w_b\mathbf{h}$. In practice, an identified CGM can only provide noisy estimates

$$\hat{\mathbf{h}}_{(b)} = w_b\mathbf{h} + \boldsymbol{\epsilon}_b, \tag{4.48}$$

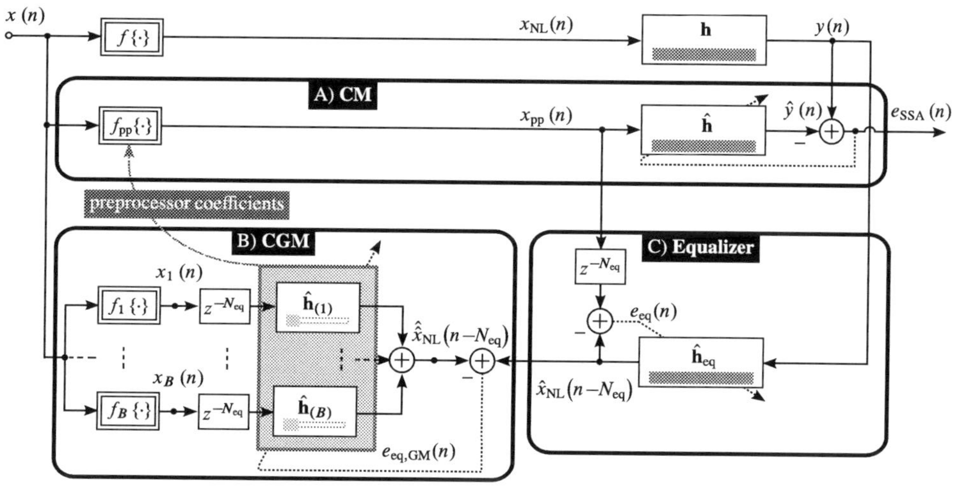


**FIGURE 4.8**

Proposed novel SSA filtering: an estimate $\hat{x}_{\mathrm{NL}}(n)$ of the nonlinearly distorted input signal is obtained in Block C by equalizing the system output signal $y(n)$ with a linear equalizer, which allows in Block B to assess the nonlinear distortions by a very short CGM, which in turn is employed for estimating the parameters of the nonlinear preprocessor $f_{\mathrm{pp}}\{\cdot\}$ of Block A.

where $\boldsymbol{\epsilon}_b$ is the coefficient error vector. As shown in [28], a least squares estimate for $w_b$ can be obtained by

$$\hat{w}_b^{(\mathrm{LS})} = \frac{\left\langle \mathbf{h}_{(b)}, \mathbf{h}_{(b_{\mathrm{ref}})} \right\rangle}{\left\langle \mathbf{h}_{(b_{\mathrm{ref}})}, \mathbf{h}_{(b_{\mathrm{ref}})} \right\rangle} \tag{4.49}$$

w.r.t. a reference branch $\mathbf{h}_{(b_{\mathrm{ref}})}$, for which $\hat{w}_{b_{\mathrm{ref}}}^{(\mathrm{LS})} = 1$. Note that $\hat{w}_{b_{\mathrm{ref}}}^{(\mathrm{LS})} = 1$ is not a model restriction—it simply removes the scaling ambiguity inherent to the cascade model by shifting the actual $w_{b_{\mathrm{ref}}}$ as gain factor into the estimate $\hat{\mathbf{h}}$ to be identified after the preprocessor. Without loss of generality, $b_{\mathrm{ref}} = 1$ and $f_1\{\cdot\} = (\cdot)$ will be assumed from now on. A preprocessor with $w_b = 0 \,\forall\, b > 1$ will be referred to as linearly configured preprocessor.

## 4.4.2 SERIAL SIGNIFICANCE-AWARE CASCADE MODELS

SSA-CMs employ the SSA decomposition from Fig. 4.6 and generalize the Serial SA HMs (SSA-HMs) from [32] by allowing branch nonlinearities with memory. The structure of an SSA-CM is depicted in Fig. 4.8. Initially, the preprocessor in Block A in Fig. 4.8, providing the signal $x_{\mathrm{pp}}(n)$, is configured to be linear. After this initialization, all subsystems in Fig. 4.8 are adapted in parallel. In Block C in Fig. 4.8, an adaptive linear equalizer implements the SSA decomposition. The equalizer filters the unknown system's output $y(n)$ to produce an estimate $\hat{x}_{\mathrm{NL}}(n - N_{\mathrm{eq}})$ of the delayed nonlinearly distorted input signal. For adaptation, $\hat{x}_{\text{NL}}(n-N_{\text{eq}})$ is matched to $x_{\text{pp}}(n-N_{\text{eq}})$.[5] Ideally, $x(n-N_{\text{eq}})$ and $\hat{x}_{\text{NL}}(n-N_{\text{eq}})$ are time-aligned as well and related by $f\{\cdot\}$ (*cf.* Fig. 4.6). Thus, the nonlinear relationship between $x(n-N_{\text{eq}})$ and $\hat{x}_{\text{NL}}(n-N_{\text{eq}})$ can be estimated by a very short adaptive CGM of $L = L_{\text{SA}}$ taps in Block B in Fig. 4.8. Although $L_{\text{SA}} = 1$ would be sufficient in the case of a perfect equalization of $\mathbf{h}$, a very small $L_{\text{SA}}$ of about $L_{\text{SA}} = 3$ is reasonable in practice due to the adaptive, imperfect equalizer. Between Blocks B and A in Fig. 4.8, the identified CGM and Eq. (4.49) are employed to estimate the coefficients of a preprocessor which combines the CGM's branch signals according to Eq. (4.8). Analogously to [32], the preprocessor coefficients $\hat{w}_b$ are estimated on a frame-wise basis as

$$\hat{w}_b(\kappa) = \gamma_{\text{w}}\hat{w}_b(\kappa-1) + (1-\gamma_{\text{w}})\tilde{w}_b(\kappa), \tag{4.50}$$

where $\gamma_{\text{w}}, 0 \le \gamma_{\text{w}} < 1$, is a recursive-smoothing constant ($\gamma_{\text{w}} = 0.95$ will be used by default for the simulations in Section 4.5) and $\tilde{w}_b(\kappa)$ is computed according to Eq. (4.49) from the instantaneous estimates of the branch signal vectors of the CGM of Block B. The preprocessor with weights $\hat{w}_b(\kappa)$ is used in a CM with an adaptive linear subsystem to model the entire unknown system (Block A in Fig. 4.8).

The SSA-CM structure according to Fig. 4.8 is efficient, because it requires only two adaptive linear filters with long temporal support (in Blocks A and C), both of which will be assumed to have length $L$, and a CGM with a very low number of taps $L_{\text{SA}}$. A number of $L_{\text{SA}} = 3$ taps will be assumed by default in the following. The long adaptive filters can be realized efficiently using a PB-FNLMS according to Section 4.3.1.2 in a block processing scheme and the CGM can be implemented due to its short length at very low computational effort by a time-domain NLMS according to Section 4.3.1.1. Hence, the computational effort for an SSA-CM is roughly twice as high as that for a linear filter. The relative complexity compared with a CGM, $R_{\text{CGM}}^{\text{SSA-CM}}$, is about

$$\tilde{R}_{\text{CGM}}^{\text{SSA-CM}} = \frac{2 \cdot P}{B \cdot P} = \frac{2}{B},$$

where the contribution of the length-$L_{\text{SA}}$ CGM has been disregarded. For $B = 5$ branches and $P = 4$ partitions, this leads to $\tilde{R}_{\text{CGM}}^{\text{SSA-CM}} = 0.40$.

### 4.4.3 PARALLEL SIGNIFICANCE-AWARE CASCADE GROUP MODELS

PSA-CGMs generalize the SA-HGMs from [28,31] by allowing branch nonlinearities with memory and are structurally identical otherwise. The structure of such a PSA-CGM is depicted in Fig. 4.9. For an explanation of the general concept, assume that the temporal support of the unknown system's dominant region (*cf.* Fig. 4.7) is known. In this case, all adaptive systems in Fig. 4.9 can be adapted simultaneously.

Block A of Fig. 4.9 contains a CM with adaptive linear subsystem (minimizing the error signal $e_{\text{CM}}(n)$). The nonlinear parametric preprocessor $f_{\text{pp}}\{\cdot\}$ is initially configured as linear function and is

[5] As opposed to the equalization-based SA-HM [32], the equalizer matches its output with $x_{\text{pp}}(n-N_{\text{eq}})$ instead of the input signal $x(n)$ itself. Thereby, the error signal $e_{\text{eq}}(n)$ used for adapting the equalizer can actually approach zero when identifying the preprocessor correctly and consequently leads to less gradient noise in the adaptation process.

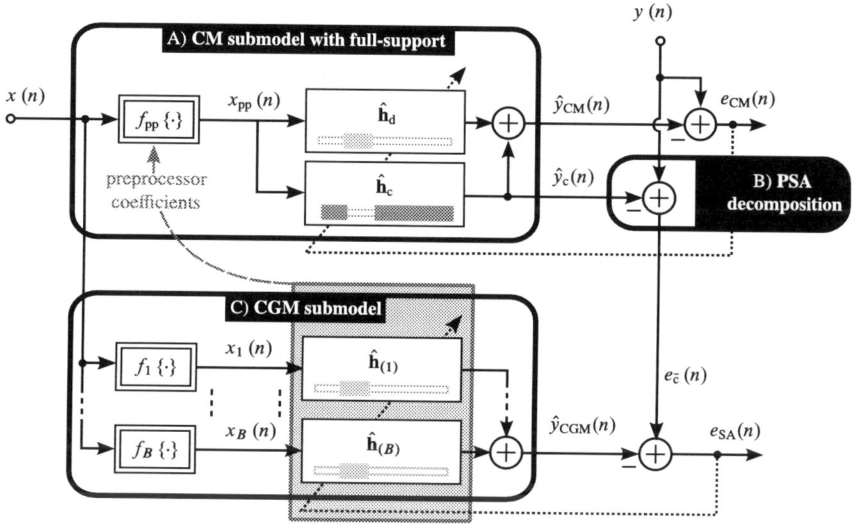


**FIGURE 4.9**

Basic structure of a PSA-CGM: a $B$-branch CGM models only the dominant-part propagation (Block C), whereas the long IR is covered by a simple preprocessor system (Block A).

refined separately from the IR. The CM's IR estimate is partitioned into a dominant region $\hat{\mathbf{h}}_d$ (upper branch in Block A) and a complementary region $\hat{\mathbf{h}}_c$ (lower branch in Block A). This allows one to apply the PSA decomposition (*cf.* Fig. 4.7) to the unknown system in Block B in Fig. 4.9: subtracting the complementary-region estimate $\hat{y}_c(n)$ from the unknown system's output $y(n)$ results in the dominant-region dominated error signal $e_{\bar{c}}(n)$. This enables in Block C of Fig. 4.9 the identification of the unknown system's dominant region alone by a CGM with short temporal support (as short as $\hat{\mathbf{h}}_d$). The CGM's IRs $\hat{\mathbf{h}}_{(b)}$ allow one to extract the preprocessor coefficients for the CM in Block A of Fig. 4.9 analogously to Section 4.4.2. It is worth noting that a PSA-CGM offers two different error signals: $e_{CM}(n)$, resulting from a CM output signal estimate $\hat{y}_{CM}(n)$, and $e_{SA}(n)$, resulting from $\hat{y}_{SA}(n) = \hat{y}_c(n) + \hat{y}_{CGM}(n)$ (not shown in Fig. 4.9). The estimate $\hat{y}_{SA}(n)$ is obtained with a CGM as dominant-path model and represents a more general model than the CM in Block A of Fig. 4.9. If the unknown system has a CM structure, $\hat{y}_{CM}(n)$ may be slightly more accurate than $\hat{y}_{SA}(n)$, because the CGM has more degrees of freedom than necessary and is more affected by gradient noise. If the unknown system is more complex than a CM, the additional degrees of freedom of the CGM may render $\hat{y}_{SA}(n)$ more accurate. For this reason, the PSA-CGM will appear in the evaluation in Section 4.5 twice.

For a highly efficient low-delay realization of the PSA-CGM, a PB-FNLMS implementation has been proposed in [31]. This method identifies partitions $\hat{\mathbf{h}}_\kappa^{(p)}$ of the linear subsystem of the CM in Block A of Fig. 4.9 by a PB-FNLMS (Section 4.3.1.2, $B = 1$ branches and $P > 1$ partitions). Employing this partitioning, the dominant region is modeled by a single partition $\hat{\mathbf{h}}_\kappa^{(p_d)}$ with index $p_d$ and the complementary region is composed of all other partitions, where $p \neq p_d$. As a consequence, the nonlinear dominant-region model of Block C in Fig. 4.9 is a single-partition CGM and adapted according to Section 4.3.1.2 ($B > 1$ branches and $P = 1$ partition).

If the temporal support of the dominant region is unknown, only $\hat{\mathbf{h}}_\kappa^{(p)}$ are adapted in an initial convergence phase. Afterwards, the fixed-length partitions $\hat{\mathbf{h}}_\kappa^{(p)}$ allow one to detect a dominant partition $\hat{\mathbf{h}}_\kappa^{(p_\mathrm{d})}$ as the partition with maximum energy. Compared to a full-length CGM with $P$ partitions in each branch, the relative computational complexity $R_{\mathrm{CGM}}^{\mathrm{PSA\text{-}CGM}}$ of a PSA-CGM is approximately

$$\tilde{R}_{\mathrm{CGM}}^{\mathrm{PSA\text{-}CGM}} = \frac{B + P}{B \cdot P}.$$

Obviously, the PSA-CGM is very efficient for large $P$ and $B$. Without block partitioning ($P = 1$), however, the described PSA-CGM does not provide any advantage. To give an example, $B = 5$ and $P = 4$ results in $\tilde{R}_{\mathrm{CGM}}^{\mathrm{PSA\text{-}CGM}} = 0.45$.

### 4.4.4 PARALLEL SIGNIFICANCE-AWARE FILTERED-X ADAPTATION

In this section, a novel SA filtering concept will be introduced. This concept exploits the PSA decomposition for the direct FX adaptation of bilinear CMs from Section 4.2.3 (see Fig. 4.7) and will be denoted Parallel SA Filtered-X CM (PSAFX-CM). The block diagram of the PSAFX-CM matches the block diagram of the PSA-CGM in Fig. 4.9, except for Block C. Like for the PSA-CGM, a CM with an adaptive linear subsystem (Block A in Fig. 4.9) enables a PSA decomposition (Block B in Fig. 4.9); in an initialization phase, the CM preprocessor is configured as linear function and only the IR is adapted. Afterwards, the CM's IR is split into a dominant and a complementary region. The complementary-region submodel is used (as in Block B in Fig. 4.9) to compute an error signal $e_{\bar{\mathrm{c}}}(n) = y(n) - \hat{y}_\mathrm{c}(n) \approx y_\mathrm{d}(n)$. This error signal mainly contains components $y_\mathrm{d}(n)$, which are caused by the unknown system's dominant region (see also Fig. 4.7).

As an alternative to the PSA-CGM, the novel PSAFX-CM implements the adaptive nonlinear model for the dominant region (Block C) according to Fig. 4.10. Therein, the PSAFX-CM employs a dominant-region CM with $\hat{\mathbf{h}}_\mathrm{d}$ from Block A in Fig. 4.9 and FX-adapted preprocessor weights $\hat{w}_b$. As it is characteristic for SA filtering, the preprocessor determined for the dominant region is then applied in the CM with the full-length adaptive linear subsystem (Block A in Fig. 4.9). Note that the complementary-region CM from Block A in Fig. 4.9 and the dominant-region CM from Fig. 4.10 seamlessly combine to the overall CM of Block A in Fig. 4.9 again. This implies $e_{\mathrm{FX}}(n) = e_{\mathrm{CM}}(n)$. Thus, the only difference from an ordinary FX implementation is that the FX branch signals are generated with the most significant fragment of $\hat{\mathbf{h}}$.

Pursuing a PB-FNLMS implementation as in Section 4.3.2, the adaptation procedures are identical up to Eq. (4.31), and Eq. (4.31) simplifies to the single partition

$$\underline{\mathring{\mathbf{x}}}_{\mathrm{FX},(b),\kappa} = \underline{\mathbf{x}}_{(b),\kappa-p} \odot \underline{\hat{\mathbf{h}}}_{\kappa-1}^{(p_\mathrm{d})} =: \underline{\mathring{\mathbf{x}}}_{\mathrm{SAFX},(b),\kappa} \tag{4.51}$$

and is therefore independent of the length of the actually modeled unknown system. Unlike for the FX algorithm of Section 4.3.2, the filtering effort for generating the FX signals is not proportional to $B \cdot P$ anymore, but only to $B$. As a consequence, the relative complexity $R_{\mathrm{CGM}}^{\mathrm{PSAFX\text{-}CM}}$ of a PSAFX-CM w.r.t.

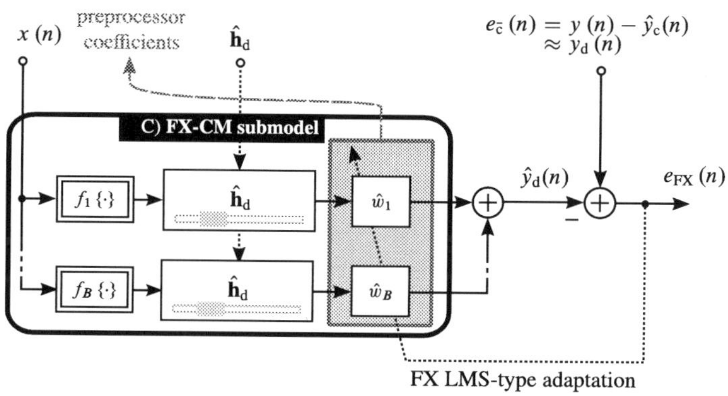


**FIGURE 4.10**

Substitution for Block C in Fig. 4.9 to form the novel PSAFX-CM: The preprocessor coefficients $\hat{w}_b$ of a dominant-region CM are adapted using an FX algorithm with $\hat{\mathbf{h}}_\mathrm{d}$ from Block A in Fig. 4.9 as linear subsystem of the CM and with $e_{\bar{c}}(n) \approx y_\mathrm{d}(n)$ from the PSA decomposition in Block B in Fig. 4.9 as target signal.

a full-length CGM can be estimated by the number of DFTs and yields

$$\tilde{R}_{\mathrm{CGM}}^{\mathrm{PSAFX\text{-}CM}} = \frac{B+2+2P}{B+2+2BP} < \tilde{R}_{\mathrm{CGM}}^{\mathrm{FX\text{-}CM}}. \tag{4.52}$$

The overhead $\varepsilon$ from Eq. (4.47), caused by non-DFT operations with a complexity proportional to $B \cdot P$, does not appear in Eq. (4.52) as the actual filtering effort is reduced to a complexity proportional to $B$ only. To give an example, $P = 4$ and $B = 5$ leads to $\tilde{R}_{\mathrm{CGM}}^{\mathrm{PSAFX\text{-}CM}} \approx 0.32$ and thereby indicates a computationally very inexpensive nonlinear model, which is even less complex than the other SA filters.

As for the other SA models in Sections 4.4.2 and 4.4.3, the nonlinear system identification has been split into synergetic subsystems, where the first models the possibly long memory of the linear subsystem (Block A in Fig. 4.9) and the second estimates the nonlinearities without modeling the possibly long memory of the system (Fig. 4.10).

## 4.5 EXPERIMENTS AND EVALUATION

In this section, the adaptive LIP nonlinear filters from Sections 4.3 and 4.4 will be evaluated exemplarily for the challenging application of AEC (*cf.* Section 4.1). To this end, the evaluation method will be introduced in Section 4.5.1 and the experiments will be presented and discussed in Section 4.5.2.

Note that, in principle, the presented methods can also be employed for many applications aside from AEC. In particular, the identification of the path between the discrete-time loudspeaker signal and a recording of the radiated sound can also be the first step towards a loudspeaker linearization [7–9] or nonlinear active noise control [68,53]. Furthermore, the presented algorithms can enable the identification of nonlinear analog signal processors (*e.g.*, guitar amplifier and effect processor devices [69, 70]) to emulate the identified processors in digital hardware afterwards. Moreover, the joints of robots can be modeled as HMs as well [71] and may be identified with the methods of Section 4.3. Lifting the latter example to a broad industrial scope, the adaptive algorithms presented in this chapter may be employed for digital twin modeling (also termed cybertwin) of nonlinear mechanical, electrical, acoustical or chemical systems in digitalized production plants for Industry 4.0 [72]—as long as the underlying nonlinear systems can be described well by CMs. However, the longer the linear subsystem of an HM or CM is, the more computational benefit can be expected by SA modeling. In case of short linear subsystems, block partitioning and frequency-domain processing may be unnecessary and time-domain adaptation of all involved adaptive linear filters should be pursued (see [28] for a time-domain SA filter).

### 4.5.1 EVALUATION METRICS

For system identification, normalized misalignment [24] and projection misalignment measures [73] evaluate how well particular components of a parametrized system are identified. However, these measures require the exact knowledge of the true system parameters—knowledge which is typically not available for actual physical systems to be identified. In this case, signal-based measures have to be adopted. In particular, the modeling accuracy can be measured by the ratio of the variance of the unknown system's output $y(n)$ and of the residual error $e(n) = y(n) - \hat{y}(n)$.

For the special system identification case of AEC, where the primary objective is the removal of the loudspeaker signal components (echoes) from the microphone signal, this ratio of variances is widely known as Echo-Return Loss Enhancement (ERLE) and is usually considered in the logarithmic domain according to

$$\mathrm{ERLE} = 10\log_{10}\left(\frac{\mathcal{E}\left\{|y(n)|^2\right\}}{\mathcal{E}\left\{|e(n)|^2\right\}}\right)\,\mathrm{dB}, \tag{4.53}$$

assuming noise-free "single talk" situations (no speaker or noise on the near end, only an echo signal). A higher ERLE measure corresponds to a better performance of the AEC system [74]. When estimating the ERLE in short time frames employing the instantaneous energies of $y(n)$ and $e(n)$ and continuously adapting the model, even severe overadaptation to the particular signal may remain unnoticed and be misinterpreted as accurate system identification because of the high ERLE values. To minimize this undesired effect, filtering a long interval of speech with frozen filter coefficients after convergence is possible. The resulting ERLE can be used as system identification performance measure. An ERLE computed in this way also evaluates the primary use case of an AEC system, as it captures the amount of echo reduction in the presence of near-end speech (double talk), when the adaptive filters cannot be adapted and the recording cannot be muted (no half-duplex but full-duplex communication) [74]. The ERLE measure for frozen filter coefficients will be employed in the following to evaluate and compare the adaptive filtering algorithms described in Sections 4.3 and 4.4.

### 4.5.2 EXPERIMENTS

In this section, the ERLE performance of the adaptive algorithms from Sections 4.3 and 4.4 will be compared in AEC experiments. To this end, the nonlinear echo paths of different devices will be considered. Device 1 is a smartphone in hands-free mode, Device 2 is a low-quality electrodynamic loudspeaker and Device 3 is a cellphone manufactured in the year 2000. About 80 s of speech (both

**Table 4.4 Computational complexity of linear and nonlinear models relative to a CGM with $B=5$ branches and $P=4$ partitions. The approximate values are a summary of the numbers in the individual algorithms' sections, whereas the number of FLOPS are computed analogously to [32] for a frameshift of $M=512$**

| | linear | PSAFX-CM | FX-CM | SSA-CM | PSA-CGM | CGM |
|---|---|---|---|---|---|---|
| approximate | 0.23 | 0.32 | 0.32+$\varepsilon$ | 0.40 | 0.45 | 1 |
| FLOPs | 0.23 | 0.32 | 0.37 | 0.50 | 0.58 | 1 |

male and female speech) have been played back at relatively high volumes of about 80 dBA at 1 m distance and recorded with the playback device itself for Device 1 and with a high-quality measurement microphone at 1 m distance for Devices 2 and 3. All AEC experiments were conducted at a sampling rate of $f_S = 16$ kHz and in three acoustic environments, denoted Scenarios 1, 2 and 3. The actual recordings were performed under low-reverberation conditions, which corresponds to Scenario 1. The data for Scenarios 2 and 3 are synthesized from the low-reverberation recordings by convolving the recordings with measured acoustic IRs from real lab environments with reverberation times of $T_{60} \approx 250$ ms and $T_{60} \approx 400$ ms, respectively. The adaptive filters employed for the identification of the acoustic echo path will be realized, as specified earlier in this chapter, as partitioned-block adaptive filters in the frequency domain. For all scenarios and devices, a linear filter (Section 4.3.1.2 with $B=1$ branches and $f_1\{\cdot\} = (\cdot)$), a CM with FX adaptation tailored to memoryless preprocessors (Section 4.3.2), a CM with PSAFX adaptation (Section 4.4.4), a PSA-CGM (Section 4.4.3), an SSA-CM (Section 4.4.2) and a CGM with full temporal support (Section 4.3.1.2) will be compared. All adaptive filters are identically parametrized by step sizes of $\mu_b = \mu_{FX} = 0.1$, block processing is done at a frame shift of $M = 512$ samples and a frame size of $N = 2M$ and the block-partitioned filters consist of $P = 4$ partitions of length $M$. Furthermore, the equalizer delay for the SSA-CM is chosen as $N_{eq} = 2M$ samples. All nonlinear models have $B = 5$ branches and preprocessor coefficients, respectively, and employ odd-order Legendre polynomials of orders 1 to 9 as branch nonlinearities $f_b\{\cdot\}$. Thereby, all CGMs are HGMs and all CMs are HMs. The computational complexity of the individual models relative to the complexity of a CGM is listed in Table 4.4. Therein, the first line ("approximate") corresponds to the relative complexity estimates introduced with the algorithms in Sections 4.3 and 4.4. The second line ("FLOPS") contains the relative numbers of FLoating Point Operations (FLOPS) of the particular method in comparison to a CGM. To this end, the FLOPS have been determined analogously to [32]. Operations counted as FLOPS are real-valued additions, multiplications and divisions. The complex-valued DFT-domain operations are considered by mapping complex-valued additions, multiplications and divisions to multiple real-valued operations. In particular, a complex-valued addition is counted as 2 real-valued FLOPS, a complex-valued multiplication is counted as 6 real-valued FLOPS and the division of a complex number by a real-valued number is counted as 2 real-valued FLOPS. Accumulating the FLOPS of all equations necessary for processing a signal frame of $M$ samples by a particular filter structure corresponds to the algorithm's number of FLOPS. Normalizing these FLOPS to the FLOPS of a CGM leads to the relative FLOPS in line two of Table 4.4.

Obviously, the qualitative rating of the different algorithms in Table 4.4 is similar by both complexity estimation methods: the further right an algorithm is in Table 4.4, the more computationally demanding it is.

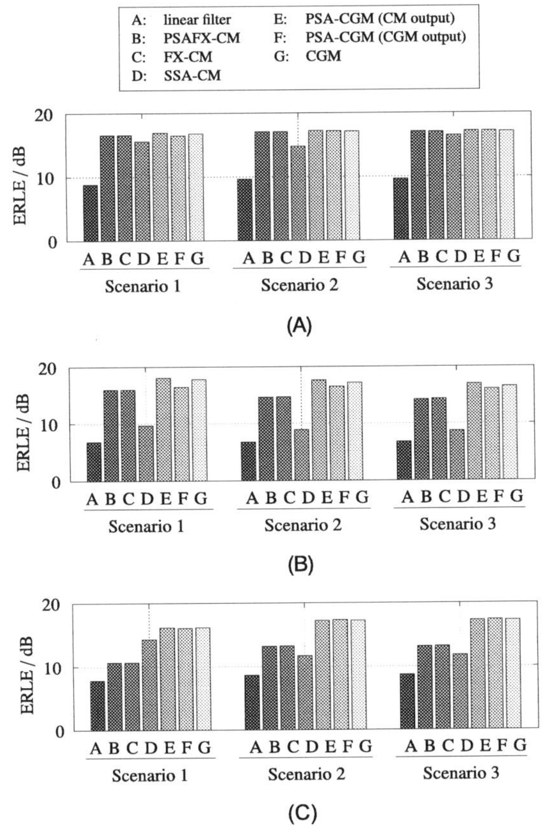


**FIGURE 4.11**

ERLE performance: the efficient SA-filtering approaches allow for nonlinear AEC and may even reach the performance of the filters without the SA extension (SAFX-CM/FX-CM and SA-CGM/CGM). (A) Smartphone. (B) Electrodynamic Loudspeaker. (C) Cellphone.

The ERLE achievable with the different adaptive filter structures with frozen parameters (like during double talk) is listed in Figs. 4.11A to 4.11C for Devices 1 to 3, respectively. Obviously, linear AEC (A in Fig. 4.11) suffers a lot from the severe nonlinearities, leading to ERLE below 10 dB under all reverberation conditions and to particularly bad performance under the low-reverberation conditions of Scenario 1. All nonlinear models outperform the linear model for all devices and scenarios. The CGM (G in Fig. 4.11), being the most complex model, shows consistently high performance. However, in particular the PSA-CGM (E and F in Fig. 4.11) approaches or even outperforms the more complex CGM. Especially the PSA-CGM's CM output (E in Fig. 4.11) performs very well. This confirms that the actual acoustic echo paths can be approximated well by CMs. However, the FX adaptation of a CM (C in Fig. 4.11) closely approaches the performance of the CGM only for Device 1. Still, introducing further efficiency into the FX adaptation by a PSA decomposition (B in Fig. 4.11) does not lead to any noticeable disadvantage. The SSA-CM performance (D in Fig. 4.11) resembles the performance of other nonlinear approaches for Device 1, but performs worse for the other devices (while still outperforming the linear filter).

To sum up, all SA filters successfully alleviate the computational burden for nonlinear AEC. The performance of a general CGM can be reached in the experiments by an efficient PSA-CGM. An even less complex (but also less effective) direct CM adaptation by an FX algorithm can benefit in terms of computational complexity from PSA filtering without losing ERLE performance.

## 4.6 OUTLOOK ON MODEL STRUCTURE ESTIMATION

All methods described in this chapter identify the parameters of nonlinear models adaptively. Yet, the algorithms do not choose an optimum model structure, such as a particular base function set, the order of a polynomial nonlinearity, or the memory length of linear subsystems. However, knowledge about the model structure is highly desirable for two reasons. First, minimizing the number of parameters leads to computational efficiency and second, unnecessarily estimated and redundant parameters introduce gradient noise (estimation errors) and consequently reduce the accuracy of the model.

Both disadvantages can be tackled by employing convex filter combinations [75], as described in Chapter 11. To briefly illustrate this concept for the scope of this chapter, consider two competing models A and B with the respective outputs $\hat{y}^{(A)}(n)$ and $\hat{y}^{(B)}(n)$, modeling the same system with output $y(n)$. Although modeling the same system, the models A and B are assumed to have a different structure (*e.g.*, memory length of linear subsystems) and consequently different numbers and values of parameters (*e.g.*, filter coefficients of linear subsystems). Convex filter combinations superimpose the outputs $y^{(A)}(n)$ and $y^{(B)}(n)$ according to

$$\hat{y}(n) = (1-\eta)\cdot\hat{y}^{(A)}(n) + \eta\cdot\hat{y}^{(B)}(n),$$

where the mixing weight $\eta$, $0 \leq \eta \leq 1$, is determined adaptively. In practice, this can perform like an adaptive switch indicating the more suitable of the two competing models and thereby the more suitable model structure. Convex combination-based methods for structure estimation can be applied to all kinds of CGMs (*e.g.*, Volterra, Legendre, Chebyshev and Fourier nonlinear filters). As the structure estimation methods are described for the example of Volterra filters in the original publications, the methods will be explained here first for Volterra filters. To this end, consider Volterra filters, which consist of kernels of orders $o$, $o \in \{1, \ldots, O\}$, with maximum order $O$. Each kernel consists of diagonals with maximum relative time lags $R_o$ in the branch nonlinearities of the $o$th kernel ($R_o$ is also known as diagonal radius) and has linear subsystems of length $L_o$ in kernel $o$. As $O$, $R_o$ and $L_o$ determine the model structure, they will be referred to as the structure parameters of the Volterra model. These structure parameters determine which of the filter coefficients of a Volterra filter are modeled in practice.

As proposed in [43], an identified model can automatically be trimmed by an additional adaptive structure, where a Volterra kernel is clustered into disjoint groups of diagonals and each group's output signal $y_g(n)$, where $g$ is the group index, is replaced by a convex combination of the group's output $y_g(n)$ with 0 (the result of filtering with an all-zero kernel region). Thereby, irrelevant diagonals, which are dominated by noise and would increase the modeling error, can be trimmed for the final computation of the model output. Thereby, the modeling accuracy is increased at the expense of a slight increase in computational complexity. In this case, the mixing variables of the convex combinations take the role of additional structure parameters. In a similar way, the FLAFs in [48] disable nonlinear model parts as well. Another related approach is the so-called EVOLVE concept [76], which approaches the estimation of a Volterra filter structure in a bottom-up way by starting with a small filter and successively growing the model until the optimal size is reached. Filters which are evolutionary w.r.t. a structure parameter dimension $S \in \{O, R_o, L_o\}$ perform a convex combination of a filter with a small model size (a low structure parameter value $S_1$) and of a filter with a larger model size (a larger structure parameter value $S_2 > S_1$). In the case of a clear preference of one of both models, both models are grown or shrunk into the preferred direction. Note that, for reasons of efficiency, the larger model can also be "virtualized" by simply augmenting the small model with additional filter coefficients [76,77]. As described in [76], an evolution in $R_o$ and $L_o$ and bypassing kernel orders is possible by cascading multiple convex combinations, each of which decides for the activation of another region of filter taps of the Volterra filter. This concept can, of course, be applied to CGMs with other branch nonlinearities as well: Legendre-based CGMs (*cf.*, [45,46] or Section 4.5.2), where $i$th powers in the Volterra filters' branch nonlinearities are replaced by $i$th-order Legendre polynomials, have the same set of structure parameters $\{O, R_o, L_o\}$ and can be grown the same way as Volterra filters in an EVOLVE fashion. Also, the model trimming from [43] can directly be applied to such Legendre-based CGMs. As opposed to [43], EVOLVE [76] starts with a low-complexity model and successively grows the model if required. Thereby, not only gradient noise in irrelevant filter coefficients is prevented, but also the computational complexity is only as high as the complexity of the optimum filter structure found by the algorithm, plus some overhead for the virtualized models of increased size.

Like in SA filtering, efficient EVOLVE implementations can employ a block partitioning suitable for PB-FNLMS adaptation. This allows one to grow and shrink the memory lengths of the diagonals in steps of $M$ taps by adding or removing entire length-$M$ partitions. However, the efficiency and accuracy are achieved differently from SA filtering. SA filtering estimates the nonlinear characteristics of the unknown system after decomposing the system into a nonlinear system with short nonzero temporal support and the estimated nonlinear characteristics are then extrapolated efficiently using CMs with parametric preprocessors. In EVOLVE, such a model decomposition and extrapolation does not take place, so SA filtering locally (at particular filter coefficients) observes the nonlinearity and extrapolates it to a global system representation, assuming a fixed model structure (CM structure). As opposed to this, EVOLVE locally grows the model (at particular diagonals in a particular Volterra kernel), aiming for the best-performing global model structure and parametrization as final result. Yet, the SA and the EVOLVE approach do not contradict each other but can jointly be implemented with the same block partitioning scheme. The resulting evolutionary SA filters may, in the future, dynamically grow their structure (*e.g.*, number of partitions $P$ and the number of branches $B$) while extrapolating nonlinearities with a preprocessor system, whenever the computational complexity restricts additional partitions or a direct estimation of the particular CGM partitions is disadvantageous due to adaptation noise.

## 4.7 SUMMARY

In this chapter, recent advances in LIP nonlinear filters have been presented with the focus on the concept of Significance-Aware (SA) filtering and nonlinear AEC as a challenging application in the background. After a unified description of nonlinear models as variants of CGMs and LIP CMs, where the output is a bilinear function of the model parameters, the parameter estimation methods for both structures have been revisited, showing that some of the recent developments can be described as FX algorithms. This lead to the proposal of a novel, highly efficient FX algorithm for the adaptation of CMs with parametric nonlinear preprocessors. Afterwards, the SA-filtering concept has been described in two variants: the SSA-CM and the PSA-CGM, which gain efficiency by equalizing parts of or partly compensating the unknown system before estimating the system's nonlinearity. Additionally, a novel PSA FX algorithm has been introduced as well. The algorithms have been compared in AEC experiments with recordings from three different physical devices, which have shown the high efficacy of SA filtering and its potential to save computational complexity without sacrificing modeling accuracy. Finally, model structure estimation based on convex filter combinations has been summarized and potential synergies with SA filters in a joint partitioned-block signal processing framework have been pointed out.

## ACKNOWLEDGMENTS

This book chapter could not have been written without the dedicated work of many former students and collaborators in the authors' research group, especially Alexander Stenger, Fabian Küch, Marcus Zeller and Luis A. Azpicueta-Ruiz.

## REFERENCES

[1] M.J. Korenberg, I.W. Hunter, The identification of nonlinear biological systems: Volterra kernel approaches, Annals of Biomedical Engineering 24 (2) (March 1996) 250–268.

[2] Y. Chen, I.W. Hunter, Nonlinear stochastic system identification of skin using Volterra kernels, Annals of Biomedical Engineering 41 (4) (April 2013) 847–862.

[3] S.A. Billings, Identification of nonlinear systems—a survey, IEEE Proc. Control Theory and Applications 127 (6) (November 1980) 272–285.

[4] S. Benedetto, E. Biglieri, R. Daffara, Modeling and performance evaluation of nonlinear satellite links—a Volterra series approach, IEEE Transactions on Aerospace and Electronic Systems AES-15 (4) (July 1979) 494–507.

[5] K.V. Peddanarappagari, M. Brandt-Pearce, Volterra series transfer function of single-mode fibers, Journal of Lightwave Technology 15 (12) (December 1997) 2232–2241.

[6] G.M. Raz, B.D. van Veen, Baseband Volterra filters for implementing carrier based nonlinearities, IEEE Transactions on Signal Processing 46 (1) (January 1998) 103–114.

[7] W.A. Frank, An efficient approximation to the quadratic Volterra filter and its application in real-time loudspeaker linearization, Signal Processing 45 (1) (July 1995) 97–113.

[8] S. Kinoshita, Y. Kajikawa, Y. Nomura, Volterra filters using multirate signal processing and their application to loudspeaker systems, in: IEEE International Conference on Acoustics, Speech, and Signal Processing, ICASSP, vol. 6, Salt Lake City, USA, May 2001, pp. 3497–3500.

[9] W. Frank, R. Reger, U. Appel, Loudspeaker nonlinearities—analysis and compensation, in: Asilomar Conference on Signals, Systems, and Computers, vol. 2, Pacific Grove, California, USA, October 1992, pp. 756–760.

[10] A. Stenger, L. Trautmann, R. Rabenstein, Nonlinear acoustic echo cancellation with 2nd order adaptive Volterra filters, in: IEEE International Conference on Acoustics, Speech, and Signal Processing, ICASSP, vol. 2, Phoenix, Arizona, USA, March 1999, pp. 877–880.

[11] A. Fermo, A. Carini, G. Sicuranza, Analysis of different low complexity nonlinear filters for acoustic echo cancellation, in: Proceedings of the First International Workshop on Image and Signal Processing and Analysis, IWISPA, Pula, Croatia, June 2000, pp. 261–266.

[12] S. Kinoshita, Y. Kajikawa, Y. Nomura, Adaptive Volterra filters using multirate signal processing and their application to identification of loudspeaker systems, Electronics and Communications in Japan (Part III: Fundamental Electronic Science) 87 (7) (July 2004) 45–54.

[13] F. Küch, W. Kellermann, Partitioned block frequency-domain adaptive second-order Volterra filter, IEEE Transactions on Signal Processing 53 (2) (February 2005) 564–575.

[14] L.A. Azpicueta-Ruiz, M. Zeller, J. Arenas-García, W. Kellermann, Novel schemes for nonlinear acoustic echo cancellation based on filter combinations, in: IEEE International Conference on Acoustics, Speech, and Signal Processing, ICASSP, Taipei, Taiwan, April 2009, pp. 193–196.

[15] L.A. Azpicueta-Ruiz, M. Zeller, A. Figueiras-Vidal, J. Arenas-García, W. Kellermann, Adaptive combination of Volterra kernels and its application to nonlinear acoustic echo cancellation, IEEE Transactions on Audio, Speech, and Language Processing 19 (1) (January 2011) 97–110.

[16] M. Zeller, L.A. Azpicueta-Ruiz, J. Arenas-García, W. Kellermann, Adaptive Volterra filters with evolutionary quadratic kernels using a combination scheme for memory control, IEEE Transactions on Signal Processing 59 (4) (April 2011) 1449–1464.

[17] H. Buchner, J. Benesty, T. Gänsler, W. Kellermann, Robust extended multidelay filter and double-talk detector for acoustic echo cancellation, IEEE Transactions on Audio, Speech, and Language Processing 14 (5) (September 2006) 1633–1644.

[18] K. Shi, X. Ma, G. Zhou, A mutual information based double-talk detector for nonlinear systems, in: Annual Conference on Information Sciences and Systems, CISS, March 2008, pp. 356–360.

[19] V.J. Mathews, G.L. Sicuranza, Polynomial Signal Processing, Wiley, New York (NY), USA, May 2000.

[20] K. Shi, G. Zhou, M. Viberg, Hammerstein system linearization with application to nonlinear acoustic echo cancellation, in: Digital Signal Processing Workshop, 12th—Signal Processing Education Workshop, 4th, September 2006, pp. 183–186.

[21] K. Shi, G. Zhou, M. Viberg, Compensation for nonlinearity in a Hammerstein system using the coherence function with application to nonlinear acoustic echo cancellation, IEEE Transactions on Signal Processing 55 (12) (December 2007) 5853–5858.

[22] A. Birkett, R. Goubran, Limitations of handsfree acoustic echo cancellers due to nonlinear loudspeaker distortion and enclosure vibration effects, in: IEEE ASSP Workshop on Applications of Signal Processing to Audio and Acoustics, WASPAA, New Paltz, NY, USA, October 1995, pp. 103–106.

[23] A. Stenger, R. Rabenstein, An acoustic echo canceller with compensation of nonlinearities, in: European Signal Processing Conference, EUSIPCO, Island of Rhodes, Greece, September 1998, pp. 969–972.

[24] A. Stenger, W. Kellermann, Adaptation of a memoryless preprocessor for nonlinear acoustic echo cancelling, Signal Processing 80 (9) (September 2000) 1747–1760.

[25] A. Stenger, Kompensation akustischer Echos unter Einfluß von nichtlinearen Audiokomponenten, Berichte aus der Kommunikations- und Informationstechnik, vol. 23, Shaker Verlag, Aachen, April 2001, in German.

[26] M. Scarpiniti, D. Comminiello, R. Parisi, A. Uncini, Comparison of Hammerstein and Wiener systems for nonlinear acoustic echo cancelers in reverberant environments, in: 17th International Conference on Digital Signal Processing, DSP, July 2011, pp. 1–6.

[27] S. Shimauchi, Y. Haneda, Nonlinear acoustic echo cancellation based on piecewise linear approximation with amplitude threshold decomposition, in: International Workshop on Acoustic Signal Enhancement, IWAENC, Aachen, Germany, September 2012.

[28] C. Hofmann, C. Huemmer, W. Kellermann, Significance-aware Hammerstein group models for nonlinear acoustic echo cancellation, in: IEEE International Conference on Acoustics, Speech, and Signal Processing, ICASSP, Florence, Italy, May 2014, pp. 5934–5938.

[29] S. Van Vaerenbergh, L.A. Azpicueta-Ruiz, Kernel-based identification of Hammerstein systems for nonlinear acoustic echo-cancellation, in: IEEE International Conference on Acoustics, Speech, and Signal Processing, ICASSP, Florence, Italy, May 2014, pp. 3739–3743.

[30] C. Huemmer, C. Hofmann, R. Maas, W. Kellermann, The significance-aware EPFES to estimate a memoryless preprocessor for nonlinear acoustic echo cancellation, in: IEEE Global Conference on Signal Information Processing, GlobalSIP, Atlanta, GA, December 2014, pp. 557–561.

[31] C. Hofmann, M. Guenther, C. Huemmer, W. Kellermann, Efficient nonlinear acoustic echo cancellation by partitioned-block significance-aware Hammerstein group models, in: European Signal Processing Conference, EUSIPCO, Budapest, Hungary, August 2016, pp. 1783–1787.

[32] C. Hofmann, C. Huemmer, M. Guenther, W. Kellermann, Significance-aware filtering for nonlinear acoustic echo cancellation, EURASIP Journal on Advances in Signal Processing 2016 (1) (2016) 113.

[33] F. Küch, W. Kellermann, Orthogonalized power filters for nonlinear acoustic echo cancellation, Signal Processing 86 (6) (June 2006) 1168–1181.
[34] D. Comminiello, M. Scarpiniti, R. Parisi, A. Uncini, A functional link based nonlinear echo canceller exploiting sparsity, in: International Workshop on Acoustic Echo and Noise Control, IWAENC, Tel Aviv, Israel, August 2010.
[35] D. Comminiello, L.A. Azpicueta-Ruiz, M. Scarpiniti, A. Uncini, J. Arenas-García, Functional link based architectures for nonlinear acoustic echo cancellation, in: Joint Workshop on Hands-Free Speech Communication and Microphone Arrays, HSCMA, Edinburgh, UK, May 2011, pp. 180–184.
[36] M. Mossi, C. Yemdji, N. Evans, C. Beaugeant, Non-linear acoustic echo cancellation using online loudspeaker linearization, in: IEEE Workshop on Applications of Signal Processing to Audio and Acoustics, WASPAA, New Paltz, NY, USA, October 2011, pp. 97–100.
[37] M.I. Mossi, C. Yemdji, N. Evans, C. Beaugeant, P. Degry, Robust and low-cost cascaded non-linear acoustic echo cancellation, in: IEEE International Conference on Acoustics, Speech, and Signal Processing, ICASSP, Prague, Czech Republic, IEEE, May 2011, pp. 89–92.
[38] F. Chang, R. Luus, A noniterative method for identification using Hammerstein model, IEEE Transactions on Automatic Control 16 (5) (October 1971) 464–468.
[39] P. Gallman, An iterative method for the identification of nonlinear systems using a Uryson model, IEEE Transactions on Automatic Control 20 (6) (December 1975) 771–775.
[40] F. Küch, A. Mitnacht, W. Kellermann, Nonlinear acoustic echo cancellation using adaptive orthogonalized power filters, in: IEEE International Conference on Acoustics, Speech, and Signal Processing, ICASSP, vol. 3, Philadelphia, PA, USA, March 2005, pp. 105–108.
[41] M.I. Mossi, C. Yemdji, N. Evans, C. Beaugeant, A Cascaded Non-linear Acoustic Echo Canceller Combining Power Filtering and Clipping Compensation, Technical Report RR-11-258, EURECOM Research Report, August 2011.
[42] E. Thomas, Some considerations on the application of the Volterra representation of nonlinear networks to adaptive echo cancellers, The Bell System Technical Journal 50 (8) (October 1971) 2797–2805.
[43] L.A. Azpicueta-Ruiz, M. Zeller, A. Figueiras-Vidal, W. Kellermann, J. Arenas-García, Enhanced adaptive Volterra filtering by automatic attenuation of memory regions and its application to acoustic echo cancellation, IEEE Transactions on Signal Processing 61 (11) (June 2013) 2745–2750.
[44] A. Carini, G. Sicuranza, Even mirror Fourier nonlinear filters, in: IEEE International Conference on Acoustics, Speech, and Signal Processing, ICASSP, Vancouver, BC, Canada, May 2013, pp. 5608–5612.
[45] A. Carini, S. Cecchi, M. Gasparini, G. Sicuranza, Introducing Legendre nonlinear filters, in: IEEE International Conference on Acoustics, Speech, and Signal Processing, ICASSP, Florence, Italy, May 2014, pp. 7939–7943.
[46] A. Carini, S. Cecchi, L. Romoli, G.L. Sicuranza, Perfect periodic sequences for Legendre nonlinear filters, in: European Signal Processing Conference, EUSIPCO, September 2014, pp. 2400–2404.
[47] A. Carini, G.L. Sicuranza, A study about Chebyshev nonlinear filters, Signal Processing 122 (May 2016) 24–32.
[48] D. Comminiello, M. Scarpiniti, L.A. Azpicueta-Ruiz, J. Arenas-García, A. Uncini, Functional link adaptive filters for nonlinear acoustic echo cancellation, IEEE Transactions on Audio, Speech, and Language Processing 21 (7) (July 2013) 1502–1512.
[49] D. Comminiello, M. Scarpiniti, L.A. Azpicueta-Ruiz, J. Arenas-García, A. Uncini, Nonlinear acoustic echo cancellation based on sparse functional link representations, IEEE/ACM Transactions on Audio, Speech, and Language Processing 22 (7) (July 2014) 1172–1183.
[50] D. Comminiello, M. Scarpiniti, L.A. Azpicueta-Ruiz, J. Arenas-García, A. Uncini, A nonlinear architecture involving a combination of proportionate functional link adaptive filters, in: European Signal Processing Conference, EUSIPCO, August 2015, pp. 2869–2873.
[51] K. Narendra, P. Gallman, An iterative method for the identification of nonlinear systems using a Hammerstein model, IEEE Transactions on Automatic Control 11 (3) (July 1966) 546–550.
[52] E. Eskinat, S.H. Johnson, W.L. Luyben, Use of Hammerstein models in identification of nonlinear systems, American Institute of Chemical Engineers (AIChE) Journal 37 (2) (February 1991) 255–268.
[53] G. Sicuranza, A. Carini, On the accuracy of generalized Hammerstein models for nonlinear active noise control, in: IEEE Instrumentation and Measurement Technology Conference, IMTC, April 2006, pp. 1411–1416.
[54] E.-W. Bai, K. Li, Convergence of the iterative algorithm for a general Hammerstein system identification, Automatica 46 (11) (November 2010) 1891–1896.
[55] S. Malik, G. Enzner, Fourier expansion of Hammerstein models for nonlinear acoustic system identification, in: IEEE International Conference on Acoustics, Speech, and Signal Processing, ICASSP, Prague, Czech Republic, May 2011, pp. 85–88.

[56] C. Huemmer, C. Hofmann, R. Maas, A. Schwarz, W. Kellermann, The elitist particle filter based on evolutionary strategies as novel approach for nonlinear acoustic echo cancellation, in: IEEE International Conference on Acoustics, Speech, and Signal Processing, ICASSP, Florence, Italy, May 2014, pp. 1315–1319.
[57] S. Haykin, Adaptive Filter Theory, 4th edition, Prentice Hall, Upper Saddle River (NJ), USA, 2002.
[58] J.-S. Soo, K. Pang, Multidelay block frequency domain adaptive filter, IEEE Transactions on Acoustics, Speech and Signal Processing 38 (2) (February 1990) 373–376.
[59] H. Buchner, J. Benesty, T. Gänsler, W. Kellermann, An outlier-robust extended multidelay filter with application to acoustic echo cancellation, in: International Workshop on Acoustic Echo and Noise Control, IWAENC, Kyoto, Japan, September 2003, pp. 19–22.
[60] M. Zeller, W. Kellermann, Self-configuring system identification via evolutionary frequency-domain adaptive filters, in: International Workshop on Acoustic Echo and Noise Control, IWAENC, Tel Aviv, Israel, August 2010.
[61] F. Küch, E. Mabande, G. Enzner, State-space architecture of the partitioned-block-based acoustic echo controller, in: IEEE International Conference on Acoustics, Speech, and Signal Processing, ICASSP, Florence, Italy, May 2014, pp. 1295–1299.
[62] G. Enzner, P. Vary, A soft-partitioned frequency-domain adaptive filter for acoustic echo cancellation, in: IEEE International Conference on Acoustics, Speech, and Signal Processing, ICASSP, vol. 5, Hong Kong, April 2003, pp. V-393–V-396 (cancelled due to the SARS epidemic).
[63] P.C.W. Sommen, Adaptive Filtering Methods, PhD thesis, Technische Universiteit Eindhoven, 1992.
[64] S.M. Kuo, D.R. Morgan, Active Noise Control Systems: Algorithms and DSP Implementations, Wiley, February 1996.
[65] B. Widrow, D. Shur, S. Shaffer, On adaptive inverse control, in: Asilomar Conference on Circuits, Systems, and Computers, Pacific Grove, California, USA, September 1981, pp. 185–189.
[66] Y.A. Huang, J. Skoglund, A. Luebs, Practically efficient nonlinear acoustic echo cancellers using cascaded block RLS and FLMS adaptive filters, in: IEEE International Conference on Acoustics, Speech, and Signal Processing, ICASSP, New Orleans, LA, USA, March 2017, pp. 596–600.
[67] E.L.O. Batista, O.J. Tobias, R. Seara, A sparse-interpolated scheme for implementing adaptive Volterra filters, IEEE Transactions on Signal Processing 58 (4) (April 2010) 2022–2035.
[68] G.L. Sicuranza, A. Carini, Nonlinear active noise control, in: European Signal Processing Conference, EUSIPCO, Vienna, Austria, September 2004, pp. 1887–1890.
[69] L. Tronchin, The emulation of nonlinear time-invariant audio systems with memory by means of Volterra series, Journal of the Audio Engineering Society 60 (12) (December 2012) 984–996.
[70] L. Tronchin, V.L. Coli, Further investigations in the emulation of nonlinear systems with Volterra series, Journal of the Audio Engineering Society 63 (9) (September 2015) 671–683.
[71] A. Krzyżak, J.Z. Sasiadek, S. Ulrich, Nonparametric identification of robot flexible joint space manipulator, in: 17th International Conference on Methods Models in Automation Robotics, MMAR, Międzyzdroje, Poland, August 2012, pp. 172–177.
[72] J. Lee, B. Bagheri, H.-A. Kao, A cyber-physical systems architecture for Industry 4.0-based manufacturing systems, Manufacturing Letters 3 (Supplement C) (January 2015) 18–23.
[73] G. Enzner, T.C. Kranemann, P. Thüne, Evaluation of estimated Hammerstein models via normalized projection misalignment of linear and nonlinear subsystems, in: IEEE International Conference on Acoustics, Speech, and Signal Processing, ICASSP, March 2016, pp. 4234–4238.
[74] C. Breining, P. Dreiseitel, E. Hänsler, A. Mader, B. Nitsch, H. Puder, T. Schertler, G. Schmidt, J. Tilp, Acoustic echo control. An application of very-high-order adaptive filters, IEEE Signal Processing Magazine 16 (4) (July 1999) 42–69.
[75] J. Arenas-García, A. Figueiras-Vidal, A. Sayed, Mean-square performance of a convex combination of two adaptive filters, IEEE Transactions on Signal Processing 54 (3) (March 2006) 1078–1090.
[76] M. Zeller, Generalized Nonlinear System Identification Using Adaptive Volterra Filters with Evolutionary Kernels, Dr. Hut Verlag, München, 2013.
[77] M. Zeller, L.A. Azpicueta-Ruiz, W. Kellermann, Adaptive FIR filters with automatic length optimization by monitoring a normalized combination scheme, in: IEEE Workshop on Applications of Signal Processing to Audio and Acoustics, WASPAA, New Paltz, NY, USA, October 2009, pp. 149–152.

PART

# 2

# ADAPTIVE ALGORITHMS IN THE REPRODUCING KERNEL HILBERT SPACE

CHAPTER

# MAXIMUM CORRENTROPY CRITERION–BASED KERNEL ADAPTIVE FILTERS

# 5

**Badong Chen***, **Xin Wang***, **Jose C. Principe***,†

**Institute of Artificial Intelligence and Robotics, Xi'an Jiaotong University, Xi'an, China* †*University of Florida, Gainesville, FL, United States*

## CONTENTS



## CHAPTER POINTS

- Kernel adaptive filters (KAFs) are powerful in online nonlinear system modeling.
- Correntropy and generalized correntropy are local similarity measures, and the maximum correntropy criterion (MCC) is robust to impulsive non-Gaussian noises.
- Combining KAFs and MCC provides an efficient way to deal with nonlinear modeling in the presence of non-Gaussian noises.

**Adaptive Learning Methods for Nonlinear System Modeling. DOI: 10.1016/B978-0-12-812976-0.00007-5**

## 5.1 INTRODUCTION

The features of nonlinearity, nonstationarity and non-Gaussianity in many practical signals (such as EEG and EMG signals) pose great challenges for signal processing methods [1–5]. To deal with nonlinearities, a lot of nonlinear models were proposed in the literature. Well-known examples include neural networks (*e.g.*, MLP, RBF), linear-in-the-parameter nonlinear models (*e.g.*, Volterra series, kernel methods, extreme learning machines) and block-oriented nonlinear models (*e.g.*, Wiener, Hammerstein) [6–10]. To handle nonstationarities, a common approach is to build the models in an online manner, where the parameters or structures of the models are updated adaptively to track the changes in a time varying process. For example, in adaptive filtering domains, various adaptive algorithms have been developed so far to learn the models adaptively, such as the least mean square (LMS), recursive least squares (RLS), affine projection algorithms (APA) and their variants [10–13]. To cope with non-Gaussianity, optimization criteria must go beyond the second-order framework, and statistical measures incorporating higher-order information should be used. In particular, information theoretic measures such as entropy, mutual information and divergences have been shown to be very efficient for extracting the non-Gaussian features of the processed data [14–17]. Recently, the concept of correntropy in information theoretic learning (ITL) has been successfully used for robust learning and signal processing in the presence of heavy-tailed non-Gaussian noises [18–27].

KAFs have recently attracted great attention due to their powerful learning capability for nonlinear and nonstationary system modeling [10,28–39]. The KAF algorithms are a family of online kernel learning methods, which are implemented as linear adaptive filters with input data being transformed into a high-dimensional feature space induced by a Mercer kernel [10]. Combining kernel methods with standard (linear) adaptive filtering algorithms, the KAFs have some desirable features such as universal approximation capability, the absence of local minima during adaptation and reasonable computational resources. To date, a number of kernel adaptive filtering algorithms have been developed, including kernel LMS (KLMS) [28], kernel RLS (KRLS) [29], kernel APA (KAPA) [36] and kernel Kalman filter [39]. Generally a KAF algorithm builds a growing RBF-like network with data-centered hidden nodes. In the literature, there are several approaches to constrain the network size of KAFs [10,31,40].

Most KAFs were developed under the popular mean square error (MSE) criterion, which is computationally simple and mathematically tractable but may suffer severe performance degradation in the presence of non-Gaussian noises. To improve the robustness of KAFs with respect to non-Gaussian noises, many efforts have been made to develop new KAF algorithms based on nonMSE (nonsecond-order) optimization criteria, such as the MCC [21,38], minimum error entropy (MEE) criterion [41], least mean $p$-power (LMP) criterion [42], least mean mixed norm (LMMN) criterion [43] and least mean kurtosis (LMK) criterion [44]. Increasing attention has recently been paid to the MCC criterion that aims at maximizing the correntropy between model output and desired response. Correntropy is a generalized correlation measure in kernel space and closely related to the Parzen estimator of Rényi's entropy [14], and its name comes from the words "correlation" and "entropy". When used as an optimality criterion, correntropy has some desirable properties: smoothness, robustness and invexity [45, 46]. Combining KAFs with MCC provides an efficient way to handle nonlinearity, nonstationarity and non-Gaussianity in a unified manner.

In this chapter, we give a brief introduction to the MCC-based KAFs. In Section 5.2, we briefly review two basic KAF algorithms. In Section 5.3, we introduce the concepts of correntropy and MCC. In Section 5.4, we present two KAF algorithms under the generalized MCC. Experimental results are shown in Section 5.5 and, finally, a conclusion is given in Section 5.6.

## 5.2 KERNEL ADAPTIVE FILTERS

Kernel methods are a powerful nonparametric modeling tool, which transform the input data into a high-dimensional feature space via a reproducing kernel such that the inner product operation in the feature space can be computed effectively through the kernel evaluation [47]. Using the linear structure of this feature space to implement a well-established linear adaptive filtering algorithm, one can obtain a KAF algorithm [10]. Among various KAF algorithms, the KLMS and KRLS are perhaps the most popular and widely used ones [28,29]. The KLMS is an LMS-type stochastic gradient–based learning algorithm which is computationally very simple yet efficient [28]. The KRLS is an RLS-type recursive learning algorithm which can outperform the KLMS in many practical situations but requires higher computational resources [10].

### 5.2.1 KLMS ALGORITHM

Consider the learning of a nonlinear function $y = f(\boldsymbol{x})$ based on the sample sequence $\{\boldsymbol{x}_i \in \mathbb{R}^d, y_i \in \mathbb{R}\}$, $(i = 1, 2, ...)$. First, we transform the input data into a feature space $\mathbb{F}$ through a nonlinear mapping $\varphi(.)$ induced by a Mercer kernel $\kappa(.,.)$. Then applying the well-known LMS algorithm [12] on the transformed sample sequence $\{\varphi(\boldsymbol{x}_i), y_i\}$, we have

$$\begin{aligned}\boldsymbol{\Omega}_i &= \boldsymbol{\Omega}_{i-1} - \mu \frac{\partial}{\partial \boldsymbol{\Omega}_{i-1}} e_i^2 \\ &= \boldsymbol{\Omega}_{i-1} + \eta e_i \varphi(\boldsymbol{x}_i),\end{aligned} \tag{5.1}$$

where $\boldsymbol{\Omega}_i$ denotes the weight vector in feature space at iteration $i$, $\eta = 2\mu$ is the step size parameter and $e_i$ stands for the prediction error, that is,

$$e_i = y_i - \boldsymbol{\Omega}_{i-1}^T \varphi(\boldsymbol{x}_i). \tag{5.2}$$

Identifying $\varphi(\boldsymbol{x}) = \kappa(.,\boldsymbol{x})$ and $\boldsymbol{\Omega}_i = f_i$, where $f_i$ denotes the estimated function at iteration $i$, and setting the initial function $f_0 = 0$, we can obtain the following update equations for KLMS [28]:

$$\begin{aligned}f_0 &= 0, \\ e_i &= y_i - f_{i-1}(\boldsymbol{x}_i), \\ f_i &= f_{i-1} + \eta e_i \kappa(\boldsymbol{x}_i, .).\end{aligned} \tag{5.3}$$

Note that, for the reproducing kernel $\kappa(.,.)$, we have $\langle \kappa(.,\boldsymbol{x}), f_i \rangle_{\mathbb{F}} = f_i(\boldsymbol{x})$. With a radial basis kernel (*e.g.*, Gaussian kernel), the KLMS creates a growing RBF-like network [10], with the input vectors as the centers and prediction errors (scaled by $\eta$) as the output weights, as shown in Fig. 5.1.

### 5.2.2 KRLS ALGORITHM

Next, we present the KRLS algorithm [29]. Consider the following optimization cost function:

$$\ell(\boldsymbol{\Omega}) = \sum_{j=1}^{i} \varsigma^{i-j} e_j^2 + \varsigma^i \gamma \|\boldsymbol{\Omega}\|^2, \tag{5.4}$$

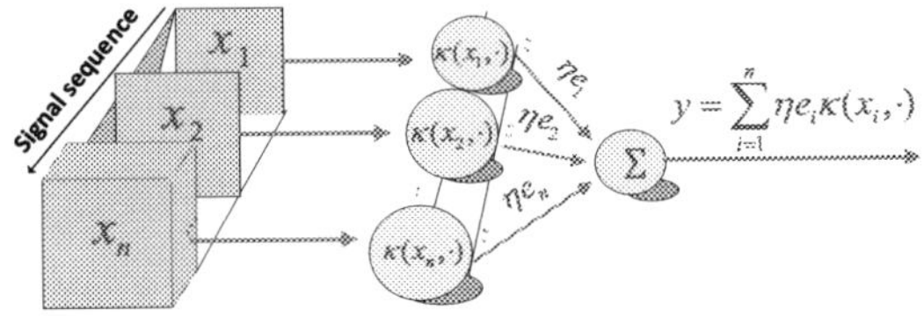


**FIGURE 5.1**

Learning process of KLMS.

where $\zeta$ is the forgetting factor and $\gamma$ is the regularization factor. Setting $\frac{\partial}{\partial \mathbf{\Omega}}\ell(\mathbf{\Omega})=0$, one can derive

$$\mathbf{\Omega}=\left(\mathbf{\Phi}_i\mathbf{B_i}\mathbf{\Phi}_i^T+\gamma\zeta^i\boldsymbol{I}\right)^{-1}\mathbf{\Phi}_i\mathbf{B_i}\boldsymbol{y}_i, \tag{5.5}$$

where $\mathbf{\Phi}_i=\left[\boldsymbol{\varphi}_1,\cdots,\boldsymbol{\varphi}_i\right]$ with $\boldsymbol{\varphi}_i$ being $\boldsymbol{\varphi}_i=\varphi(\boldsymbol{x}_i)$, $\boldsymbol{y}_i=\left[y_1,\cdots,y_i\right]$, $\mathbf{I}$ is an identity matrix and $\mathbf{B_i}=diag\left\{\zeta^{i-1},\zeta^{i-2},\cdots,1\right\}$ is a diagonal matrix. Using the matrix inversion lemma [10], we have

$$\mathbf{\Omega}=\mathbf{\Phi}_i\left(\mathbf{\Phi}_i^T\mathbf{\Phi}_i+\gamma\zeta^i\mathbf{B_i^{-1}}\right)^{-1}y_i. \tag{5.6}$$

Let $\mathbf{a}_i=\left(\mathbf{\Phi}_i^T\mathbf{\Phi}_i+\gamma\zeta^i\mathbf{B_i^{-1}}\right)^{-1}y_i$ be a coefficient vector. Then the weight vector $\mathbf{\Omega}$ can be expressed explicitly as a linear combination of the transformed data, that is, $\mathbf{\Omega}=\sum_{j=1}^{i}(\mathbf{a}_i)_j\boldsymbol{\varphi}_j$, where $(\mathbf{a}_i)_j$ denotes the $j$th component of $\mathbf{a}_i$. Denote $\boldsymbol{Q}_i=\left(\mathbf{\Phi}_i^T\mathbf{\Phi}_i+\gamma\zeta^i\mathbf{B_i^{-1}}\right)^{-1}$. We have

$$\boldsymbol{Q}_i^{-1}=\begin{bmatrix}\boldsymbol{Q}_{i-1}^{-1} & \mathbf{h}_i\\ \mathbf{h}_i^T & \boldsymbol{\varphi}_i^T\boldsymbol{\varphi}_i+\gamma\zeta^i\end{bmatrix}, \tag{5.7}$$

where $\mathbf{h}_i=\mathbf{\Phi}_{i-1}^T\boldsymbol{\varphi}_i$. Using the block matrix inversion identity [10], we have the following update rule for $\boldsymbol{Q}_i$:

$$\boldsymbol{Q}_i=r_i^{-1}\begin{bmatrix}\boldsymbol{Q}_{i-1}r_i+\mathbf{z}_i\mathbf{z}_i^T & -\mathbf{z}_i\\ -\mathbf{z}_i^T & 1\end{bmatrix}, \tag{5.8}$$

where $\mathbf{z}_i=\boldsymbol{Q}_{i-1}\mathbf{h}_i$ and $r_i=\gamma\zeta^i+\boldsymbol{\varphi}_i^T\boldsymbol{\varphi}_i-\mathbf{z}_i^T\mathbf{h}_i$. Therefore, the coefficient vector $\mathbf{a}_i$ can be updated as

$$\mathbf{a}_i=\begin{bmatrix}\mathbf{a}_{i-1}-\mathbf{z}_i r_i^{-1}e_i\\ r_i^{-1}e_i\end{bmatrix}. \tag{5.9}$$

From the above derivation, the KRLS algorithm can be briefly described in Algorithm 1.

Similar to the KLMS, the KRLS also produces a growing RBF-like network. In order to reduce the computational cost and memory requirements of KAFs, several approaches have been proposed to

**Algorithm 1** KRLS Algorithm.

**Input:** samples $\{\boldsymbol{x}_i, y_i\}, i = 1, 2, ...,$ Mercer kernel $\kappa$, forgetting factor $\zeta$, regularization factor $\gamma$

**Output:** estimated function $f_i = \sum_{j=1}^{i} (\mathbf{a}_i)_j \kappa(\boldsymbol{x}_j, .)$

Initialization: $\begin{cases} Q_1 = (\kappa(\boldsymbol{x}_1, \boldsymbol{x}_1) + \gamma\zeta)^{-1}, \\ \mathbf{a}_1 = Q_1 y_1 \end{cases}$

**while** $\{\boldsymbol{x}_i, y_i\}$ $i > 1$ **do**

  $e_i = y_i - f_{i-1}(\boldsymbol{x}_i)$

  $\mathbf{h}_i = \boldsymbol{\Phi}_{i-1}{}^T \boldsymbol{\varphi}_i$

  $\mathbf{z}_i = Q_{i-1}\mathbf{h}_i$

  $r_i = \gamma\zeta^i + \boldsymbol{\varphi}_i{}^T \boldsymbol{\varphi}_i - \mathbf{z}_i{}^T \mathbf{h}_i$

  $Q_i = r_i{}^{-1} \begin{bmatrix} Q_{i-1} r_i + \mathbf{z}_i \mathbf{z}_i{}^T & -\mathbf{z}_i \\ -\mathbf{z}_i{}^T & 1 \end{bmatrix}$

  $\mathbf{a}_i = \begin{bmatrix} \mathbf{a}_{i-1} - \mathbf{z}_i r_i{}^{-1} e_i \\ r_i{}^{-1} e_i \end{bmatrix}$

**end while**

curb the network growth [10]. Particularly, a simple online quantization approach was proposed in [31]. The basic idea of the quantization approach is to represent the input data with a smaller data set and quantize the original centers to the nearest code vector in the quantization code book (or dictionary).

# 5.3 MAXIMUM CORRENTROPY CRITERION

## 5.3.1 CORRENTROPY

First, we give some background on information theoretic learning (ITL) [14,15]. There is a close relationship between kernel methods and ITL. Most ITL cost functions, when estimated using the Parzen kernel estimator, can be expressed in terms of inner products in kernel space $\mathbb{F}$. For example, the Parzen estimator of the quadratic information potential (QIP) from samples $x_1, x_2, \cdots, x_N \in \mathbb{R}$ can be obtained as [14]

$$QIP(X) = \frac{1}{N^2\sqrt{2\pi}\sigma} \sum_{i=1}^{N}\sum_{j=1}^{N} \kappa_\sigma\left(x_i - x_j\right), \tag{5.10}$$

where $\kappa_\sigma(.)$ denotes the translation-invariant Gaussian kernel with bandwidth $\sigma$, given by

$$\kappa_\sigma(x - y) = \exp\left(-\frac{(x-y)^2}{2\sigma^2}\right). \tag{5.11}$$

Since $\kappa_\sigma(x, y) = \langle\varphi(x), \varphi(y)\rangle_{\mathbb{F}}$, we have

$$QIP(X) = \frac{1}{\sqrt{2\pi}\sigma}\left\|\frac{1}{N}\sum_{i=1}^{N}\varphi(x_i)\right\|_{\mathbb{F}}^2 = \frac{1}{\sqrt{2\pi}\sigma}\left\langle\frac{1}{N}\sum_{i=1}^{N}\varphi(x_i), \frac{1}{N}\sum_{j=1}^{N}\varphi(x_j)\right\rangle_{\mathbb{F}}, \tag{5.12}$$

where $\|.\|_{\mathbb{F}}$ stands for the norm in $\mathbb{F}$. Thus, the estimated QIP represents the squared mean of the transformed data in kernel space.

The intrinsic link between ITL and kernel methods enlightens researchers to define new similarity measures in kernel space. As a local similarity measure in ITL, the correntropy between two random variables $X$ and $Y$ is defined [18] by

$$V(X, Y) = \mathbf{E}[\kappa_\sigma(X - Y)] = \int \kappa_\sigma(x - y) dF_{XY}(x, y), \tag{5.13}$$

where the expectation value is over the joint distribution $F_{XY}(x, y)$. Correntropy can easily be estimated from data as

$$V(X, Y) = \frac{1}{N}\sum_{i=1}^{N}\kappa_\sigma(x_i - y_i). \tag{5.14}$$

Correntropy can be expressed in terms of an inner product as

$$V(X, Y) = \mathbf{E}\left[\langle\varphi(X), \varphi(Y)\rangle_{\mathbb{F}}\right], \tag{5.15}$$

which is a generalized correlation measure in kernel space and itself is also positive definite, *i.e.*, it defines a new kernel space for inference. In addition, by a simple Taylor series expansion on the kernel, one can see that correntropy provides a number that is the sum of all the statistical moments expressed by the kernel. In many applications this sum may be sufficient to quantify better than correlation the relationships of interest and it is much simpler to estimate than the higher-order statistical moments. Therefore, it can be considered as a new type of statistical descriptor. To date correntropy has been successfully applied to develop new adaptive filters for non-Gaussian noises [21–24], a correntropy MACE filter for pattern matching [48], robust PCA methods [19], new ICA algorithms [49], pitch detection [50] and nonlinear Granger causality analysis [51] methods with very good results. It is worth noting that many other similarity measures can be defined in kernel space such as the centered correntropy [14], correntropy coefficient [14], kernel risk sensitive loss [52] and kernel mean p-power error [53].

Based on the correntropy, one can define the correntropic loss (C-Loss) [52] as

$$CL(X, Y) = 1 - \mathbf{E}[\kappa_\sigma(X - Y)], \tag{5.16}$$

which essentially is the MSE in kernel space since we have

$$CL(X, Y) = \frac{1}{2}\mathbf{E}\left[\|\varphi(X) - \varphi(Y)\|_{\mathbb{F}}^2\right]. \tag{5.17}$$

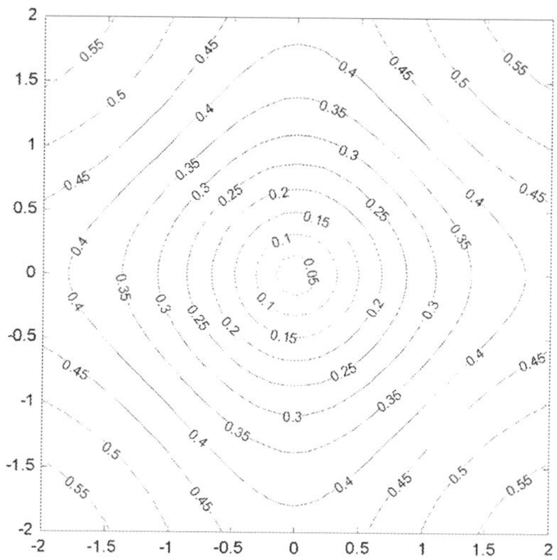


**FIGURE 5.2**

Contours of CIM$(X, 0)$ in 2D vector space (kernel size is set to 1.0).

The empirical C-Loss can be calculated by

$$\hat{CL}(X, Y) = 1 - \frac{1}{N}\sum_{i=1}^{N}\kappa_\sigma(x_i - y_i), \tag{5.18}$$

which is a function of the sample vectors $\boldsymbol{X} = [x_1, x_2, \cdots, x_N]^T$ and $\boldsymbol{Y} = [y_1, y_2, \cdots, y_N]^T$. If no confusion can arise, we denote the empirical C-Loss by $\hat{CL}(\boldsymbol{X}, \boldsymbol{Y})$. The square root of $\hat{CL}(\boldsymbol{X}, \boldsymbol{Y})$ is also called the Correntropy Induced Metric (CIM) between $\boldsymbol{X}$ and $\boldsymbol{Y}$ since it defines a metric in the sample space [18].

Fig. 5.2 shows the contours of distance from $\boldsymbol{e} = \boldsymbol{X} - \boldsymbol{Y}$ to the origin in a two-dimensional space. The interesting observation from the figure is as follows. When two points are close, CIM behaves like an $L_2$ norm (which is clear from the Taylor expansion of the kernel function) and we call this area the Euclidean zone; outside of the Euclidean zone CIM then behaves like an $L_1$, which is named the Transition zone; eventually in the Rectification zone as two points are further apart, the metric is much less sensitive to the change of the distance and saturates. Specifically, note that the CIM can be used in at least two respects for robust signal processing: (1) the relative insensitivity to very large values and the $L_1$ equivalence in the transition region enables its use as an outlier-robust error measure and (2) its concave contours for large-norm vectors in the rectification regime can be used to further impose sparsity in adaptive system coefficients, which is typically done using an $L_p (p \leq 1)$ norm penalization of the weight-vector norm. In both cases, the CIM has the useful property that for small-signal/weight situations the traditional quadratic measures are approximated, while for large-signal/weight conditions, it makes a smooth but sharp transition to desirable and robust $L_p (p \leq 1)$ penalization. The kernel size controls the speed of transition and can be optimally determined to best fit the data through the kernel density estimation connection. It is worth noting that there is a single parameter in correntropy, the kernel bandwidth, which controls the behavior of the CIM and MCC. Intuitively, if the data size is large and the problem dimension is low, a small kernel size can be used so that MCC searches with high precision (small bias) for the maximum position of the error PDF. However, if the data size is small and/or the problem dimension large, the kernel size has to be chosen so that the bulk of the data are inside the bandwidth to ensure efficient data usage while the outliers stay outside to enable an effective attenuation (small variance).

The kernel function in correntropy can easily be extended to the generalized Gaussian function and the new measure is called the generalized correntropy [23]. Specifically, the generalized correntropy between random variables $X$ and $Y$ is defined by

$$V_{\alpha,\beta}(X,Y) = \mathbf{E}[G_{\alpha,\beta}(X-Y)], \tag{5.19}$$

with $G_{\alpha,\beta}(.)$ being the generalized Gaussian density (GGD) function [23]:

$$G_{\alpha,\beta}(x) = \frac{\alpha}{2\beta\Gamma(1/\alpha)} \exp\left(-\left|\frac{x}{\beta}\right|^{\alpha}\right) = \gamma_{\alpha,\beta} \exp\left(-\lambda |x|^{\alpha}\right), \tag{5.20}$$

where $\Gamma(.)$ denotes the gamma function, $\alpha > 0$ is the shape parameter, $\beta > 0$ is the scale parameter, $\lambda = 1/\beta^{\alpha}$ is the kernel parameter and $\gamma_{\alpha,\beta} = \alpha/(2\beta\Gamma(1/\alpha))$ is the normalization constant. Obviously, when $\alpha = 2$, the generalized correntropy is equivalent to the original correntropy. Note that the kernel function in the generalized correntropy is a Mercer kernel only when $0 < \alpha \leq 2$.

Similarly, one can define the generalized C-Loss (GC-Loss) and generalized CIM (GCIM). The GCIM also behaves like different norms (from $L_{\alpha}$ to $L_0$) of the data in different regions (see [23] for more details).

### 5.3.2 MAXIMUM CORRENTROPY CRITERION

The correntropy and generalized correntropy can be used as an optimization cost in estimation-related problems. Given two random variables $X \in \mathbb{R}$, an unknown real-valued parameter to be estimated, and $Y \in \mathbb{R}^m$, a vector of observations (measurements), an estimator of $X$ can be defined as a function of $Y{:}\hat{X} = g(Y)$, where $g$ is solved by optimizing a certain cost function. Under the MCC, the estimator $g$ can be solved by maximizing the correntropy (or minimizing the C-Loss) between $X$ and $\hat{X}$, that is [45],

$$g_{MCC} = \arg\max_{g\in\mathbf{G}} V(X,\hat{X}) = \arg\max_{g\in\mathbf{G}} \mathbf{E}[\kappa_{\sigma}(X - g(Y))], \tag{5.21}$$

where $\mathbf{G}$ stands for the collection of all measurable functions of $Y$. If $\forall y \in \mathbb{R}^m$, $X$ has a conditional PDF $p_{X|Y}(x\,|y)$, then the estimation error $e = X - \hat{X}$ has PDF

$$p_e(\varepsilon) = \int_{\mathbb{R}^m} p_{X|Y}(\varepsilon + g(y)|y)dF_Y(y), \tag{5.22}$$

where $F_Y(y)$ is the distribution function of $Y$. It follows that

$$g_{MCC} = \arg\max_{g\in G} \int_{\mathbb{R}} \kappa_{\sigma}(\varepsilon) \int_{\mathbb{R}^m} p_{X|Y}(\varepsilon + g(y)|y)dF_Y(y)d\varepsilon. \tag{5.23}$$

In [45], it has been proven that the MCC estimate is a smoothed maximum *a posteriori* probability (MAP) estimate, which equals the mode (at which the PDF attains its maximum value). Similar results hold for the generalized MCC (GMCC) estimation.

**Theorem 1.** *The GMCC estimator can be expressed as*

$$g_{GMCC}(y) = \arg\max_{x\in\mathbb{R}} \rho_{\alpha,\beta}(x|y), \forall y \in \mathbb{R}^m, \tag{5.24}$$

*where* $\rho_{\alpha,\beta}(x|y) = G_{\alpha,\beta}(x) * p_{X|Y}(x|y)$ *("$*$" denotes the convolution operator with respect to $x$).*

*Proof.* It is easy to derive

$$\begin{aligned}
V_{\alpha,\beta}(X,\hat{X}) &= \textstyle\int_{\mathbb{R}} G_{\alpha,\beta}(\varepsilon) \int_{\mathbb{R}^m} p_{X|Y}(\varepsilon + g(y)|y) dF_Y(y) d\varepsilon \\
&= \textstyle\int_{\mathbb{R}^m} \left\{ \int_{-\infty}^{\infty} G_{\alpha,\beta}(\varepsilon) p_{X|Y}(\varepsilon + g(y)|y) d\varepsilon \right\} dF_Y(y) \\
&= \textstyle\int_{\mathbb{R}^m} \left\{ \int_{-\infty}^{\infty} G_{\alpha,\beta}(\varepsilon' - g(y)) p_{X|Y}(\varepsilon'|y) d\varepsilon' \right\} dF_Y(y) \\
&\overset{(a)}{=} \textstyle\int_{\mathbb{R}^m} \left\{ \int_{-\infty}^{\infty} G_{\alpha,\beta}(g(y) - \varepsilon') p_{X|Y}(\varepsilon'|y) d\varepsilon' \right\} dF_Y(y) \\
&= \textstyle\int_{\mathbb{R}^m} \left\{ \left( G_{\alpha,\beta}(.) * p_{X|Y}(.|y) \right) (g(y)) \right\} dF_Y(y) \\
&= \textstyle\int_{\mathbb{R}^m} \rho_{\alpha,\beta}(g(y)|y) dF_Y(y),
\end{aligned} \tag{5.25}$$

where $\varepsilon' = \varepsilon + g(y)$ and (a) comes from the symmetry of $G_{\alpha,\beta}(.)$. Thus

$$\begin{aligned}
& g_{GMCC} = \arg\max_{g\in\mathbf{G}} \textstyle\int_{\mathbb{R}^m} \rho_{\alpha,\beta}(g(y)|y) dF_Y(y) \\
& \Rightarrow g_{GMCC}(y) = \arg\max_{x\in\mathbb{R}} \rho_{\alpha,\beta}(x|y), \forall y \in \mathbb{R}^m
\end{aligned} \tag{5.26}$$

This completes the proof.

*Remark.* Considering the conditional PDF $p_{X|Y}(x|y)$ as an *a posteriori* PDF given the observation $y$, the function $\rho_{\alpha,\beta}(x|y)$ is a smoothed (by convolution) *a posteriori* PDF. Therefore, the GMCC estimate is a smoothed MAP estimate.

**Corollary 1.** *When $\beta \to 0+$ (or $\lambda \to \infty$), the GMCC estimation becomes the MAP estimation.*

*Proof.* As $\beta \to 0+$, the GGD function $G_{\alpha,\beta}(.)$ will approach the Dirac delta function and the smoothed conditional PDF $\rho_{\alpha,\beta}(x|y)$ will reduce to the original conditional PDF $p_{X|Y}(x|y)$. In this case, the GMCC estimation will be equivalent to the MAP estimation.

**Theorem 2.** *The GMCC estimator can also be expressed as*

$$g_{GMCC} = \arg\max_{g\in G} p_e^{\alpha,\beta}(0), \tag{5.27}$$

*where $p_e^{\alpha,\beta}(x) = p_e(x) * G_{\alpha,\beta}(x)$ is the smoothed error PDF.*

*Proof.* It is easy to derive

$$\begin{aligned}\mathbf{E}\left[G_{\alpha,\beta}(e)\right] &= \textstyle\int_{\mathbb{R}} G_{\alpha,\beta}(\varepsilon)p_e(\varepsilon)d\varepsilon \\ &= \left(\textstyle\int_{\mathbb{R}} G_{\alpha,\beta}(x-\varepsilon)p_e(\varepsilon)d\varepsilon\right)_{x=0} \\ &= \left(p_e(x) * G_{\alpha,\beta}(x)\right)_{x=0} \\ &= p_e^{\alpha,\beta}(0).\end{aligned} \tag{5.28}$$

Hence

$$g_{GMCC} = \arg\max_{g\in\mathbf{G}} \mathbf{E}\left[G_{\alpha,\beta}(e)\right] = \arg\max_{g\in\mathbf{G}} p_e^{\alpha,\beta}(0). \tag{5.29}$$

**Theorem 3.** *When $\beta \to \infty$ (or $\lambda \to 0+$), the GMCC estimation will be equivalent to the LMP estimation with $p = \alpha$.*

*Proof.* Under the LMP criterion, the optimal estimator is solved by minimizing the mean $p$-power of the error:

$$g_{LMP} = \arg\min_{g\in\mathbf{G}} \mathbf{E}\left[|e|^p\right]. \tag{5.30}$$

On the other hand, we have $V_{\alpha,\beta}(X,\hat{X}) \approx \gamma_{\alpha,\beta}\left(1 - \lambda\mathbf{E}\left[|e|^\alpha\right]\right)$ as $\lambda \to 0+$. It follows that

$$\max \mathrm{V}_{\alpha,\beta}(X,\hat{X}) \sim \min \mathbf{E}\left[|e|^\alpha\right], \tag{5.31}$$

that is, as $\lambda \to 0+$, the GMCC estimation will be, approximately, equivalent to the LMP estimation with $p = \alpha$.

## 5.4 KERNEL ADAPTIVE FILTERS UNDER GENERALIZED MCC

Most of the KAFs, such as KLMS and KRLS, have been derived under the well-known MSE criterion. The MSE is computationally simple and mathematically tractable but may perform very poorly under non-Gaussian noise (especially impulsive noise) disturbance. An efficient approach to improve the robustness to non-Gaussian noises is to use the MCC-based costs [21–24]. In the following, we present two KAF algorithms under the generalized MCC [54,55]. Of course, one can derive many other KAFs under GMCC.

### 5.4.1 GENERALIZED KERNEL MAXIMUM CORRENTROPY

In a way very similar to the derivation of KLMS, applying the GMCC on the transformed sample sequence $\{\varphi(\boldsymbol{x}_i), y_i\}$, one can derive

$$\begin{aligned}\boldsymbol{\Omega}_i &= \boldsymbol{\Omega}_{i-1} + \mu\frac{\partial}{\partial\boldsymbol{\Omega}_{i-1}}\exp(-\lambda|e_i|^\alpha) \\ &= \boldsymbol{\Omega}_{i-1} + \mu\lambda\alpha\exp(-\lambda|e_i|^\alpha)|e_i|^{\alpha-1}sign(e_i)\varphi(\boldsymbol{x}_i) \\ &= \boldsymbol{\Omega}_{i-1} + \eta\exp(-\lambda|e_i|^\alpha)|e_i|^{\alpha-1}sign(e_i)\varphi(\boldsymbol{x}_i),\end{aligned} \tag{5.32}$$

where the step size $\eta = \mu\lambda\alpha$. The above algorithm is called the generalized kernel maximum correntropy (GKMC) [54]. From (5.32), we have the following observations:
1) When $\alpha = 2$, the GKMC becomes

$$\boldsymbol{\Omega}_i = \boldsymbol{\Omega}_{i-1} + \eta \exp(-\lambda e_i^2) e_i \varphi(\boldsymbol{x}_i), \tag{5.33}$$

which is the kernel maximum correntropy (KMC) algorithm developed in [21].
2) When $\lambda \to 0+$, the GKMC becomes

$$\boldsymbol{\Omega}_i = \boldsymbol{\Omega}_{i-1} + \eta |e_i|^{\alpha-1} sign(e_i) \varphi(\boldsymbol{x}_i), \tag{5.34}$$

which is the kernel LMP (KLMP) algorithm developed in [42].
3) When $\alpha = 2$ and $\lambda \to 0+$, the GKMC will reduce to the KLMS algorithm

$$\boldsymbol{\Omega}_i = \boldsymbol{\Omega}_{i-1} + \eta e_i \varphi(\boldsymbol{x}_i). \tag{5.35}$$

In addition, one can rewrite (5.32) as

$$\boldsymbol{\Omega}_i = \boldsymbol{\Omega}_{i-1} + \eta_i e_i \varphi(\boldsymbol{x}_i), \tag{5.36}$$

where $\eta_i = \eta \exp(-\lambda |e_i|^{\alpha}) |e_i|^{\alpha-2}$ is a variable step size (VSS). Thus, the GKMC can be viewed as a KLMS with VSS $\eta_i$. Since $\eta_i \to 0$ as $|e_i| \to \infty$, the learning rate of GKMC will approach zero when a large error occurs. This property ensures the robustness of the GKMC with respect to large outliers.

In summary, the GKMC algorithm is summarized in Algorithm 2, where $\mathbf{a}_i$ denotes the coefficient vector at iteration $i$, $(\mathbf{a}_i)_j$ is the $j$th component of $\mathbf{a}_i$ and $\boldsymbol{C}_i$ is the center set (or dictionary).

The network size of the GKMC grows linearly with the number of input data. One can use a simple quantization approach to curb the network growth [31]. In this way, the weight update Eq. (5.32) becomes

$$\boldsymbol{\Omega}_i = \boldsymbol{\Omega}_{i-1} + \eta \exp(-\lambda |e_i|^{\alpha}) |e_i|^{\alpha-1} sign(e_i) \varphi(Q[\boldsymbol{x}_i]), \tag{5.37}$$

---

**Algorithm 2** GKMC Algorithm.

**Input:** samples $\{\boldsymbol{x}_i, y_i\}, i = 1, 2, ...$, step size $\eta$, Mercer kernel $\kappa$, generalized correntropy shape parameter $\alpha$, kernel parameter $\lambda$

**Output:** estimated function $f_i = \sum_{j=1}^{i} (\mathbf{a}_i)_j \kappa(\boldsymbol{x}_j, .)$.

Initialization: $\begin{cases} \mathbf{a}_1 = \{\eta \exp(-\lambda |y_1|^{\alpha}) |y_1|^{\alpha-1} sign(y_1)\}, \\ \boldsymbol{C}_1 = \{\boldsymbol{x}_1\} \end{cases}$

**while** $\{\boldsymbol{x}_i, y_i\}$ $i > 1$ **do**

  $f_{i-1}(\boldsymbol{x}_i) = \sum_{j=1}^{i-1} (\mathbf{a}_{i-1})_j \kappa(\boldsymbol{x}_j, \boldsymbol{x}_i)$

  $e_i = y_i - f_{i-1}(\boldsymbol{x}_i)$

  $\boldsymbol{C}_i = \{\boldsymbol{C}_{i-1}, \boldsymbol{x}_i\}$

  $\mathbf{a}_i = \{\mathbf{a}_{i-1}, \eta \exp(-\lambda |e_i|^{\alpha}) |e_i|^{\alpha-1} sign(e_i)\}$

**end while**

---

where $Q[.]$ denotes the quantization operator [31]:

$$Q[\boldsymbol{x}_i] = \left\{ \begin{array}{ll} \boldsymbol{C}_{j^*}(i-1) & dis(\boldsymbol{x}_i, \boldsymbol{C}_{i-1}) \leq \varepsilon \\ \boldsymbol{x}_i & dis(\boldsymbol{x}_i, \boldsymbol{C}_{i-1}) > \varepsilon \end{array} \right\}, \tag{5.38}$$

with $\boldsymbol{C}_{i-1}$ being the codebook at iteration $i-1$, $\varepsilon$ the quantization size and $dis(\boldsymbol{x}_i, \boldsymbol{C}_{i-1})$ the distance between $\boldsymbol{x}_i$ and $\boldsymbol{C}_{i-1}$. We have

$$dis(\boldsymbol{x}_i, \boldsymbol{C}_{i-1}) = \min_{1 \leq j \leq size(\boldsymbol{C}_{i-1})} \left\| \boldsymbol{x}_i - (\boldsymbol{C}_{i-1})_j \right\|, \tag{5.39}$$

where $(\boldsymbol{C}_{i-1})_j$ denotes the $j$th element (code vector) of the codebook $\boldsymbol{C}_{i-1}$, $j^*$ is the index of the closest center

$$j^* = \arg\min_{1 \leq j \leq size(\boldsymbol{C}(i-1))} \left\| \boldsymbol{x}_i - (\boldsymbol{C}_{i-1})_j \right\|. \tag{5.40}$$

The quantized GKMC (QGKMC) algorithm is summarized in Algorithm 3.

---

**Algorithm 3** QGKMC Algorithm.

**Input:** samples $\{\boldsymbol{x}_i, y_i\}, i = 1, 2, \ldots$, step size $\eta$, Mercer kernel $\kappa$, quantization size $\varepsilon$, generalized correntropy shape parameter $\alpha$, kernel parameter $\lambda$

**Output:** estimated function $f_i = \sum_{j=1}^{size(\boldsymbol{C}_i)} (\mathbf{a}_i)_j \kappa(\boldsymbol{x}_j, .)$

Initialization: $\left\{ \begin{array}{l} \mathbf{a}_1 = \{\eta \exp(-\lambda |y_1|^{\alpha}) |y_1|^{\alpha-1} sign(y_1)\}, \\ \boldsymbol{C}_1 = \{\boldsymbol{x}_1\} \end{array} \right.$

**while** $\{\boldsymbol{x}_i, y_i\}$ $i > 1$ **do**

  $f_{i-1}(\boldsymbol{x}_i) = \sum_{j=1}^{size(\boldsymbol{C}_{i-1})} (\mathbf{a}_{i-1})_j \kappa((\boldsymbol{C}_{i-1})_j, \boldsymbol{x}_i)$

  $e_i = y_i - f_{i-1}(\boldsymbol{x}_i)$

  $dis(\boldsymbol{x}_i, \boldsymbol{C}_{i-1}) = \min_{1 \leq j \leq size(\boldsymbol{C}_{i-1})} \left\| \boldsymbol{x}_i - (\boldsymbol{C}_{i-1})_j \right\|$

  $j^* = \arg\min_{1 \leq j \leq size(\boldsymbol{C}_{i-1})} \left\| \boldsymbol{x}_i - (\boldsymbol{C}_{i-1})_j \right\|$

  **if** $dis(\boldsymbol{x}_i, \boldsymbol{C}_{i-1}) \leq \varepsilon$ **then**

    $\boldsymbol{C}_i = \boldsymbol{C}_{i-1}, \mathbf{a}_i = \mathbf{a}_{i-1}$

    $(\mathbf{a}_i)_{j^*} = (\mathbf{a}_i)_{j^*} + \eta \exp(-\lambda |e_i{}^{\alpha}|) |e_i|^{\alpha-1} sign(e_i)$

  **else**

    $\boldsymbol{C}_i = \{\boldsymbol{C}_{i-1}, \boldsymbol{x}_i\}$

    $\mathbf{a}_i = \{\mathbf{a}_{i-1}, \eta \exp(-\lambda |e_i|^{\alpha}) |e_i|^{\alpha-1} sign(e_i)\}$

  **end if**

**end while**

---

### 5.4.2 GENERALIZED KERNEL RECURSIVE MAXIMUM CORRENTROPY

Next, we derive the generalized kernel recursive maximum correntropy (GKRMC) algorithm [55]. A weighted and regularized cost function under GMCC is proposed as

$$\ell(\mathbf{\Omega}) = \sum_{j=1}^{i} \zeta^{i-j} G_{\alpha,\beta}\left(y_j - \mathbf{\Omega}^{\mathrm{T}}\boldsymbol{\varphi}_j\right) - \frac{1}{2}\zeta^i \gamma \|\mathbf{\Omega}\|^2, \tag{5.41}$$

where $\zeta$ is the forgetting factor, satisfying $0 \le \zeta \le 1$, $\gamma$ is the regularization factor and $G_{\alpha,\beta}(y_j - \mathbf{\Omega}^{\mathrm{T}}\boldsymbol{\varphi}_j) = \frac{\alpha}{2\beta\Gamma(1/\alpha)} \exp\left(-\left|\frac{y_j - \mathbf{\Omega}^T\boldsymbol{\varphi}_j}{\beta}\right|^{\alpha}\right)$. Setting $\frac{\partial}{\partial \mathbf{\Omega}}\ell(\mathbf{\Omega}) = 0$, one can obtain the solution

$$\Omega_i = (\Phi_i \mathbf{B}_i \Phi_i{}^{\mathrm{T}} + \gamma\zeta^i\beta^{\alpha}\mathrm{I})^{-1}\Phi_i \mathbf{B}_i y_i, \tag{5.42}$$

where

$$\mathbf{B}_i = diag\{b_1, b_2, \ldots, b_i\}$$

with

$$b_j = \zeta^{i-j} \times \left(y_j - \Omega^{\mathrm{T}}\varphi_j\right)^{\alpha-2} \times \left(\frac{\alpha^2}{2\beta\Gamma(1/\alpha)}\right) \times \exp\left(-\left|\frac{y_j - \Omega^T\varphi_j}{\beta}\right|^{\alpha}\right), \quad j = 1, 2, \ldots i.$$

With the matrix inversion lemma, we have

$$(\Phi_i \mathbf{B}_i \Phi_i{}^{\mathrm{T}} + \gamma\zeta^i\beta^{\alpha}\mathrm{I})^{-1}\Phi_i \mathbf{B}_i = \Phi_i(\Phi_i{}^{\mathrm{T}}\Phi_i + \gamma\zeta^i\beta^{\alpha}\mathbf{B}_i{}^{-1})^{-1}. \tag{5.43}$$

Thus, (5.42) can be rewritten as

$$\Omega_i = \Phi_i(\Phi_i{}^{\mathrm{T}}\Phi_i + \gamma\zeta^i\beta^{\alpha}\mathbf{B}_i{}^{-1})^{-1} y_i. \tag{5.44}$$

The above weight vector can be expressed as a linear combination of the transformed feature vectors, where the combination coefficients can be calculated recursively. We denote $\Omega_i = \Phi_i \mathbf{a}_i$ with $\mathbf{a}_i = (\Phi_i{}^{\mathrm{T}}\Phi_i + \gamma\zeta^i\beta^{\alpha}\mathbf{B}_i{}^{-1})^{-1} y_i$ being the coefficient vector. In addition, let $\boldsymbol{Q}_i = (\mathbf{\Phi}_i{}^{\mathrm{T}}\mathbf{\Phi}_i + \gamma\zeta^i\beta^{\alpha} \times \mathbf{B}_i{}^{-1})^{-1}$. Then we have

$$\boldsymbol{Q}_i = \begin{bmatrix} \mathbf{\Phi}_{i-1}{}^{\mathrm{T}}\mathbf{\Phi}_{i-1} + \gamma\zeta^i\beta^{\alpha}\mathbf{B}_{i-1}{}^{-1} & \mathbf{\Phi}_{i-1}{}^{\mathrm{T}}\boldsymbol{\varphi}_i \\ \boldsymbol{\varphi}_i{}^{\mathrm{T}}\mathbf{\Phi}_{i-1} & \boldsymbol{\varphi}_i{}^{T}\boldsymbol{\varphi}_i + \gamma\zeta^i\beta^{\alpha}\theta_i \end{bmatrix}^{-1}, \tag{5.45}$$

where $\theta_i = \left(y_i - \Omega^{\mathrm{T}}\boldsymbol{\varphi}_i\right)^{\alpha-2} \times \left(\frac{\alpha^2}{2\beta\Gamma(1/\alpha)}\right) \times \exp\left(-\left|\frac{y_i - \Omega^T\varphi_i}{\beta}\right|^{\alpha}\right)$. The above result can be converted to

$$\boldsymbol{Q}_i{}^{-1} = \begin{bmatrix} \boldsymbol{Q}_{i-1}{}^{-1} & \mathbf{h}_i \\ \mathbf{h}_i{}^{\mathrm{T}} & \boldsymbol{\varphi}_i{}^{T}\boldsymbol{\varphi}_i + \gamma\zeta^i\beta^{\alpha}\theta_i \end{bmatrix}, \tag{5.46}$$

where $\mathbf{h}_i = \Phi_{i-1}{}^{\mathrm{T}}\boldsymbol{\varphi}_i$. Based on the following block matrix inversion identity,

$$\begin{bmatrix} \mathbf{A} & \mathbf{B} \\ \mathbf{C} & \mathbf{D} \end{bmatrix}^{-1} = \begin{bmatrix} (\mathbf{A}-\mathbf{B}\mathbf{D}^{-1}\mathbf{C})^{-1} & -\mathbf{A}^{-1}\mathbf{B}(\mathbf{D}-\mathbf{C}\mathbf{A}^{-1}\mathbf{B})^{-1} \\ -\mathbf{D}^{-1}\mathbf{C}(\mathbf{A}-\mathbf{B}\mathbf{D}^{-1}\mathbf{C})^{-1} & (\mathbf{D}-\mathbf{C}\mathbf{A}^{-1}\mathbf{B})^{-1} \end{bmatrix}, \tag{5.47}$$

one can derive

$$\boldsymbol{Q}_i = {r_i}^{-1} \begin{bmatrix} \boldsymbol{Q}_{i-1} r_i + \mathbf{z}_i {\mathbf{z}_i}^{\mathrm{T}} & -\mathbf{z}_i \\ -{\mathbf{z}_i}^{\mathrm{T}} & 1 \end{bmatrix}, \tag{5.48}$$

where $r_i = \gamma \zeta^i \beta^\alpha \theta_i + {\boldsymbol{\varphi}_i}^T \boldsymbol{\varphi}_i - {\mathbf{z}_i}^{\mathrm{T}} \mathbf{h}_i$ and $\mathbf{z}_i = \boldsymbol{Q}_{i-1} \mathbf{h}_i$. Therefore, the expansion coefficients are

$$\begin{aligned} \mathbf{a}_i &= \boldsymbol{Q}_i y_i \\ &= {r_i}^{-1} \begin{bmatrix} \boldsymbol{Q}_{i-1} r_i + \mathbf{z}_i {\mathbf{z}_i}^{\mathrm{T}} & -\mathbf{z}_i \\ -{\mathbf{z}_i}^{\mathrm{T}} & 1 \end{bmatrix} \begin{bmatrix} y_{i-1} \\ y_i \end{bmatrix} \\ &= \begin{bmatrix} \mathbf{a}_{i-1} - \mathbf{z}_i {r_i}^{-1} e_i \\ {r_i}^{-1} e_i \end{bmatrix}, \end{aligned} \tag{5.49}$$

where $e_i = y_i - \Omega^{\mathrm{T}} \boldsymbol{\varphi}_i$. Then we have obtained the GKRMC algorithm, whose pseudocode is given in Algorithm 4.

---

**Algorithm 4** GKRMC Algorithm.

**Input:** samples $\{\boldsymbol{x}_i, y_i\}, i = 1, 2, ...,$ Mercer kernel $\kappa$, forgetting factor $\zeta$, regularization factor $\gamma$, generalized correntropy shape parameter $\alpha$ and scale parameter $\beta$.

**Output:** estimated function $f_i = \sum_{j=1}^{i} (\mathbf{a}_i)_j \kappa(\boldsymbol{x}_j, .)$

Initialization: $\begin{cases} \boldsymbol{Q}_1 = (\zeta \gamma \beta^\alpha + \kappa(\boldsymbol{x}_1, \boldsymbol{x}_1))^{-1}, \\ \mathbf{a}_1 = \boldsymbol{Q}_1 y_1 \end{cases}$

**while** $\{\boldsymbol{x}_i, y_i\}$ $i > 1$ **do**

$h_i = [\kappa(\boldsymbol{x}_i, \boldsymbol{x}_1), \kappa(\boldsymbol{x}_i, \boldsymbol{x}_2), ..., \kappa(\boldsymbol{x}_i, \boldsymbol{x}_{i-1})]^T$

$y_i = {h_i}^T \mathbf{a}_{i-1}$

$e_i = y_i - \boldsymbol{\Omega}^T \boldsymbol{\varphi}_i$

$\mathbf{z}_i = \boldsymbol{Q}_{i-1} h_i$

$\theta_i = (y_i - \Omega^{\mathrm{T}} \boldsymbol{\varphi}_i)^{\alpha-2} \times \left(\frac{\alpha^2}{2\beta\Gamma(1/\alpha)}\right) \times \exp\left(-\left|\frac{y_i - \Omega^T \boldsymbol{\varphi}_i}{\beta}\right|^\alpha\right)$

$r_i = \gamma \zeta^i \beta^\alpha \theta_i + {\boldsymbol{\varphi}_i}^T \boldsymbol{\varphi}_i - {\mathbf{z}_i}^{\mathrm{T}} \mathbf{h}_i$

$\boldsymbol{Q}_i = {r_i}^{-1} \begin{bmatrix} \boldsymbol{Q}_{i-1} r_i + \mathbf{z}_i {\mathbf{z}_i}^{\mathrm{T}} & -\mathbf{z}_i \\ -{\mathbf{z}_i}^{\mathrm{T}} & 1 \end{bmatrix}$

$\mathbf{a}_i = \begin{bmatrix} \mathbf{a}_{i-1} - \mathbf{z}_i {r_i}^{-1} e_i \\ {r_i}^{-1} e_i \end{bmatrix}$

**end while**

---

**Table 5.1 Testing MSEs at final iteration in frequency doubling**

| | KLMS | GKMC $\alpha = 2$ | GKMC $\alpha = 4$ | GKMC $\alpha = 6$ |
|---|---|---|---|---|
| Gaussian | $0.2250 \pm 0.1896$ | $0.0353 \pm 0.0168$ | $0.0196 \pm 0.0101$ | $0.0260 \pm 0.0121$ |
| Uniform | $0.1781 \pm 0.1842$ | $0.0137 \pm 0.0057$ | $0.0037 \pm 0.0016$ | $0.0030 \pm 0.0022$ |

*Remark.* When $\alpha = 2$, the GKRMC algorithm will reduce to the kernel recursive maximum correntropy (KRMC) developed in [38].

## 5.5 SIMULATION RESULTS

In the following, we present some simulation results to demonstrate the performance of the GKMC and GKRMC algorithms.

### 5.5.1 FREQUENCY DOUBLING PROBLEM

First, consider the frequency-doubling problem, where the input signal is a sine wave with 20 samples per period. The desired signal is another sine wave with double frequency, disturbed by an additive noise $v_i$. With a time embedding dimension of 5, the input vector is $\boldsymbol{x}_i = [x_{i-4}, x_{i-3}, ..., x_i]$. In this example, we consider a mixture noise model which combines two independent noise processes, given by

$$v_i = (1 - \delta_i)A_i + \delta_i B_i, \tag{5.50}$$

where $\delta_i$ is binary distributed over $\{0, 1\}$, with probability $p(\delta_i = 1) = c$, $p(\delta_i = 0) = 1 - c$, where $0 \leq c \leq 1$ controls the occurrence probability of the noise processes $A_i$ and $B_i$. In general, $A_i$ represents the "background noise" with small variance, while $B_i$ stands for large outliers that occur occasionally with larger variance. In the simulations below, without explicit mention we set $c = 0.05$. $B_i$ is assumed to be a white Gaussian process with zero mean and variance 15. The Gaussian kernel is adopted as the Mercer kernel in kernel adaptive filters and the bandwidth is set to $\sigma = 2.0$. The step sizes are chosen such that all the algorithms achieve almost the same initial convergence speed.

We choose 4000 noisy samples for training and 200 clean samples for testing (the filters are fixed during the testing). The average learning curves of the KLMS and GKMC over 100 Monte Carlo runs in terms of the testing MSEs with different values of $\alpha$ ($\lambda = 2.0$) and distributions of $A_i$ (Gaussian and uniform) are illustrated in Fig. 5.3, and the corresponding testing MSEs at the final iteration are summarized in Table 5.1. From the simulation results we observe: (i) the GKMC can outperform the KLMS significantly, showing much stronger robustness with respect to the impulsive noises; (ii) with different values of $\alpha$, the GKMC can achieve difference performances, and the best performance may be obtained when $\alpha \neq 2.0$. Note that the GKMC will reduce to the KMC algorithm when $\alpha = 2.0$.

Further, we compare the performance of the GKMC and QGKMC. We set $\varepsilon = 0.03$, $\alpha = 4$, $\lambda = 2$ and $\eta = 0.3$ and assume that $A_i$ is of Gaussian distribution with variance 0.04. The average learning curves over 100 runs are shown in Fig. 5.4, and the network size growth curve of the QGKMC is presented in Fig. 5.5. One can see that both algorithms achieve almost the same convergence performance

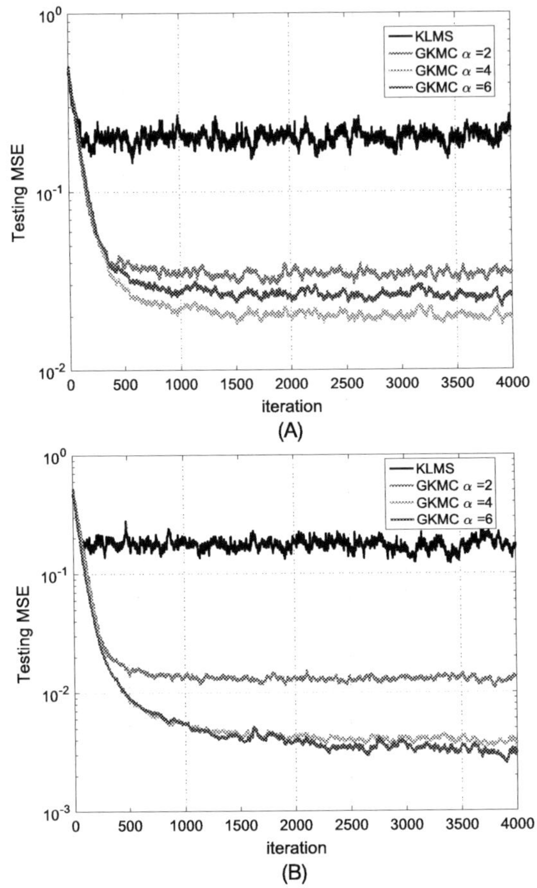


**FIGURE 5.3**

Average learning curves with different values of $\alpha$ ($\alpha = 2, 4, 6$) and distributions of $A_i$: (A) Gaussian distribution with zero mean and variance 0.25; (B) uniform distribution over $[-0.5, 0.5]$.

but the network size of QGKMC has been significantly curbed (only slightly larger than 30 at steady state).

## 5.5.2 ONLINE NONLINEAR SYSTEM IDENTIFICATION

Next, we show the performance of the GKMC and GKRMC algorithms in online nonlinear system identification. Suppose the nonlinear system is of the form [41]

$$\begin{aligned} y_i &= (0.8 - 0.5exp(-y_{i-1}^2))y_{i-1} \\ &-(0.3 + 0.9\exp(-y_{i-1}^2))y_{i-2} \\ &+0.1\sin(3.1415926y_{i-1}) + v_i, \end{aligned} \tag{5.51}$$

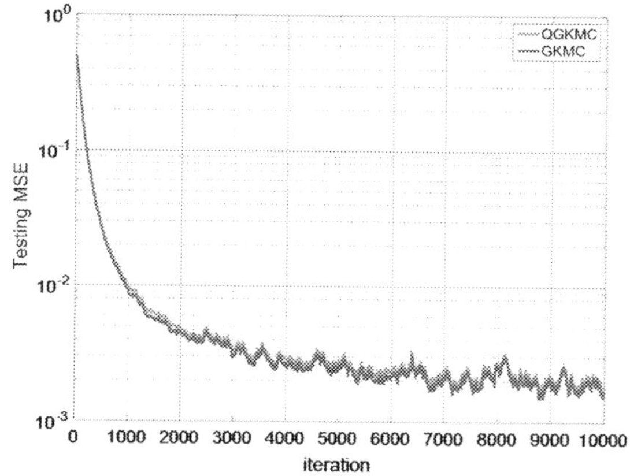


**FIGURE 5.4**

Average learning curves of QGKMC and GKMC.

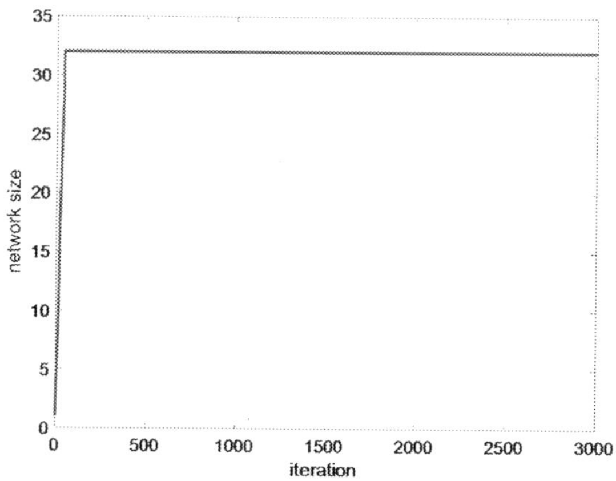


**FIGURE 5.5**

Network size growth curve of QGKMC.

where $v_i$ is a disturbance noise described by (5.50). The average learning curves in terms of testing MSEs over 100 Monte Carlo runs with different values of $\alpha$ and distributions of $A_i$ (uniform and alpha-stable) are shown in Fig. 5.6, and the corresponding testing MSEs at final iteration are listed in Table 5.2. In the simulation, 1000 noisy samples are used for training and 100 noise-free samples are used for testing (the filters are fixed during the testing). The parameters of the GKMC and GKRMC are experimentally chosen such that both achieve a desirable performance. It can be clearly seen that the GKRMC can outperform the GKMC, with faster convergence speed and lower testing MSE. However, we should note that the GKRMC is computationally more expensive than the GKMC.

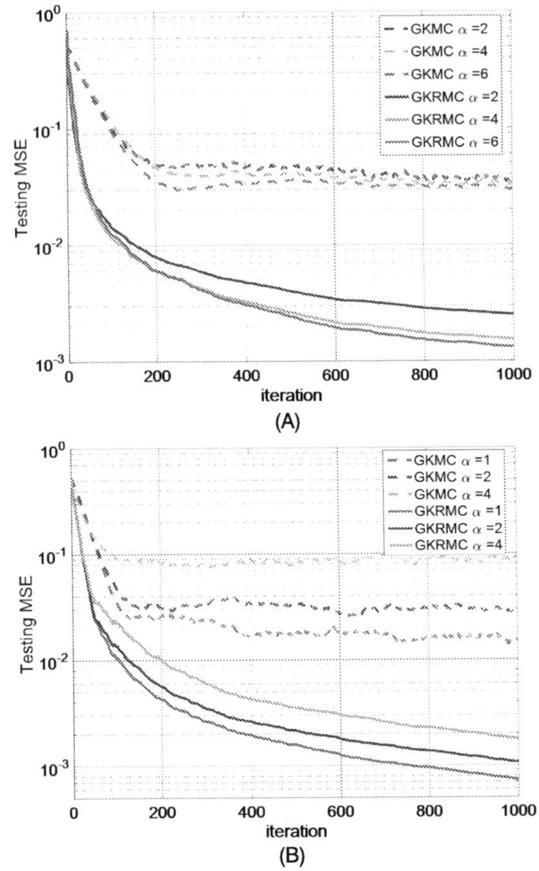


**FIGURE 5.6**

Average learning curves with different values of $\alpha$ and distributions of $A_i$: (A) uniform distribution over $[-0.5, 0.5]$; (B) alpha-stable distribution with shape parameter equal to 0.7 and scale parameter equal to 0.05.

**Table 5.2 Testing MSEs at the final iteration in online nonlinear system identification**

| | GKMC $\alpha=2$ | GKMC $\alpha=4$ | GKMC $\alpha=6$ | GKRMC $\alpha=2$ | GKRMC $\alpha=4$ | GKRMC $\alpha=6$ |
|---|---|---|---|---|---|---|
| Uniform | $0.0383 \pm 0.0250$ | $0.0330 \pm 0.0145$ | $0.0298 \pm 0.0132$ | $0.0024 \pm 0.0009$ | $0.0014 \pm 0.0007$ | $0.0012 \pm 0.0006$ |
| | GKMC $\alpha=1$ | GKMC $\alpha=2$ | GKMC $\alpha=4$ | GKRMC $\alpha=1$ | GKRMC $\alpha=2$ | GKRMC $\alpha=4$ |
| Alpha-stable | $0.0152 \pm 0.0110$ | $0.0286 \pm 0.0194$ | $0.0907 \pm 0.0556$ | $0.0007 \pm 0.0004$ | $0.0011 \pm 0.0005$ | $0.0017 \pm 0.0009$ |

## 5.5.3 NOISE CANCELLATION

In this part, we consider a noise cancellation problem in signal processing. The basic structure of a noise cancellation system is shown in Fig. 5.7. The primary signal is $s_i$, whose noisy measurement $d_i$ serves as the desired signal of the system. The term $\rho_i$ represents an unknown white noise process and $x_i$ is its reference measurement. The signal $x_i$ can be used as an input of an adaptive filter such that the

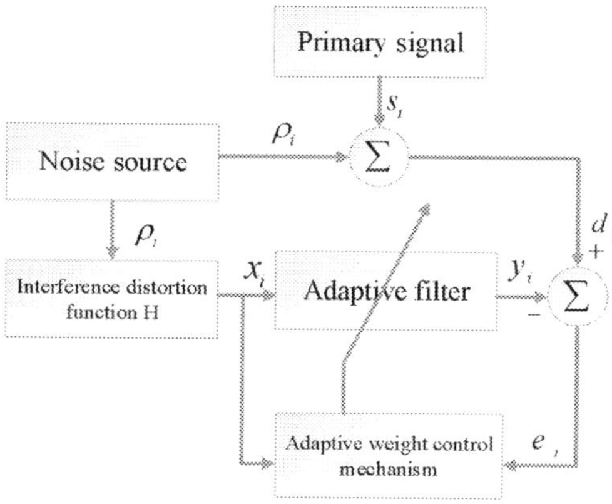


**FIGURE 5.7**

Basic structure of a noise cancellation system.

**Table 5.3 Performance comparison of KLMS, KRLS and GKRMC in noise cancellation**

| Algorithm | KLMS | KRLS | GKMC $\alpha = 2$ | GKMC $\alpha = 4$ | GKMC $\alpha = 6$ |
|---|---|---|---|---|---|
| NR (dB) | $5.7356 \pm 0.6441$ | $11.6057 \pm 1.3710$ | $27.4908 \pm 1.2422$ | $30.5206 \pm 1.4516$ | $33.5343 \pm 1.7440$ |

filter output is an estimate of the noise $\rho_i$. Clearly, the signal-to-noise ratio (SNR) can be improved by subtracting the estimated noise from $d_i$. In the simulation, the noise source is assumed to be uniformly distributed over $[-0.5, 0.5]$. The interference distortion function is

$$x_i = \rho_i - 0.2x_{i-1} - x_{i-1}\rho_{i-1} + 0.1\rho_{i-1} + 0.4x_{i-2}, \tag{5.52}$$

which is nonlinear with an infinite impulsive response. It is impossible to recover $\rho_i$ from a finite time delay embedding of $x_i$. Actually, one can rewrite the distortion function as

$$\rho_i = x_i + 0.2x_{i-1} - 0.4x_{i-2} + (x_{i-1} - 0.1)\rho_{i-1}. \tag{5.53}$$

Obviously, the present value of the noise source $\rho_i$ not only depends on the reference noise measurements $\left[x_i, x_{i-1}, x_{i-2}\right]$, but also on the past value $\rho_{i-1}$. However, the filter output $y_{i-1}$ can be used as a surrogate signal. In this case, the input vector of the adaptive filter is $\left[x_i, x_{i-1}, x_{i-2}, y_{i-1}\right]$. We assume the primary signal $s_i = 0$ during the training phase. The system tries to reconstruct the noise source from the reference measurements. We use three nonlinear filters trained with KLMS, KRLS and GKRMC, respectively. In the simulation, the step size and the Gaussian kernel bandwidth for KLMS are 2.0 and 0.15. The regularization parameter for KRLS and GKRMC is 0.1. The Gaussian kernel bandwidth for KRLS and GKRMC are 0.17 and 0.1, respectively. The parameter $\beta$ for GKRMC is set to 0.5. The ensemble learning curves over 100 Monte Carlo runs for different algorithms are illustrated in Fig. 5.8. In addition, the noise reduction factor (NR) defined by $10\log_{10}\left\{\mathbf{E}\left[\rho_i^2\right]/\mathbf{E}\left[\rho_i - y_i\right]^2\right\}$ is given in Table 5.3. In this example, the GKRMC achieves the best performance when $\alpha = 6$.

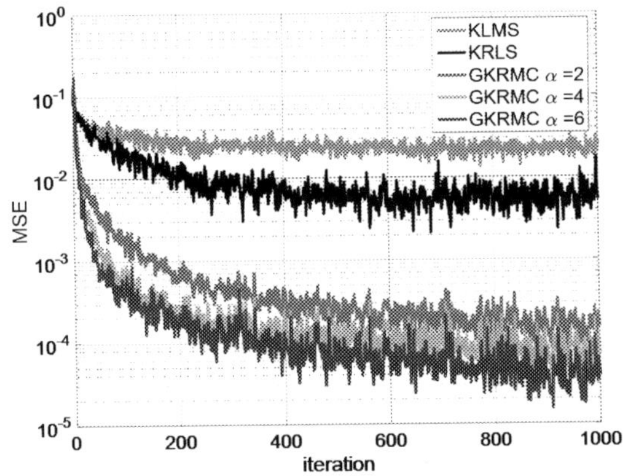


**FIGURE 5.8**

Learning curves of noise cancellation.

## 5.6 CONCLUSION

In this chapter, we characterized a unified and efficient approach to deal with online nonlinear modeling with non-Gaussian noises, which combines the powerful modeling capability of the KAFs and the robustness of the MCC with respect to heavy-tailed non-Gaussian noises. We focused mainly on the KLMS- and KRLS-type algorithms but the results can easily be extended to other KAFs. Particularly the GKMC and GKRMC algorithms were presented. Simulation results on frequency doubling, nonlinear system identification and noise cancellation have been presented to illustrate the benefits of the new approach.

## REFERENCES


[1] R.A. Gonzalo, Nonlinear Signal Processing: A Statistical Approach, J. Wiley and Sons Editors, 2005.

[2] W.J. Fitzgerald, R.L. Smith, A.T. Walden, P.C. Young, Nonlinear and Nonstationary Signal Processing, Cambridge University Press, 2000.

[3] S. Haykin, L. Li, Nonlinear adaptive prediction of nonstationary signals, IEEE Transactions on Signal Processing 43 (2) (1995) 526–535.

[4] M.S. Arulampalam, S. Maskell, N. Gordon, T. Clapp, A tutorial on particle filters for online nonlinear/non-Gaussian Bayesian tracking, IEEE Transactions on Signal Processing 50 (2) (2002) 174–188.

[5] S.C. Schwartz, J.B. Thomas, E.J. Wegman, Topics in Non-Gaussian Signal Processing, Springer-Verlag, 1989.

[6] D. Robert Nowak, Nonlinear system identification, Circuits, Systems, and Signal Processing 21 (1) (2002) 109–122.

[7] Oliver Nelles, Nonlinear System Identification: From Classical Approaches to Neural Networks and Fuzzy Models, Springer Science and Business Media, 2013.

[8] Er-WeiBai FouadGiri, Block-Oriented Nonlinear System Identification, Springer, London, 2010.

[9] G. Huang, Y. Qin, Siew Chee-Kheong, Extreme learning machine: theory and applications, Neurocomputing 70 (1) (2006) 489–501.

[10] W. Liu, J.C. Principe, S. Haykin, Kernel adaptive filtering: a comprehensive introduction, in: Neural Networks (IJCNN), the 2013 International Joint Conference on, IEEE, 2011, pp. 1–6.

[11] Ali H. Sayed, Fundamentals of adaptive filtering, Control Systems IEEE 25 (4) (2003) 77–79.

[12] S. Haykin, B. Widrow, Least-Mean-Square Adaptive Filters, vol. 31, J. Wiley and Sons, 2003.
[13] Paulo S.R. Diniz, Adaptive Filtering: Algorithms and Practical Implementation, Springer Science and Business Media, 2012.
[14] J.C. Principe, Information Theoretic Learning: Renyi's Entropy and Kernel Perspectives, Springer Publishing Company, Incorporated, 2010.
[15] J.C. Principe, D. Xu, J. Fisher, Information theoretic learning, in: Unsupervised Adaptive Filtering, vol. 1, 2000, pp. 265–319.
[16] E. Kenneth Hild, D. Erdogmus, K. Torkkola, J.C. Principe, Feature extraction using information-theoretic learning, IEEE Transactions on Pattern Analysis and Machine Intelligence 28 (9) (2006) 1385–1392.
[17] B. Chen, Y. Zhu, J. Hu, J.C. Principe, System Parameter Identification: Information Criteria and Algorithms, Elsevier, 2013.
[18] W. Liu, P.P. Pokharel, J.C. Principe, Correntropy: properties and applications in non-Gaussian signal processing, IEEE Transactions on Signal Processing 55 (11) (2007) 5286–5298.
[19] R. He, B. Hu, W. Zheng, X. Kong, Robust principal component analysis based on maximum correntropy criterion, IEEE Transactions on Image Processing 20 (6) (2011) 1485–1494.
[20] R. He, W. Zheng, B. Hu, Maximum correntropy criterion for robust face recognition, IEEE Transactions on Pattern Analysis and Machine Intelligence 33 (8) (2011) 1561–1576.
[21] S. Zhao, B. Chen, J.C. Principe, Kernel adaptive filtering with maximum correntropy criterion, in: Neural Networks (IJCNN), the 2011 International Joint Conference on, IEEE, 2011.
[22] B. Chen, L. Xing, J. Liang, N. Zheng, J.C. Principe, Steady-state mean-square error analysis for adaptive filtering under the maximum correntropy criterion, IEEE Signal Processing Letters 21 (7) (2014) 880–884.
[23] B. Chen, L. Xing, H. Zhao, N. Zheng, J.C. Principe, Generalized correntropy for robust adaptive filtering, IEEE Transactions on Signal Processing 64 (13) (2016) 3376–3387.
[24] W. Ma, H. Qu, G. Gui, L. Xu, J. Zhao, B. Chen, Maximum correntropy criterion based sparse adaptive filtering algorithms for robust channel estimation under non-Gaussian environments, Journal of the Franklin Institute 352 (7) (2015) 2708–2727.
[25] B. Chen, X. Liu, H. Zhao, J.C. Principe, Maximum correntropy Kalman filter, Automatica 76 (2017) 70–77.
[26] X. Liu, B. Chen, B. Xu, Z. Wu, P. Honeine, Maximum correntropy unscented filter, International Journal of Systems Science 48 (8) (2017) 1607–1615.
[27] Z. Wu, S. Peng, B. Chen, H. Zhao, Robust Hammerstein adaptive filtering under maximum correntropy criterion, Entropy 17 (10) (2015) 7149–7166.
[28] W. Liu, P.P. Pokharel, J.C. Principe, The kernel least-mean-square algorithm, IEEE Transactions on Signal Processing 56 (2) (2008) 543–554.
[29] Y. Engel, S. Mannor, R. Meir, The kernel recursive least-squares algorithm, IEEE Transactions on Signal Processing 52 (8) (2004) 2275–2285.
[30] C. Richard, J.C.M. Bermudez, P. Honeine, Online prediction of time series data with kernels, IEEE Transactions on Signal Processing 57 (3) (Mar. 2009) 1058–1067.
[31] B. Chen, S. Zhao, P. Zhu, J.C. Principe, Quantized kernel least mean square algorithm, IEEE Transactions on Neural Networks and Learning Systems 23 (1) (Jan. 2012) 22–32.
[32] B. Chen, S. Zhao, P. Zhu, J.C. Principe, Quantized kernel recursive least squares algorithm, IEEE Transactions on Neural Networks and Learning Systems 24 (9) (2013) 1484–1491.
[33] S. Zhao, B. Chen, P. Zhu, J.C. Principe, Fixed budget quantized kernel least-mean-square algorithm, Signal Processing 93 (9) (2013) 2759–2770.
[34] M. Yukawa, Multikernel adaptive filtering, IEEE Transactions on Signal Processing 60 (9) (2012) 4672–4682.
[35] W. Liu, I. Park, Y. Wang, J.C. Principe, Extended kernel recursive least squares algorithm, IEEE Transactions on Signal Processing 57 (10) (2009) 3801–3814.
[36] W. Liu, J.C. Príncipe, Kernel affine projection algorithms, EURASIP Journal on Advances in Signal Processing 2008 (1) (2008) 1–12.
[37] V.V. Steven, M. Lazaro-Gredilla, I. Santamaría, Kernel recursive least-squares tracker for time-varying regression, IEEE Transactions on Neural Networks and Learning Systems 23 (8) (2002) 1313–1326.
[38] Z. Wu, J. Shi, X. Zhang, W. Ma, B. Chen, Kernel recursive maximum correntropy, Signal Processing 117 (2015) 11–16.
[39] P. Zhu, B. Chen, J.C. Principe, Learning nonlinear generative models of time series with a Kalman filter in RKHS, IEEE Transactions on Signal Processing 62 (2014) 141–155.
[40] W. Liu, I. Park, J.C. Principe, An information theoretic approach of designing sparse kernel adaptive filters, IEEE Transactions on Neural Networks 20 (12) (2009) 1950–1961.

[41] B. Chen, Zejian Yuan, Nanning Zheng, Jose C. Principe, Kernel minimum error entropy algorithm, Neurocomputing 121 (2013) 160–169.

[42] W. Ma, J. Duan, W. Man, H. Zhao, B. Chen, Robust kernel adaptive filters based on mean p-power error for noisy chaotic time series prediction, Engineering Applications of Artificial Intelligence 58 (2017) 101–110.

[43] J. Liu, H. Qu, B. Chen, W. Ma, Kernel robust mixed-norm adaptive filtering, in: Neural Networks (IJCNN), 2014 International Joint Conference on, IEEE, 2014, pp. 3021–3024.

[44] H. Qu, W. Ma, J. Zhao, B. Chen, Kernel least mean kurtosis based online chaotic time series prediction, Chinese Physics Letters 30 (11) (2013) 110505.

[45] B. Chen, J.C. Principe, Maximum correntropy estimation is a smoothed MAP estimation, IEEE Signal Processing Letters 19 (8) (2012) 491–494.

[46] M.N. Syed, P.M. Pardalos, J.C. Principe, On the optimization properties of the correntropic loss function in data analysis, Optimization Letters 3 (8) (2014) 823–839.

[47] N. Cristianini, J. Shawe-Taylor, An Introduction to Support Vector Machines and Other Kernel-Based Learning Methods, Cambridge University Press, 2000.

[48] K.H. Jeong, W. Liu, S. Han, E. Hasanbelliu, J.C. Principe, The correntropy MACE filter, Pattern Recognition 42 (5) (2009) 871–885.

[49] R. Li, W. Liu, J.C. Principe, A unifying criterion for instantaneous blind source separation based on correntropy, Signal Processing 87 (8) (2007) 1872–1881.

[50] J. Xu, J.C. Principe, A pitch detector based on a generalized correlation function, IEEE Transactions on Audio, Speech, and Language Processing 16 (8) (2008) 1420–1432.

[51] I. Park, J.C. Principe, Correntropy based Granger causality, in: Acoustics, Speech and Signal Processing, 2008, ICASSP 2008, IEEE International Conference on, IEEE, 2008, pp. 3605–3608.

[52] B. Chen, L. Xing, B. Xu, H. Zhao, N. Zheng, J.C. Principe, Kernel risk-sensitive loss: definition, properties and application to robust adaptive filtering, IEEE Transactions on Signal Processing 65 (11) (2017) 2888–2901.

[53] B. Chen, L. Xing, X. Wang, J. Qin, N. Zheng, Robust learning with kernel mean p-power error loss, arXiv preprint, arXiv:1612.07019, 2016.

[54] Y. He, F. Wang, J. Yang, B. Chen, Kernel adaptive filtering under generalized maximum correntropy criterion, in: Neural Networks (IJCNN), 2016 International Joint Conference on, IEEE, 2016, pp. 1738–1745.

[55] C. Zheng, L. Xing, T. Li, H. Yang, J. cao, B. chen, Z. Zhou, L. zhang, Developing a robust colorectal cancer (CRC) risk predictive model with the big genetic and environment related CRC data, in: Bioinformatics and Biomedicine (BIBM), 2016 IEEE International Conference on, IEEE, 2016, pp. 1885–1893.

CHAPTER

# KERNEL SUBSPACE LEARNING FOR PATTERN CLASSIFICATION

# 6

**Yinan Yu*, Konstantinos Diamantaras†, Tomas McKelvey*, S.Y. Kung‡**

**Chalmers University of Technology, Electrical Engineering, Gothenburg, Sweden †TEI of Thessaloniki, Thessaloniki, Greece ‡Princeton University, Princeton, NJ, United States*

## CONTENTS



Adaptive Learning Methods for Nonlinear System Modeling. DOI: 10.1016/B978-0-12-812976-0.00008-7

## CHAPTER POINTS

- How to use kernel methods in practice.
- How to control the model complexity.
- Kernel approximation as a subspace learning technique for feature construction.
- Learning criteria for adaptive kernel approximation.
- Infrastructure for kernel methods.

## 6.1 INTRODUCTION

Kernel methods are nonlinear transformation techniques that map a given input set into an implicit high-dimensional feature space by utilizing a positive-definite function called the kernel function. In practice, kernel methods are nonparametric learning techniques where the learning process is restricted to the subspace spanned by the training feature vectors. It implies that the complexity of the model grows with the training size. More specifically, given $N$ training data points, the computational complexity and storage depend on an $N \times N$ Gram matrix called the "kernel matrix" $\mathbf{K}$. The elements in the matrix are pairwise operations on all training data.

To reduce this complexity, kernel approximation techniques are commonly applied. There are mainly two types of approximation: the random Fourier features [1] and the Nyström methods [2–4]. Briefly speaking, random Fourier features construct a vector by randomly sampling in the frequency domain. They are computationally efficient, but do not always provide satisfying performance due to the fact that they are not data-dependent in general. Another category is the Nyström methods, where the kernel approximation is posed as a subspace learning problem. That is, one tries to approximate $\mathbf{K}$ using a low-rank matrix $\hat{\mathbf{K}}$, with $rank(\hat{\mathbf{K}}) \ll N$. This is usually implemented by sampling a subset of the training data. The actual computational complexity of some operations, such as matrix inversion and multiplications, can be reduced from being a function of $N$ to a function of $rank(\hat{\mathbf{K}})$. In its basic version [2], Nyström approximates the subspace by random sampling. More sophisticated strategies for sampling are proposed to achieve different objectives, such as control over the model complexity [5], low approximation error [6,4], high class separability [7], efficient memory usage [8]. These algorithms can be naturally computed in an adaptive fashion for streaming data acquisition.

There are typically two ways of representing the approximation outcome: (i) the kernel matrix, which is used in most standard kernel models; or (ii) explicit feature vectors, where one applies a subspace transformation of the feature space induced by the kernel function, which is sometimes more straightforward and explicit compared with the kernel matrix formulation. This is illustrated in Fig. 6.1. Given a training data set and a kernel function, the transformation $\phi : \mathcal{X} \to \mathbb{R}^m$ can be constructed in an iterative fashion. For a data point $\mathbf{x} \in \mathcal{X}$, $\phi(\mathbf{x})$ can be used as the input to the subsequent learning modules, such as regression models, deep neural networks and support vector machines.

This chapter serves as a practical guide to the utilization of subspace learning techniques applied to kernel methods.

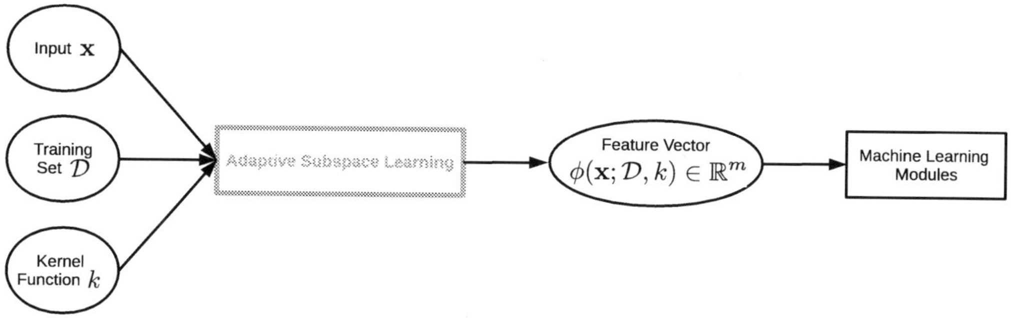


**FIGURE 6.1**

The transformation from input data $\mathbf{x}$ to the feature vector as inputs to the subsequent machine learning module.

## 6.2 KERNEL METHODS

Kernel methods [9,10] are popular feature extraction techniques that:

- have the ability to handle nonnumerical/nonvectorial input data;
- are able to deal with high-dimensional data;
- can be formulated as convex optimization problems;
- have well-established theoretical properties.

Essentially, kernel methods map input data onto a high (possibly infinite)-dimensional space for a better data representation. In this section, we introduce kernel methods as a nonparametric learning technique, where the model complexity grows with respect to an increasing training size, which is essentially the dimension of the feature space. We then show that the extracted features can be represented in two different ways. Although being essentially equivalent, one may choose one or the other depending on the application. Issues related to feature dimension and model complexity are then discussed, leading to the necessity of kernel approximation.

### 6.2.1 NOTATIONS

Given a nonempty input set $\mathcal{X}$ that could be a string of text, a collection of websites, medical records, etc., let $\mathcal{H}$ be the Reproducing Kernel Hilbert Space (RKHS) [9] associated with a kernel function $k : \mathcal{X} \times \mathcal{X} \to \mathbb{R}$ and $k(\mathbf{x}, \mathbf{z}) = \langle \mathbf{x}, \mathbf{z} \rangle_{\mathcal{H}}$ for $\mathbf{x}, \mathbf{z} \in \mathcal{X}$, where $\langle\rangle_{\mathcal{H}}$ denotes the inner-product in $\mathcal{H}$. Denote the mapping $\varphi : \mathcal{X} \to \mathcal{H}$ and the feature vectors $\boldsymbol{\varphi} := \varphi(\mathbf{x})$, $\forall \mathbf{x} \in \mathcal{X}$, where $d$ is the dimension of $\mathcal{X}$. In a classification setup, let $y_i$ denote the label information corresponding to $\mathbf{x}_i$, where $y_i \in 1, \cdots, C$ with $C$ being the number of classes. Given a training set $\mathcal{D} = \{(\mathbf{x}_i, y_i)\}_{i=1}^{N}$, $N \in \mathbb{N}$ and a subset $\mathcal{M} \subseteq \mathcal{D}$, the set of indices is denoted $\mathcal{I}_{\mathcal{M}} = \{i : (\mathbf{x}_i, y_i) \in \mathcal{M}\}$. Moreover, let $\mathbf{X}_{\mathcal{M}} = [\mathbf{x}_i]_{i \in \mathcal{I}_{\mathcal{M}}}$ be a matrix with columns $\mathbf{x}_i$'s. Given a kernel function $k$, for $\mathcal{M}, \mathcal{N} \subseteq \mathcal{D}$, denote

$\mathbf{K}_{\mathcal{MN}} = k(\mathbf{X}_{\mathcal{M}}, \mathbf{X}_{\mathcal{N}}) = \left[k\left(\mathbf{x}_i, \mathbf{x}_j\right)\right]_{i\in\mathcal{I}_{\mathcal{M}}, j\in\mathcal{I}_{\mathcal{N}}}$ and $\mathbf{K}_{\mathcal{M}} = \left[k\left(\mathbf{x}_i, \mathbf{x}_j\right)\right]_{i\in\mathcal{I}_{\mathcal{M}}, j\in\mathcal{I}_{\mathcal{M}}}$. In particular, we write $\mathbf{K} = \left[k\left(\mathbf{x}_i, \mathbf{x}_j\right)\right]_{i\in\mathcal{I}_{\mathcal{D}}, j\in\mathcal{I}_{\mathcal{D}}}$.

### 6.2.2 NONPARAMETRIC LEARNING MODEL

There are many machine learning techniques that are closely related to kernel methods, such as Gaussian Processes (GPs) [11] and Support Vector Machines (SVMs) [12]. The focus of kernel methods is mainly on the feature representation instead of the learning objective itself. However, the feature representation is strongly tied to the learning objective via the Representer Theorem and the empirical risk given below [9].

#### Empirical Risk

The learning objective used in most kernel techniques is the regularized empirical risk with the following structure:

$$\mathcal{R}_e = \mathcal{L}(f \mid \mathcal{D}) + \lambda \,||f||\,, \tag{6.1}$$

where $f$ is a point in the Hilbert space, $\mathcal{L}$ is any loss function and $\lambda\,||f||$ is the regularization term.

#### Representer Theorem

One of the most important results in kernel methods is the **Representer Theorem** [9], which states that, given an empirical risk (*cf.* Eq. (6.1)), the optimal solution $f^*$ lies in the **subspace spanned by** $\{\varphi(\mathbf{x}_1), \cdots, \varphi(\mathbf{x}_N)\}$, denoted by

$$\mathcal{F}_{\mathcal{D}} \triangleq \operatorname{span}\left(\{\varphi(\mathbf{x}_1), \cdots, \varphi(\mathbf{x}_N)\}\right). \tag{6.2}$$

This indicates the nonparametric nature of kernel techniques. A nonparametric model does not make strong assumptions on the data structure, but the model is instead chosen to fit the training data, where regularizations on the model complexity are usually applied to achieve a reasonable generalization ability on unseen datasets. Such an approach is commonly adopted when there are a lot of training data without sufficient prior knowledge available.

### 6.2.3 TRAINING WITH KERNEL METHODS

#### *6.2.3.1 Feature Representations*

The main task of kernel methods is to find a good feature representation that is adapted to the training data, which can be used as an input for the subsequent training model that, for example, tries to find a $f^*$ that minimizes Eq. (6.1).

From Eq. (6.2), we know that the solution $f^*$ can be found by projecting data onto $\mathcal{F}_{\mathcal{D}}$. In practice, for a given training set $\{(\mathbf{x}_1, y_1), \cdots, (\mathbf{x}_N, y_N)\}$, a kernel function $k : \mathcal{X} \times \mathcal{X} \to \mathbb{R}$, the projection $\mathcal{T}_{\mathcal{D}} : \mathcal{H} \to \mathcal{F}_{\mathcal{D}}$ and an empirical risk $\mathcal{R}_e$, there are mainly two ways of representing the features: (i) by the kernel matrix or (ii) by constructing an explicit feature vector.

**(i)** Representation by inner-product and the kernel matrix $\mathbf{K}$:
One way to represent data is by computing the kernel matrix projected onto $\mathcal{F}_{\mathcal{D}}$, which is computed by

$$\mathbf{K}_{\mathcal{D}} = \left[\langle \mathcal{T}_{\mathcal{D}}\varphi(\mathbf{x}_i), \mathcal{T}_{\mathcal{D}}\varphi(\mathbf{x}_j)\rangle\right]_{1\le i,j\le N} = \begin{bmatrix} k(\mathbf{x}_1,\mathbf{x}_1) & \cdots & k(\mathbf{x}_1,\mathbf{x}_N) \\ \vdots & \ddots & \vdots \\ k(\mathbf{x}_N,\mathbf{x}_1) & \cdots & k(\mathbf{x}_N,\mathbf{x}_N) \end{bmatrix} = \mathbf{K}. \tag{6.3}$$

Eq. (6.3) shows that, by projecting feature vectors onto the subspace spanned by the training set, the kernel matrix remains unchanged.
Note that the training data can be entirely represented by $\mathbf{K}$.

**(ii)** Representation by an explicit feature vector $\phi(\mathbf{x})$:
Alternatively, instead of using an $N \times N$ Gram matrix $\mathbf{K}$, one can construct an explicit feature map $\phi : \mathcal{X} \to \mathbb{R}^N$ to represent data. More specifically, for any data vector $\mathbf{x} \in \mathcal{X}$, the explicit feature vector is computed as

$$\phi(\mathbf{x}) = \mathbf{A}^{\mathrm{T}} \begin{bmatrix} k(\mathbf{x}_1,\mathbf{x}) \\ \vdots \\ k(\mathbf{x}_N,\mathbf{x}), \end{bmatrix} \tag{6.4}$$

where $\mathbf{A} \in \mathbb{R}^{N\times N}$ is a transformation matrix such that $\mathbf{A}^T\mathbf{K}\mathbf{A} = \mathbf{I}_{N\times N}$.

#### 6.2.3.2 Kernelization of Linear Models

Given a linear model, kernelization refers to introducing nonlinearity to the linear model by using the aforementioned feature representations. This transformation can be implemented using Eq. (6.3) or Eq. (6.4). These two alternative representations are essentially equivalent, in the sense that, for any $\mathbf{x}, \mathbf{z} \in \mathcal{X}$,

$$\phi(\mathbf{x})^T\phi(\mathbf{z}) = < \mathcal{T}_{\mathcal{D}}\varphi(\mathbf{x}), \mathcal{T}_{\mathcal{D}}\varphi(\mathbf{z}) > . \tag{6.5}$$

It is more common to use Eq. (6.3), where the learning model is formulated using the kernel matrix, such as SVMs, Kernel Ridge Regression (KRR), Kernel Principal Component Analysis (KPCA) and GPs. This alternative requires the follow-up learning algorithm to be formulated using only inner-products to represent input data. However, compared with Eq. (6.3), the explicit feature vector (*cf.* Eq. (6.4)) provides a more flexible representation. One can simply apply the nonlinear transformation in Eq. (6.4) and use $\phi(\mathbf{x}) \in \mathbb{R}^N$ as the new feature vector to input the next learning module. The advantages are:

**(i)** Instead of requiring a reformulation using inner-products, any linear model can be applied directly in $\mathbb{R}^N$, and hence all the theoretical properties of well-studied linear models can be adopted.
**(ii)** One can explicitly control the complexity of feature space using subspace transformations.

### 6.2.4 MODEL COMPLEXITY

One of the main challenges of kernel methods, or nonparametric modeling techniques in general, is the control of the model complexity, where a low complexity leads to an oversimplified description of the world, which results in a low variance but a high bias and *vice versa*. Particularly, for nonparametric learning techniques, where the model complexity is a function of the training size, this might cause issues discussed in this section.

#### 6.2.4.1 Computational Complexity and Memory Usage

One of the major challenges of kernel methods is the computational complexity and memory usage. As shown in Eq. (6.3), to project the features onto the subspace $\mathcal{F}_\mathcal{D}$ (*cf.* Eq. (6.2)), one has to store and invert an $N \times N$ kernel matrix which has a computational complexity of $\mathcal{O}(N^3)$ and $\mathcal{O}(N^2)$ memory. It is usually not possible to run on a single node machine with limited RAM when $N$ is very large.

#### 6.2.4.2 Robustness and Overfitting

Another issue related to the model complexity is overfitting. Since the feature space is essentially the subspace spanned by $\{\varphi(\mathbf{x}_1), \cdots, \varphi(\mathbf{x}_N)\}$, the "quality" of the training dataset is crucial to construct a robust feature space. That is, noisy training data might result in a relatively arbitrary feature representation. Besides trying to increase the Signal-to-Noise Ratio (SNR) of the measurement, which has limitations on its own, one commonly adopted solution is to find a more robust and invariant subspace of the feature space.

### 6.2.5 KERNEL APPROXIMATION AND RELATED WORK

With the rapidly growing data sizes in modern machine learning applications, model reduction techniques are developed to reduce the computational complexity, the storage requirement and the model complexity. To this end, in kernel methods, approximation techniques are introduced. There are mainly two types of kernel approximation techniques: the random Fourier features [1] and the Nyström methods. A comparison between the original versions of these two techniques can be found in [13,14].

Briefly speaking, random Fourier features are explicitly constructed by a random sampling in the Fourier domain. Although computations of such techniques can be quite efficient [15], the main limitation is that the basis functions of the feature construction are in general not data-dependent. Moreover, in its original form, random Fourier features require numerical input sets.

Another popular technique is the Nyström method, where a subspace approximation for the kernel matrix is learned from data. This technique is motivated in Section 6.3.1. Nyström is generally more computationally costly compared with random Fourier features, but due to its data-dependent nature, it typically provides a better performance in data driven problems. Moreover, with Nyström techniques, one can iteratively update the approximated subspace by including/rejecting new data points with respect to some criteria, which induces a nice adaptive framework.

#### 6.2.5.1 Random Fourier Features

Random Fourier feature [1] techniques construct an explicit feature map such that the inner-product of the explicit feature vectors can be used to approximate the kernel function. The kernel functions are required to be shift-invariant, *i.e.*, $k(\mathbf{x}, \mathbf{z}) = k(\mathbf{x} - \mathbf{z})$ for any $\mathbf{x}, \mathbf{z} \in \mathcal{X}$. The construction is carried out by drawing independent and identically distributed samples in the frequency domain of the kernel function. The main theorem to ensure the existence of random Fourier features is shown here.

**Theorem 1** ([16]). *A continuous shift-invariant kernel* $k(\mathbf{x}, \mathbf{z})$ *on* $\mathbb{R}^d$ *is positive definite* **if and only if** $k(\boldsymbol{\delta})$ *is the Fourier transform of a nonnegative measure,* i.e.,

$$k(\mathbf{x}, \mathbf{z}) = \int_{\mathcal{X}} p(\boldsymbol{\omega}) e^{j\boldsymbol{\omega}^T(\mathbf{x}-\mathbf{z})} d\boldsymbol{\omega} = \mathbb{E}_{\boldsymbol{\omega}}(\zeta_{\boldsymbol{\omega}}(\mathbf{x})\zeta_{\boldsymbol{\omega}}(\mathbf{z})^*), \tag{6.6}$$

*where* $\zeta_{\boldsymbol{\omega}}(\mathbf{x}) \triangleq e^{j\boldsymbol{\omega}^T\mathbf{x}} = j\sin(\boldsymbol{\omega}^T\mathbf{x}) + \cos(\boldsymbol{\omega}^T\mathbf{x})$ *and* $^*$ *denotes the complex conjugate and* $j$ *is the imaginary unit.*

Theorem 1 shows that $\zeta_{\boldsymbol{\omega}}(\mathbf{x})\zeta_{\boldsymbol{\omega}}(\mathbf{z})^*$ is an unbiased estimator of $k(\mathbf{x}, \mathbf{z})$ for shift-invariant kernels. Since $k(\mathbf{x}, \mathbf{z})$ is a real number, we have $\zeta_{\boldsymbol{\omega}}(\mathbf{x}) = \cos(\boldsymbol{\omega}^\mathrm{T}\mathbf{x})$.

The idea is then to construct an explicit feature map $z : \mathcal{X} \to \mathbb{R}^m$, such that

$$\begin{aligned} z_{\boldsymbol{\omega}}(\mathbf{x})^T z_{\boldsymbol{\omega}}(\mathbf{z}) &= \zeta_{\boldsymbol{\omega}}(\mathbf{x})\zeta_{\boldsymbol{\omega}}(\mathbf{z}) && (6.7)\\ &= \cos(\boldsymbol{\omega}^\mathrm{T}\mathbf{x} - \boldsymbol{\omega}^\mathrm{T}\mathbf{z}) && (6.8)\\ &= \sin(\boldsymbol{\omega}^\mathrm{T}\mathbf{x})\sin(\boldsymbol{\omega}^\mathrm{T}\mathbf{z}) + \cos(\boldsymbol{\omega}^\mathrm{T}\mathbf{x})\cos(\boldsymbol{\omega}^\mathrm{T}\mathbf{z}). && (6.9) \end{aligned}$$

Empirically, the feature map can then be constructed using

$$\phi(\mathbf{x}) = \sqrt{\frac{1}{m}} \begin{bmatrix} \cos(\boldsymbol{\omega}_1^\mathrm{T}\mathbf{x}) & \cdots & \cos(\boldsymbol{\omega}_m^\mathrm{T}\mathbf{x}) & \sin(\boldsymbol{\omega}_1^\mathrm{T}\mathbf{x}) & \cdots & \sin(\boldsymbol{\omega}_m^\mathrm{T}\mathbf{x}) \end{bmatrix}^\mathrm{T},$$

where $\boldsymbol{\omega}_1, \cdots, \boldsymbol{\omega}_m$ are independent and identically distributed samples drawn from $p(\boldsymbol{\omega})$.

The algorithm is summarized in Algorithm 1. Details will be addressed in Chapter 7.

#### 6.2.5.2 Nyström Methods: Subspace Learning

Another important family of kernel approximation techniques are the Nyström methods, where the effective rank of the kernel matrix is exploited in a more direct fashion. That is, one assumes an underlying subspace structure in the kernel matrix, *i.e.*, $\mathbf{K} \approx \hat{\mathbf{K}}$ for some $\hat{\mathbf{K}}$ such that $rank(\hat{\mathbf{K}}) = m < N$. The task is then to use the rank-deficient matrix $\hat{\mathbf{K}}$ to approximate the kernel matrix.

In its basic form, Nyström randomly selects a subset of the training data $\mathcal{G} \subset \mathcal{D}$ and constructs the approximation using

$$\hat{\mathbf{K}} = \underbrace{k(\mathbf{X}_\mathcal{D}, \mathbf{X}_\mathcal{G})}_{\in\mathbb{R}^{N\times m}} \underbrace{k(\mathbf{X}_\mathcal{G}, \mathbf{X}_\mathcal{G})^{-1}}_{\in\mathbb{R}^{m\times m}} \underbrace{k(\mathbf{X}_\mathcal{G}, \mathbf{X}_\mathcal{D})}_{\in\mathbb{R}^{m\times N}} = \mathbf{K}_{\mathcal{D}\mathcal{G}}\mathbf{K}_\mathcal{G}^{-1}\mathbf{K}_{\mathcal{G}\mathcal{D}}.$$

The random subset selection is computationally simple and in many cases sufficient to capture the underlying data structure in many applications. Many more advanced sampling strategies have been proposed for different purposes. Most techniques cast the approximation problem as an estimation of the top $m$ singular values of $\mathbf{K}$. Essentially, they attempt to approximate the subspace as well as possible while trying to reduce the computational complexity and/or storage usage. For example, some of these techniques include subspace approximation using the incomplete Cholesky decomposition [17]; identifying representative points by clustering [18]; enabling parallelization and performance boosting by

**Algorithm 1:** Random Fourier Features.

**Input**:
- shift-invariant kernel function $k$, where $k(\mathbf{x}, \mathbf{z}) = k(\mathbf{x} - \mathbf{z})$
- data vector $\mathbf{x}$

**Output**:
- output feature vector $\phi(\mathbf{x}) \in \mathbb{R}^m$

**Procedure**:

Compute the Fourier transform $p$ of the kernel function $k$. We have

$$p(\boldsymbol{\omega}) = \frac{1}{2\pi} \int e^{-j\omega^{\mathrm{T}}\boldsymbol{\Delta}} k(\boldsymbol{\Delta}) d\boldsymbol{\Delta}, \tag{6.10}$$

where $\boldsymbol{\omega} \in \mathbb{R}^d$;

Draw $m$ independent and identically distributed samples $\boldsymbol{\omega}_1, \cdots, \boldsymbol{\omega}_m$ from $p$;

Construct the feature vector as[a]

$$\phi(\mathbf{x}) = \sqrt{\frac{1}{m}} \begin{bmatrix} \cos(\boldsymbol{\omega}_1^{\mathrm{T}}\mathbf{x}) & \cdots & \cos(\boldsymbol{\omega}_m^{\mathrm{T}}\mathbf{x}) & \sin(\boldsymbol{\omega}_1^{\mathrm{T}}\mathbf{x}) & \cdots & \sin(\boldsymbol{\omega}_m^{\mathrm{T}}\mathbf{x}) \end{bmatrix}. \tag{6.11}$$

[a] Variations on the feature map construction can be found in [1].

the ensemble Nyström method. Some other methods explore the intrinsic structure of the kernel matrix to obtain a more compact and efficient memory storage, such as the Memory-Efficient Kernel Approximation (MEKA) [8], Clustered Low-Rank Approximation (CLRA) [19] and the CLAss-Specific Kernel (CLASK) subspace approximation [7]. These techniques more or less share the same base algorithm with different sampling criteria and/or feature representation strategy. In the following sections, we explore the underlying framework of kernel subspace approximation and its adaptive learning algorithms.

## 6.3 KERNEL SUBSPACE APPROXIMATION

### 6.3.1 MOTIVATION

Briefly speaking, kernel subspace approximation applies a subspace transformation to the feature vector space $\mathcal{F}_{\mathcal{D}}$. The transformed data vectors with a lower dimensionality are then used as the new input features to the subsequent module.

That is,

$$\mathcal{T} : \mathcal{F}_{\mathcal{D}} \rightarrow \mathcal{F}, \tag{6.12}$$

where $\mathcal{F}$ has a lower dimension than $\mathcal{F}_{\mathcal{D}}$. By applying this transformation, the complexity is then reduced to be a function of $dim(\mathcal{F})$. The flowchart of this process can be found in Fig. 6.2.

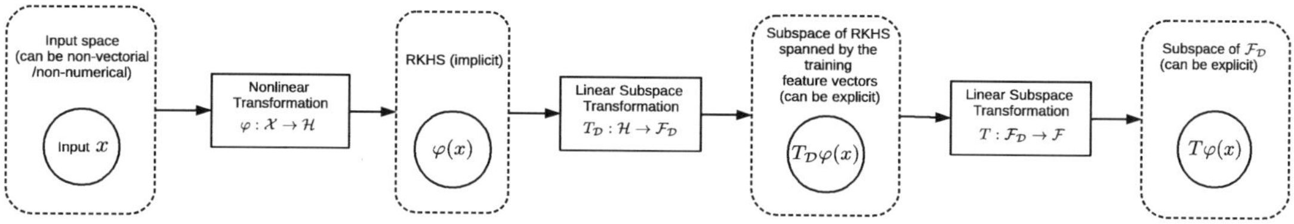


**FIGURE 6.2**

Transformations in kernel methods.

Given a training set $\mathcal{D}$ and a kernel function $k$, the idea is to find a subset $\mathcal{G} \subseteq \mathcal{D}$ with cardinality $|\mathcal{G}| \ll N$ and a transformation matrix $\mathbf{T} \in \mathbb{R}^{N_{\mathcal{G}} \times m}$ with $m \leq |\mathcal{G}|$, such that

$$\mathbf{T}^{\mathrm{T}} k(\mathbf{X}_{\mathcal{G}}, \mathbf{X}_{\mathcal{G}}) \mathbf{T} = \mathbf{I}_{m \times m}, \tag{6.13}$$

which gives a transformation matrix $k(\cdot, \mathbf{X}_{\mathcal{G}})\mathbf{T}$ with orthonormal columns.

## 6.3.2 TRAINING COMPLEXITY WITH APPROXIMATION

In kernel methods, during the training process, to find the exact or approximated optimal solution of unknown parameters in the learning model, the computational complexity is typically dominated by the operation $(\mathbf{K} + \lambda \mathbf{I})^{-1}$, where $\mathbf{I}$ is the $N \times N$ identity matrix; $\lambda \in \mathbb{R}^{+}$ is the ridge parameter to handle the ill-conditioned matrix inversion [20]. In practice, this matrix inversion requires $\mathcal{O}(N^{2.8})$ [21] to $\mathcal{O}(N^3)$ operations depending on the algorithm.[1] Kernel approximations transform the learning model onto a lower dimensional feature space and subsequently reduce the computational complexity. In this section, we discuss the approximation for the two different feature representations in Eq. (6.3) and Eq. (6.4).

### 6.3.2.1 Approximated Kernel Matrix

One computational advantage of kernel methods is that all computations are carried out by inner-products. The kernel matrix $\mathbf{K}$ (*cf.* Eq. (6.3)) that holds the pair-wise inner-product of the training data is then a key component of kernel methods. Essentially, the subspace approximation can be interpreted as weighted inner-products, which transforms the kernel matrix as

$$\mathbf{K} \approx \tilde{\mathbf{K}} = \underbrace{\begin{bmatrix} k(\mathbf{X}_{\mathcal{G}}, \mathbf{x}_1)^{\mathrm{T}} \\ \vdots \\ k(\mathbf{X}_{\mathcal{G}}, \mathbf{x}_N)^{\mathrm{T}} \end{bmatrix}}_{\mathbf{K}_{\mathcal{D}\mathcal{G}} \in \mathbb{R}^{N \times |\mathcal{G}|}} \mathbf{T}\mathbf{T}^{\mathrm{T}} \underbrace{\begin{bmatrix} k(\mathbf{X}_{\mathcal{G}}, \mathbf{x}_1) & \cdots & k(\mathbf{X}_{\mathcal{G}}, \mathbf{x}_N) \end{bmatrix}}_{\mathbf{K}_{\mathcal{G}\mathcal{D}} \in \mathbb{R}^{|\mathcal{G}| \times N}} \in \mathbb{R}^{N \times N}, \tag{6.14}$$

[1] Throughout the paper, we assume that the complexity for matrix inversion is $\mathcal{O}(N^3)$.

where $k(\mathbf{X}_{\mathcal{G}}, \mathbf{x}) = \begin{bmatrix} k(\mathbf{x}_{i_1}, \mathbf{x}) \\ \vdots \\ k(\mathbf{x}_{i_{|\mathcal{G}|}}, \mathbf{x}) \end{bmatrix}$ and indices $\{i_j : i_j \in \mathcal{I}_{\mathcal{G}},\ j = 1 \cdots |\mathcal{G}|\}$ denote the index set of the subset $\mathcal{G}$. The weighting matrix $\mathbf{T}\mathbf{T}^{\mathrm{T}}$ (*cf.* Eq. (6.13)) is rank-deficient.

After kernel approximation, the size of the kernel matrix remains the same and the approximated kernel matrix $\tilde{\mathbf{K}}$ becomes rank-deficient, *i.e.*, rank($\tilde{\mathbf{K}}$)$= m$. However, in combination with the matrix inversion lemma, the kernel matrix inversion can be reduced to

$$(\mathbf{K} + \lambda\mathbf{I})^{-1} \approx \left(\tilde{\mathbf{K}} + \lambda\mathbf{I}\right)^{-1} = \frac{1}{\lambda}\mathbf{I} - \frac{1}{\lambda^2}\mathbf{K}_{\mathcal{D}\mathcal{G}}\mathbf{T}\underbrace{\left(\mathbf{I} + \frac{1}{\lambda}\mathbf{T}^{\mathrm{T}}\mathbf{K}_{\mathcal{G}\mathcal{D}}\mathbf{K}_{\mathcal{D}\mathcal{G}}\mathbf{T}\right)^{-1}}_{\in\mathbb{R}^{m\times m}}\mathbf{T}^{\mathrm{T}}\mathbf{K}_{\mathcal{G}\mathcal{D}}. \tag{6.15}$$

The computational complexity is listed as follows:

- Matrix multiplication $\mathbf{K}_{\mathcal{D}\mathcal{G}}\mathbf{T}$: $\mathcal{O}(N|\mathcal{G}|m)$.
- Matrix multiplication $\left(\mathbf{K}_{\mathcal{D}\mathcal{G}}\mathbf{T}\right)^{\mathrm{T}}\mathbf{K}_{\mathcal{D}\mathcal{G}}\mathbf{T}$: $\mathcal{O}(Nm^2)$.
- Matrix inversion: $\mathcal{O}(m^3)$.

Due to the fact that $N \gg |\mathcal{G}| \geq m$, we have the overall complexity $\mathcal{O}(N|\mathcal{G}|m)$.

#### 6.3.2.2 Approximated Sample Covariance Matrix Using Explicit Feature Vectors

Kernel approximation can be readily applied to create an explicit feature vector by a map $\tilde{\phi} : \mathcal{X} \to \mathbb{R}^m$. Specifically,

$$\tilde{\phi}(\mathbf{x}) = \mathbf{T}^{\mathrm{T}} \begin{bmatrix} k\left(\mathbf{x}_{i_1}, \mathbf{x}\right) \\ \vdots \\ k(\mathbf{x}_{i_{|\mathcal{G}|}}, \mathbf{x}) \end{bmatrix} \in \mathbb{R}^m. \tag{6.16}$$

Eq. (6.16) can then be used as the feature vector for the succeeding learning model. When applying Eq. (6.16) to linear learning techniques, the sample covariance matrix typically needs to be inverted. Let us assume that the feature vector $\tilde{\phi}(\mathbf{x})$ is always centered (*cf.* Appendix 6.A.1), the (scaled) sample covariance matrix is defined as

$$\mathbf{C} = \left[\tilde{\phi}(\mathbf{x}_1) \quad \cdots \quad \tilde{\phi}(\mathbf{x}_N)\right] \begin{bmatrix} \tilde{\phi}(\mathbf{x}_1)^{\mathrm{T}} \\ \vdots \\ \tilde{\phi}(\mathbf{x}_N)^{\mathrm{T}} \end{bmatrix} \tag{6.17}$$

$$= \mathbf{T}^{\mathrm{T}}\left[k(\mathbf{X}_{\mathcal{G}}, \mathbf{x}_1) \quad \cdots \quad k(\mathbf{X}_{\mathcal{G}}, \mathbf{x}_N)\right] \begin{bmatrix} k(\mathbf{X}_{\mathcal{G}}, \mathbf{x}_1)^{\mathrm{T}} \\ \vdots \\ k(\mathbf{X}_{\mathcal{G}}, \mathbf{x}_N)^{\mathrm{T}} \end{bmatrix} \mathbf{T} \tag{6.18}$$

$$\in \mathbb{R}^{m\times m}.$$

The computational complexity of $\mathbf{C}^{-1}$ is composed of:

- Matrix multiplication: $\mathcal{O}(N|\mathcal{G}|m)$.
- Matrix inversion: $\mathcal{O}(m^3)$.

Hence, we have the overall computational complexity $\mathcal{O}(N|\mathcal{G}|m)$.

In the next section, we focus on an adaptive algorithmic framework and its instantiations to construct desired subspaces.

## 6.4 ADAPTIVE KERNEL SUBSPACE APPROXIMATION ALGORITHM

### 6.4.1 ALGORITHMIC FRAMEWORK

With an online data acquisition, the subspace approximation can be implemented in an adaptive fashion. One generic framework is summarized in Algorithm 2.

The matrix inversion lemma [22] is applied to obtain an iterative update of the kernel matrix inversion. The actual sampling strategy is embedded in the "criteria", which are discussed in the upcoming section.

### 6.4.2 ALGORITHM DESIGN

The key design components in Algorithm 2 are (1) the sampling criteria to find $\mathcal{G}$ and (2) the construction of the transformation matrix $\mathbf{T}$. In this section, we discuss the principles of finding $\mathcal{G}$ and $\mathbf{T}$ and how they may affect the performance of the algorithm.

#### *6.4.2.1 Design Criteria*

Generally speaking, we are interested in building a model (*i.e.*, the approximated subspace) with a high generalization ability, which is measured by the performance on unseen datasets. The generalization ability is closely related to the model complexity. That is, a model with a high complexity tends to result in a high performance on the training data, but it might perform poorly on testing datasets and *vice versa*. It is referred to as the "overfitting" problem.

This can be observed by the bias-variance trade-off, where the performance statistics are measured when the model is trained using different training sets with a fixed training size. More precisely, when the model complexity is high, the search space of the model parameters is large and hence the bias of the model tends to be low. On the other hand, since the model is derived from the training data, the higher complexity the model has, the higher variance as a result we find.

The strategy for achieving a high generalization is twofold: one needs to (1) reduce the model complexity and (2) maintain a low empirical risk on the training data. By combining an intuitive illustration in Fig. 6.3 and the actual model representation in Eq. (6.16), we see that the model complexity (the "size" of the circle) is related to (1) the size of the subset $\mathcal{G}$ and (2) the number of columns in $\mathbf{T}$; and the empirical risk (the "location" of the circle) depends on (1) the sampling criteria for $\mathcal{G}$ and (2) the construction strategy for the transformation matrix $\mathbf{T}$. Given the analysis, in this section, we demonstrate some design criteria for Algorithm 2.

**Algorithm 2:** Generic Kernel Subspace Approximation Framework.

**Input**:

- kernel function $k$
- new training batch $\mathcal{D}$
- hyperparameters $\eta$ and $m$
- initial values:
  - initial set $\mathcal{G}$ (if none, need to initialize)
  - initial kernel matrix $\mathbf{K}_{\mathcal{G}} = k\left(\mathbf{X}_{\mathcal{G}}, \mathbf{X}_{\mathcal{G}}\right)$
  - $\mathbf{invK}_{\mathcal{G}} = \left(\mathbf{K}_{\mathcal{G}}\right)^{-1}$

**Output**:

- approximated subset $\mathcal{G}$
- transformation matrix $\mathbf{T} \in \mathbb{R}^{|\mathcal{G}| \times m}$
- output kernel matrix $\mathbf{invK}_{\mathcal{G}} = \left(\mathbf{K}_{\mathcal{G}}\right)^{-1}$ and/or $\mathbf{K}_{\mathcal{G}}$
- transformation matrix $\mathbf{T}$, such that $\mathbf{T}^{\mathrm{T}}\mathbf{K}_{\mathcal{G}}\mathbf{T} = \mathbf{I}$

**Procedure**:

**for** $j \in \mathcal{I}_{\mathcal{D}}$ **do**

  Preprocessing (centering, normalization, etc.);

  **for** $i \in \mathcal{I}_{\mathcal{G}}$ **do**

$$\mathbf{k}_{\mathcal{G}j}(i, j) = k(\mathbf{x}_i, \mathbf{x}_j) \tag{6.19}$$

$$\mathbf{k}_j = \mathbf{invK}_{\mathcal{G}}\mathbf{k}_{\mathcal{G}j} \tag{6.20}$$

  **if** *Criteria*[a] *(for given* $\eta$*) fulfilled* **then**

    Compute $\Delta_j = \frac{1}{k(\mathbf{x}_j, \mathbf{x}_j) - \mathbf{k}_{\mathcal{G}j}k_j}$;

    Update and store[b]:

$$\mathbf{K}_{\mathcal{G}} \leftarrow \begin{bmatrix} \mathbf{K}_{\mathcal{G}} & k(\mathbf{X}_{\mathcal{G}}, \mathbf{x}_j) \\ k(\mathbf{x}_j, \mathbf{X}_{\mathcal{G}}) & k(\mathbf{x}_j, \mathbf{x}_j) \end{bmatrix}^{c}$$

$$\mathbf{invK}_{\mathcal{G}} \leftarrow \begin{bmatrix} \mathbf{invK}_{\mathcal{G}} + \delta_j \mathbf{k}_j \mathbf{k}_j^{\mathrm{T}} & -\mathbf{k}_j \Delta_j \\ -\Delta_j \mathbf{k}_j^{\mathrm{T}} & \Delta_j \end{bmatrix} \tag{6.21}$$

$$\mathcal{I}_{\mathcal{G}} \leftarrow \mathcal{I}_{\mathcal{G}} \cup j \tag{6.22}$$

  **else**

    Continue;

Construction of the transformation matrix $\mathbf{T}$;

[a] The design of the criteria is discussed in the next section.

[b] The update might vary. The presented update is one example adopted by many algorithms.

[c] This matrix does not need to be stored if not used as an output.

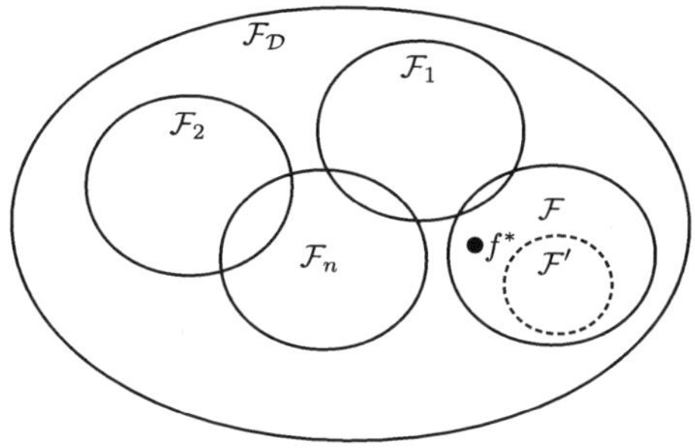


**FIGURE 6.3**

Illustration of the algorithm design criteria. The size of the circles indicates the model complexity. Circles $\mathcal{F}$, $\mathcal{F}_1, \cdots, \mathcal{F}_n$ are subspaces with the same complexity and $\mathcal{F}'$ has a lower complexity compared with the rest. The dot $f^*$ indicates the optimal solution that minimizes the empirical risk given the subspace $\mathcal{F}_\mathcal{D}$ that is spanned by the whole training dataset (*cf.* Eq. (6.1)). The goal of the algorithm design is to find a subspace with an appropriate size (model complexity) with a reasonable location (which hopefully contains $f^*$) before the training process.

**Rejecting samples with a low innovation:** The "innovation" of a data point $\mathbf{x}$ is the projection error from $\varphi(\mathbf{x})$ to the existing subspace. The sample is only accepted if the innovation is large enough, *i.e.*,

$$k(\mathbf{x}, \mathbf{x}) - \mathbf{k}_{\mathcal{G}j}^{\mathrm{T}} \mathbf{invK}_{\mathcal{G}} \mathbf{k}_{\mathcal{G}j} \geq \eta \in (0, 1). \tag{6.23}$$

When Eq. (6.23) is violated, it means that it is not necessary to include $\varphi(\mathbf{x})$. Hence, the larger $\eta$ is, the less data points are included in the subspace construction.

**Dimension reduction by choosing principal components:** Principal Component Analysis (PCA) [23] belongs to the most commonly used linear techniques for dimension reduction purposes. From Eq. (6.16), we see that, given the subset $\mathcal{G}$ and a hyperparameter $m < |\mathcal{G}|$, the transformation matrix $\mathbf{T} \in \mathbb{R}^{|\mathcal{G}| \times m}$ can be constructed by applying PCA on the dataset $\mathcal{U} = \left\{ \left[ k\left(\mathbf{x}_{i_1}, \mathbf{x}\right) \quad \cdots \quad k(\mathbf{x}_{i_{|\mathcal{G}|}}, \mathbf{x}) \right]^{\mathrm{T}} \}, \ \forall \mathbf{x} \in \mathcal{D} \right\}$. That is, let $\tilde{\mathbf{T}}$ be the matrix that contains the sorted principal vectors computed from $\mathcal{U}$ as its columns. The matrix $\mathbf{T}$ is obtained by truncation. We have

$$\mathbf{T} = \tilde{\mathbf{T}}(:, 1 : m). \tag{6.24}$$

Since $m < |\mathcal{G}|$, the model complexity is then further reduced.

**Class Separability:** In classification problems, the reconstruction error is not the only criterion to measure the quality of the subspace. When label information is available, the discriminative ability is preferably taken into consideration. This information can be encoded in various ways. For example, the CLAss-specific Subspace Kernel (CLASK) representation [7] encodes the information into the feature vector by constructing one subspace for each class. The sampling criteria depend on the between-class projection error. Fisher discriminant analysis [24] finds a transformation matrix $\mathbf{T}$ that maximizes the Fisher Quotient, which measures the class separability. Other constrained optimizations can be applied to finding $\mathbf{T}$ that results in a good class separation, which can be cast as metric learning [25–27] problems.

**Ensemble techniques:** One efficient way of reducing the variance is to learn multiple models from subsets of training data and the final estimate is based on an ensemble decision. One such example would be the ensemble Nyström method [28], where

$$\hat{\mathbf{K}} = \sum_{r=1}^{P} \mu_r \hat{\mathbf{K}}_r \tag{6.25}$$

and each $\hat{\mathbf{K}}_r$ is estimated from a subset of $\mathcal{D}$ and $\mu_r$ are the weights for $r = 1 \cdots P$, $P \in \mathbb{N}^+$.

### 6.4.2.2 Empirical Risk

Given a model $\mathcal{F}$ designed by the aforementioned criteria, empirical risks are measured to ensure that the "location" of the circle in Fig. 6.3 is not way off. Typically, with the same model complexity, the model corresponding to a lower empirical risk is preferred. Typically, there are two empirical risks to be considered.

**Reconstruction error:** The approximated subspace $\mathcal{F}$ (*cf.* Eq. (6.12)) has to be representative and is evaluated by the following reconstruction error on the training data. Let $\hat{\boldsymbol{\varphi}}_n$ be the projection of $\varphi(\mathbf{x}_n)$ onto $\mathcal{F}$. The empirical risk of the reconstruction error (on the training set $\mathcal{D}$) is defined as

$$\mathcal{E}_{\mathcal{F}} = \frac{1}{N}\sum_{n=1}^{N} \frac{\left|\left|\boldsymbol{\varphi}_n - \sum_{i=1}^{|\mathcal{G}|} a^*_{n,i}\hat{\boldsymbol{\varphi}}_n\right|\right|_2^2}{||\boldsymbol{\varphi}_n||^2} \tag{6.26}$$

$$= \frac{1}{N}\sum_{n=1}^{N}\left(1 - \frac{\left|\left|\sum_{j=1}^{|\mathcal{G}|} a^*_{n,i} k(\mathbf{x}_n, \mathbf{x}_i)\mathbf{T}\mathbf{T}^{\mathrm{T}} k(\mathbf{x}_n, \mathbf{x}_i)\right|\right|_2^2}{k(\mathbf{x}_n, \mathbf{x}_n)}\right), \tag{6.27}$$

where $a^*_{n,i}$'s are found by

$$a^*_{n,i} = \underset{a_{1,1},\cdots,a_{N,m}}{\arg\min}\ \mathcal{E}_{\mathcal{F}}(a_{1,1}, \cdots, a_{N,m}).$$

We see that, by setting the hyperparameters $\eta$ (*cf.* Eq. (6.23)) to a small value, the reconstruction error $\mathcal{E}_{\mathcal{F}}$ becomes small. On the other hand, a small number of columns in $\mathbf{T}$ results in a large $\mathcal{E}_{\mathcal{F}}$ due to the truncation (*cf.* Eq. (6.24)).

**Classification error:** In classification problems, the ultimate risk function is the 0–1 loss, which is defined as

$$\mathcal{L}_{\mathcal{F}} = \frac{1}{N}\sum_{n=1}^{N} \mathbf{1}\left(f^*_{\mathcal{F}}(\hat{\boldsymbol{\varphi}}_n) \neq y_n\right), \tag{6.28}$$

where $y_n$ are the labels for data $\mathbf{x}_n$, $\mathbf{1}$ is the indicator function and $f_{\mathcal{F}}^*$ is found by minimizing $\mathcal{L}_{\mathcal{F}}$. Note that, given a training set, this empirical risk depends not only on the subspace approximation algorithm, but also on the classification learning model.

## 6.5 INFRASTRUCTURES

The computational complexity of kernel methods mainly depends on the dimension $d$ of the input space, the training size $N$ and the approximated rank $m$. Given a large training size $N$ and a complex intrinsic data structure, to fully explore the capability of the representative power of kernel methods, we need to take advantage of the modern computational infrastructures and frameworks.

### 6.5.1 SPEEDUP: GPU/CUDA

To speed up the kernel algorithms, one can take advantage of the graphics processing units (GPUs) for general computational purposes [29]. This is readily applicable due to the fact that computations in kernel methods mainly involve a large amount of simple operations that can easily be parallelized.

In recent years graphics cards have evolved into GPUs offering computing resources for general purpose computations. Typically, a GPU card contains hundreds or even thousands of simple processors (cores) allowing the concurrent execution of multiple threads. GPU cores, unlike the cores in a multicore CPU, are not full-blown processors. They do not have their own memory but share memory with other cores and they allow the execution of simple tasks, such as mathematical operations, shared memory access and access to the global GPU memory. The *Computer-Unified Device Architecture* (CUDA) by NVIDIA is a library enabling the use of the GPU cores for general purpose parallel computing. Threads are organized into blocks that run on groups of cores called streaming multiprocessors. CUDA is compatible with the data-parallel model of execution. All threads execute the same function, called a "kernel". Each thread, however, operates on different parts of the data depending on its ID. This programming paradigm is also known as the Single Instruction–Multiple Threads (SIMT) model [30]. The basic idea behind CUDA is that the GPU is operating separately and independently from the host. The CPU submits a job to the GPU copying the necessary data onto the global GPU memory, the GPU does the processing and returns the result in the CPU so that the results can be printed, saved on disk, etc. The GPU device has no access to the system I/O and it essentially operates as a slave accelerator. Most popular machine learning packages, such as Tensorflow and Theano, can take advantage of CUDA in order to accelerate computations.

#### Example: Computing Kernel Functions

We see from Algorithm 2 that the computational complexity of the algorithm is dominated by kernel function computations and vector-matrix multiplication. In this example, we show the capacity of the speedup achieved by using CUDA on a GPU compared with the CPU performance. The setup of the experiment is specified as follows.

**Hardware specs:** GPU/CPU specifications

- GPU: GeForce GTX 960M with 640 CUDA cores
- CPU: Intel i7-6700HQ processor 2.6 GHz 4 cores 8 threads

**Data:** two synthetic datasets $\{\mathbf{x}_1 \cdots, \mathbf{x}_{N1}\}$ and $\{\mathbf{z}_1 \cdots, \mathbf{z}_{N2}\}$, where each element in each data vector is generated independently from a normal distribution.

**Computations involved:** allocate memory and compute the entries for a $N1 \times N2$ matrix, where each element is $\mathbf{M}(i, j) = k(\mathbf{x}_i, \mathbf{z}_j)$ for $i = 1 \cdots N1$, $j = 1 \cdots N2$ and $\mathbf{x}_i, \mathbf{z}_j \in \mathbb{R}^d$. To vary the number of operations per thread, we compute and compare the run time for the following kernels:

- Single kernel: polynomial kernel,

$$k(\mathbf{x}_i, \mathbf{z}_j) = (\mathbf{x}_i^{\mathrm{T}} \mathbf{z}_j + 1)^2.$$

- Multiple-kernel [31],

$$\begin{aligned} k(\mathbf{x}_i, \mathbf{z}_j) &= \sum_{l=1}^{3} k_l(\mathbf{x}_i, \mathbf{z}_j) \\ &= (\mathbf{x}_i^{\mathrm{T}} \mathbf{z}_j + 1)^2 + \exp(-\frac{1}{2}||\mathbf{x}_i - \mathbf{z}_j||^2) + \exp(-\frac{1}{2}||\mathbf{x}_i - \mathbf{z}_j||). \end{aligned} \tag{6.29}$$

**Variables:** to compare the performance and gain a better understanding, we vary the following parameters:

- Data-related: $N1$, $N2$, $d$
- Operation related: single kernel computation versus multiple-kernel computation
- CUDA-related: `THREAD_PER_BLOCK`
- CUDA block grid: `dim3 blocks(`$N1$, $\frac{N2}{\texttt{THREAD\_PER\_BLOCK}}$`,1)`

The design of the CUDA algorithm can be further optimized. For simplicity, we choose to compute one element $k(\mathbf{x}_i, \mathbf{z}_j)$ per thread.

The run time comparison can be found in Fig. 6.4, Fig. 6.5 and Fig. 6.6, where the run time is measured in milliseconds using the C function `clock()` with an output type `clock_t`. With the aforementioned setup, we observe a major speedup using GPU under most circumstances with various GPU parameters. However, note that, when the number of operations per thread is too low ($d = 10$), CPU can outperform GPU, since the overhead of CUDA, such as data transfer, will dominate the computational complexity.

### 6.5.2 SCALING: SPARK

Apache Spark [32–34] is a data-parallel computing framework offering scalability, fault tolerance and increased performance due to the use of in-memory operations. Spark operates on top of various cluster managers and is agnostic of the architecture of the underlying cluster. Compared to Hadoop MapReduce [35], Spark can be 10 to 100 times faster. A central data structure concept in the Spark programming model is the Resilient Distributed Dataset (RDD) [36], which is a fault-tolerant collection of data partitioned across nodes. An RDD can be rebuilt automatically in case of data loss because of a node failure. RDDs can be created by various actions, for instance reading a text file from a distributed file system or by transforming an existing RDD. A typical transformation is the map operation which applies the same function on each element of the RDD via closure. Reductions of RDDs into single objects are also possible. For example, an RDD can be reduced into a number by summing up its elements, by counting its elements, etc.

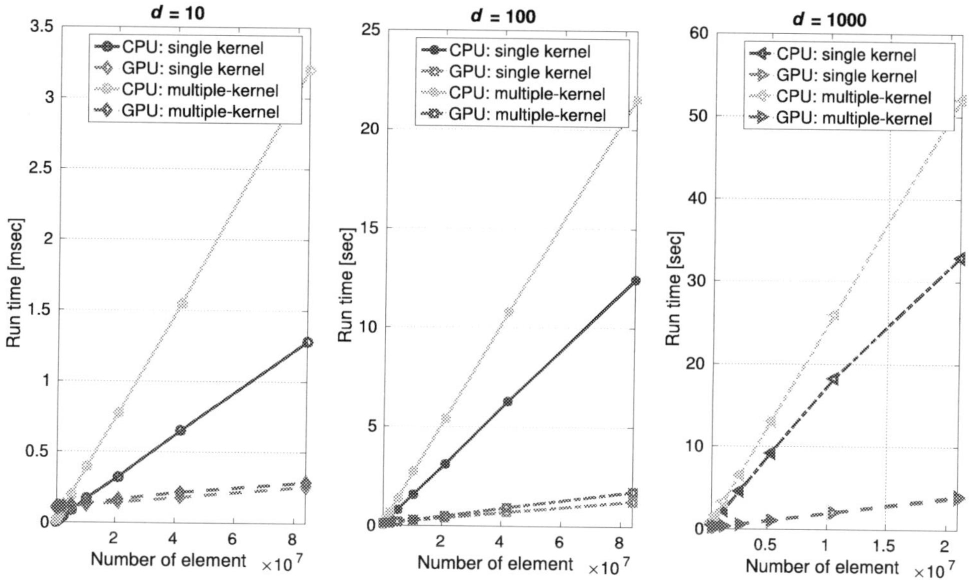


**FIGURE 6.4**

CUDA performance measurement I. Run time comparison for different $d$. The number of elements is $N1 \times N2$.

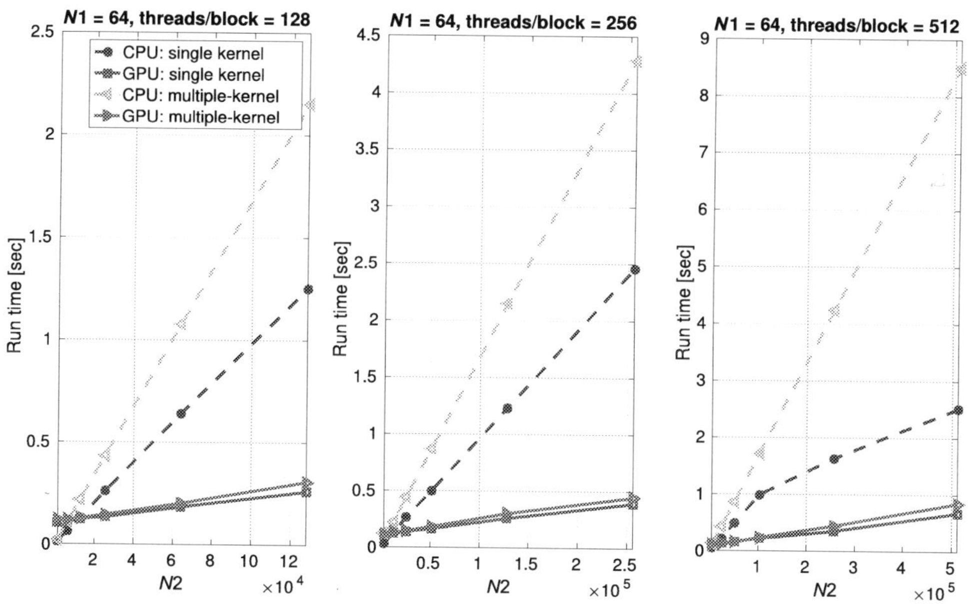


**FIGURE 6.5**

CUDA performance measurement II(1). THREAD_PER_BLOCK is varied with $N1 = 64$.

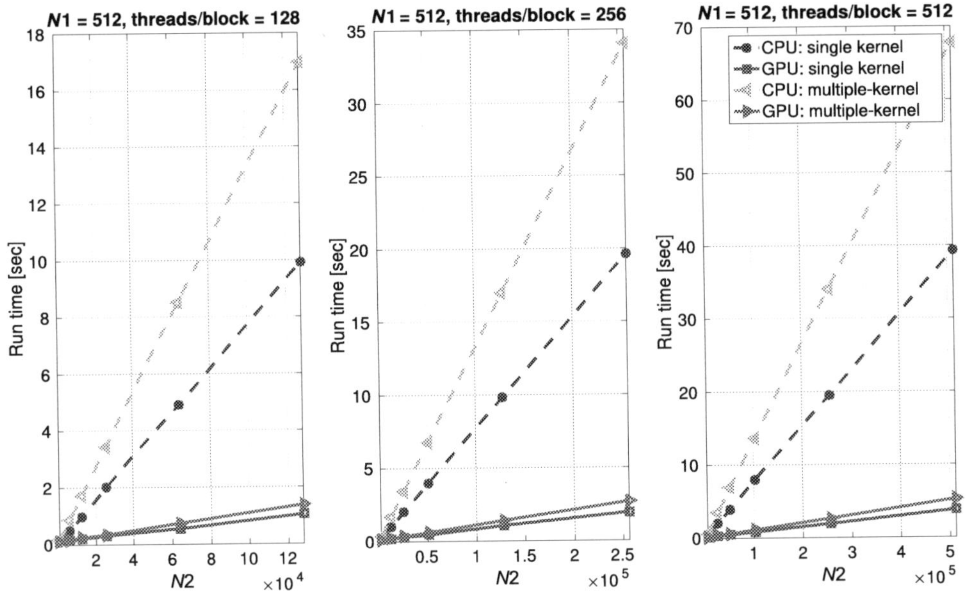


**FIGURE 6.6**

CUDA performance measurement II(2). THREAD_PER_BLOCK is varied for $N1 = 512$.

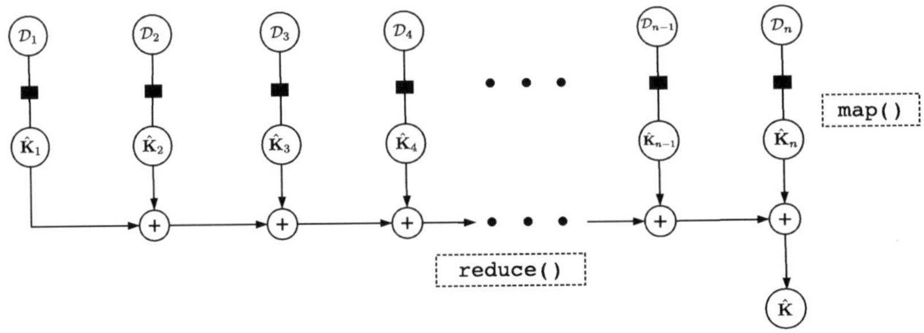


**FIGURE 6.7**

Ensemble Nyström (*cf.* Eq. (6.25)). A diagram illustrating how the Ensemble Nyström can be implemented using the Spark map() and reduce() functions.

### Example: Ensemble Nyström

The Spark framework can be adopted for ensemble algorithms, such as the ensemble Nyström in Eq. (6.25) [28] (*cf.* Fig. 6.7). The training set is divided into $n$ disjoint subsets denoted by $\mathcal{D} = \mathcal{D}_1 \cup \mathcal{D}_2 \cup \cdots \cup \mathcal{D}_n$, for $n \in \mathbb{N}^+$. In Fig. 6.7, each solid black node denotes the kernel subspace approximation in Algorithm 2. The framework is composed of two parts:

- `map()`: The `map` function takes a collection of training sets $\{\mathcal{D}_1, \mathcal{D}_2, \cdots, \mathcal{D}_n\}$ and applies Algorithm 2 to each element $\mathcal{D}_i$, for $i = 1, \cdots, n$. The output of each operation is the kernel matrix $\hat{\mathbf{K}}_i$.
- `reduce()`: The `reduce` function takes the collection of outputs $\left\{\hat{\mathbf{K}}_1, \cdots, \hat{\mathbf{K}}_n\right\}$ and applies a weighted sum (*cf.* Eq. (6.25)) to obtain the final result $\hat{\mathbf{K}}$.

The Spark framework scales linearly with respect to the number of nodes. Moreover, one can enable the NVIDIA CUDA GPU on Spark to accelerate the computations.

These frameworks and libraries are updated frequently with the development of new technologies around them. However, when the underlying principle is taken into account, this rapid evolution is not as overwhelming as it seems. Nevertheless, learning how to use these tools is an essential part of the algorithm design at hand.

## 6.6 CONCLUSION

Kernel techniques are powerful nonlinear feature extraction techniques that handle nonvectorial input sets with well-established theoretical foundations. Given its nonparametric nature, the model complexity of kernel techniques grows with respect to an increasing training size. High model complexity might lead to prohibitive computational complexity and memory usage. Kernel approximation techniques are typically applied to address this issue. In this chapter, we have mainly discussed two types of kernel approximation strategies, the Random Fourier Features and the Nyström subspace approximation. We have focused on the Nyström due to its data-dependent nature and ease of modification to an adaptive learning framework. Designs of the basic Nyström and its extensions have been discussed and analyzed. Moreover, to fully take advantage of the large-scale training size, we have presented possibilities of using modern computing infrastructures and frameworks, e.g. GPU with CUDA library, the map-reduce framework, etc., to scale and speed up kernel methods. Furthermore, the advantage of theoretical properties can be beneficial in combination with other machine learning techniques such as deep learning [37–41], where kernel methods can be used as a preprocessing or a classification module, which can be implemented using popular APIs such as Torch [42] and Tensorflow [43]. The interested reader is referred to the example code from this chapter for further investigation.

## APPENDIX 6.A

### 6.A.1 CENTERING

One preprocessing technique is to center the feature vectors as follows:

$$\begin{aligned} k(\mathbf{x}, \mathbf{z}) &\leftarrow \langle \varphi(\mathbf{x}) - \mathbb{E}(\varphi(\mathbf{x})), \varphi(\mathbf{z}) - \mathbb{E}(\varphi(\mathbf{z})) \rangle \\ &= \langle \varphi(\mathbf{x}), \varphi(\mathbf{z}) \rangle - \langle \mathbb{E}(\varphi(\mathbf{x})), \varphi(\mathbf{z}) \rangle - \langle \varphi(\mathbf{x}), \mathbb{E}(\varphi(\mathbf{z})) \rangle + \langle \mathbb{E}(\varphi(\mathbf{x})), \mathbb{E}(\varphi(\mathbf{z})) \rangle, \end{aligned}$$

where $\mathbb{E}(\varphi(\cdot))$ can be estimated from the training data $\{\mathbf{x}_1, \cdots, \mathbf{x}_N\}$. We have

$$k(\mathbf{x}, \mathbf{z}) \leftarrow k(\mathbf{x}, \mathbf{z}) - \frac{1}{N}\sum_{i=1}^{N} k(\mathbf{x}_i, \mathbf{x}) - \frac{1}{N}\sum_{i=1}^{N} k(\mathbf{x}_i, \mathbf{z}) + \frac{1}{N^2}\sum_{i=1}^{N}\sum_{j=1}^{N} k(\mathbf{x}_i, \mathbf{x}_j). \tag{6.30}$$

After centering the feature vectors, the kernel function only depends on the relative position between $\varphi(\mathbf{x})$ and $\varphi(\mathbf{z})$.

### 6.A.2 NORMALIZATION

The implementation of normalization on the feature vector $\varphi(\mathbf{x})$ is as follows:

$$k(\mathbf{x}, \mathbf{z}) \leftarrow \frac{k(\mathbf{x}, \mathbf{z})}{\sqrt{k(\mathbf{x}, \mathbf{x})}\sqrt{k(\mathbf{z}, \mathbf{z})}}. \tag{6.31}$$

With normalization, the kernel function essentially represents the angle between two feature vectors. The effect of the absolute length of the vectors is dismissed. This leads to a more robust result in many applications.

## REFERENCES

[1] Ali Rahimi, Benjamin Recht, Random features for large-scale kernel machines, in: Neural Information Processing Systems, 2007, pp. 1–10.

[2] Christopher Williams, Matthias Seeger, Using the Nyström method to speed up kernel machines, in: Advances in Neural Information Processing Systems 13, MIT Press, 2001, pp. 682–688.

[3] Sanjiv Kumar, Mehryar Mohri, Ameet Talwalkar, Sampling methods for the Nyström method, Journal of Machine Learning Research 13 (2012) 981–1006.

[4] Petros Drineas, Michael W. Mahoney, On the Nyström method for approximating a gram matrix for improved kernel-based learning, Journal of Machine Learning Research 6 (2005) 2153–2175.

[5] Mu Li, James T. Kwok, Bao-Liang Lu, Making large-scale Nyström approximation possible, in: Johannes Fürnkranz, Thorsten Joachims (Eds.), ICML, Omnipress, 2010, pp. 631–638.

[6] Ming Lin, Fei Wang, Changshui Zhang, Large-scale eigenvector approximation via Hilbert space embedding Nyström, Pattern Recognition 48 (5) (2015) 1904–1912.

[7] Yinan Yu, I. Diamantaras Konstantinos, Tomas McKelvey, S.Y. Kung, Class-specific subspace kernel representations and adaptive margin slack minimization for large scale classification, IEEE Transaction on Neural Network Learn System 29 (2) (2016) 440–456.

[8] Si Si, Cho-Jui Hsieh, Inderjit S. Dhillon, Memory efficient kernel approximation, in: ICML, in: JMLR Workshop and Conference Proceedings, vol. 32, 2014, pp. 701–709, JMLR.org.

[9] Bernhard Schölkopf, J. Smola Alexander, Learning with Kernels: Support Vector Machines, Regularization, Optimization, and Beyond, The MIT Press, 2001.

[10] S.Y. Kung, Kernel Methods and Machine Learning, Cambridge University Press, 2014.

[11] Carl Edward Rasmussen, Christopher K.I. Williams, Gaussian Processes for Machine Learning, The MIT Press, 2006.

[12] Corinna Cortes, Vladimir Vapnik, Support vector networks, Machine Learning 20 (1995) 273–297.

[13] Tianbao Yang, Yu-feng Li, Mehrdad Mahdavi, Rong Jin, Zhi-hua Zhou, San Ramon, East Lansing, Nyström method vs random Fourier features: a theoretical and empirical comparison, in: Neural Information Processing Systems, 2012, pp. 1–9.

[14] Steven C.H. Hoi, Jialei Wang, Peilin Zhao, Jinfeng Zhuang, Zhiyong Liu, Large scale online kernel classification, in: Francesca Rossi (Ed.), IJCAI, IJCAI/AAAI, 2013, pp. 1750–1756.

[15] Ali Rahimi, Benjamin Recht, Weighted sums of random kitchen sinks: replacing minimization with randomization in learning, in: Daphne Koller, Dale Schuurmans, Yoshua Bengio, Léon Bottou (Eds.), NIPS, Curran Associates, Inc., 2008, pp. 1313–1320.
[16] Walter Rudin, Fourier Analysis on Groups, reprint edition, Wiley-Interscience, New York, 1994.
[17] Francis R. Bach, Michael I. Jordan, Computer Science Division, Predictive low-rank decomposition for kernel methods, in: NIPS, 2005.
[18] Kai Zhang, James T. Kwok, Clustered Nyström method for large scale manifold learning and dimension reduction, IEEE Transactions on Neural Networks 21 (10) (2010) 1576–1587.
[19] Berkant Savas, Inderjit S. Dhillon, Clustered low rank approximation of graphs in information science applications, in: SDM, SIAM/Omnipress, 2011, pp. 164–175.
[20] A.E. Hoerl, R.W. Kennard, Ridge regression: biased estimation for nonorthogonal problems, Technometrics 12 (1970) 55–67.
[21] Steven Skiena, The Algorithm Design Manual, 2nd ed., Springer, 2008.
[22] William W. Hager, Updating the inverse of a matrix, SIAM Review 31 (2) (1989) 221–239.
[23] H. Abdi, L.J. Williams, Principal component analysis, Computational Statistics 2 (2010) 433–459.
[24] Sebastian Mika, Gunnar Rätsch, Weston Jason, Schölkopf Bernhard, Müller Klaus-Robert, Fisher discriminant analysis with kernels, in: Neural Networks for Signal Processing IX, IEEE, 1999, pp. 3–10.
[25] Eric P. Xing, Andrew Y. Ng, Michael I. Jordan, Stuart J. Russell, Distance metric learning with application to clustering with side-information, in: NIPS'02, 2002, pp. 505–512.
[26] Kilian Q. Weinberger, Metric learning, 2010.
[27] Dor Kedem, Stephen Tyree, Kilian Weinberger, Fei Sha, Gert Lanckriet, Non-linear metric learning, in: P. Bartlett, F.C.N. Pereira, C.J.C. Burges, L. Bottou, K.Q. Weinberger (Eds.), Advances in Neural Information Processing Systems 25, 2012, pp. 2582–2590.
[28] Sanjiv Kumar, Mehryar Mohri, Ameet Talwalkar, Ensemble Nyström method, in: Yoshua Bengio, Dale Schuurmans, John D. Lafferty, Christopher K.I. Williams, Aron Culotta (Eds.), NIPS, Curran Associates, Inc., 2009, pp. 1060–1068.
[29] CUDA: http://www.nvidia.com/object/cuda_home_new.html.
[30] NVIDIA, Whitepaper – NVIDIA's next generation CUDA compute architecture: Fermi, 2009.
[31] Mehmet Gönen, Ethem Alpaydin, Multiple kernel learning algorithms, Journal of Machine Learning Research 12 (2011) 2211–2268.
[32] SPARK: http://spark.apache.org/.
[33] Matei Zaharia, Mosharaf Chowdhury, Michael J. Franklin, Scott Shenker, Ion Stoica, Spark: cluster computing with working sets, in: HotCloud 10, 2010, p. 10.
[34] apache-spark™: Lightning-fast cluster computing, http://spark.apache.org/. (Accessed 31 December 2016).
[35] Hadoop MapReduce: https://hadoop.apache.org/docs/r1.2.1/mapred_tutorial.html.
[36] Matei Zaharia, Mosharaf Chowdhury, Tathagata Das, Ankur Dave, Justin Ma, Murphy McCauley, Michael J. Franklin, Scott Shenker, Ion Stoica, Resilient distributed datasets: a fault-tolerant abstraction for in-memory cluster computing, in: Proceedings of the 9th USENIX Conference on Networked Systems Design and Implementation, USENIX Association, 2012, p. 2.
[37] Andrew Gordon Wilson, Zhiting Hu, Ruslan Salakhutdinov, Eric P. Xing, Deep kernel learning, CoRR, arXiv:1511.02222, 2015.
[38] Youngmin Cho, Lawrence K. Saul, Kernel methods for deep learning, in: Yoshua Bengio, Dale Schuurmans, John D. Lafferty, Christopher K.I. Williams, Aron Culotta (Eds.), NIPS, Curran Associates, Inc., 2009, pp. 342–350.
[39] Grégoire Montavon, Mikio L. Braun, Klaus-Robert Müller, Kernel analysis of deep networks, Journal of Machine Learning Research 12 (2011) 2563–2581.
[40] Grégoire Montavon, Mikio L. Braun, Klaus-Robert Müller, Layer-wise analysis of deep networks with Gaussian kernels, in: John D. Lafferty, Christopher K.I. Williams, John Shawe-Taylor, Richard S. Zemel, Aron Culotta (Eds.), NIPS, Curran Associates, Inc., 2010, pp. 1678–1686.
[41] Brendan McCane, Lech Szymanski, Deep radial kernel networks: approximating radially symmetric functions with deep networks, CoRR, arXiv:1703.03470, 2017.
[42] TORCH: http://torch.ch/.
[43] TENSORFLOW: https://www.tensorflow.org/.

CHAPTER

# 7

# A RANDOM FOURIER FEATURES PERSPECTIVE OF KAFS WITH APPLICATION TO DISTRIBUTED LEARNING OVER NETWORKS

**Pantelis Bouboulis*, Sergios Theodoridis*, Symeon Chouvardas†**

**National and Kapodistrian University of Athens, Athens, Greece* †*Capital Fund Management, Paris, France*

## CONTENTS



## CHAPTER POINTS

- Provide an overview of the random Fourier features rationale, as a tool to approximate the kernel.
- Provide a review of existing online kernel-based schemes (*e.g.*, KLMS, KRLS, PEGASOS), using the random Fourier features procedure.
- Propose elegant and efficient online distributed kernel-based schemes for learning over networks, using the random Fourier features rationale.

**Adaptive Learning Methods for Nonlinear System Modeling. DOI: 10.1016/B978-0-12-812976-0.00009-9**

## 7.1 INTRODUCTION

During the last decade, there has been an increasing interest for online learning schemes based on the rationale of reproducing kernels both for regression and classification. This has been sparked by the great success of the Support Vectors Machines (SVMs) and other related kernel methods [1–3] in addressing nonlinear tasks in a batch setting. Similar to the way that the *kernel trick* generalizes the classical linear SVM to the nonlinear case, these methods exploit the inner product structure of a *Reproducing Kernel Hilbert Space (RKHS)*, $\mathbb{H}$, to modify well-established linear learning methods (such as the LMS and the RLS) to treat nonlinear tasks [4–7]. In essence, the data are transformed to a high-dimensional space equipped with an inner product (induced by the selected kernel) and then standard linear learning methods are applied to the transformed data. This is equivalent to nonlinear processing in the original space.

In general, these methods consider that the training data points $\{(\mathbf{x}_n, y(n)),\ n = 1, 2, \dots\}$, $\mathbf{x}_n \in \mathbb{R}^d$, are generated by a nonlinear model of the form $y = g(\mathbf{x})$, where $g$ is an unknown nonlinear function, and that the data pairs arrive in a sequential (online) fashion. The objective is to learn a function $f \in \mathbb{H}$, to minimize a certain cost, *i.e.*, $\mathcal{L}(\mathbf{x}_n, y(n), f)$. Typically, a specific kernel function, $\kappa$, is selected and each data pair (arriving at the $n$th time instant) is mapped to $(\Phi(\mathbf{x}_n), y(n))$, using the *feature map* $\Phi : \mathbb{R}^d \to \mathbb{H} : \Phi(\mathbf{x}) = \kappa(\mathbf{x}, \cdot)$. Consequently, the system's output is modeled as $\langle \Phi(\mathbf{x}_n), f \rangle_{\mathbb{H}}$ and the data pair is used to update the previously available estimate of $f$. The result is that (after $n$ iterations) the solution, at time instant $n$, is given in terms of all past training points, *i.e.*,

$$f_n = \sum_{i=1}^{n} w_i \Phi(\mathbf{x}_i) = \sum_{i=1}^{n} w_i \kappa(\mathbf{x}_i, \cdot). \tag{7.1}$$

Since $n$ can grow indefinitely, it is evident that this scheme soon becomes impractical. This becomes very serious in big data applications, where online algorithms are the only viable possibility, *e.g.*, [3]. Usually, some sort of pruning criterion is adopted to make the aforementioned linear expansion sparse. Thus, a dictionary is formed, which contains all valid training points (centers and coefficients) and the specific criterion is used to determine whether a new point will enter the dictionary or not. For example, if the *novelty* criterion is adopted, new arriving points enter the expansion only if the respective center, $\mathbf{x}_n$, is far away from all other centers of the dictionary. If the quantization criterion is adopted, the centers are quantized and each new data pair is used to update the coefficient of the dictionary's center with the same quantization value.

In this chapter, we propose a different approach. Based on the original idea of Rahimi and Recht [8], which has also been applied in [9,10] in a similar context, we approximate the kernel function using randomly selected Fourier features and transform the problem from the original RKHS, $\mathbb{H}$, to another Euclidean space $\mathbb{R}^D$. Thus, we can model the solution as a *fixed-size* vector in $\mathbb{R}^D$ and directly apply any desired linear learning method. More importantly, we can use this rationale to develop learning methods over distributed networks. These are decentralized networks, which are represented by different nodes; each node collects the observations in a sequential fashion and communicates with a subset of nodes, which define its neighborhood, in order to reach a consensus regarding the solution. In such a setting, although the noise and observations, in each node, may follow different statistics, the parameters defining the generating model are assumed to be common for all nodes. The problem in this case is that, if one follows the rationale of Eq. (7.1), the typical sparsification methods become increasingly complicated, as each node has to transmit the entire sum at each time instant to its neighbors and the node has to, somehow, fuse together all received sums to compute the updated estimate. This is probably the reason that (until now) no practical solutions for online learning with kernels in distributed networks have been proposed. On the contrary, we will show that the newly proposed scheme can be directly applied to distributed environments and that the standard established linear methods can be exploited.

The rest of the chapter is organized as follows. Section 7.2 presents the kernel approximation rationale of Rahimi and Recht. In Section 7.3, it is shown how the aforementioned kernel approximation rationale can be exploited to generate fixed-budget kernel-based online learning methods (kernel adaptive filters). The section contains fixed-budget implementations of the KLMS, the KRLS and the kernel-PEGASOS algorithms, as well as simulations to demonstrate the effectiveness of the methods. Moreover, in Section 7.4 the kernel approximation rationale is employed to generate elegant and efficient distributed learning methods over networks, based on the combine-then-adapt strategy. Finally, Section 7.5 presents some concluding remarks.

## 7.2 APPROXIMATING THE KERNEL

All kernel-based learning techniques usually demand a large number of kernel computations between training samples. For example, the popular SVM algorithm requires the computation of a large kernel matrix consisting of the kernel evaluations between pairs of all possible combinations of the training points. Hence, to alleviate the computational burden, one common line of research suggests to use some sort of approximation of the kernel matrix. This can be done by the celebrated Nyström method [11,12] or by other similar techniques. In this chapter, we are focusing on the Fourier features approach proposed in [8,13], as this fits more naturally to the online setting. The following theorem plays a key role in this procedure.

**Theorem 1.** *Consider a shift-invariant positive definite kernel $\kappa(\mathbf{x}-\mathbf{y})$ defined on $\mathbb{R}^d$ and its Fourier transform, $p(\boldsymbol{\omega}) = \frac{1}{(2\pi)^d}\int_{\mathbb{R}^d}\kappa(\boldsymbol{\delta})e^{-i\boldsymbol{\omega}^T\boldsymbol{\delta}}d\boldsymbol{\delta}$, which (according to Bochner's theorem) can be regarded as a* ***probability density*** *function. Then, defining $z_{\boldsymbol{\omega},b}(\mathbf{x}) = \sqrt{2}\cos(\boldsymbol{\omega}^T\mathbf{x}+b)$, it turns out that*

$$\kappa(\mathbf{x}-\mathbf{y}) = E_{\boldsymbol{\omega},b}[z_{\boldsymbol{\omega},b}(\mathbf{x})z_{\boldsymbol{\omega},b}(\mathbf{y})], \tag{7.2}$$

*where $\boldsymbol{\omega}$ is drawn from $p$, $b$ from the uniform distribution on $[0,2\pi]$ and $E[\cdot]$ denotes the related expectation.*

Following Theorem 1, we choose to approximate $\kappa(\mathbf{x}_n-\mathbf{x}_m)$ using $D$ random Fourier features, $\boldsymbol{\omega}_1,\boldsymbol{\omega}_2,\ldots,\boldsymbol{\omega}_D$, (drawn from $p$) and $D$ random numbers, $b_1,b_2,\ldots,b_D$ (drawn uniformly from $[0,2\pi]$) that define a sample average:

$$\kappa(\mathbf{x}_n-\mathbf{x}_m) \approx \frac{1}{D}\sum_{i=1}^{D} z_{\boldsymbol{\omega}_i,b_i}(\mathbf{x}_m)z_{\boldsymbol{\omega}_i,b_i}(\mathbf{x}_n). \tag{7.3}$$

Hence, instead of relying on the implicit map, $\Phi$, provided by the kernel trick, we can map the input data to a finite-dimensional Euclidean space (with dimension lower than $\mathbb{H}$, but still larger than the input space) using the randomized feature map $\mathbf{z}_{\mathbf{\Omega}} : \mathbb{R}^d \to \mathbb{R}^D$:

$$\mathbf{z}_{\mathbf{\Omega}}(\mathbf{x}) = \sqrt{\frac{2}{D}} \begin{pmatrix} \cos(\boldsymbol{\omega}_1^T \mathbf{x} + b_1) \\ \vdots \\ \cos(\boldsymbol{\omega}_D^T \mathbf{x} + b_D) \end{pmatrix}, \tag{7.4}$$

where $\mathbf{\Omega}$ is the $(d+1) \times D$ matrix defining the random Fourier features of the respective kernel, *i.e.*,

$$\mathbf{\Omega} = \begin{pmatrix} \boldsymbol{\omega}_1 & \boldsymbol{\omega}_2 & \dots & \boldsymbol{\omega}_D \\ b_1 & b_2 & \dots & b_D \end{pmatrix}, \tag{7.5}$$

so that the kernel evaluations can be approximated as $\kappa(\mathbf{x}_n, \mathbf{x}_m) \approx \mathbf{z}_{\mathbf{\Omega}}(\mathbf{x}_n)^T \mathbf{z}_{\mathbf{\Omega}}(\mathbf{x}_m)$. Details on the quality of this approximation, as well as other theoretical results, can be found in [8,13–15]. We note that for the Gaussian kernel, *i.e.*, $\kappa_\sigma(\mathbf{x}, \mathbf{y}) = \exp\left(-\|\mathbf{x} - \mathbf{y}\|^2/\sigma^2\right)$, which is employed throughout the paper, the respective Fourier transform becomes

$$p(\boldsymbol{\omega}) = \left(\sigma/\sqrt{2\pi}\right)^D e^{-\frac{\sigma^2 \|\boldsymbol{\omega}\|^2}{2}}, \tag{7.6}$$

which is actually the multivariate Gaussian distribution with mean $\mathbf{0}_D$ and covariance matrix $\frac{1}{\sigma^2}\mathbf{I}_D$.

## 7.3 ONLINE KERNEL-BASED LEARNING: A RANDOM FOURIER FEATURES PERSPECTIVE

Inspired by the randomized feature map, we propose an alternative approach for online learning in RKHS. The procedure can be summarized as follows:

- Map all data pairs, to $(\mathbf{z}_{\mathbf{\Omega}}(\mathbf{x}_n), y(n))$.
- Approximate the desired output as $f(\mathbf{x}) \approx \boldsymbol{\theta}^T \mathbf{z}_{\mathbf{\Omega}}(\mathbf{x})$, for some $\boldsymbol{\theta} \in \mathbb{R}^D$.
- Estimate $\boldsymbol{\theta}$ using any standard linear learning method (*e.g.*, LMS, RLS) on the transformed data.

In the following, we will demonstrate in details three applications of the proposed procedure: (a) the Kernel LMS (KLMS), (b) the Kernel RLS (KRLS) and (c) the Primal Estimated sub-GrAdient SOlver for non linear SVM (kernel PEGASOS).

### 7.3.1 RFF-KLMS

Assuming a sequentially arriving data set, $\{(\mathbf{z}_{\mathbf{\Omega}}(\mathbf{x}_n), y(n)), n = 1, 2, \dots\}$, the typical LMS algorithm models the desired output as $f(\mathbf{x}) = \hat{y} = \boldsymbol{\theta}^T \mathbf{z}_{\mathbf{\Omega}}(\mathbf{x})$ and estimates $\boldsymbol{\theta}$, so that the MSE cost function, $\mathcal{L}(\mathbf{x}_n, y(n), \boldsymbol{\theta}) = E\left[\left(y(n) - \boldsymbol{\theta}^T \mathbf{z}_{\mathbf{\Omega}}(\mathbf{x}_n)\right)^2\right]$ is minimized. The minimization is carried out via a gradient descent rationale, where the gradient of the cost, is approximated by the current measurement, *i.e.*,

**Algorithm 1** The Random Fourier Features Kernel LMS algorithm.

Set $\boldsymbol{\theta} = \mathbf{0}$. Select the step update $\mu$, the dimension of the new space, $D$, and the parameter of the kernel ($\sigma$).
Generate the matrix $\Omega$ as in Eq. (7.5).
$\boldsymbol{\theta}_0 \leftarrow \mathbf{0}_D$ ▷ Initialization
**for** $n = 1, 2, 3, \ldots$ **do**
  Map the current data pair, $(\mathbf{x}_n, y(n))$, to $(\mathbf{z}_\Omega(\mathbf{x}_n), y(n))$.
  $\hat{y}(n) \leftarrow \boldsymbol{\theta}_{n-1}^T \mathbf{z}_\Omega(\mathbf{x}_n)$. ▷ System output
  $e(n) \leftarrow y(n) - \hat{y}(n)$. ▷ Estimation error
  $\boldsymbol{\theta}_n \leftarrow \boldsymbol{\theta}_{n-1} + \mu e(n) \mathbf{z}_\Omega(\mathbf{x}_n)$. ▷ Step update

$$\begin{aligned}\nabla_{\boldsymbol{\theta}} \mathcal{L}(\mathbf{x}_n, y(n), \boldsymbol{\theta}) &= -2E\left[y(n) - \boldsymbol{\theta}^T \mathbf{z}_\Omega(\mathbf{x}_n)\right] \mathbf{z}_\Omega(\mathbf{x}_n) \\ &\approx -2\left(y(n) - \boldsymbol{\theta}^T \mathbf{z}_\Omega(\mathbf{x}_n)\right) \mathbf{z}_\Omega(\mathbf{x}_n).\end{aligned}$$

This method (RFF-KLMS) has been introduced independently in [9] and [10]. Algorithm 1 presents the steps in details. It is a matter of elementary algebra to see that after $n-1$ steps, the estimation of the solution will be $\boldsymbol{\theta}_n = \mu \sum_{i=1}^{n-1} e(i) \mathbf{z}_\Omega(\mathbf{x}_i)$, which leads us to conclude that the RFF-KLMS scheme will produce, approximately, the same output as that of the standard KLMS (provided that $D$ is sufficiently large):

$$\hat{y}(n) = \mu \sum_{i=1}^{n-1} e(i) \mathbf{z}_\Omega(\mathbf{x}_i)^T \mathbf{z}_\Omega(\mathbf{x}_n) \approx \mu \sum_{i=1}^{n-1} e(i) \kappa(\mathbf{x}_i, \mathbf{x}_n), \tag{7.7}$$

where for the last relation (which gives the output of the standard KLMS [7,4]) we used Theorem 1. The major difference is that the RFF-KLMS provides a single vector $\boldsymbol{\theta}$ of fixed dimensions, instead of a growing expansion of kernel functions, hence no pruning mechanisms are required.

Working similar to the case of the standard LMS, we can study convergence properties of RFFKLMS. Henceforth, we will assume that the data pairs are generated by

$$y(n) = \sum_{m=1}^{M} \alpha_m \kappa(\mathbf{c}_m, \mathbf{x}_n) + \eta(n), \tag{7.8}$$

where $\mathbf{c}_1, \ldots, \mathbf{c}_M$ are fixed centers, $\mathbf{x}_n$ are zero-mean independent and identically distributed samples drawn from the Gaussian distribution with covariance matrix $\sigma_X^2 \mathbf{I}_d$ and $\eta(n)$ are independent and identically distributed noise samples drawn from $\mathcal{N}(0, \sigma_\eta^2)$. We note that the parameters $\sigma_X$ and $\sigma_\eta$ are the variances of the input and the noise, respectively, and they are not in any way related to the kernel parameter $\sigma$. Applying the approximation rationale of Theorem 1, we obtain

$$\kappa(\mathbf{c}_m, \mathbf{x}_n) = E_{\boldsymbol{\omega}, b}[z_{\boldsymbol{\omega}, b}(\mathbf{c}_m) z_{\boldsymbol{\omega}, b}(\mathbf{x}_n)] = \mathbf{z}_\Omega(\mathbf{c}_m)^T \mathbf{z}_\Omega(\mathbf{x}_n) + \epsilon_m(n),$$

where $\epsilon_m(n)$ is the error between the actual value of the kernel function and the approximated one. Thus, if we define

$$\mathbf{Z_C} = (\mathbf{z}_\Omega(\mathbf{c}_1), \ldots, \mathbf{z}_\Omega(\mathbf{c}_M)), \boldsymbol{\alpha} = (\alpha_1, \ldots, \alpha_M)^T, \ \boldsymbol{\theta}_o = \mathbf{Z}_\Omega \boldsymbol{\alpha},$$

we get

$$y(n) = \boldsymbol{\alpha}^T \mathbf{Z}_\Omega^T \mathbf{z}_\Omega(\mathbf{x}_n) + \epsilon(n) + \eta(n) = \boldsymbol{\theta}_o^T \mathbf{z}_\Omega(\mathbf{x}_n) + \epsilon(n) + \eta(n),$$

where $\epsilon(n)$ stands for the approximation error between the noise-free component of $y(n)$ (evaluated only by the linear kernel expansion of Eq. (7.8)) and the approximation of this component using random Fourier features, *i.e.*,

$$\epsilon(n) = \sum_{m=1}^{M} \alpha_m \kappa(\mathbf{c}_m, \mathbf{x}_n) - \boldsymbol{\theta}_o^T \mathbf{z}_\Omega(\mathbf{x}_n) = \sum_{m=1}^{M} a_m \epsilon_m(n).$$

Let $\mathbf{x} \in \mathbb{R}^d$, $y \in \mathbb{R}$, be the random variables that generate the measurements and $\mathbf{R}_{zz} = E[\mathbf{z}_\Omega(\mathbf{x})\mathbf{z}_\Omega^T(\mathbf{x})]$ the corresponding autocorrelation matrix. We can prove that $\mathbf{R}_{zz}$ is positive definite, provided that all the random features are different from each other, *i.e.*, $\boldsymbol{\omega}_i \neq \boldsymbol{\omega}_j$, for all $i \neq j$ [16]. Thus, the cross-correlation vector takes the form

$$E[\mathbf{z}_\Omega(\mathbf{x})y] = E\left[\mathbf{z}_\Omega(\mathbf{x})\left(\mathbf{z}_\Omega(\mathbf{x})^T\boldsymbol{\theta}_o + \epsilon + \eta\right)\right] = \mathbf{R}_{zz}\boldsymbol{\theta}_o + E[\mathbf{z}_\Omega(\mathbf{x})\epsilon],$$

where for the last relation we have used the fact that $\eta$ is a zero-mean variable representing noise and that $\mathbf{z}_\Omega(\mathbf{x})$ and $\eta$ are independent. For large enough $D$, we can assume that the approximation error, $\epsilon$, approaches 0 [14], hence the optimal solution becomes

$$\boldsymbol{\theta}_* = E\left[\left(y - \mathbf{z}_\Omega(\mathbf{x})^T\boldsymbol{\theta}_o\right)^2\right] = \mathbf{R}_{zz}^{-1}(\mathbf{R}_{zz}\boldsymbol{\theta}_o + E[\mathbf{z}_\Omega(\mathbf{x})\epsilon]) = \boldsymbol{\theta}_o + \mathbf{R}_{zz}^{-1}E[\mathbf{z}_\Omega(\mathbf{x})\epsilon] \approx \boldsymbol{\theta}_o.$$

To summarize, the proposed rationale assumes that, for sufficiently large $D$, Eq. (7.8) can be closely approximated by $y(n) \approx \boldsymbol{\alpha}^T\mathbf{Z}_\mathbf{C}^T\mathbf{z}_\Omega(\mathbf{x}_n) + \eta(n)$. The major difference with the standard LMS case is that the transformed inputs, $\mathbf{z}_\Omega(\mathbf{x}_n)$, can no longer be assumed to be generated by the Gaussian distribution. Applying similar assumptions as in the case of the standard LMS (*e.g.*, independence between $\mathbf{x}_n, \mathbf{x}_m$, for $n \neq m$ and between $\mathbf{x}_n, \eta(n)$), we get several convergence properties, where the eigenvalues of the positive definite matrix $\mathbf{R}_{zz}$, *i.e.*, $0 < \lambda_1 \leq \lambda_2 \leq \cdots \leq \lambda_D$, play a pivotal role:

1. If $0 < \mu < 2/\lambda_D$, then RFFKLMS converges in the mean, *i.e.*, $E[\boldsymbol{\theta}_n - \boldsymbol{\theta}_o] \to \mathbf{0}$.
2. The optimal MSE is given by

$$J(n)^{\text{opt}} = \sigma_\eta^2 + E[\epsilon(n)] - E[\epsilon(n)\mathbf{z}_\Omega(\mathbf{x}_n)]\mathrm{R}_{zz}^{-1}E[\epsilon(n)\mathbf{z}_\Omega(\mathbf{x}_n)^T].$$

   For large enough $D$, we have $J(n)^{\text{opt}} \approx \sigma_\eta^2$.
3. The excess MSE is given by $J(n)^{\text{ex}} = J(n) - J(n)^{\text{opt}} = \operatorname{tr}(\mathbf{R}_{zz}\mathbf{A}_n)$, where $\mathbf{A}_n = E[(\boldsymbol{\theta}_n - \boldsymbol{\theta}_o)(\boldsymbol{\theta}_n - \boldsymbol{\theta}_o)^T]$.
4. If $0 < \mu < 1/\lambda_D$, then $\mathbf{A}_n$ converges.

**Algorithm 2** The Random Fourier Features Kernel RLS algorithm.

Set $\boldsymbol{\theta} = \mathbf{0}$. Select the dimension of the new space, $D$, the regularization parameter $\lambda$, the forgetting factor $\beta$ and the parameter of the kernel ($\sigma$).

Generate the matrix $\Omega$ as in (7.5).

$\boldsymbol{\theta}_0 \leftarrow \mathbf{0}_D$, $\mathbf{P}_0^{-1} \leftarrow \lambda^{-1}\mathbf{I}_D$ ▷ Initialization

**for** $n = 1, 2, 3, \ldots$ **do**

Map the current data pair, $(\mathbf{x}_n, y(n))$, to $(\mathbf{z}_\Omega(\mathbf{x}_n), y(n))$.

$\hat{y}(n) \leftarrow \boldsymbol{\theta}_{n-1}^T \mathbf{z}_\Omega(\mathbf{x}_n)$. ▷ System output

$e(n) \leftarrow y(n) - \hat{y}(n)$. ▷ Estimation error

$\mathbf{P}_n^{-1} \leftarrow \beta^{-1}\mathbf{P}_{n-1}^{-1} - \frac{\beta^{-2}\mathbf{P}_{n-1}^{-1}\mathbf{z}_\Omega(\mathbf{x}_n)\mathbf{z}_\Omega(\mathbf{x}_n)^T\mathbf{P}_{n-1}^{-1}}{1+\beta^{-1}\mathbf{z}_\Omega(\mathbf{x}_n)^T\mathbf{P}_{n-1}^{-1}\mathbf{z}_\Omega(\mathbf{x}_n)}$.

$\boldsymbol{\theta}_n \leftarrow \boldsymbol{\theta}_{n-1} + \frac{\beta^{-1}\mathbf{P}_{n-1}^{-1}\mathbf{z}_\Omega(\mathbf{x}_n)}{1+\beta^{-1}\mathbf{z}_\Omega(\mathbf{x}_n)^T\mathbf{P}_{n-1}^{-1}\mathbf{z}_\Omega(\mathbf{x}_n)} e(n)$. ▷ Step update

### 7.3.2 RFF-KRLS

Contrary to the stochastic approach employed by the LMS, the RLS utilizes information of all past data to estimate $\boldsymbol{\theta}$ by minimizing the regularized risk cost function, *i.e.*,

$$\mathcal{L}(\mathbf{x}_n, y(n), \boldsymbol{\theta}) = \sum_{k=1}^{n} \beta^{n-k}\left(y(n) - \boldsymbol{\theta}^T \mathbf{z}_\Omega(\mathbf{x}_n)\right)^2 + \lambda\beta^n \|\boldsymbol{\theta}\|^2,$$

for some chosen $\lambda > 0$, where $\beta$ is a weighting factor that is usually added as a forgetting mechanism and which makes the algorithm more sensitive to recent data. Solving $\boldsymbol{\theta}_* = \min_{\boldsymbol{\theta}} \mathcal{L}(\mathbf{x}_n, y(n), \boldsymbol{\theta})$ requires the inversion of an $n \times n$ matrix at each step. However, employing the matrix inversion lemma, it is possible to compute this solution recursively. Algorithm 2 presents the details. The procedure is identical to that of the standard RLS, except that we replace $\mathbf{x}_n$ with the transformed data $\mathbf{z}_\Omega(\mathbf{x}_n)$. Moreover, we should note that, unlike the LMS case, the proposed scheme provides a solution that is different from other implementations of KRLS (like the ones presented in [6] and [4]). Although all KRLS-like implementations, provided in the respective literature, estimate the system's output as a linear expansion of kernel functions centered at the past data, the estimation of the expansion's coefficients differs in the case presented here.

### 7.3.3 RFF-PEGASOS

The methods presented in Sections 7.3.1 and 7.3.2 are suitable for online regression problems. In this section, we present an online classification method based on the PEGASOS scheme (see [17]). Consider a training set of the form $\{(\mathbf{x}_n, y(n)),\ n = 1, 2, \ldots N\}$, where $\mathbf{x}_n \in \mathbb{R}^d$ and $y(n) = \pm 1$. PEGASOS is an SVM-like algorithm that processes the data sequentially as follows. On iteration $t$, the algorithm selects a random training example $(\mathbf{x}_{n_t}, y(n_t))$ by picking an index $n_t$ uniformly at random from $\{1, 2, \ldots, N\}$. The scheme follows a stochastic gradient rationale, adopting the regularized hinge loss function, *i.e.*,

$$\boldsymbol{\theta}_* = \max\{0, 1 - y\langle f, \Phi(\mathbf{x})\rangle_{\mathbb{H}}\} + \frac{\lambda}{2}\|f\|_{\mathbb{H}}^2,$$

**Algorithm 3** The Random Fourier Features PEGASOS algorithm.

Set $\boldsymbol{\theta} = \mathbf{0}$. Select the dimension of the new space, $D$, the regularization parameter $\lambda$ and the parameter of the kernel ($\sigma$).
Generate the matrix $\boldsymbol{\Omega}$ as in (7.5).
$\boldsymbol{\theta}_0 \leftarrow \mathbf{0}_D$ ▷ Initialization
**for** $n = 1, 2, 3, \ldots$ **do**
  $\mu_n \leftarrow \frac{1}{\lambda \cdot n}$
  Map the current data pair, $(\mathbf{x}_n, y(n))$, to $(\mathbf{z}_{\Omega}(\mathbf{x}_n), y(n))$.
  $\hat{y}(n) \leftarrow \boldsymbol{\theta}_{n-1}^T \mathbf{z}_{\Omega}(\mathbf{x}_n)$. ▷ System output
  **if** $y(n)\hat{y}(n) < 1$ **then**
    $\boldsymbol{\theta}_n \leftarrow (1 - \mu_n \lambda)\boldsymbol{\theta}_{n-1} + \mu_n y(n) \mathbf{z}_{\Omega}(\mathbf{x}_n)$.
  **else**
    $\boldsymbol{\theta}_n \leftarrow (1 - \mu_n \lambda)\boldsymbol{\theta}_{n-1}$.

where $\mathbb{H}$ is the RKHS associated with a preselected kernel $\kappa$ and $\Phi$ the respective feature map. Thus, the step update equation becomes:

$$f_t = \left(1 - \frac{1}{t}\right) f_{t-1} + \mathbb{1}_+ \left(1 - y(n_t)\langle f_{t-1}, \Phi(\mathbf{x}_{n_t})\rangle_{\mathbb{H}}\right) \frac{y_{n_t}}{\lambda t} \Phi(\mathbf{x}_{n_t}). \tag{7.9}$$

After a predetermined number of iterations (say $T$), the algorithm provides the solution, which is given as a sparse linear expansion of kernel functions, similar to Eq. (7.1). The algorithm is implemented by keeping in memory an $N$-size vector that comprises all the coefficients of the expansion. During each iteration, at most one of these coefficients might change. The performance of the scheme has been shown to improve with data reuse, *i.e.*, when the same data are used again and again for training.

Evidently, since the solution provided by the algorithm depends on the dataset's size, PEGASOS cannot be considered as an online scheme (where the actual size of the dataset is not known beforehand). However, employing the random Fourier features rationale, we can model the output as $f(\mathbf{x}) = \boldsymbol{\theta}^T \mathbf{z}_{\Omega}(\mathbf{x})$, where $\boldsymbol{\theta} \in \mathbb{R}^D$, and thus rewrite the step update equation as follows (see Algorithm 3):

$$\boldsymbol{\theta}_n = \left(1 - \frac{1}{n}\right) \boldsymbol{\theta}_{n-1} + \mathbb{1}_+ \left(1 - y(n)\boldsymbol{\theta}_{n-1}^T \mathbf{z}_{\Omega}(\mathbf{x}_n)\right) \frac{y(n)}{\lambda n} \mathbf{z}_{\Omega}(\mathbf{x}_n), \tag{7.10}$$

where we have assumed that the data arrive sequentially. Similar to the LMS case, the proposed algorithm can be seen as a linear PEGASOS on the dataset $\{(\mathbf{z}_{\Omega}(\mathbf{x}_n), y(n)),\ n = 1, 2, \ldots\}$, hence all convergence properties of [17] hold in this case too. We will call this scheme RFF-PEGASOS.

We should note that the proposed algorithm features several significant advantages with respect to the original form of kernel PEGASOS presented in [17]. Firstly, the complexity per iteration is $\mathcal{O}(Dd)$ and does not depend on the training database size, in contrast with the kernel-PEGASOS, which has complexity $\mathcal{O}(Md)$, $M$ being the number of the support vectors. Moreover, RFF-PEGASOS solves the problem of the growing sum and thus it can work in a truly online fashion (where the dataset is not known beforehand), while kernel-PEGASOS was designed for batch processing (*i.e.*, fixed datasets). Finally, it is evident that for datasets with many support vectors (where $M >> D$) RFF-PEGASOS is

**Table 7.1 The parameters of the KLMS and KRLS variants**

| QKLMS | RFF-KLMS | E-KRLS | RFF-KRLS |
|---|---|---|---|
| $\mu = 1, \epsilon = 5$ | $\mu = 1$ | $\nu = 0.0005$ | $\lambda = 0.00001,\ w = 1$ |

significantly faster. However, in databases with a low number of support vectors ($M << D$), it runs slower.

### 7.3.4 SIMULATIONS—REGRESSION

In this section, we provide some simple simulations, which are indicative of the behavior of the proposed algorithms, compared to their typical implementations. For the regression case, we generate 5000 data pairs using the following model: $y(n) = \sum_{m=1}^{M} a_m \kappa(\mathbf{c}_m, \mathbf{x}_n) + \eta(n)$, where $\mathbf{x}_n \in \mathbb{R}^5$ are drawn from $\mathcal{N}(\mathbf{0}, \mathbf{I}_5)$ and the noise are independent and identically distributed Gaussian samples with $\sigma_\eta = 0.1$. The parameters of the expansion (*i.e.*, $a_1, \ldots, a_M$) are drawn from $\mathcal{N}(0, 25)$ and the kernel parameter $\sigma$ is set to 5. The specific parameters of the algorithms are shown in Table 7.1. We compare RFF-KLMS and RFF-KRLS with a standard implementation of KLMS using the quantization pruning technique and the KRLS version of [6]. Fig. 7.1 shows the evolution of the MSE for 100 realizations of the experiment for several values of $D$ (*i.e.*, the total number of random Fourier features used), where the number of the free parameters of the model is set to $M = 100$. As this is a kernel approximation scheme, it is expected that the performance (in terms of MSE) would be somewhat reduced. However, we can see in Fig. 7.1 that the performance of both RFF-KLMS and RFF-KRLS is very close to their typical implementations. In this particular example, both algorithms can achieve similar performance with their standard counterparts earlier in time (see Table 7.2 and Figs. 7.1B, 7.1D for the KLMS and KRLS, respectively). Fig. 7.2 shows the evolution of the MSE for the same experiment using $M = 400$ parameters in the design model. In this case, it is evident that a larger number of Fourier features is required to achieve the same approximation quality. For example, the RFF-KRLS requires more than $D = 200$ random features and the RFF-KLMS requires $D = 2000$ features to achieve approximately the same steady-state MSE as their typical variants. To conclude:

- Both RFF-KLMS and RFF-KRLS can efficiently approximate their typical counterparts for large enough values of $D > D_0$.
- The value of $D$ needed (so that the RFF-KLMS and the KRLS approximate their counterparts) depends on the model complexity (it also depends on the size of the input space, $d$, for obvious reasons).
- The RFF-KRLS can achieve the same performance with the KRLS using significantly smaller $D$ than the one needed by the RFF-KLMS for comparable performance to KLMS. Possibly, this is due to the fact that the KRLS mechanism utilizes all past data in each iteration.

### 7.3.5 SIMULATIONS—CLASSIFICATION

For the classification case, we have performed two sets of experiments. In the first set, we have compared the performances of the standard kernel-PEGASOS and the RFF-PEGASOS on four datasets downloaded from Leon Bottou's LASVM web page [18]. The chosen datasets are (a) the Adult dataset,

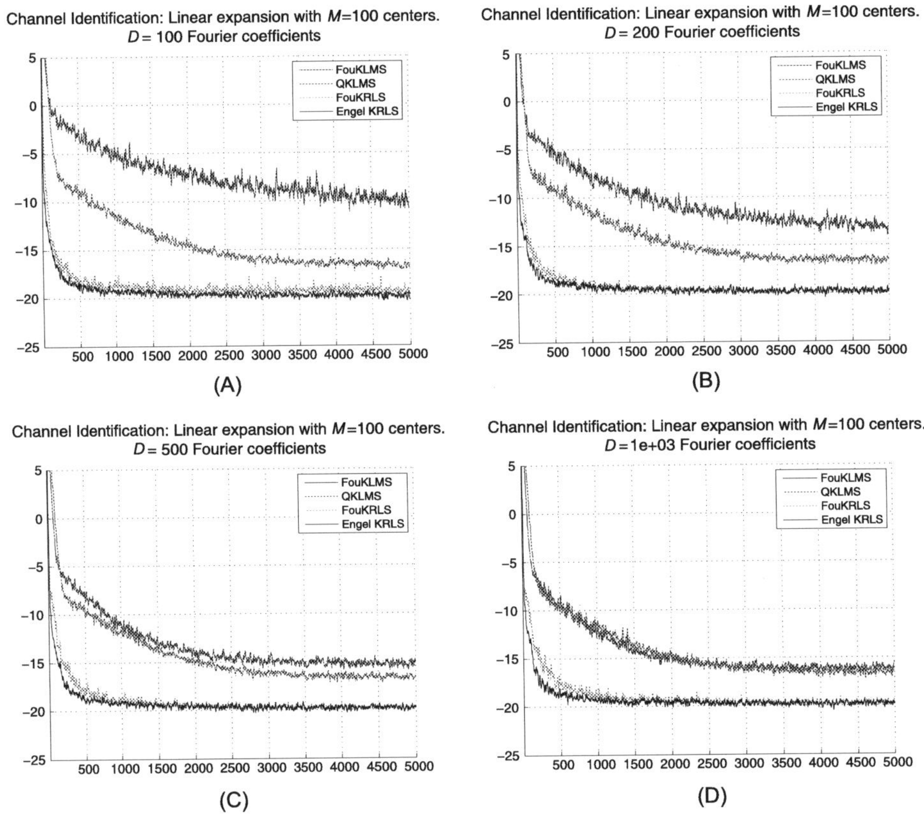


**FIGURE 7.1**

Comparing the performances of KLMS and KRLS versus their random Fourier features variants (FouKLMS and FouKRLS, respectively). (A) $D = 100$. (B) $D = 200$. (C) $D = 500$. (D) $D = 1000$.

**Table 7.2 Mean times needed for the KLMS and KRLS on a typical core i5 machine for $M = 100$ and $M = 400$**

| | $M = 100$ | | | | $M = 400$ | | | |
|---|---|---|---|---|---|---|---|---|
| | $D = 100$ | $D = 200$ | $D = 500$ | $D = 1000$ | $D = 100$ | $D = 200$ | $D = 1000$ | $D = 2000$ |
| RFF-KLMS | 0.03 s | 0.07 s | 0.14 s | 0.25 s | 0.03 s | 0.06 s | 0.26 s | 0.41 s |
| QKLMS | 0.27 s | 0.27 s | 0.27 s | 0.27 s | 0.27 s | 0.27 s | 0.27 s | 0.27 s |
| RFF-KRLS | 0.35 s | 1.00 s | 11.5 s | 58.2 s | 0.39 s | 1.0 s | 53.2 s | 140.2 s |
| KRLS | 3.03 s | 3.2 s | 3.2 s | 3.2 s | 3.2 s | 3.2 s | 3.2 s | 3.2 |

(b) the Banana dataset (where we have used the first 4000 points as training data and the remaining 1300 as testing data), (c) the Waveform dataset (where we have used the first 4000 points as training data and the remaining 1000 as testing data) and (d) the MNIST dataset (for the task of classifying the digit 8 versus the rest). The sizes of the datasets are given in Table 7.4. Table 7.5 reports the

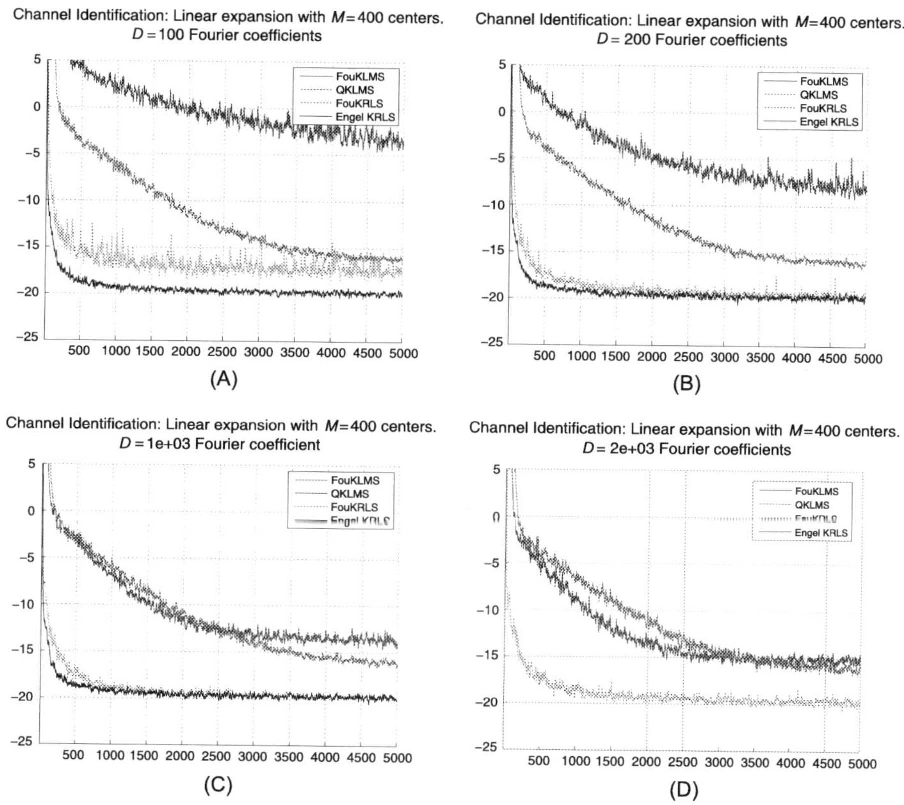


**FIGURE 7.2**

Comparing the performances of KLMS and KRLS versus their random Fourier features variants (FouKLMS and FouKRLS, respectively) on a more complex input–output model. (A) $D = 100$. (B) $D = 200$. (C) $D = 1000$. (D) $D = 2000$.

test errors for each dataset and each method, along with the total number of support vectors after training (only for the kernel-PEGASOS) and the respective time in seconds (in parentheses). The experiments were performed on an i7-3770 machine using a MatLab implementation. The parameters used in each method are reported in Table 7.3. The parameters were selected after extensive trials to give the best possible accuracy. The parentheses (after the title of the method) indicate the number of times the algorithm passed through the entire dataset. As expected from the theoretical analysis of the corresponding complexities, for datasets with a large number of support vectors (such as Adult and Banana), RFF-PEGASOS requires significantly less time than its standard implementation. For datasets with a medium number of SVs (such as Waveform), this difference is diminishing, while for datasets with very few SVs (*e.g.*, MNIST) kernel-PEGASOS can be significantly faster. We note that, for the MNIST database, RFF-PEGASOS attains a 0.47% test error percentage after 20 reruns (19000 s).

In the second set of experiments we tested the performance of RFF-PEGASOS and compared it with that of the LA-SVM [19]. LA-SVM is one of the most elegant and fast SVM training implemen-

**Table 7.3 Parameters for each classification method**

| Method | Adult | Banana | Waveform | MNIST |
|---|---|---|---|---|
| Kernel-PEGASOS | $\sigma = \sqrt{10}$<br>$\lambda = 0.0000307$ | $\sigma = 0.7$<br>$\lambda = \frac{1}{316}$ | $\sigma = \sqrt{10}$<br>$\lambda = 0.001$ | $\sigma = 4$<br>$\lambda = 10^{-7}$ |
| RFF-PEGASOS | $\sigma = \sqrt{10}$<br>$\lambda = 0.0000307$<br>$D = 2000$ | $\sigma = 0.7$<br>$\lambda = \frac{1}{316}$<br>$D = 200$ | $\sigma = \sqrt{10}$<br>$\lambda = 0.001$<br>$D = 2000$ | $\sigma = 4$<br>$\lambda = 10^{-7}$<br>$D = 100000$ |

**Table 7.4 Dataset Information**

| Method | Adult | Banana | Waveform | MNIST |
|---|---|---|---|---|
| Training size | 32562 | 4000 | 4000 | 60000 |
| Testing size | 16282 | 1300 | 1000 | 10000 |
| dimensions | 123 | 2 | 21 | 784 |

**Table 7.5 Comparing the performances of kernel-PEGASOS and RFF-PEGASOS**

| Method | Adult | Banana | Waveform | MNIST |
|---|---|---|---|---|
| kernel-PEGASOS (1) | 16.52% (8754 sv/129 s) | 9.46% (1510 sv/0.42 s) | 9.2% (1069 sv/0.52 s) | 0.85% (940 sv/182 s) |
| kernel-PEGASOS (2) | 15.89% (11009 sv/464 s) | 9.46% (1696 sv/1.11 s) | 9.8% (1237 sv/1.38 s) | 0.56% (1194 sv/548 s) |
| kernel-PEGASOS (5) | 15% (12513 sv/1660 s) | 9.38% (1760 sv/3.22 s) | 8.8% (1306 sv/4.05 s) | 0.52% (1432 sv/1947 s) |
| kernel-PEGASOS (10) | 14.83% (13078 sv/3804 s) | 9.46% (1784 sv/6.77 s) | 8.1% (1336 sv/8.61 s) | 0.43% (1555 sv/4692 s) |
| RFF-PEGASOS (1) | 16.55% (2.53 s) | 9.61% (0.06 s) | 9.3% (0.26 s) | 0.96% (1908 s) |
| RFF-PEGASOS (2) | 15.84% (5.02 s) | 9.55% (0.09 s) | 10.1% (0.51 s) | 0.69% (3800 s) |
| RFF-PEGASOS (5) | 14.95% (12.59 s) | 9.69% (0.19 s) | 8.9% (1.24 s) | 0.55% (9463 s) |
| RFF-PEGASOS (10) | 14.75% (25 s) | 9.62% (0.36 s) | 8.7% (2.45 s) | 0.51% (18986 s) |

**Table 7.6 Comparing the performances of LA-SVM and RFF-PEGASOS**

| Method | Adult | Banana | Waveform | MNIST |
|---|---|---|---|---|
| RFF-Peg (1) | 16.34% (10 s) | 9.62% ($<$ 1 s) | 10% (1 s) | 0.98% (1.5 h) |
| RFF-Peg (2) | 15.67% (21 s) | 9.46% ($<$ 1 s) | 8.9% (1 s) | 0.76% (3 h) |
| RFF-Peg (5) | 14.97% (52 s) | 9.46% ($\approx$1 s) | 8.7% (2 s) | 0.62% (7 h) |
| RFF-Peg (10) | 14.76% (104 s) | 9.62% ($\approx$1 s) | 8.8% (5 s) | 0.52% (14 h) |
| LA-SVM | 14.94% (276 s) | 9.92% (3 s) | 8.8% (1 s) | 0.46% (313 s) |

tations. In order to compare running times we used a C implementation of the respective algorithms (as the C implementation of LA-SVM is publicly available on the web). The experiments were performed on an i5 650 machine running at 3.20 GHz with 6 Gb of RAM. The parameters for LA-SVM (shown in Table 7.3) have been taken from [19]. The results are shown in Table 7.6. Once more, we observe that for datasets with a large number of support vectors RFF-PEGASOS is faster. This is not the case however for datasets with a lower number of support vectors, where RFF-PEGASOS can be significantly slower.

*Remark* 1. We stress that the second set of experiments concerns classification tasks in a batch mode. The LA-SVM method is one of the most powerful SVM solvers, but it is not designed for online tasks,

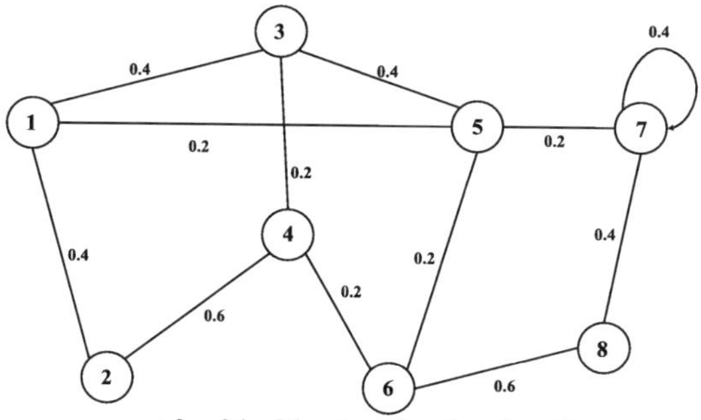


$$\mathbf{A} = \begin{pmatrix} 0 & 0.4 & 0.4 & 0 & 0.2 & 0 & 0 & 0 \\ 0.4 & 0 & 0 & 0.6 & 0 & 0 & 0 & 0 \\ 0.4 & 0 & 0 & 0.2 & 0.4 & 0 & 0 & 0 \\ 0 & 0.6 & 0.2 & 0 & 0 & 0.2 & 0 & 0 \\ 0.2 & 0 & 0.4 & 0 & 0 & 0.2 & 0 & 0 \\ 0 & 0 & 0 & 0.2 & 0.2 & 0 & 0 & 0.6 \\ 0 & 0 & 0 & 0 & 0.2 & 0 & 0.4 & 0.4 \\ 0 & 0 & 0 & 0 & 0 & 0.6 & 0.4 & 0 \end{pmatrix}$$

**FIGURE 7.3**

A network topology (with 8 nodes) and the respective weight matrix.

as is the case for the RFF-Pegasos. The provided experiments demonstrate that in many cases the RFF-Pegasos can be used in a batch mode as well.

## 7.4 ONLINE DISTRIBUTED LEARNING WITH KERNELS

Although the RFF methods, presented in Section 7.3, constitute a competitive alternative to existing online learning techniques in RKHS, the obtained results do not offer any significant real benefits from a practical point of view. In this section, we discuss the scenario of online distributed learning in RKHS, where it will be shown that the random Fourier features rationale can solve major open problems and lead to simple and elegant solutions.

In the following, we consider $K$ connected nodes, labeled $k \in \mathcal{N} = \{1, 2, \dots K\}$, which operate in cooperation to solve a specific learning task. This is a more general learning scenario compared to the one considered in Section 7.3, as the nodes are able to communicate with each other and exchange valuable information. The nodes comprising the network and the corresponding topology are represented as an undirected *connected* graph, consisting of $K$ vertices (*i.e.*, the nodes) and a set of edges that represent the communication between those nodes (see Fig. 7.3). The nodes that interact with node $k$ are called the "neighbors" of $k$ and the respective subset (that comprises the neighbors of $k$) is denoted as $\mathcal{N}_k \subseteq \mathcal{N}$. Moreover, we assign a nonnegative weight $a_{k,l}$ to the edge connecting node $k$ to $l$, which is used by $k$ to scale the data transmitted from $l$ and *vice versa* (*i.e.*, the confidence that node $k$ assigns to node $l$). We collect all those weights into a $K \times K$ symmetric matrix $\mathbf{A} = (a_{k,l})$, such that the entries of the $k$th row of $\mathbf{A}$ contain the coefficients used by node $k$ to scale the data arriving from its neighbors. Additionally, we assume that $\mathbf{A}$ is doubly stochastic, so that the weights of all incoming and outgoing "transmissions" sum to 1. Such a matrix can be generated using the *Metropolis rule*:

$$a_{k,l} = \begin{cases} \frac{1}{\max\{|\mathcal{N}_k|,|\mathcal{N}_l|\}}, & \text{if } l \in \mathcal{N}_k \text{ and } l \neq k, \\ 1 - \sum_{i \in \mathcal{N}_k \setminus k} a_{k,i}, & \text{if } l = k, \\ 0, & \text{otherwise.} \end{cases}$$

Note that the topology considered here is similar to that of Chapter 9, which also deals with nonlinear learning over networks. Finally, we assume that each node, $k$, receives streaming data $\{(\mathbf{x}_{k,n}, y_k(n)), n = 1, 2, \ldots\}$ that are generated by $y_k(n) = g(\mathbf{x}_{k,n}) + \eta_k(n)$, where $\mathbf{x}_{k,n} \in \mathbb{R}^d$, $y_k(n) \in \mathbb{R}$ and $\eta_k(n)$ represents the respective noise, for the regression task. For classification, we assume that $y_k(n) = \phi(f(\boldsymbol{x}_{k,n}))$, where $\phi$ is a thresholding function, such that $y_k(n) \in \{-1, 1\}$. In both tasks, the goal of each node is to obtain an estimate of the nonlinear function $g$, working in cooperation with its neighbors, so that a specific cost function is minimized.

In the case where $g$ is linear, *i.e.*, it can be modeled as $g(\mathbf{x}) = \boldsymbol{\theta}^T \mathbf{x}$, for some vector $\boldsymbol{\theta} \in \mathbb{R}^d$, the problem has been studied extensively and several learning methods have been proposed based on the diffusion scheme (*e.g.*, the diffusion LMS and the diffusion RLS). In a nutshell, these methods can be summarized in the following steps (this rationale is usually called Combine-Then-Adapt or CTA for short):

- Each node, $k$, receives the new data pair, *i.e.*, $(\mathbf{x}_{k,n}, y_k(n))$ as well as the current estimates, $\boldsymbol{\theta}_{l,n-1}$, from all neighboring nodes, *i.e.*, for all $l \in \mathcal{N}_k$.
- Consequently, the node combines these estimates to a single one, $\boldsymbol{\psi}_{k,n-1}$, taking into account the weights between the nodes, for example, $\boldsymbol{\psi}_{k,n-1} = \sum_{l \in \mathcal{N}_k} a_{k,l} \boldsymbol{\theta}_{l,n-1}$.
- Finally, the node exploits the combined estimate, $\boldsymbol{\psi}_{k,n-1}$, as well as the newly received data to update the estimation. For example in diffusion LMS, where the mean square cost is minimized using a gradient descent rationale, the respective step update equation becomes $\boldsymbol{\theta}_{k,n} = \boldsymbol{\psi}_{k,n-1} + \mu e_{k,n} \mathbf{x}_{k,n}$, where $e_k(n) = y_k(n) - \boldsymbol{\psi}_{k,n-1}^T \mathbf{x}_{k,n}$ is the estimation error.

The implementation of such an approach in the context of RKHS presents significant challenges. The big difference lies in the fact that the estimation of the solution at each node is not a simple vector, but instead it is a function, which is expressed as a growing sum of kernel evaluations centered at the points observed by the specific node, *i.e.*,

$$f_{k,n} = \sum_{i=1}^{n} w_{k,i} \kappa(\cdot, \mathbf{x}_{k,i}). \tag{7.11}$$

Hence, the implementation of a straightforward CTA strategy would require from each node, $k$, to transmit its entire growing sum (*i.e.*, the so-called dictionary, which comprises the coefficients $w_{k,i}$ as well as the respective centers $\mathbf{x}_{k,i}$, for all $i = 1, 2, \ldots, n$) to all neighbors. This would significantly increase both the communication load among the nodes, as well as the computational cost at each node, since the size of the transmitted data would become increasingly larger as time evolves (as for every time instant, they gather the centers transmitted by all neighbors). Evidently, after a specific point in time this scheme would become so demanding that no machine would be able to carry on indefinitely. This is the rationale adopted in [20–22] for the case of KLMS. Alternatively, one could follow a pruning method similar to those applied for the standard KLMS and devise an efficient method to sparsify the solution at each node. However, this would require that, at each node, to somehow merge all the respective dictionaries (*i.e.*, Eq. (7.11)) transmitted by its neighbors, at each time instant. For example, each node would have to search all the dictionaries that it receives from the neighboring nodes, for similar centers, and treat them as a single one. Alternatively, one could adopt a single prearranged dictionary (*i.e.*, a specific set of centers) for all nodes and then fuse each observed point with the best-suited center. However, no such strategy has appeared in the respective literature, perhaps due to its increased complexity and lack of theoretical elegance.

In the present chapter, we propose an alternative approach, which addresses the aforementioned problems. Once again, our starting point is the observation that the input–output relationship, $y = f(\mathbf{x})$, can be closely approximated using the random Fourier features rationale (similar to Section 7.3). In particular, we see that

$$\begin{aligned} f_{k,n}(\mathbf{x}) &= \sum_{i=1}^{n} w_{k,i}\kappa(\mathbf{x}, \mathbf{x}_{k,i}) \approx \sum_{i=1}^{n} w_{k,i}\mathbf{z}_{\Omega}(\mathbf{x})^T \mathbf{z}_{\Omega}(\mathbf{x}_{k,i}) \\ &= \left(\sum_{i=1}^{n} w_{k,i}\mathbf{z}_{\Omega}(\mathbf{x}_{k,i})\right)^T \mathbf{z}_{\Omega}(\mathbf{x}). \end{aligned}$$

Inspired by the last relation, we propose to approximate the desired input–output relationship as $y = \boldsymbol{\theta}^T \mathbf{z}_{\Omega}(\mathbf{x})$ and follow a two-step procedure: (a) we map each observed point $(\mathbf{x}_{k,n}, y_k(n))$ to $(\mathbf{z}_{\Omega}(\mathbf{x}_{k,n}), y_k(n))$ and then (b) we adopt a simple linear CTA diffusion strategy on the transformed points, where each node aims to estimate the vector $\boldsymbol{\theta} \in \mathbb{R}^D$ minimizing a specific (convex) cost function, $\mathcal{L}(\mathbf{x}, y, \boldsymbol{\theta})$. We can summarize the proposed scheme for each node by the following relations:

$$\boldsymbol{\psi}_{k,n-1} = \sum_{l \in \mathcal{N}_k} a_{k,l}\boldsymbol{\theta}_{l,n-1}, \tag{7.12}$$

$$\boldsymbol{\theta}_{k,n} = \boldsymbol{\psi}_{k,n-1} - \mu_{k,n}\nabla_{\boldsymbol{\theta}}\mathcal{L}(\mathbf{z}_{\Omega}(\mathbf{x}_{k,n}), y_k(n), \boldsymbol{\psi}_{k,n-1}), \tag{7.13}$$

where $\mu_{k,n}$ is the (possibly time varying) learning rate at the $n$th time instant on node $k$ and $\nabla_{\boldsymbol{\theta}}\mathcal{L}(\mathbf{z}_{\Omega}(\mathbf{x}_{k,n}), y_{k,n}, \boldsymbol{\psi}_{k,n-1})$ is the gradient, or any subgradient of $\mathcal{L}(\mathbf{x}, y, \boldsymbol{\theta})$ (with respect to $\boldsymbol{\theta}$), if the loss function is not differentiable. Eq. (7.12) refers to the fact that each node collects the past estimates from its neighbors to produce the combined estimate, $\boldsymbol{\psi}_{k,n-1}$, while in Eq. (7.13) the most recent pair of observations is exploited to update the new estimate vector $\boldsymbol{\theta}_{k,n}$, employing a gradient descent–based rationale. We emphasize that $\mathcal{L}$ need not be differentiable. Hence, a large family of loss functions can be adopted. However, in this section, we focus on the following two types of cost functions:

- Mean squared error (for Regression): $\mathcal{L}(\mathbf{x}, y, \boldsymbol{\theta}) = E[(y - \boldsymbol{\theta}^T\mathbf{x})^2]$.
- Hinge loss (for Classification): $\mathcal{L}(\mathbf{x}, y, \boldsymbol{\theta}) = \max(0, 1 - y\boldsymbol{\theta}^T\mathbf{x})$.

We can write Eq. (7.12) and Eq. (7.13) more compactly using block-matrices as follows:

$$\underline{\boldsymbol{\theta}}_n = \underline{\mathbf{A}\boldsymbol{\theta}}_{n-1} - \underline{\boldsymbol{M}}_n\underline{\boldsymbol{G}}_n, \tag{7.14}$$

**Algorithm 4** The Random Fourier Features Diffusion KLMS algorithm.

Set $\boldsymbol{\theta} = \mathbf{0}$. Select the dimension of the new space, $D$, the learning rate $\mu$ and the parameter of the kernel ($\sigma$).
Generate the matrix $\boldsymbol{\Omega}$ as in (7.5).
**for** each node $k$ **do**
  $\boldsymbol{\theta}_{k,0} \leftarrow \mathbf{0}_D$ ▷ Initialization
**for** $n = 1, 2, 3, \ldots$ **do**
  **for** each node $k$ **do**
    $\boldsymbol{\psi}_{k,n-1} \leftarrow \sum_{l \in \mathcal{N}_k} a_{k,l} \boldsymbol{\theta}_{l,n-1}$,
    Map the current data pair, $(\mathbf{x}_{k,n}, y_k(n))$, to $(\mathbf{z}_{\boldsymbol{\Omega}}(\mathbf{x}_{k,n}), y_k(n))$.
    $\hat{y}_k(n) \leftarrow \boldsymbol{\psi}_{k,n-1}^T \mathbf{z}_{\boldsymbol{\Omega}}(\mathbf{x}_{k,n})$. ▷ System output
    $\varepsilon_k(n) = y_k(n) - \hat{y}_k(n)$
    $\boldsymbol{\theta}_{k,n} \leftarrow \boldsymbol{\psi}_{k,n-1} + \mu \varepsilon_k(n) \mathbf{z}_{\boldsymbol{\Omega}}(\mathbf{x}_{k,n})$.

where $\underline{\boldsymbol{\theta}}_n := (\boldsymbol{\theta}_{1,n}^T, \ldots, \boldsymbol{\theta}_{K,n}^T)^T \in \mathbb{R}^{KD}$ is the collection of the estimations from each node, $\underline{\boldsymbol{M}}_n := \operatorname{diag}\{\mu_{1,n}, \ldots, \mu_{K,n}\} \otimes \mathbf{I}_D$ is the collection of learning rates, $\underline{\mathbf{A}} := \mathbf{A} \otimes \mathbf{I}_D$ and $\underline{\boldsymbol{G}}_n := [\mathbf{u}_{1,n}^T, \ldots, \mathbf{u}_{K,n}^T]^T$, where $\mathbf{u}_{k,n} = \nabla \mathcal{L}(\mathbf{z}_{\boldsymbol{\Omega}}(\mathbf{x}_{k,n}), y_k(n), \boldsymbol{\psi}_{k,n})$ comprises all (sub)gradients. Here, the symbol $\otimes$ denotes the matrix tensor product.

The advantage of the proposed scheme is that each node transmits a single vector (*i.e.*, its current estimate, $\boldsymbol{\theta}_{k,n}$) to its neighbors, instead of an entire dictionary of centers (and their respective coefficients) while the merging of the currently available estimates requires only a straightforward summation. However, we should point out that the new vector space, $\mathbb{R}^D$, usually has a significantly larger dimension than the original input space $\mathbb{R}^d$. This may pose some communication obstacles; to this end, various suboptimal techniques can be used and have been proposed, *e.g.*, [23].

### 7.4.1 DIFFUSION KLMS

Adopting the mean square error in place of $\mathcal{L}$ and estimating the gradient by its current measurement, the Random Fourier Features Diffusion KLMS (RFF-DKLMS) results, where Eq. (7.12) can be recast as

$$\boldsymbol{\theta}_{k,n} = \boldsymbol{\psi}_{k,n-1} + \mu \varepsilon_k(n) \mathbf{z}_{\Omega}(\mathbf{x}_{k,n}), \tag{7.15}$$

where $\varepsilon_k(n) = y_n - \boldsymbol{\psi}_{k,n-1}^T \mathbf{z}_{\Omega}(\mathbf{x}_{k,n})$. Algorithm 4 describes the proposed method in detail.

Under certain general conditions, we can establish consensus (*i.e.*, that all nodes converge to the same solution), [16], following the results of the standard Diffusion LMS (*e.g.*, [24,25]). To this end, we will assume that the data pairs are generated by

$$y_k(n) = \sum_{m=1}^{M} \alpha_m \kappa(\mathbf{c}_m, \mathbf{x}_{k,n}) + \eta_k(n), \tag{7.16}$$

where $\mathbf{c}_1, \ldots, \mathbf{c}_M$ are fixed centers, $\mathbf{x}_{k,n}$ are zero-mean i.i.d, samples drawn from the Gaussian distribution with covariance matrix $\sigma_X^2 \mathbf{I}_d$ and $\eta_k(n)$ are independent and identically distributed noise samples drawn from $\mathcal{N}(0, \sigma_\eta^2)$. Hence, we have

$$\begin{aligned} y_k(n) &= \sum_{m=1}^{M} \alpha_m E_{\omega,b}[z_{\omega,b}(\mathbf{c}_m) z_{\omega,b}(\mathbf{x}_{k,n})] + \eta_k(n) \\ &\approx \boldsymbol{\alpha}^T \mathbf{Z}_\Omega^T \mathbf{z}_\Omega(\mathbf{x}_{k,n}) + \epsilon_k(n) + \eta_k(n) \\ &= \boldsymbol{\theta}_o^T \mathbf{z}_\Omega(\mathbf{x}_{k,n}) + \epsilon_k(n) + \eta_k(n), \end{aligned}$$

where $\mathbf{Z}_\Omega = (\mathbf{z}_\Omega(\mathbf{c}_1), \ldots, \mathbf{z}_\Omega(\mathbf{c}_M))$, $\boldsymbol{\alpha} = (\alpha_1, \ldots, \alpha_M)^T$, $\boldsymbol{\theta}_o = \mathbf{Z}_\Omega \boldsymbol{\alpha}$ and $\epsilon_k(n)$ is the approximation error between the noise-free component of $y_k(n)$ (evaluated only by the linear kernel expansion of Eq. (7.16)) and the approximation of this component using random Fourier features, *i.e.*, $\epsilon_k(n) = \sum_{m=1}^{M} \alpha_m \kappa(\mathbf{c}_m, \mathbf{x}_{k,n}) - \boldsymbol{\theta}_o^T \mathbf{z}_\Omega(\mathbf{x}_{k,n})$. Using block-matrices we can derive the following, more compact, formulation for the entire network:

$$\underline{\mathbf{y}}_n = \underline{\boldsymbol{V}}_n^T \underline{\boldsymbol{\theta}}_o + \underline{\boldsymbol{\epsilon}}_n + \underline{\boldsymbol{\eta}}_n, \tag{7.17}$$

where

- $\underline{\mathbf{y}}_n := (y_1(n), y_2(n), \ldots, y_K(n))^T$,
- $\underline{\boldsymbol{V}}_n := \text{diag}(\mathbf{z}_\Omega(\mathbf{x}_{1,n}), \mathbf{z}_\Omega(\mathbf{x}_{2,n}), \ldots, \mathbf{z}_\Omega(\mathbf{x}_{K,n}))$, is a $DK \times K$ matrix,
- $\underline{\boldsymbol{\theta}}_o = \left(\boldsymbol{\theta}_o^T, \boldsymbol{\theta}_o^T, \ldots, \boldsymbol{\theta}_o^T\right)^T \in \mathbb{R}^{DK}$,
- $\underline{\boldsymbol{\epsilon}}_n = (\epsilon_1(n), \epsilon_2(n), \ldots, \epsilon_K(n))^T \in \mathbb{R}^K$,
- $\underline{\boldsymbol{\eta}}_n = (\eta_1(n), \eta_2(n), \ldots, \eta_K(n))^T \in \mathbb{R}^K$.

Let $\mathbf{x}_1, \ldots, \mathbf{x}_K \in \mathbb{R}^d$, $\underline{\mathbf{y}} \in \mathbb{R}^K$, be the random variables that generate the measurements of the nodes and

$$\underline{\boldsymbol{V}} = \text{diag}(\mathbf{z}_\Omega(\mathbf{x}_1), \mathbf{z}_\Omega(\mathbf{x}_2), \ldots, \mathbf{z}_\Omega(\mathbf{x}_K))$$

be the $DK \times K$ matrix that collects the transformed random variables for the whole network. Moreover, let $\underline{\mathbf{R}} = E[\underline{\boldsymbol{V}}\,\underline{\boldsymbol{V}}^T]$ be the $DK \times DK$ respective autocorrelation matrix. It is not difficult to see that $\underline{\mathbf{R}}$ can be given in a block form as $\underline{\mathbf{R}} = E[\underline{\boldsymbol{V}}\,\underline{\boldsymbol{V}}^T] = \text{diag}(\mathbf{R}_{zz}, \mathbf{R}_{zz} \ldots, \mathbf{R}_{zz})$, where $\mathbf{R}_{zz} = E[\mathbf{z}_\Omega(\mathbf{x}_k)\mathbf{z}_\Omega(\mathbf{x}_k)^T]$, for all $k = 1, 2, \ldots, K$. We can prove that, under certain general conditions, the autocorrelation matrix $\mathbf{R}$ is invertible (see [16] for the proof).

**Lemma 1.** *Consider a selection of samples $\boldsymbol{\omega}_1, \boldsymbol{\omega}_2, \ldots, \boldsymbol{\omega}_D$, drawn from (7.6) such that $\boldsymbol{\omega}_i \neq \boldsymbol{\omega}_j$, for any $i \neq j$. Then both $\mathbf{R}_{zz}$ and $\underline{\mathbf{R}} = E[\underline{\boldsymbol{V}}\,\underline{\boldsymbol{V}}^T]$ are strictly positive definite (hence invertible) matrices.*

Moreover, as $\mathbf{R}_{zz}$ is positive definite (this has been also discussed in Section 7.4.1), the respective eigenvalues satisfy $0 < \lambda_1 \leq \lambda_2 \leq \cdots \leq \lambda_D$. In this case, the optimal solution is given by

$$\begin{aligned} \underline{\boldsymbol{\theta}}_* &= \text{argmin}_{\boldsymbol{\theta}} E[\|\underline{\mathbf{y}} - \underline{\boldsymbol{V}}^T \underline{\boldsymbol{\theta}}\|^2] \\ &= E[\underline{\boldsymbol{V}}\,\underline{\boldsymbol{V}}^T]^{-1} E[\underline{\boldsymbol{V}}\underline{\mathbf{y}}] \\ &= E[\underline{\boldsymbol{V}}\,\underline{\boldsymbol{V}}^T]^{-1} \left(E[\underline{\boldsymbol{V}}(\underline{\boldsymbol{V}}^T \underline{\boldsymbol{\theta}}_o + \underline{\boldsymbol{\epsilon}} + \underline{\boldsymbol{\eta}})]\right) \end{aligned}$$

$$= E[\underline{V}\underline{V}^T]^{-1}\left(E[\underline{V}\underline{V}^T]\underline{\theta}_o + E[\underline{V}\epsilon] + E[\underline{V}\underline{\eta}]\right)$$
$$= E[\underline{V}\underline{V}^T]^{-1}\left(E[\underline{V}\underline{V}^T]\underline{\theta}_o + E[\underline{V}\epsilon]\right),$$

where for the last relation we have used the notion that $\eta$ is a zero mean vector representing noise and that $\underline{V}$ and $\underline{\eta}$ are independent. For large enough $D$, the approximation error vector $\underline{\epsilon}$ approaches $\mathbf{0}_K$, hence the optimal solution becomes

$$\underline{\theta}_* = \underline{\theta}_o + E[\underline{V}\underline{V}^T]^{-1}E[\underline{V}\epsilon] \approx \underline{\theta}_o.$$

Since we work under the assumption that $\underline{\epsilon}$ approaches $\mathbf{0}_K$, we can say that Eq. (7.17) can be closely approximated by $\underline{\mathbf{y}}_n \approx \underline{V}_n\underline{\theta}_o + \underline{\eta}_n$. This actually means that the RFF-DKLMS is nothing more than the standard diffusion LMS applied to the data pairs $\{(\mathbf{z}_\Omega(\mathbf{x}_{k,n}), y_k(n), k = 1, \ldots, K, n = 1, 2 \ldots\}$. However, the difference is that the input vectors $\mathbf{z}_\Omega(\mathbf{x}_{k,n})$ may have nonzero mean and do not follow, necessarily, the Gaussian distribution. In fact we can prove that, if $\mathbf{x}_{k,n} \sim \mathcal{N}(\mathbf{0}, \sigma_X \mathbf{I}_d)$, then the entries of $\mathbf{R}_{zz}$ can be evaluated as

$$r_{i,j} = \frac{1}{2}\exp\left(\frac{-\|\boldsymbol{\omega}_i - \boldsymbol{\omega}_j\|^2\sigma_X^2}{2}\right)\cos(b_i - b_j)$$
$$+ \frac{1}{2}\exp\left(\frac{-\|\boldsymbol{\omega}_i + \boldsymbol{\omega}_j\|^2\sigma_X^2}{2}\right)\cos(b_i + b_j).$$

Consequently, the available results regarding convergence and stability of diffusion LMS (*e.g.*, [24,26]) cannot be applied here directly, as in these works the inputs are assumed to be zero mean Gaussian (to simplify the formulas related to stability). However, it is possible to end up with similar conclusions (the proofs are given in [16]).

**Proposition 1.** *If the step update $\mu$ satisfies: $0 < \mu < \frac{2}{\lambda_D}$, where $\lambda_D$ is the maximum eigenvalue of $R_{zz}$, then the RFF-DKLMS achieves asymptotic consensus in the mean,* i.e.,

$$\lim_n E[\boldsymbol{\theta}_{k,n} - \boldsymbol{\theta}_o] = \mathbf{0}_D, \text{ for all } k = 1, 2, \ldots, K.$$

**Proposition 2.** *For stability in the mean square sense, we must ensure that both $\mu$ and $\mathbf{A}$ satisfy*

$$\rho\left(\mathbf{I}_{D^2K^2} - \mu\left(\underline{\mathbf{R}} \boxtimes \mathbf{I}_{DK} + \mathbf{I}_{DK} \boxtimes \underline{\mathbf{R}}\right)(\underline{\mathbf{A}} \boxtimes \underline{\mathbf{A}})\right) < 1,$$

*where $\boxtimes$ denotes the unbalanced block Kronecker product (see [27]) and $\rho(\cdot)$ denotes the spectral radius of the respective matrix.*

### 7.4.2 DIFFUSION PEGASOS

For the case of classification tasks, we can adopt the regularized hinge loss function, for a specific value of the regularization parameter, $\lambda$, and get a diffusion version of PEGASOS, in its full online version (see Algorithm 5). In this case, the gradient becomes

$$\nabla_{\boldsymbol{\theta}}\mathcal{L}(\mathbf{x}, y, \boldsymbol{\theta}) = \lambda\boldsymbol{\theta} - \boldsymbol{I}_+(1 - y\boldsymbol{\theta}^T\mathbf{z}_\Omega(\mathbf{x}))y\mathbf{z}_\Omega(\mathbf{x}),$$

**Algorithm 5** The Random Fourier Features Diffusion PEGASOS algorithm.

Set $\boldsymbol{\theta} = \mathbf{0}$. Select the dimension of the new space, $D$, the regularization parameter $\lambda$ and the parameter of the kernel ($\sigma$).
Generate the matrix $\Omega$ as in (7.5).
**for** each node $k$ **do**
  $\boldsymbol{\theta}_{k,0} \leftarrow \mathbf{0}_D$ ▷ Initialization
**for** $n = 1, 2, 3, \ldots$ **do**
  **for** each node $k$ **do**
    $\mu_n \leftarrow \frac{1}{\lambda \cdot n}$
    $\boldsymbol{\psi}_{k,n-1} \leftarrow \sum_{l \in \mathcal{N}_k} a_{k,l} \boldsymbol{\theta}_{l,n-1}$,
    Map the current data pair, $(\mathbf{x}_{k,n}, y_k(n))$, to $(\mathbf{z}_\Omega(\mathbf{x}_{k,n}), y_k(n))$.
    $\hat{y}_k(n) \leftarrow \boldsymbol{\psi}_{k,n-1}^T \mathbf{z}_\Omega(\mathbf{x}_{k,n})$. ▷ System output
    **if** $y_k(n)\hat{y}_k(n) < 1$ **then**
      $\boldsymbol{\theta}_{k,n} \leftarrow (1 - \mu_n \lambda)\boldsymbol{\psi}_{k,n} + \mu_n y_k(n) \mathbf{z}_\Omega(\mathbf{x}_{k,n})$.
    **else**
      $\boldsymbol{\theta}_{k,n} \leftarrow (1 - \mu_n \lambda)\boldsymbol{\psi}_{k,n}$.

where $\boldsymbol{I}_+$ is the indicator function of $(0, +\infty)$, which takes a value of 1, if its argument belongs in $(0, +\infty)$, and zero otherwise. Hence the step update equation of the algorithm becomes

$$\boldsymbol{\theta}_{k,n} = (1 - \frac{1}{n})\boldsymbol{\psi}_{k,n-1} + \boldsymbol{I}_+(1 - y_n \boldsymbol{\psi}_{k,n-1}^T \mathbf{z}_\Omega(\mathbf{x}_{k,n}))\frac{y_k(n)}{\lambda n}\mathbf{z}_\Omega(\mathbf{x}_{k,n}), \tag{7.18}$$

where, following [17], we have used a decreasing learning rate, $\mu_n = \frac{1}{\lambda n}$. Recall, however, that we can write the update scheme more compactly by Eq. (7.14), setting $\boldsymbol{M}_n = \text{diag}(1/(\lambda n), \ldots, 1/(\lambda n)) \otimes I_D$.

We can prove that all nodes converge to the same solution using the result from [16], where it is proven that the general update scheme described by Eq. (7.14) achieves asymptotic consensus, if the following assumptions are satisfied:

Assumption 1: The step size is time decaying and is bounded by the inverse square root of time, *i.e.*, $\mu_{k,n} = \mu_n \leq \mu n^{-1/2}$.

Assumption 2: The norm of the transformed input is bounded, *i.e.*, there is $U_1 > 0$, such that $\|\mathbf{z}_\Omega(\mathbf{x}_{k,n})\| \leq U_1$, $\forall k \in \mathcal{N}$, $\forall n \in \mathbb{N}$. Furthermore, $y_{k,n}$ is bounded, *i.e.*, there is $V > 0$, such that $|y_{k,n}| \leq V$, $\forall k \in \mathcal{N}$, $\forall n \in \mathbb{N}$.

Assumption 3: The estimates are bounded, *i.e.*, there is $U_2 > 0$, such that $\|\boldsymbol{\theta}_{k,n}\| \leq U_2$, $\forall k \in \mathcal{N}$, $\forall n \in \mathbb{N}$.

Assumption 4: The matrix comprising the combination weights, *i.e.*, $A$, is doubly stochastic.

Moreover, in [16], it is shown that the regret is sublinearly bounded, which means that on average the algorithm performs at least as well as the best fixed strategy.

**Proposition 3.** *Under Assumptions 1–4 the network-wise regret is bounded by*

$$\sum_{i=1}^{N}\sum_{k \in \mathcal{N}}(\mathcal{L}(\mathbf{x}_{k,i}, y_{k,i}, \boldsymbol{\psi}_{k,i}) - \mathcal{L}(\mathbf{x}_{k,i}, y_{k,i}, \mathbf{g})) \leq \gamma\sqrt{N} + \delta,$$

*for all* $\mathbf{g} \in \mathcal{B}_{[\mathbf{0}_D, U_2]}$, *where* $\gamma$, $\delta$ *are positive constants and* $\mathcal{B}_{[\mathbf{0}_D, U_2]}$ *is the closed ball with center* $\mathbf{0}_D$ *and radius* $U_2$.

### 7.4.3 SIMULATIONS—DIFFUSION KLMS

In order to demonstrate that the estimation provided by the cooperative strategy is better than having each node working alone, we follow a similar experimental setup as in Section 7.3.4. Each realization of the experiments uses a different random connected graph with $K = 10$ nodes and with a probability of attachment per node equal to 0.25 (*i.e.*, there is a 25% probability that a specific node $k$ is connected to any other node $l$). The graphs are generated using MIT's random_graph routine (see [28]) and their adjacency matrices, $\mathbf{A}$, using the Metropolis rule (resulting in graphs with mean algebraic connectivity around 0.69). For the noncooperative case, we simply used a graph that connects each node to itself, *i.e.*, $\mathbf{A} = \mathbf{I}_{10}$. All parameters were optimized (after trials) to give the lowest MSE. The algorithms were implemented in Matlab and the experiments were performed on an i7-3770 machine running at 3.4 GHz with 32 Mb of RAM. Similar to Section 7.3.4, we generated 5000 data pairs for each node using the following model:

$$y_k(n) = \sum_{m=1}^{M} a_m \kappa(\mathbf{c}_m, \mathbf{x}_{k,n}) + \eta_k(n),$$

where $\mathbf{x}_{k,n} \in \mathbb{R}^5$ are drawn from $\mathcal{N}(\mathbf{0}, \mathbf{I}_5)$ and the noise are independent and identically distributed Gaussian samples with $\sigma_\eta = 0.1$. The parameters of the expansion (*i.e.*, $a_1, \ldots, a_M$) are drawn from $\mathcal{N}(0, 25)$, the kernel parameter $\sigma$ is set to 5 and the learning rate is set to $\mu = 1$. Fig. 7.4 shows the evolution of the MSE over all network nodes for 100 realizations of the experiment (for various values of $D$ and $M$). We note that the selected value of the step size satisfies the conditions of Proposition 1.

### 7.4.4 SIMULATIONS—DIFFUSION PEGASOS

We have tested the performance of Diffusion-PEGASOS versus the noncooperative kernel Pegasus on the same four datasets that were used in Section 7.3.5. In all experiments, we generated random graphs (using MIT's random_graph routine) and compared the proposed diffusion method versus the respective noncooperative strategy (where each node works independent of the rest). Similar to Section 7.4.3, a different random connected graph with $K = 5$ or $K = 20$ nodes was generated, for each realization of the experiments. The probability of attachment per node was set equal to 0.2 and the respective adjacency matrix, $\mathbf{A}$, of each graph was generated using the Metropolis rule. For the noncooperative strategies, we used a graph that connects each node to itself, *i.e.*, $\mathbf{A} = \mathbf{I}_5$ or $\mathbf{A} = \mathbf{I}_{20}$, respectively. Moreover, for each realization, the corresponding dataset was randomly split into $K$ subsets of equal size (one for every node). All parameters were optimized (after trials) to give the lowest number of test errors. Their values are reported in Table 7.9. The algorithms were implemented in Matlab and the experiments were performed on an i7-3770 machine running at 3.4 GHz with 32 Gb of RAM. Tables 7.7 and 7.8 report the mean test errors obtained by both procedures. The number inside the parentheses indicates the times of data reuse (*i.e.*, running the algorithm again over the same data, albeit with a continuously decreasing step size $\mu_n$), which has been suggested to improve the classification accuracy of PEGASOS (see [17]). For example, the number 2 indicates that the algorithm runs

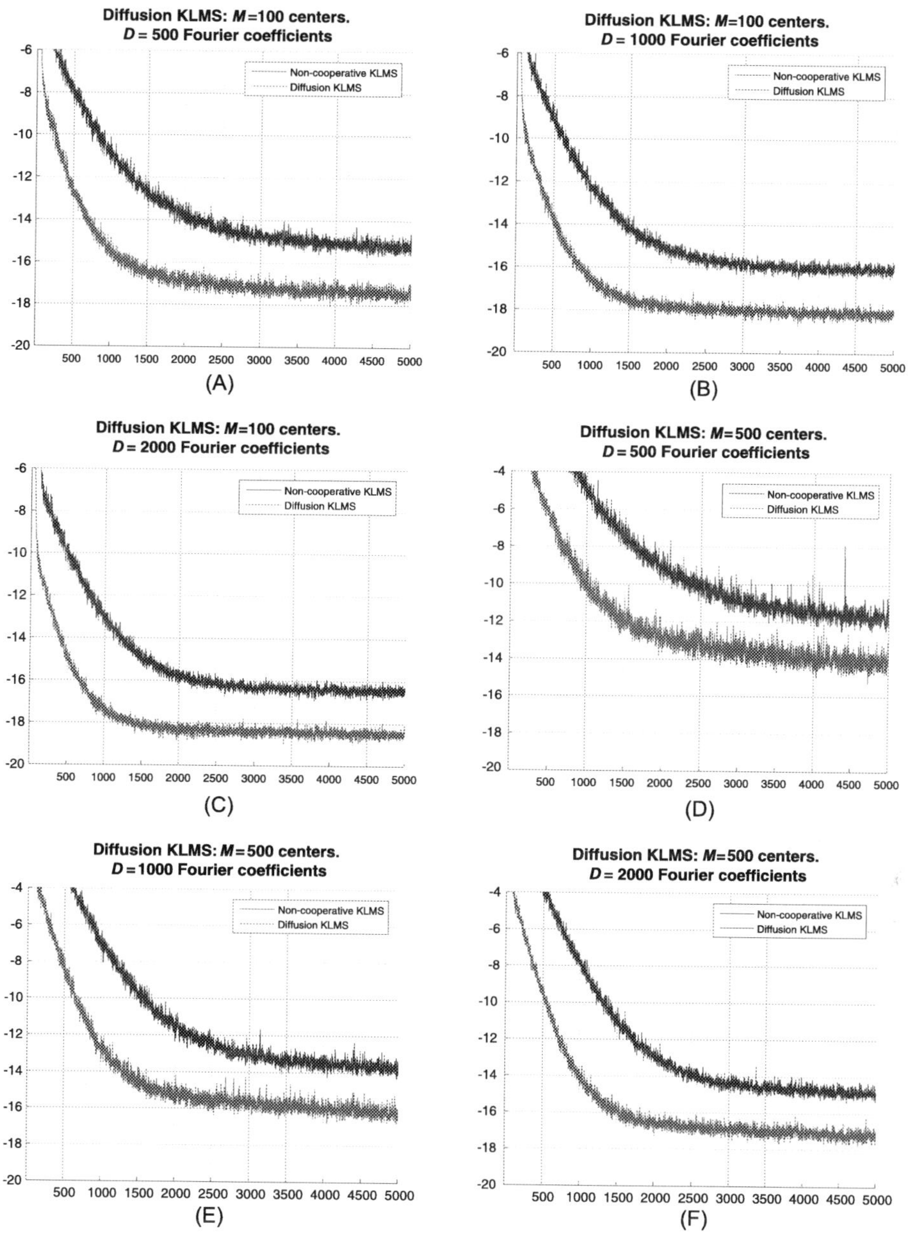


**FIGURE 7.4**

Comparing the performances of diffusion KLMS versus the uncooperative KLMS. (A) $D=500$, $M=100$. (B) $D=1000$, $M=100$. (C) $D=2000$, $M=100$. (D) $D=500$, $M=500$. (E) $D=1000$, $M=500$. (F) $D=2000$, $M=500$.

**Table 7.7 Comparing the performances of the Diffusion PEGASOS versus the noncooperative PEGASOS for graphs with $K = 5$ nodes**

| Method | Adult | Banana | Waveform | MNIST |
|---|---|---|---|---|
| Diffusion-PEGASOS (1) | 19% | 11.80% | 11.82% | 0.79% |
| Diffusion-PEGASOS (2) | 17.43% | 10.84% | 10.49% | 0.68% |
| Diffusion-PEGASOS (5) | 15.87% | 10.34% | 9.56% | 0.59% |
| Noncooperative-PEGASOS (1) | 19.11% | 14.52% | 13.75% | 1.42% |
| Noncooperative-PEGASOS (2) | 18.31% | 12.52% | 12.59% | 1.19% |
| Noncooperative-PEGASOS (5) | 17.29% | 11.32% | 11.86% | 1.01% |

**Table 7.8 Comparing the performances of the Distributed PEGASOS versus the noncooperative PEGASOS for graphs with $K = 20$ nodes**

| Method | Adult | Banana | Waveform | MNIST |
|---|---|---|---|---|
| Diffusion-PEGASOS (1) | 24.04% | 16.38% | 16.26% | 1.03% |
| Diffusion-PEGASOS (2) | 22.34% | 13.23% | 13.93% | 0.77% |
| Diffusion-PEGASOS (5) | 18.94% | 10.83% | 11.20% | 0.57% |
| Noncooperative-PEGASOS (1) | 20.81% | 21.74% | 18.40% | 2.93% |
| Noncooperative-PEGASOS (2) | 20.52% | 18.64% | 16.54% | 2.19% |
| Noncooperative-PEGASOS (5) | 19.88% | 15.96% | 14.86% | 1.87% |

**Table 7.9 Parameters for each method**

| Method | Adult | Banana | Waveform | MNIST |
|---|---|---|---|---|
| Kernel-PEGASOS | $\sigma = \sqrt{10}$<br>$\lambda = 0.0000307$ | $\sigma = 0.7$<br>$\lambda = \frac{1}{316}$ | $\sigma = \sqrt{10}$<br>$\lambda = 0.001$ | $\sigma = 4$<br>$\lambda = 10^{-7}$ |
| RFF-PEGASOS | $\sigma = \sqrt{10}$<br>$\lambda = 0.0000307$<br>$D = 2000$ | $\sigma = 0.7$<br>$\lambda = \frac{1}{316}$<br>$D = 200$ | $\sigma = \sqrt{10}$<br>$\lambda = 0.001$<br>$D = 2000$ | $\sigma = 4$<br>$\lambda = 10^{-7}$<br>$D = 100000$ |

over a dataset of double size, that contains the same data pairs twice. For the three first datasets (Adult, Banana, Waveform) we have run 100 realizations of the experiment, while for the fourth (MNIST) we have run only 10 (to save time). Besides the ADULT dataset, all other simulations show that the distributed implementation significantly outperforms the noncooperative one. For that particular dataset, we observe that for a single run the noncooperative strategy behaves better (for $K = 20$), but as data reuse increases the distributed implementation reaches lower error floors.

## 7.5 CONCLUSIONS

In this chapter, we presented an alternative path for designing online kernel-based learning methods. Instead of relying on the typical growing sum, the presented approach suggests to approximate the solution using randomly sampled Fourier features of the kernel. This is equivalent to transforming the training data to a Euclidean space of larger dimension, thus leading to linear fixed-budget strategies.

Moreover, the proposed approach can be exploited to derive online distributed schemes for learning over networks, where the nodes cooperate to find a combined solution, based on the combine-then-adapt rationale. We stress out that this is the first practical scheme for kernel-based distributed learning that has been proposed in the literature so far. The provided simulations verify the validity of the proposed approach.

## REFERENCES


[1] V.N. Vapnik, The Nature of Statistical Learning Theory, Springer Verlag, 1999.

[2] B. Scholkopf, A.J. Smola, Learning with Kernels: Support Vector Machines, Regularization, Optimization and Beyond, MIT Press, 2002.

[3] Sergios Theodoridis, Machine Learning: A Bayesian and Optimization Perspective, Academic Press, 2015.

[4] W. Liu, J.C. Principe, S. Haykin, Kernel Adaptive Filtering, Wiley, Hoboken, NJ, 2010.

[5] W. Liu, P. Pokharel, J.C. Principe, The kernel least-mean-square algorithm, IEEE Transactions on Signal Processing 56 (2) (Feb. 2008) 543–554.

[6] Y. Engel, S. Mannor, R. Meir, The kernel recursive least-squares algorithm, IEEE Transactions on Signal Processing 52 (8) (Aug. 2004) 2275–2285.

[7] K. Slavakis, P. Bouboulis, S. Theodoridis, Online learning in reproducing kernel Hilbert spaces, in: Rama Chellappa, Sergios Theodoridis (Eds.), Academic Press Library in Signal Processing, Signal Processing Theory and Machine Learning, Academic Press, 2014, pp. 883–987.

[8] A. Rahimi, B. Recht, Random features for large scale kernel machines, in: Neural Information Processing Systems, vol. 20, 2007.

[9] A. Singh, N. Ahuja, P. Moulin, Online learning with kernels: overcoming the growing sum problem, in: IEEE International Workshop on Machine Learning for Signal Processing, September 2012.

[10] P. Bouboulis, S. Pougkakiotis, S. Theodoridis, Efficient KLMS and KRLS algorithms: a random Fourier feature perspective, in: IEEE Workshop on Statistical Signal Processing, 2016.

[11] C. Williams, M. Seeger, Using the Nystrom method to speed up kernel machines, in: Neural Information Processing Systems, vol. 14, 2001, pp. 682–688.

[12] P. Drineas, M.W. Mahoney, On the Nystrom method for approximating a gram matrix for improved kernel-based learning, Journal of Machine Learning Research 6 (2005) 2153–2175.

[13] A. Rahimi, B. Recht, Weighted sums of random kitchen sinks: replacing minimization with randomization in learning, in: Neural Information Processing Systems, vol. 22, 2009, pp. 1313–1320.

[14] D.J. Sutherland, J. Schneider, On the error of random Fourier features, in: Conference on Uncertainty in Artificial Intelligence, 2015.

[15] T. Yang, Y-F. Li, M. Mahdavi, J. Rong, Z-H. Zhou, Nyström method vs random Fourier features: a theoretical and empirical comparison, in: Neural Information Processing Systems, vol. 25, 2012, pp. 476–484.

[16] P. Bouboulis, S. Chouvardas, S. Theodoridis, Online distributed learning over networks in RKH spaces using random Fourier features, IEEE Transactions on Signal Processing 66 (7) (2018) 1920–1932.

[17] S. Shalev-Shwartz, Y. Singer, N. Srebro, A. Cotter, Pegasos: primal estimated sub-gradient solver for SVM, Mathematical Programming 127 (1) (2011) 3–30.

[18] L. Bottou, http://leon.bottou.org/projects/lasvm.

[19] A. Bordes, S. Ertekin, J. Weston, L. Bottou, Fast kernel classifiers with online and active learning, Journal of Machine Learning Research 6 (2005) 1579–1619.

[20] R. Mitra, V. Bhatia, The diffusion-KLMS algorithm, in: ICIT, 2014, Dec 2014, pp. 256–259.

[21] W. Gao, J. Chen, C. Richard, J. Huang, Diffusion adaptation over networks with kernel least-mean-square, in: IEEE International Workshop on Computational Advances in Multi-Sensor Adaptive Processing, 2015.

[22] Chouvardas Symeon, Draief Moez, A diffusion kernel LMS algorithm for nonlinear adaptive networks, in: International Conference on Acoustics, Speech, and Signal Processing, 2016.

[23] Symeon Chouvardas, Yannis Kopsinis, Sergios Theodoridis, Robust subspace tracking with missing entries: the set-theoretic approach, IEEE Transactions on Signal Processing 63 (19) (2015) 5060–5070.

[24] Cassio G. Lopes, Ali H. Sayed, Diffusion least-mean squares over adaptive networks: formulation and performance analysis, IEEE Transactions on Signal Processing 56 (7) (July 2008) 3122–3136.

[25] A.H. Sayed, Adaptive networks, Proceedings of the IEEE 102 (4) (2014) 460–497.

[26] F.S. Cattiveli, A.H. Sayed, Diffusion LMS strategies for distributed estimation, IEEE Transactions on Signal Processing 58 (3) (2010) 1035–1048.

[27] R.H. Koning, H. Neudecker, Block Kronecker products and the vecb operator, Linear Algebra and Its Applications 149 (1991) 165–184.

[28] MIT, Strategic Engineering Research Group, MatLab Tools for Network analysis, http://strategic.mit.edu/.

CHAPTER

# KERNEL-BASED INFERENCE OF FUNCTIONS OVER GRAPHS

# 8

**Vassilis N. Ioannidis*, Meng Ma*, Athanasios N. Nikolakopoulos*, Georgios B. Giannakis*, Daniel Romero†**

**University of Minnesota, Minneapolis, MN, United States* †*University of Agder, Grimstad, Norway*

## CONTENTS



## 8.1 INTRODUCTION

Numerous applications arising in diverse disciplines involve *inference over networks* [1]. Modeling nodal attributes as signals that take values over the vertices of the underlying graph allows the associated inference tasks to leverage node dependencies captured by the graph structure. In many real settings one often affords to work with only a limited number of node observations due to inherent restrictions particularly to the inference task at hand. In social networks, for example, individuals may be reluctant to share personal information; in sensor networks the nodes may report observations sporadically in order to save energy; in brain networks acquiring node samples may involve invasive procedures (*e.g.*, electrocorticography). In this context, a frequently encountered challenge that often emerges is that of *inferring the attributes for every node in the network given the attributes for a*

Adaptive Learning Methods for Nonlinear System Modeling. DOI: 10.1016/B978-0-12-812976-0.00010-5

*subset of nodes*. This is typically formulated as the task of reconstructing a function defined on the nodes [1–6], given information about some of its values.

Reconstruction of functions over graphs has been studied by the machine learning community, in the context of *semisupervised learning* under the term of *transductive* regression and classification [6–8]. Existing approaches assume "smoothness" with respect to the graph—in the sense that neighboring vertices have similar values—and devise *nonparametric* methods [2,3,6,9] targeting primarily the task of reconstructing binary-valued signals. Function estimation has also been investigated recently by the community of signal processing on graphs (SPoG) under the term *signal reconstruction* [10–17]. Most of such approaches commonly adopt *parametric* estimation tools and rely on *bandlimitedness*, by which the signal of interest is assumed to lie in the span of the $B$ principal eigenvectors of the graph's Laplacian (or adjacency) matrix.

This chapter cross-pollinates ideas arising from both communities and presents a unifying framework for tackling signal reconstruction problems both in the traditional *time-invariant*, as well as in the more challenging *time varying* setting. We begin by a comprehensive presentation of kernel-based learning for solving problems of signal reconstruction over graphs (Section 8.2). Data-driven techniques are then presented based on multikernel learning (MKL) that enables combining optimally the kernels in a given dictionary and simultaneously estimating the graph function by solving a single optimization problem (Section 8.2.3). For the case where prior information is available, semiparametric estimators are discussed that can incorporate seamlessly structured prior information into the signal estimators (Section 8.2.4). We then move to the problem of reconstructing time evolving functions on dynamic graphs (Section 8.3). The kernel-based framework is now extended to accommodate the time evolving setting building on the notion of *graph extension*, specific choices of which can lend themselves to a reduced complexity online solver (Section 8.3.1). Next, a more flexible model is introduced that captures multiple forms of time dynamics, and kernel-based learning is employed to derive an online solver that effects online MKL by selecting the optimal combination of kernels on-the-fly (Section 8.3.2). Our analytical exposition, in both parts, is supplemented with a set of numerical tests based on both real and synthetic data that highlight the effectiveness of the methods, while providing examples of interesting realistic problems that they can address.

**Notation**: Scalars are denoted by lowercase characters, vectors by bold lowercase, matrices by bold uppercase; $(A)_{m,n}$ is the $(m,n)$th entry of matrix $\mathbf{A}$; superscripts $^T$ and $^\dagger$ respectively denote transpose and pseudoinverse. If $\mathbf{A} := [\mathbf{a}_1, \ldots, \mathbf{a}_N]$, then $\text{vec}\{\mathbf{A}\} := [\mathbf{a}_1^T, \ldots, \mathbf{a}_N^T]^T := \mathbf{a}$. With $N \times N$ matrices $\{\mathbf{A}_t\}_{t=1}^T$ and $\{\mathbf{B}_t\}_{t=2}^T$ satisfying $\mathbf{A}_t = \mathbf{A}_t^T \ \forall t$, $\text{btridiag}\{\mathbf{A}_1, \ldots, \mathbf{A}_T; \mathbf{B}_2, \ldots, \mathbf{B}_T\}$ represents the symmetric block tridiagonal matrix. Symbols $\odot$, $\otimes$ and $\oplus$, respectively, denote the element-wise (Hadamard) matrix product, Kronecker product and Kronecker sum, the latter being defined for $\mathbf{A} \in \mathbb{R}^{M \times M}$ and $\mathbf{B} \in \mathbb{R}^{N \times N}$ as $\mathbf{A} \oplus \mathbf{B} := \mathbf{A} \otimes \mathbf{I}_N + \mathbf{I}_M \otimes \mathbf{B}$. The $n$th column of the identity matrix $\mathbf{I}_N$ is represented by $\mathbf{i}_{N,n}$. If $\mathbf{A} \in \mathbb{R}^{N \times N}$ is positive definite and $\mathbf{x} \in \mathbb{R}^N$, then $||\mathbf{x}||_{\mathbf{A}}^2 := \mathbf{x}^T \mathbf{A}^{-1} \mathbf{x}$ and $||\mathbf{x}||_2 := ||\mathbf{x}||_{\mathbf{I}_N}$. The cone of $N \times N$ positive definite matrices is denoted by $\mathbb{S}_+^N$. Finally, $\delta[\cdot]$ stands for the Kronecker delta and $\mathbb{E}$ for expectation.

## 8.2 RECONSTRUCTION OF FUNCTIONS OVER GRAPHS

Before giving the formal problem statement, it is instructive to start with the basic definitions that will be used throughout this chapter.

**Definitions**: A graph can be specified by a tuple $\mathcal{G} := (\mathcal{V}, \mathbf{A})$, where $\mathcal{V} := \{v_1, \ldots, v_N\}$ is the vertex set and $\mathbf{A}$ is the $N \times N$ adjacency matrix, whose $(n, n')$th entry, $A_{n,n'} \geq 0$, denotes the nonnegative edge weight between vertices $v_n$ and $v_{n'}$. For simplicity, it is assumed that the graph has no self-loops, *i.e.*, $A_{n,n} = 0,\ \forall v_n \in \mathcal{V}$. This chapter focuses on *undirected* graphs, for which $A_{n',n} = A_{n,n'}\ \forall v_n, v_{n'} \in \mathcal{V}$. A graph is said to be *unweighted* if $A_{n,n'}$ is either 0 or 1. The edge set is defined as $\mathcal{E} := \{(v_n, v_{n'}) \in \mathcal{V} \times \mathcal{V} : A_{n,n'} \neq 0\}$. Two vertices $v_n$ and $v_{n'}$ are *adjacent, connected* or *neighbors* if $(v_n, v_{n'}) \in \mathcal{E}$. The *Laplacian* matrix is defined as $\mathbf{L} := \text{diag}\{\mathbf{A1}\} - \mathbf{A}$ and is symmetric and positive semidefinite [1, Chapter 2]. A real-valued function (or signal) on a graph is a map $f : \mathcal{V} \to \mathbb{R}$. The value $f(v)$ represents an attribute or feature of $v \in \mathcal{V}$, such as age, political alignment or annual income of a person in a social network. Signal $f$ is thus represented by $\mathbf{f} := [f(v_1), \ldots, f(v_N)]^T$.

**Problem statement.** Suppose that a collection of noisy samples (or observations) $\{y_s | y_s = f(v_{n_s}) + e_s\}_{s=1}^S$ is available, where $e_s$ models noise and $\mathcal{S} := \{n_1, \ldots, n_S\}$ contains the indices $1 \leq n_1 < \cdots < n_S \leq N$ of the sampled vertices, with $S \leq N$. Given $\{(n_s, y_s)\}_{s=1}^S$ and assuming knowledge of $\mathcal{G}$, the goal is to estimate $f$. This will provide estimates of $f(v)$ both at observed and unobserved vertices. By defining $\mathbf{y} := [y_1, \ldots, y_S]^T$, the observation model is summarized as

$$\mathbf{y} = \mathbf{Sf} + \mathbf{e}, \tag{8.1}$$

where $\mathbf{e} := [e_1, \ldots, e_S]^T$ and $\mathbf{S}$ is a known $S \times N$ binary sampling matrix with entries $(s, n_s)$, $s = 1, \ldots, S$, set to one and the rest set to zero.

### 8.2.1 KERNEL REGRESSION

Kernel methods constitute the "workhorse" of machine learning for nonlinear function estimation [18]. Their popularity can be attributed to their simplicity, flexibility and good performance. Here, we present kernel regression as a unifying framework for graph signal reconstruction along with the so-called representer theorem.

Kernel regression seeks an estimate of $f$ in a reproducing kernel Hilbert space (RKHS) $\mathcal{H}$, which is the space of functions $f : \mathcal{V} \to \mathbb{R}$ defined as

$$\mathcal{H} := \left\{ f : f(v) = \sum_{n=1}^{N} \alpha_n \kappa(v, v_n),\ \alpha_n \in \mathbb{R} \right\}, \tag{8.2}$$

where the *kernel map* $\kappa : \mathcal{V} \times \mathcal{V} \to \mathbb{R}$ is any function defining a symmetric and positive semidefinite $N \times N$ matrix with entries $[\mathbf{K}]_{n,n'} := \kappa(v_n, v_{n'})$ [19]. Intuitively, $\kappa(v, v')$ is a basis function in (8.2) measuring similarity between the values of $f$ at $v$ and $v'$. (For a more detailed treatment of RKHS, see, *e.g.*, [3].)

Note that, for signals over graphs, the expansion in (8.2) is finite since $\mathcal{V}$ is finite-dimensional. Thus, any $f \in \mathcal{H}$ can be expressed in compact form

$$\mathbf{f} = \mathbf{K}\boldsymbol{\alpha} \tag{8.3}$$

for some $N \times 1$ vector $\boldsymbol{\alpha} := [\alpha_1, \ldots, \alpha_N]^T$.

Given two functions $f(v) := \sum_{n=1}^{N} \alpha_n \kappa(v, v_n)$ and $f'(v) := \sum_{n=1}^{N} \alpha'_n \kappa(v, v_n)$, their RKHS inner product is defined as[1]

$$\langle f, f' \rangle_{\mathcal{H}} := \sum_{n=1}^{N} \sum_{n'=1}^{N} \alpha_n \alpha'_{n'} \kappa(v_n, v_{n'}) = \boldsymbol{\alpha}^T \mathbf{K} \boldsymbol{\alpha}', \tag{8.4}$$

where $\boldsymbol{\alpha}' := [\alpha'_1, \ldots, \alpha'_N]^T$ and the reproducing property has been employed that suggests $\langle \kappa(\cdot, v_{n_0}), \kappa(\cdot, v_{n'_0}) \rangle_{\mathcal{H}} = \mathbf{i}_{n_0}^T \mathbf{K} \mathbf{i}_{n'_0} = \kappa(v_{n_0}, v_{n'_0})$. The RKHS norm is defined by

$$||f||_{\mathcal{H}}^2 := \langle f, f \rangle_{\mathcal{H}} = \boldsymbol{\alpha}^T \mathbf{K} \boldsymbol{\alpha} \tag{8.5}$$

and will be used as a regularizer to control overfitting and to cope with the under-determined reconstruction problem. As a special case, setting $\mathbf{K} = \mathbf{I}_N$ recovers the standard inner product $\langle f, f' \rangle_{\mathcal{H}} = \mathbf{f}^T \mathbf{f}'$ and the Euclidean norm $||f||_{\mathcal{H}}^2 = ||\mathbf{f}||_2^2$. Note that when $\mathbf{K} \succ \mathbf{0}$, the set of functions of the form (8.3) coincides with $\mathbb{R}^N$.

Given $\{y_s\}_{s=1}^S$, RKHS-based function estimators are obtained by solving functional minimization problems formulated as (see also, *e.g.*, [18–20])

$$\hat{f} := \arg\min_{f \in \mathcal{H}} \mathcal{L}(\mathbf{y}, \bar{\mathbf{f}}) + \mu \Omega(||f||_{\mathcal{H}}), \tag{8.6}$$

where the loss $\mathcal{L}$ measures how the estimated function $f$ at the observed vertices $\{v_{n_s}\}_{s=1}^S$, collected in $\bar{\mathbf{f}} := [f(v_{n_1}), \ldots, f(v_{n_S})]^T = \mathbf{S}\mathbf{f}$, deviates from the data $\mathbf{y}$. The so-called *square loss* $\mathcal{L}(\mathbf{y}, \bar{\mathbf{f}}) := (1/S) \sum_{s=1}^S \left[ y_s - f(v_{n_s}) \right]^2$ constitutes a popular choice for $\mathcal{L}$. The increasing function $\Omega$ is used to promote smoothness with typical choices including $\Omega(\zeta) = |\zeta|$ and $\Omega(\zeta) = \zeta^2$. The regularization parameter $\mu > 0$ controls overfitting. Substituting (8.3) and (8.5) into (8.6) shows that $\hat{\mathbf{f}}$ can be found as

$$\hat{\boldsymbol{\alpha}} := \arg\min_{\boldsymbol{\alpha} \in \mathbb{R}^N} \mathcal{L}(\mathbf{y}, \mathbf{S}\mathbf{K}\boldsymbol{\alpha}) + \mu \Omega((\boldsymbol{\alpha}^T \mathbf{K} \boldsymbol{\alpha})^{1/2}), \tag{8.7a}$$

$$\hat{\mathbf{f}} = \mathbf{K} \hat{\boldsymbol{\alpha}}. \tag{8.7b}$$

An alternative form of (8.5) that will be frequently used in the sequel results upon noting that

$$\boldsymbol{\alpha}^T \mathbf{K} \boldsymbol{\alpha} = \boldsymbol{\alpha}^T \mathbf{K} \mathbf{K}^\dagger \mathbf{K} \boldsymbol{\alpha} = \mathbf{f}^T \mathbf{K}^\dagger \mathbf{f}. \tag{8.8}$$

Thus, one can rewrite (8.6) as

$$\hat{\mathbf{f}} := \arg\min_{\mathbf{f} \in \mathcal{R}\{\mathbf{K}\}} \mathcal{L}(\mathbf{y}, \mathbf{S}\mathbf{f}) + \mu \Omega((\mathbf{f}^T \mathbf{K}^\dagger \mathbf{f})^{1/2}). \tag{8.9}$$

[1] While $f$ denotes a *function*, $f(v)$ represents the *scalar* resulting from evaluating $f$ at vertex $v$.

Although graph signals can be reconstructed from (8.7), such an approach involves optimizing over $N$ variables. Thankfully, the solution can be obtained by solving an optimization problem in $S$ variables (where typically $S \ll N$), by invoking the so-called representer theorem [19,21].

The representer theorem plays an instrumental role in the traditional infinite-dimensional setting where (8.6) cannot be solved directly; however, even when $\mathcal{H}$ comprises graph signals, it can still be beneficial to reduce the dimension of the optimization in (8.7). The theorem essentially asserts that the solution to the functional minimization in (8.6) can be expressed as

$$\hat{f}(v) = \sum_{s=1}^{S} \bar{\alpha}_s \kappa(v, v_{n_s}) \tag{8.10}$$

for some $\bar{\alpha}_s \in \mathbb{R}$, $s = 1, \ldots, S$.

The representer theorem shows the form of $\hat{f}$, but does not provide the optimal $\{\bar{\alpha}_s\}_{s=1}^{S}$, which are found after substituting (8.10) into (8.6) and solving the resulting optimization problem with respect to these coefficients. To this end, let $\bar{\boldsymbol{\alpha}} := [\bar{\alpha}_1, \ldots, \bar{\alpha}_S]^T$ and write $\boldsymbol{\alpha} = \mathbf{S}^T \bar{\boldsymbol{\alpha}}$ to deduce that

$$\hat{\mathbf{f}} = \mathbf{K}\boldsymbol{\alpha} = \mathbf{K}\mathbf{S}^T \bar{\boldsymbol{\alpha}}. \tag{8.11}$$

With $\bar{\mathbf{K}} := \mathbf{S}\mathbf{K}\mathbf{S}^T$ and using (8.7) and (8.11), the optimal $\bar{\boldsymbol{\alpha}}$ can be found as

$$\hat{\bar{\boldsymbol{\alpha}}} := \arg\min_{\bar{\boldsymbol{\alpha}} \in \mathbb{R}^S} \mathcal{L}(\mathbf{y}, \bar{\mathbf{K}}\bar{\boldsymbol{\alpha}}) + \mu \Omega((\bar{\boldsymbol{\alpha}}^T \bar{\mathbf{K}} \bar{\boldsymbol{\alpha}})^{1/2}). \tag{8.12}$$

**Kernel ridge regression**. For $\mathcal{L}$ chosen as the square loss and $\Omega(\zeta) = \zeta^2$, the $\hat{f}$ in (8.6) is referred to as the *kernel ridge regression* (RR) estimate [18]. If $\bar{\mathbf{K}}$ is full-rank, this estimate is given by $\hat{\mathbf{f}}_{\mathrm{RR}} = \mathbf{K}\mathbf{S}^T \hat{\bar{\boldsymbol{\alpha}}}$, where

$$\hat{\bar{\boldsymbol{\alpha}}} := \arg\min_{\bar{\boldsymbol{\alpha}} \in \mathbb{R}^S} \frac{1}{S} \left\| \mathbf{y} - \bar{\mathbf{K}}\bar{\boldsymbol{\alpha}} \right\|^2 + \mu \bar{\boldsymbol{\alpha}}^T \bar{\mathbf{K}} \bar{\boldsymbol{\alpha}} \tag{8.13a}$$

$$= (\bar{\mathbf{K}} + \mu S \mathbf{I}_S)^{-1} \mathbf{y}. \tag{8.13b}$$

Therefore, $\hat{\mathbf{f}}_{\mathrm{RR}}$ can be expressed as

$$\hat{\mathbf{f}}_{\mathrm{RR}} = \mathbf{K}\mathbf{S}^T (\bar{\mathbf{K}} + \mu S \mathbf{I}_S)^{-1} \mathbf{y}. \tag{8.14}$$

Section 8.2.2 shows that (8.14) generalizes a number of existing signal reconstructors upon properly selecting $\mathbf{K}$.

### 8.2.2 KERNELS ON GRAPHS

When estimating functions on graphs, conventional kernels such as the Gaussian kernel cannot be adopted because the underlying set where graph signals are defined is not a metric space. Indeed, no vertex addition $v_n + v_{n'}$, scaling $\beta v_n$ or norm $||v_n||$ can be naturally defined on $\mathcal{V}$. An alternative is to embed $\mathcal{V}$ into Euclidean space via a feature map $\phi : \mathcal{V} \to \mathbb{R}^D$ and invoke a conventional kernel afterwards. However, for a given graph it is generally unclear how to explicitly design $\phi$ or select $D$. This motivates the adoption of *kernels on graphs* [3].

**Table 8.1 Common spectral weight functions**

| Kernel name | Function | Parameters |
|---|---|---|
| Diffusion [2] | $r(\lambda) = \exp\{\sigma^2\lambda/2\}$ | $\sigma^2$ |
| $p$-step random walk [3] | $r(\lambda) = (a - \lambda)^{-p}$ | $a \geq 2,\ p \geq 0$ |
| Regularized Laplacian [3,22] | $r(\lambda) = 1 + \sigma^2\lambda$ | $\sigma^2$ |
| Bandlimited [23] | $r(\lambda) = \begin{cases} 1/\beta, & \lambda \leq \lambda_{\max}, \\ \beta, & \text{otherwise} \end{cases}$ | $\beta > 0,\ \lambda_{\max}$ |

A common approach to designing kernels on graphs is to apply a transformation function on the graph Laplacian [3]. The term *Laplacian kernel* comprises a wide family of kernels obtained by applying a certain function $r(\cdot)$ to the Laplacian matrix $\mathbf{L}$. Laplacian kernels are well motivated since they constitute the graph counterpart of the so-called *translation-invariant kernels* in Euclidean spaces [3]. This section reviews Laplacian kernels, provides beneficial insights in terms of interpolating signals and highlights their versatility in capturing information about the *graph Fourier transform* of the estimated signal.

The reason why the graph Laplacian constitutes one of the prominent candidates for regularization on graphs becomes clear upon recognizing that

$$\mathbf{f}^T\mathbf{L}\mathbf{f} = \frac{1}{2}\sum_{(n,n')\in\mathcal{E}} A_{n,n'}(f_n - f_{n'})^2, \tag{8.15}$$

where $A_{n,n'}$ denotes weight associated with edge $(n, n')$. The quadratic form of (8.15) becomes larger when function values vary a lot among connected vertices and therefore quantifies the *smoothness* of $\mathbf{f}$ on $\mathcal{G}$.

Let $0 = \lambda_1 \leq \lambda_2 \leq \ldots \leq \lambda_N$ denote the eigenvalues of the graph Laplacian matrix $\mathbf{L}$ and consider the eigendecomposition $\mathbf{L} = \mathbf{U}\boldsymbol{\Lambda}\mathbf{U}^T$, where $\boldsymbol{\Lambda} := \operatorname{diag}\{\lambda_1, \ldots, \lambda_N\}$. A Laplacian kernel matrix is defined by

$$\mathbf{K} := r^{\dagger}(\mathbf{L}) := \mathbf{U}r^{\dagger}(\boldsymbol{\Lambda})\mathbf{U}^T, \tag{8.16}$$

where $r(\boldsymbol{\Lambda})$ is the result of applying a user-selected, scalar, nonnegative map $r : \mathbb{R} \to \mathbb{R}_+$ to the diagonal entries of $\boldsymbol{\Lambda}$. The selection of map $r$ generally depends on desirable properties that the target function is expected to have. Table 8.1 summarizes some well-known examples arising for specific choices of $r$.

At this point, it is prudent to offer interpretations and insights on the operation of Laplacian kernels. Towards this objective, note first that the regularizer from (8.9) is an increasing function of

$$\boldsymbol{\alpha}^T\mathbf{K}\boldsymbol{\alpha} = \boldsymbol{\alpha}^T\mathbf{K}\mathbf{K}^{\dagger}\mathbf{K}\boldsymbol{\alpha} = \mathbf{f}^T\mathbf{K}^{\dagger}\mathbf{f} = \mathbf{f}^T\mathbf{U}r(\boldsymbol{\Lambda})\mathbf{U}^T\mathbf{f} = \check{\mathbf{f}}^T r(\boldsymbol{\Lambda})\check{\mathbf{f}} = \sum_{n=1}^{N} r(\lambda_n)|\check{f}_n|^2, \tag{8.17}$$

where $\check{\mathbf{f}} := \mathbf{U}^T\mathbf{f} := [\check{f}_1, \ldots, \check{f}_N]^T$ comprises the projections of $\mathbf{f}$ onto the eigenspace of $\mathbf{L}$ and is referred to as the *graph Fourier transform* of $\mathbf{f}$ in the SPoG parlance [4]. Consequently, $\{\check{f}_n\}_{n=1}^N$ are called *frequency components*. The so-called *bandlimited (BL) functions* in SPoG refer to those whose frequency components only exist inside some band $B$, that is, $\check{f}_n = 0, \forall n > B$.

By adopting the aforementioned SPoG notions, one can intuitively interpret the role of BL kernels. Indeed, it follows from (8.17) that the regularizer strongly penalizes those $\check{f}_n$ for which the corresponding $r(\lambda_n)$ is large, thus promoting a specific structure in this "frequency" domain. Specifically, one prefers $r(\lambda_n)$ to be large whenever $|\check{f}_n|^2$ is small and *vice versa*. The fact that $|\check{f}_n|^2$ is expected to decrease with $n$ for smooth $f$ motivates the adoption of an increasing function $r$ [3]. From (8.17) it is clear that $r(\lambda_n)$ determines how heavily $\check{f}_n$ is penalized. Therefore, by setting $r(\lambda_n)$ to be small when $n \leq B$ and extremely large when $n > B$, one can expect the result to be a BL signal.

Observe that Laplacian kernels can capture forms of prior information richer than bandlimitedness [11,13,16,17] by selecting function $r$ accordingly. For instance, using $r(\lambda) = \exp\{\sigma^2\lambda/2\}$ (diffusion kernel) accounts not only for smoothness of $\mathbf{f}$ as in (8.15), but also for the prior knowledge that $\mathbf{f}$ is generated by a process diffusing over the graph. Similarly, the use of $r(\lambda) = (\alpha - \lambda)^{-1}$ (1-step random walk) can accommodate cases where the signal captures a notion of network centrality.[2]

So far, $f$ has been assumed deterministic, which precludes accommodating certain forms of prior information that probabilistic models can capture, such as domain knowledge and historical data. Suppose without loss of generality that $\{f(v_n)\}_{n=1}^N$ are zero mean random variables. The LMMSE estimator of $\mathbf{f}$ given $\mathbf{y}$ in (8.1) is the linear estimator $\hat{\mathbf{f}}_{\text{LMMSE}}$ minimizing $\mathbb{E}||\mathbf{f} - \hat{\mathbf{f}}_{\text{LMMSE}}||_2^2$, where the expectation is over all $\mathbf{f}$ and noise realizations. With $\mathbf{C} := \mathbb{E}\left[\mathbf{f}\mathbf{f}^T\right]$, the LMMSE estimate is

$$\hat{\mathbf{f}}_{\text{LMMSE}} = \mathbf{C}\mathbf{S}^T[\mathbf{S}\mathbf{C}\mathbf{S}^T + \sigma_e^2\mathbf{I}_S]^{-1}\mathbf{y}, \tag{8.18}$$

where $\sigma_e^2 := (1/S)\mathbb{E}\left[||\mathbf{e}||_2^2\right]$ denotes the noise variance. Comparing (8.18) with (8.14) and recalling that $\bar{\mathbf{K}} := \mathbf{S}\mathbf{K}\mathbf{S}^T$, it follows that $\hat{\mathbf{f}}_{\text{LMMSE}} = \hat{\mathbf{f}}_{\text{RR}}$ if $\mu S = \sigma_e^2$ and $\mathbf{K} = \mathbf{C}$. In other words, the similarity measure $\kappa(v_n, v_{n'})$ embodied in such a kernel map is just the covariance $\text{cov}[f(v_n), f(v_{n'})]$. A related observation was pointed out in [27] for general kernel methods.

In short, one can interpret kernel ridge regression as the LMMSE estimator of a signal $\mathbf{f}$ with covariance matrix equal to $\mathbf{K}$; see also [28]. The LMMSE interpretation also suggests the usage of $\mathbf{C}$ as a kernel matrix, which enables signal reconstruction even when the graph topology is unknown. Although this discussion hinges on kernel ridge regression after setting $\mathbf{K} = \mathbf{C}$, any other kernel estimator of the form (8.7) can benefit from vertex covariance kernels too.

### 8.2.3 SELECTING KERNELS FROM A DICTIONARY

The selection of the pertinent kernel matrix is of paramount importance to the performance of kernel-based methods [23,29]. This section presents an MKL approach that effects kernel selection in graph signal reconstruction. Two algorithms with complementary strengths will be presented. Both rely on a user-specified *kernel dictionary*, and the best kernel is built from the dictionary in a data-driven way.

The first algorithm, which we call *RKHS superposition*, is motivated by the fact that one specific $\mathcal{H}$ in (8.6) is determined by some $\kappa$; therefore, kernel selection is tantamount to RKHS selection. Con-

[2] Smola et al. [3], for example, discuss the connection between $r(\lambda) = (\alpha - \lambda)^{-1}$ and PageRank [24] whereby the sought-after signal is essentially defined as the limiting distribution of a simple underlying "*random surfing*" process. For more about random surfing processes, see also [25,26].

sequently, a kernel dictionary $\{\kappa_m\}_{m=1}^M$ gives rise to an RKHS dictionary $\{\mathcal{H}_m\}_{m=1}^M$, which motivates estimates of the form[3]

$$\hat{f} = \sum_{m=1}^{M} \hat{f}_m, \quad \hat{f}_m \in \mathcal{H}_m. \tag{8.19}$$

Upon adopting a criterion that controls sparsity in this expansion, the "best" RKHSs will be selected. A reasonable approach is therefore to generalize (8.6) to accommodate multiple RKHSs. With $\mathcal{L}$ selected to be the square loss and $\Omega(\zeta) = |\zeta|$, one can pursue an estimate $\hat{f}$ by solving

$$\min_{\{f_m \in \mathcal{H}_m\}_{m=1}^M} \frac{1}{S} \sum_{s=1}^{S} \left[ y_s - \sum_{m=1}^{M} f_m(v_{n_s}) \right]^2 + \mu \sum_{m=1}^{M} \| f_m \|_{\mathcal{H}_m}. \tag{8.20}$$

Invoking the representer theorem per $f_m$ establishes that the minimizers of (8.20) can be written as

$$\hat{f}_m(v) = \sum_{s=1}^{S} \bar{\alpha}_s^m \kappa_m(v, v_{n_s}), \quad m = 1, \ldots, M \tag{8.21}$$

for some coefficients $\bar{\alpha}_s^m$. Substituting (8.21) into (8.20) suggests obtaining these coefficients as

$$\arg\min_{\{\bar{\boldsymbol{\alpha}}_m\}_{m=1}^M} \frac{1}{S} \left\| \mathbf{y} - \sum_{m=1}^{M} \bar{\mathbf{K}}_m \bar{\boldsymbol{\alpha}}_m \right\|^2 + \mu \sum_{m=1}^{M} \left( \bar{\boldsymbol{\alpha}}_m^T \bar{\mathbf{K}}_m \bar{\boldsymbol{\alpha}}_m \right)^{1/2}, \tag{8.22}$$

where $\bar{\boldsymbol{\alpha}}_m := [\bar{\alpha}_1^m, \ldots, \bar{\alpha}_S^m]^T$ and $\bar{\mathbf{K}}_m := \mathbf{S}\mathbf{K}_m\mathbf{S}^T$ with $(\mathbf{K}_m)_{n,n'} := \kappa_m(v_n, v_{n'})$. Letting $\check{\boldsymbol{\alpha}}_m := \bar{\mathbf{K}}_m^{1/2} \bar{\boldsymbol{\alpha}}_m$, Eq. (8.22) becomes

$$\arg\min_{\{\check{\boldsymbol{\alpha}}_m\}_{m=1}^M} \frac{1}{S} \left\| \mathbf{y} - \sum_{m=1}^{M} \bar{\mathbf{K}}_m^{1/2} \check{\boldsymbol{\alpha}}_m \right\|^2 + \mu \sum_{m=1}^{M} \| \check{\boldsymbol{\alpha}}_m \|_2. \tag{8.23}$$

Interestingly, (8.23) can be efficiently solved using the alternating-direction method of multipliers (ADMM) [30,31] after some necessary reformulation [23].

After obtaining $\{\check{\boldsymbol{\alpha}}_m\}_{m=1}^M$, the sought-after function estimate can be recovered as

$$\hat{\mathbf{f}} = \sum_{m=1}^{M} \mathbf{K}_m \mathbf{S}^T \bar{\boldsymbol{\alpha}}_m = \sum_{m=1}^{M} \mathbf{K}_m \mathbf{S}^T \bar{\mathbf{K}}_m^{-1/2} \check{\boldsymbol{\alpha}}_m. \tag{8.24}$$

This MKL algorithm can identify the best subset of RKHSs—and therefore kernels—but entails $MS$ unknowns (*cf.* (8.22)). Next, an alternative approach is discussed which can reduce the number of variables to $M + S$ at the price of not being able to assure a sparse kernel expansion.

[3] A sum is chosen here for tractability, but the right-hand side of (8.19) could in principle combine the functions $\{\hat{f}_m\}_m$ in different forms.

The alternative approach is to postulate a kernel of the form $\mathbf{K}(\boldsymbol{\theta}) = \sum_{m=1}^{M} \theta_m \mathbf{K}_m$, where $\{\mathbf{K}_m\}_{m=1}^{M}$ is given and $\theta_m \geq 0\ \forall m$. The coefficients $\boldsymbol{\theta} := [\theta_1, \ldots, \theta_M]^T$ can be found by jointly minimizing (8.12) with respect to $\boldsymbol{\theta}$ and $\bar{\boldsymbol{\alpha}}$ [32]. We have

$$(\boldsymbol{\theta}, \hat{\bar{\boldsymbol{\alpha}}}) := \arg\min_{\boldsymbol{\theta}, \bar{\boldsymbol{\alpha}}} \frac{1}{S} \mathcal{L}(\mathbf{v}, \mathbf{y}, \bar{\mathbf{K}}(\boldsymbol{\theta})\bar{\boldsymbol{\alpha}}) + \mu\Omega((\bar{\boldsymbol{\alpha}}^T \bar{\mathbf{K}}(\boldsymbol{\theta})\bar{\boldsymbol{\alpha}})^{1/2}), \tag{8.25}$$

where $\bar{\mathbf{K}}(\boldsymbol{\theta}) := \mathbf{S}\mathbf{K}(\boldsymbol{\theta})\mathbf{S}^T$. Except for degenerate cases, problem (8.25) is not jointly convex in $\boldsymbol{\theta}$ and $\hat{\bar{\boldsymbol{\alpha}}}$, but it is separately convex in each vector for a convex $\mathcal{L}$ [32]. Iterative algorithms for solving (8.23) and (8.25) are available in [23].

### 8.2.4 SEMIPARAMETRIC RECONSTRUCTION

The approaches discussed so far are applicable to various problems but they are certainly limited by the modeling assumptions they make. In particular, the performance of algorithms belonging to the parametric family [11,15,33] is restricted by how well the signals actually adhere to the selected model. Nonparametric models, on the other hand [2,3,6,34], offer flexibility and robustness but they cannot readily incorporate information available *a priori*.

In practice however, it is not uncommon that neither of these approaches alone suffices for reliable inference. Consider, for instance, an employment-oriented social network such as LinkedIn, and suppose the goal is to estimate the salaries of all users given information about the salaries of a few. Clearly, besides network connections, exploiting available information regarding the users' education level and work experience could benefit the reconstruction task. The same is true in problems arising in link analysis, where the exploitation of hierarchical web structure can aid the task of estimating the importance of web pages [35]. In recommender systems, inferring preference scores for every item, given the users' feedback about particular items, could be cast as a signal reconstruction problem over the item correlation graph. Data sparsity imposes severe limitations on the quality of pure collaborative filtering methods [64]. Exploiting side information about the items is known to alleviate such limitations [36], leading to considerably improved recommendation performance [37,38].

A promising direction to endow nonparametric methods with prior information relies on a *semiparametric* approach whereby the signal of interest is modeled as the superposition of a parametric and a nonparametric component [39]. While the former leverages side information, the latter accounts for deviations from the parametric part and can also promote smoothness using kernels on graphs. In this section we outline two simple and reliable semiparametric estimators with complementary strengths, as detailed in [39].

#### *8.2.4.1 Semiparametric Inference*

Function $f$ is modeled as the superposition[4]

$$/\mathbf{f} = \mathbf{f}_{\mathrm{P}} + \mathbf{f}_{\mathrm{NP}}, \tag{8.26}$$

where $\mathbf{f}_{\mathrm{P}} := [f_{\mathrm{P}}(v_1), \ldots, f_{\mathrm{P}}(v_N)]^T$ and $\mathbf{f}_{\mathrm{NP}} := [f_{\mathrm{NP}}(v_1), \ldots, f_{\mathrm{NP}}(v_N)]^T$.

[4]For simplicity here we consider only the case of semiparametric partially linear models.

The parametric term $f_P(v) := \sum_{m=1}^{M} \beta_m b_m(v)$ captures the known signal structure via the basis $\mathcal{B} := \{b_m\}_{m=1}^{M}$, while the nonparametric term $f_{NP}$ belongs to an RKHS $\mathcal{H}$, which accounts for deviations from the span of $\mathcal{B}$. The goal of this section is to provide an efficient and reliable estimation of $\mathbf{f}$ given $\mathbf{y}$, $\mathbf{S}$, $\mathcal{B}$, $\mathcal{H}$ and $\mathcal{G}$.

Since $f_{NP} \in \mathcal{H}$, vector $\mathbf{f}_{NP}$ can be represented as in (8.3). By defining $\boldsymbol{\beta} := [\beta_1, \ldots, \beta_M]^T$ and the $N \times M$ matrix $\mathbf{B}$ with entries $(\mathbf{B})_{n,m} := b_m(v_n)$, the parametric term can be written in vector form as $\mathbf{f}_P := \mathbf{B}\boldsymbol{\beta}$. The semiparametric estimates can be found as the solution of the following optimization problem:

$$\begin{aligned} \{\hat{\boldsymbol{\alpha}}, \hat{\boldsymbol{\beta}}\} &= \arg\min_{\boldsymbol{\alpha},\boldsymbol{\beta}} \frac{1}{S}\sum_{s=1}^{S} \mathcal{L}(y_s, f(v_{n_s})) + \mu \| f_{NP} \|_{\mathcal{H}}^2 \\ \text{s.t.} \quad \mathbf{f} &= \mathbf{B}\boldsymbol{\beta} + \mathbf{K}\boldsymbol{\alpha}, \end{aligned} \tag{8.27}$$

where the fitting loss $\mathcal{L}$ quantifies the deviation of $f$ from the data and $\mu > 0$ is the regularization scalar that controls overfitting of the nonparametric term. Using (8.27), the semiparametric estimates are expressed as $\hat{\mathbf{f}} = \mathbf{B}\hat{\boldsymbol{\beta}} + \mathbf{K}\hat{\boldsymbol{\alpha}}$.

Solving (8.27) entails minimization over $N + M$ variables. Clearly, when dealing with large-scale graphs this could lead to prohibitively large computational costs. To reduce complexity, the semiparametric version of the representer theorem [18,19] is employed, which establishes that

$$\hat{\mathbf{f}} = \mathbf{B}\hat{\boldsymbol{\beta}} + \mathbf{K}\mathbf{S}^T \hat{\bar{\boldsymbol{\alpha}}}, \tag{8.28}$$

where $\hat{\bar{\boldsymbol{\alpha}}} := [\hat{\bar{\alpha}}_1, \ldots, \hat{\bar{\alpha}}_S]^T$. Estimates $\hat{\bar{\boldsymbol{\alpha}}}$, $\hat{\boldsymbol{\beta}}$ are found as

$$\begin{aligned} \{\hat{\bar{\boldsymbol{\alpha}}}, \hat{\boldsymbol{\beta}}\} &= \arg\min_{\bar{\boldsymbol{\alpha}},\boldsymbol{\beta}} \frac{1}{S}\sum_{s=1}^{S} \mathcal{L}(y_s, f(v_{n_s})) + \mu \| f_{NP} \|_{\mathcal{H}}^2 \\ \text{s.t.} \quad \mathbf{f} &= \mathbf{B}\boldsymbol{\beta} + \mathbf{K}\mathbf{S}^T \bar{\boldsymbol{\alpha}}, \end{aligned} \tag{8.29}$$

where $\bar{\boldsymbol{\alpha}} := [\bar{\alpha}_1, \ldots, \bar{\alpha}_S]^T$. The RKHS norm in (8.29) is expressed as $\| f_{NP} \|_{\mathcal{H}}^2 = \bar{\boldsymbol{\alpha}}^T \bar{\mathbf{K}} \bar{\boldsymbol{\alpha}}$, with $\bar{\mathbf{K}} := \mathbf{S}\mathbf{K}\mathbf{S}^T$. Relative to (8.27), the number of optimization variables in (8.29) is reduced to the more affordable $S + M$, with $S \ll N$.

Next, two loss functions with complementary benefits will be considered: the *square* loss and the *ϵ-insensitive* loss. The square loss function is

$$\mathcal{L}(y_s, f(v_{n_s})) := \| y_s - f(v_{n_s}) \|_2^2 \tag{8.30}$$

and (8.29) admits the following closed-form solution:

$$\hat{\bar{\boldsymbol{\alpha}}} = (\mathbf{P}\bar{\mathbf{K}} + \mu \mathbf{I}_S)^{-1} \mathbf{P}\mathbf{y}, \tag{8.31a}$$

$$\hat{\boldsymbol{\beta}} = (\bar{\mathbf{B}}^T \bar{\mathbf{B}})^{-1} \bar{\mathbf{B}}^T (\mathbf{y} - \bar{\mathbf{K}} \hat{\bar{\boldsymbol{\alpha}}}), \tag{8.31b}$$

where $\bar{\mathbf{B}} := \mathbf{S}\mathbf{B}$, $\mathbf{P} := \mathbf{I}_S - \bar{\mathbf{B}}(\bar{\mathbf{B}}^T \bar{\mathbf{B}})^{-1} \bar{\mathbf{B}}^T$. The complexity of (8.31) is $\mathcal{O}(S^3 + M^3)$.

**Table 8.2 Multikernel reconstruction**

| | NMSE | RMSE[min] |
|---|---|---|
| KRR with cov. kernel | 0.34 | 3.95 |
| Multikernel, RS | 0.44 | 4.51 |
| Multikernel, KS | 0.43 | 4.45 |
| BL for B=2 | 1.55 | 8.45 |
| BL for B=3 | 32.64 | 38.72 |
| BL, cut-off | 3.97 | 13.5 |

The $\epsilon$-*insensitive* loss function is given by

$$\mathcal{L}(y_s, f(v_{n_s})) := \max(0, |y_s - f(v_{n_s})| - \epsilon), \tag{8.32}$$

where $\epsilon$ is tuned, *e.g.* via cross-validation, to minimize the generalization error and it has well-documented merits in signal estimation from quantized data [40]. Substituting (8.32) into (8.29) yields a convex nonsmooth quadratic problem that can be solved efficiently for $\bar{\alpha}$ and $\beta$ using, *e.g.*, interior-point methods [18].

## 8.2.5 NUMERICAL TESTS

This section reports on the signal reconstruction performance of different methods using real as well as synthetic data. The performance of the estimators is assessed via Monte Carlo simulation by comparing the normalized mean square error (NMSE)

$$\text{NMSE} = \mathbb{E}\left[\frac{\|\hat{\mathbf{f}} - \mathbf{f}\|^2}{\|\mathbf{f}\|^2}\right]. \tag{8.33}$$

**Multikernel reconstruction.** The first data set contains departure and arrival information for flights among U.S. airports [41], from which $3 \times 10^6$ flights in the months of July, August and September of 2014 and 2015 were selected. We construct a graph with $N = 50$ vertices corresponding to the airports with highest traffic, and whenever the number of flights between the two airports exceeds 100 within the observation window, we connect the corresponding nodes with an edge.

A signal was constructed per day averaging the arrival delay of *all* inbound flights per selected airport. A total of 184 signals were considered, of which the first 154 were used for training (July, August, September 2014 and July, August 2015), and the remaining 30 for testing (September 2015). The weights of the edges between airports were learned using the training data based on the technique described in [23].

Table 8.2 lists the NMSE and the RMSE in minutes for the task of predicting the arrival delay at 40 airports when the delay at a randomly selected collection of 10 airports is observed. The second row corresponds to the ridge regression estimator that uses the nearly optimal *estimated* covariance kernel. The next two rows correspond to the multikernel approaches in §8.2.3 with a dictionary of 30 diffusion kernels with values of $\sigma^2$ uniformly spaced between 0.1 and 7. The rest of the rows pertain to graph BL estimators. Table 8.2 demonstrates the reliable performance of covariance kernels as well as the herein discussed multikernel approaches relative to competing alternatives.

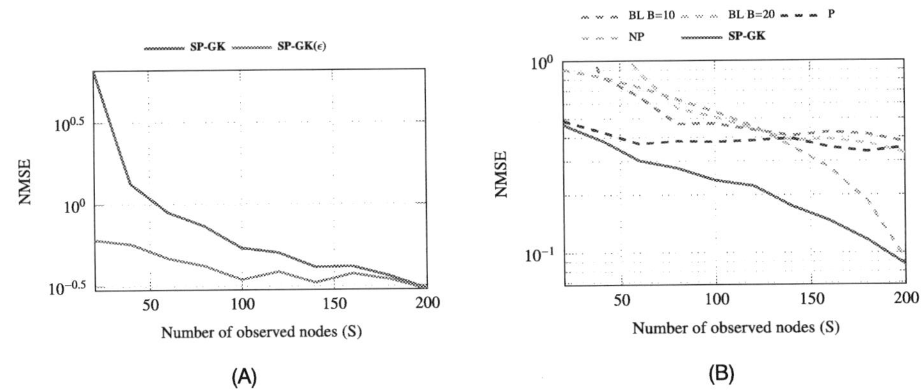


**FIGURE 8.1**

NMSE of the synthetic signal estimates. (A) $\mu = 5 \times 10^{-4}$, $\sigma = 5 \times 10^{-4}$, SNR$_e = 5$ dB. (B) $\mu = 5 \times 10^{-4}$, $\sigma = 5 \times 10^{-4}$, $\epsilon = 10^{-4}$ and SNR$_o = -5$ dB.

**Semiparametric reconstruction.** An Erdős–Rényi graph with probability of edge presence 0.6 and $N = 200$ nodes was generated, and $\mathbf{f}$ was formed by superimposing a BL signal [13,15] plus a piecewise constant signal [42]; that is,

$$\mathbf{f} = \sum_{i=1}^{10} \gamma_i \mathbf{u}_i + \sum_{i=1}^{6} \delta_i \mathbf{1}_{\mathcal{V}_c}, \tag{8.34}$$

where $\{\gamma_i\}_{i=1}^{10}$ and $\{\delta_i\}_{i=1}^{6}$ are standardized Gaussian distributed, $\{\mathbf{u}_i\}_{i=1}^{10}$ are the eigenvectors associated with the 10 smallest eigenvalues of the Laplacian matrix, $\{\mathcal{V}_i\}_{i=1}^{6}$ are the vertex sets of six clusters obtained via spectral clustering [43] and $\mathbf{1}_{\mathcal{V}_i}$ is the indicator vector with entries $(\mathbf{1}_{\mathcal{V}_i})_n := 1$, if $v_n \in \mathcal{V}_i$, and 0 otherwise. The parametric basis $\mathcal{B} = \{\mathbf{1}_{\mathcal{V}_i}\}_{i=1}^{6}$ was used by the estimators capturing the prior knowledge, and $S$ vertices were sampled uniformly at random. The subsequent experiments evaluate the performance of the semiparametric graph kernel estimators, SP-GK and SP-GK($\epsilon$), resulting from using (8.30) and (8.32) in (8.29), respectively; the *parametric* (P) that considers only the parametric term in (8.26), the *nonparametric* (NP) [2,3] that considers only the nonparametric term in (8.26) and the graph BL estimators from [13,15], which assume a BL model with bandwidth $B$. For all the experiments, the diffusion kernel (*cf.* Table 8.1) with parameter $\sigma$ is employed. First, white Gaussian noise $e_s$ of variance $\sigma_e^2$ is added to each sample $f_s$ to yield the signal-to-noise ratio SNR$_e := \|\mathbf{f}\|_2^2/(N\sigma_e^2)$. Fig. 8.1B presents the NMSE of different methods. As expected, the limited flexibility of the parametric approaches, BL and P, affects their ability to capture the true signal structure. The NP estimator achieves smaller NMSE, but only when the amount of available samples is adequate. Both semiparametric estimators were found to outperform other approaches, exhibiting reliable reconstruction even with few samples.

To illustrate the benefits of employing different loss functions (8.30) and (8.32), we compare the performance of SP-GK and SP-GK($\epsilon$) in the presence of outlying noise. Each sample $f_s$ is contaminated with Gaussian noise $o_s$ of large variance $\sigma_o^2$ with probability $p = 0.1$. Fig. 8.1A demonstrates the robustness of SP-GK($\epsilon$) which is attributed to the $\epsilon$-insensitive loss function (8.32). Further experiments using real signals can be found in [39].

## 8.3 INFERENCE OF DYNAMIC FUNCTIONS OVER DYNAMIC GRAPHS

Networks that exhibit *time varying connectivity* patterns with *time varying node attributes* arise in a plethora of network science–related applications. Sometimes these dynamic network topologies switch between a finite number of discrete states, governed by sudden changes of the underlying dynamics [44, 45]. A challenging problem that arises in this setting is that of reconstructing time evolving functions on graphs, given their values on a subset of vertices and time instants. Efficiently exploiting spatiotemporal dynamics can markedly impact sampling costs by reducing the number of vertices that need to be observed to attain a target performance. Such a reduction can be of paramount importance in certain applications, *e.g.*, in monitoring time-dependent activity of different regions of the brain through invasive electrocorticography (ECoG), where observing a vertex requires the implantation of an intracranial electrode [44].

Although one could reconstruct a time varying function per time slot using the non- or semi-parametric methods of §8.2, leveraging time correlations typically yields estimators with improved performance. Schemes tailored for time evolving functions on graphs include [46] and [47], which predict the function values at time $t$ given observations up to time $t-1$. However, these schemes assume that the function of interest adheres to a specific vector autoregressive model. Other works target time-invariant functions, but can only afford tracking sufficiently slow variations. This is the case with the dictionary learning approach in [48] and the distributed algorithms in [49] and [50]. Unfortunately, the flexibility of these algorithms to capture spatial information is also limited since [48] focuses on Laplacian regularization, whereas [49] and [50] require the signal to be BL.

Motivated by the aforementioned limitations, in what comes next we extend the framework presented in §8.2 accommodating time varying function reconstruction over dynamic graphs. But before we delve into the time varying setting, a few definitions are in order.

**Definitions**: A time varying graph is a tuple $\mathcal{G}(t) := (\mathcal{V}, \mathbf{A}_t)$, where $\mathcal{V} := \{v_1, \ldots, v_N\}$ is the vertex set and $\mathbf{A}_t \in \mathbb{R}^{N\times N}$ is the adjacency matrix at time $t$, whose $(n,n')$th entry $A_{n,n'}(t)$ assigns a weight to the pair of vertices $(v_n, v_{n'})$ at time $t$. A time-invariant graph is a special case with $\mathbf{A}_t = \mathbf{A}_{t'}\ \forall t, t'$. Adopting common assumptions made in related literature (*e.g.*, [1, Chapter 2], [4,9]), we also define $\mathcal{G}(t)$ (i) to have nonnegative weights ($A_{n,n'}(t) \geq 0\ \forall t$, and $\forall n \neq n'$), (ii) to have no self-edges ($A_{n,n}(t) = 0\ \forall n, t$) and (iii) to be undirected ($A_{n,n'}(t) = A_{n',n}(t)\ \forall n, n', t$).

A time varying function or signal on a graph is a map $f : \mathcal{V} \times \mathcal{T} \to \mathbb{R}$, where $\mathcal{T} := \{1, 2, \ldots\}$ is the set of time indices. The value $f(v_n, t)$ of $f$ at vertex $v_n$ and time $t$ can be thought of as the value of an attribute of $v_n \in \mathcal{V}$ at time $t$. The values of $f$ at time $t$ will be collected in $\mathbf{f}_t := [f(v_1, t), \ldots, f(v_N, t)]^T$.

At time $t$, vertices with indices in the time-dependent set $\mathcal{S}_t := \{n_1(t), \ldots, n_{S(t)}(t)\}$, $1 \leq n_1(t) < \cdots < n_{S(t)}(t) \leq N$, are observed. The resulting samples can be expressed as $y_s(t) = f(v_{n_s(t)}, t) + e_s(t)$, $s = 1, \ldots, S(t)$, where $e_s(t)$ models the observation error. By letting $\mathbf{y}_t := [y_1(t), \ldots, y_{S(t)}(t)]^T$, the observations can be conveniently expressed as

$$\mathbf{y}_t = \mathbf{S}_t \mathbf{f}_t + \mathbf{e}_t, \quad t = 1, 2, \ldots, \tag{8.35}$$

where $\mathbf{e}_t := [e_1(t), \ldots, e_{S(t)}(t)]^T$ and the $S(t) \times N$ sampling matrix $\mathbf{S}_t$ contains ones at positions $(s, n_s(t))$, $s = 1, \ldots, S(t)$ and zeros elsewhere.

The broad goal of this section is to "reconstruct" $f$ from the observations $\{\mathbf{y}_t\}_t$ in (8.35). Two formulations will be considered.

**Batch formulation.** In the batch reconstruction problem, one aims at finding $\{\mathbf{f}_t\}_{t=1}^T$ given $\{\mathcal{G}(t)\}_{t=1}^T$, the sample locations $\{\mathbf{S}_t\}_{t=1}^T$ and all observations $\{\mathbf{y}_t\}_{t=1}^T$.

**Online formulation.** At every time $t$, one is given $\mathcal{G}$ together with $\mathbf{S}_t$ and $\mathbf{y}_t$, and the goal is to find $\mathbf{f}_t$. The latter can be obtained possibly based on a previous estimate of $\mathbf{f}_{t-1}$, but the complexity per time slot $t$ must be independent of $t$.

To solve these problems, we will rely on the assumption that $f$ evolves smoothly over space and time, yet more structured dynamics can be incorporated if known.

## 8.3.1 KERNELS ON EXTENDED GRAPHS

This section extends the kernel-based learning framework of §8.2 to subsume time evolving functions over possibly dynamic graphs through the notion of *graph extension*, by which the time dimension receives the same treatment as the spatial dimension. The versatility of kernel-based methods to leverage spatial information [23] is thereby inherited and broadened to account for temporal dynamics as well. This vantage point also accommodates time varying sampling sets and topologies.

### 8.3.1.1 Extended Graphs

An immediate approach to reconstructing time evolving functions is to apply (8.9) separately for each $t = 1, \ldots, T$. This yields the instantaneous estimator (IE)

$$\hat{\mathbf{f}}_t^{(IE)} := \arg\min_{\mathbf{f}} \frac{1}{S(t)} \|\mathbf{y}_t - \mathbf{S}_t \mathbf{f}\|_2^2 + \mu \mathbf{f}^T \mathbf{K}_t^\dagger \mathbf{f}. \tag{8.36}$$

Unfortunately, this estimator does not account for the possible relation between, *e.g.*, $f(v_n, t)$ and $f(v_n, t-1)$. If, for instance, $f$ varies slowly over time, an estimate of $f(v_n, t)$ may as well benefit from leveraging observations $y_s(\tau)$ at time instants $\tau \neq t$. Exploiting temporal dynamics potentially reduces the number of vertices that have to be sampled to attain a target reconstruction performance, which in turn can markedly reduce sampling costs.

Incorporating temporal dynamics into kernel-based reconstruction, which can only handle a single snapshot (*cf.* §8.2), necessitates an appropriate reformulation of time evolving function reconstruction as a problem of reconstructing a time-invariant function. An appealing possibility is to replace $\mathcal{G}$ with its *extended* version $\tilde{\mathcal{G}} := (\tilde{\mathcal{V}}, \tilde{\mathbf{A}})$, where each vertex in $\mathcal{V}$ is replicated $T$ times to yield the extended vertex set $\tilde{\mathcal{V}} := \{v_n(t),\ n = 1, \ldots, N,\ t = 1, \ldots, T\}$, and the $(n + N(t-1), n' + N(t'-1))$th entry of the $TN \times TN$ extended adjacency matrix $\tilde{\mathbf{A}}$ equals the weight of the edge $(v_n(t), v_{n'}(t'))$. The time varying function $f$ can thus be replaced with its extended time-invariant counterpart $\tilde{f} : \tilde{\mathcal{V}} \to \mathbb{R}$ with $\tilde{f}(v_n(t)) = f(v_n, t)$.

**Definition 1.** Let $\mathcal{V} := \{v_1, \ldots, v_N\}$ denote a vertex set and let $\mathcal{G} := (\mathcal{V}, \{\mathbf{A}_t\}_{t=1}^T)$ be a time varying graph. A graph $\tilde{\mathcal{G}}$ with vertex set $\tilde{\mathcal{V}} := \{v_n(t),\ n = 1, \ldots, N,\ t = 1, \ldots, T\}$ and $NT \times NT$ adjacency matrix $\tilde{\mathbf{A}}$ is an extended graph of $\mathcal{G}$ if the $t$th $N \times N$ diagonal block of $\tilde{\mathbf{A}}$ equals $\mathbf{A}_t$.

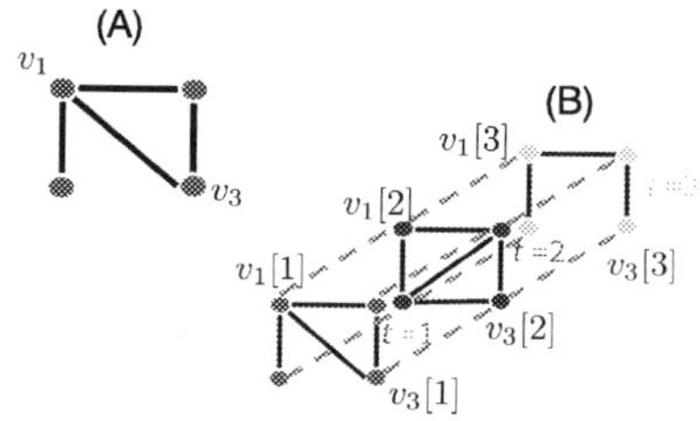


**FIGURE 8.2**

NMSE of the synthetic signal estimates. (A) Original graph and (B) extended graph $\tilde{\mathcal{G}}$ for diagonal $\mathbf{B}_t^{(\mathcal{T})}$. Edges connecting vertices at the same time instant are represented by solid lines whereas edges connecting vertices at different time instants are represented by dashed lines.

In general, the diagonal blocks $\{\mathbf{A}_t\}_{t=1}^T$ do not provide full description of the underlying extended graph. Indeed, one also needs to specify the off-diagonal block entries of $\tilde{\mathbf{A}}$ to capture the spatiotemporal dynamics of $f$.

As an example, consider an extended graph with

$$\tilde{\mathbf{A}} = \text{btridiag}\{\mathbf{A}_1, \ldots, \mathbf{A}_T; \mathbf{B}_2^{(\mathcal{T})}, \ldots, \mathbf{B}_T^{(\mathcal{T})}\}, \tag{8.37}$$

where $\mathbf{B}_t^{(\mathcal{T})} \in \mathbb{R}_+^{N\times N}$ connects $\{v_n(t-1)\}_{n=1}^N$ to $\{v_n(t)\}_{n=1}^N$, $t = 2, \ldots, T$, and btridiag$\{\mathbf{A}_1, \ldots, \mathbf{A}_T; \mathbf{B}_2, \ldots, \mathbf{B}_T\}$ represents the symmetric block tridiagonal matrix

$$\tilde{\mathbf{A}} = \begin{bmatrix} \mathbf{A}_1 & \mathbf{B}_2^T & \mathbf{0} & \ldots & \mathbf{0} & \mathbf{0} \\ \mathbf{B}_2 & \mathbf{A}_2 & \mathbf{B}_3^T & \ldots & \mathbf{0} & \mathbf{0} \\ \mathbf{0} & \mathbf{B}_3 & \mathbf{A}_3 & \ldots & \mathbf{0} & \mathbf{0} \\ \vdots & \vdots & \vdots & \ddots & \vdots & \vdots \\ \mathbf{0} & \mathbf{0} & \mathbf{0} & \ldots & \mathbf{A}_{T-1} & \mathbf{B}_T^T \\ \mathbf{0} & \mathbf{0} & \mathbf{0} & \ldots & \mathbf{B}_T & \mathbf{A}_T \end{bmatrix}.$$

For instance, each vertex can be connected to its neighbors at the previous time instant by setting $\mathbf{B}_t^{(\mathcal{T})} = \mathbf{A}_{t-1}$, or it can be connected to its replicas at adjacent time instants by setting $\mathbf{B}_t^{(\mathcal{T})}$ to be diagonal. Such a graph is depicted in Fig. 8.2.

### 8.3.1.2 Batch and Online Reconstruction via Space–Time Kernels

The extended graph enables a generalization of the estimators in §8.2 for time evolving functions. The rest of this subsection discusses two such KRR estimators.

Consider first the batch formulation, where all the $\tilde{S} := \sum_{t=1}^T S_t$ samples in $\tilde{\mathbf{y}} := [\mathbf{y}_1^T, \ldots, \mathbf{y}_T^T]^T$ are available, and the goal is to estimate $\tilde{\mathbf{f}} := [\mathbf{f}_1^T, \ldots, \mathbf{f}_T^T]^T$. Directly applying the KRR criterion in (8.9) to reconstruct $\tilde{\mathbf{f}}$ on the extended graph $\tilde{\mathcal{G}}$ yields

$$\hat{\tilde{\mathbf{f}}} := \arg\min_{\tilde{\mathbf{f}}} ||\tilde{\mathbf{y}} - \tilde{\mathbf{S}}\tilde{\mathbf{f}}||^2_{\mathbf{D}} + \mu\tilde{\mathbf{f}}^T\tilde{\mathbf{K}}^{\dagger}\tilde{\mathbf{f}}, \tag{8.38a}$$

where $\tilde{\mathbf{K}}$ is now a $TN \times TN$ "space–time" kernel matrix, $\tilde{\mathbf{S}} := \text{bdiag}\{\mathbf{S}_1, \ldots, \mathbf{S}_T\}$ and $\mathbf{D} := \text{bdiag}\{S(1)\mathbf{I}_{S(1)}, \ldots, S(T)\mathbf{I}_{S(T)}\}$. If $\tilde{\mathbf{K}}$ is invertible, (8.38a) can be solved in closed form as

$$\hat{\tilde{\mathbf{f}}} = \tilde{\mathbf{K}}\tilde{\mathbf{S}}^T(\tilde{\mathbf{S}}\tilde{\mathbf{K}}\tilde{\mathbf{S}}^T + \mu\mathbf{D})^{-1}\tilde{\mathbf{y}}. \tag{8.38b}$$

The "space–time" kernel $\tilde{\mathbf{K}}$, captures complex spatiotemporal dynamics. If the topology is time-invariant, $\tilde{\mathbf{K}}$ can be specified in a bidimensional plane of spatiotemporal frequency similar to §8.2.2.[5]

In the online formulation, one aims to estimate $\mathbf{f}_t$ after the $\tilde{S}(t) := \sum_{\tau=1}^{t} S(\tau)$ samples in $\tilde{\mathbf{y}}_t := [\mathbf{y}_1^T, \ldots, \mathbf{y}_t^T]^T$ become available. Based on these samples, the KRR estimate of $\tilde{\mathbf{f}}$, denoted as $\hat{\tilde{\mathbf{f}}}_{1:T|t}$, is clearly

$$\begin{aligned}\hat{\tilde{\mathbf{f}}}_{1:T|t} :=& \quad \arg\min_{\tilde{\mathbf{f}}} ||\tilde{\mathbf{y}}_t - \tilde{\mathbf{S}}_t\tilde{\mathbf{f}}||^2_{\mathbf{D}_t} + \mu\tilde{\mathbf{f}}^T\tilde{\mathbf{K}}^{-1}\tilde{\mathbf{f}} &(8.39a)\\ =& \quad \tilde{\mathbf{K}}\tilde{\mathbf{S}}_t^T(\tilde{\mathbf{S}}_t\tilde{\mathbf{K}}\tilde{\mathbf{S}}_t^T + \mu\mathbf{D}_t)^{-1}\tilde{\mathbf{y}}_t, &(8.39b)\end{aligned}$$

where $\tilde{\mathbf{K}}$ is assumed invertible for simplicity, $\mathbf{D}_t := \text{bdiag}\{S(1)\mathbf{I}_{S(1)}, \ldots, S(t)\mathbf{I}_{S(t)}\}$ and $\tilde{\mathbf{S}}_t := [\text{diag}\{\mathbf{S}_1, \ldots, \mathbf{S}_t\}, \mathbf{0}_{\tilde{S}(t)\times(T-t)N}] \in \{0,1\}^{\tilde{S}(t)\times TN}$.

The estimate in (8.39) comprises the per slot estimates $\{\hat{\mathbf{f}}_{\tau|t}\}_{\tau=1}^{T}$; that is, $\hat{\tilde{\mathbf{f}}}_{1:T|t} := [\hat{\mathbf{f}}_{1|t}^T, \ldots, \hat{\mathbf{f}}_{T|t}^T]^T$ with $\hat{\mathbf{f}}_{\tau|t} := [\hat{f}_1(\tau|t), \ldots, \hat{f}_N(\tau|t)]^T$, where $\hat{\mathbf{f}}_{\tau|t}$ (respectively $\hat{f}_n(\tau|t)$) is the KRR estimate of $\mathbf{f}_\tau$ ($f(v_n, \tau)$) given the observations up to time $t$. With this notation, it follows that for all $t, \tau$

$$\hat{\mathbf{f}}_{\tau|t} = (\mathbf{i}_{T,\tau}^T \otimes \mathbf{I}_N)\hat{\tilde{\mathbf{f}}}_{1:T|t}. \tag{8.40}$$

Regarding $t$ as the present, (8.39) therefore provides estimates of past, present and future values of $f$. The solution to the online problem comprises the sequence of *present* KRR estimates for all $t$, that is, $\{\hat{\mathbf{f}}_{t|t}\}_{t=1}^{T}$. This can be obtained by solving (8.39a) in closed form per $t$ as in (8.39b) and then applying (8.40). However, such an approach does not yield a desirable online algorithm since its complexity per time slot is $\mathcal{O}(\tilde{S}^3(t))$ and therefore increasing with $t$. For this reason, this approach is not satisfactory since the online problem formulation requires the complexity per time slot of the desired algorithm to be independent of $t$. An algorithm that does satisfy this requirement yet provides the exact KRR estimate is presented next when the kernel matrix is any positive definite matrix $\tilde{\mathbf{K}}$ satisfying

$$\tilde{\mathbf{K}}^{-1} = \text{btridiag}\{\mathbf{D}_1, \ldots, \mathbf{D}_T; \mathbf{C}_2, \ldots, \mathbf{C}_T\} \tag{8.41}$$

for some $N \times N$ matrices $\{\mathbf{D}_t\}_{t=1}^{T}$ and $\{\mathbf{C}_t\}_{t=2}^{T}$. Kernels in this important family are designed in [51].

If $\tilde{\mathbf{K}}$ is of the form (8.41), then the kernel Kalman filter (KKF) in Algorithm 2 returns the sequence $\{\hat{\mathbf{f}}_{t|t}\}_{t=1}^{T}$, where $\hat{\mathbf{f}}_{t|t}$ is given by (8.40). The $N \times N$ matrices $\{\mathbf{P}_\tau\}_{\tau=2}^{T}$ and $\{\boldsymbol{\Sigma}_\tau\}_{\tau=1}^{T}$ are obtained offline by Algorithm 1, and $\sigma_e^2(\tau) = \mu S(\tau)\ \forall\tau$.

[5]For general designs of space–time kernels $\tilde{\mathbf{K}}$ for time-invariant as well as time varying topologies, see [51].

**Algorithm 1:** Recursion to set the parameters of the KKF.

**Input:** $\mathbf{D}_t, t = 1, \ldots, T, \mathbf{C}_t, t = 2, \ldots, T$.

**Set** $\boldsymbol{\Sigma}_T^{-1} = \mathbf{D}_T$

**for** $t = T, T-1, \ldots, 2$ **do**

$\quad \mathbf{P}_t = -\boldsymbol{\Sigma}_t \mathbf{C}_t$

$\quad \boldsymbol{\Sigma}_{t-1}^{-1} = \mathbf{D}_{t-1} - \mathbf{P}_t^T \boldsymbol{\Sigma}_t^{-1} \mathbf{P}_t$

**Output:** $\boldsymbol{\Sigma}_t, t = 1, \ldots, T, \mathbf{P}_t, t = 2, \ldots, T$

**Algorithm 2:** Kernel Kalman filter (KKF).

**Input:** $\{\boldsymbol{\Sigma}_t \in \mathbb{S}_+^N\}_{t=1}^T, \{\mathbf{P}_t \in \mathbb{R}^{N \times N}\}_{t=2}^T$,
$\{\mathbf{y}_t \in \mathbb{R}^{S(t)}\}_{t=1}^T, \{\mathbf{S}_t \in \{0, 1\}^{S(t) \times N}\}_{t=1}^T$,
$\{\sigma_e^2(t) > 0\}_{t=1}^T$.

**Set** $\hat{\mathbf{f}}_{0|0} = \mathbf{0}, \mathbf{M}_{0|0} = \mathbf{0}, \mathbf{P}_1 = \mathbf{0}$

**for** $t = 1, \ldots, T$ **do**

$\quad \hat{\mathbf{f}}_{t|t-1} = \mathbf{P}_t \hat{\mathbf{f}}_{t-1|t-1}$

$\quad \mathbf{M}_{t|t-1} = \mathbf{P}_t \mathbf{M}_{t-1|t-1} \mathbf{P}_t^T + \boldsymbol{\Sigma}_t$

$\quad \mathbf{G}_t = \mathbf{M}_{t|t-1} \mathbf{S}_t^T (\sigma_e^2(t) \mathbf{I} + \mathbf{S}_t \mathbf{M}_{t|t-1} \mathbf{S}_t^T)^{-1}$

$\quad \hat{\mathbf{f}}_{t|t} = \hat{\mathbf{f}}_{t|t-1} + \mathbf{G}_t (\mathbf{y}_t - \mathbf{S}_t \hat{\mathbf{f}}_{t|t-1})$

$\quad \mathbf{M}_{t|t} = (\mathbf{I} - \mathbf{G}_t \mathbf{S}_t) \mathbf{M}_{t|t-1}$

**Output:** $\hat{\mathbf{f}}_{t|t}, t = 1, \ldots, T; \mathbf{M}_t, t = 1, \ldots, T$.

The KKF generalizes the probabilistic KF since the latter is recovered upon setting $\tilde{\mathbf{K}}$ to be the covariance matrix of $\tilde{\mathbf{f}}$ in the probabilistic KF. The assumptions made by the probabilistic KF are stronger than those involved in the KKF. Specifically, in the probabilistic KF, $\mathbf{f}_t$ must adhere to a linear state space model, $\mathbf{f}_t = \mathbf{P}_t \mathbf{f}_{t-1} + \boldsymbol{\eta}_t$, with known transition matrix $\mathbf{P}_t$, where the state noise $\boldsymbol{\eta}_t$ is uncorrelated over time and has known covariance matrix $\boldsymbol{\Sigma}_t$. Furthermore, the observation noise $\mathbf{e}_t$ must be uncorrelated over time and have a known covariance matrix. Correspondingly, the performance guarantees of the probabilistic KF are also stronger: the resulting estimate is optimal in the mean square error sense among all linear estimators. Furthermore, if $\boldsymbol{\eta}_t$ and $\mathbf{y}_t$ are jointly Gaussian, $t = 1, \ldots, T$, then the probabilistic KF estimate is optimal in the mean square error sense among all (not necessarily linear) estimators. In contrast, the requirements of the proposed KKF are much weaker since they only specify that $f$ must evolve smoothly with respect to a given extended graph. As expected, the performance guarantees are similarly weaker; see *e.g.* [18, Chapter 5]. However, since the KKF generalizes the probabilistic KF, the reconstruction performance of the former for judiciously selected $\tilde{\mathbf{K}}$ cannot be worse than the reconstruction performance of the latter for any given criterion. The caveat, however, is that such a selection is not necessarily easy.

For the rigorous statement and proof relating the celebrated KF [52, Chapter 17] and the optimization problem in (8.39a), see [51]. Algorithm 2 requires $\mathcal{O}(N^3)$ operations per time slot, whereas the complexity of evaluating (8.39b) for the $t$th time slot is $\mathcal{O}(\tilde{S}^3(t))$, which increases with $t$ and becomes eventually prohibitive. Since distributed versions of the Kalman filter are well studied [53], decentralized KKF algorithms can be pursued to further reduce the computational complexity.

### 8.3.2 MULTIKERNEL KRIGED KALMAN FILTERS

The following section applies the KRR framework presented in §8.2 to online data-adaptive estimators of $\mathbf{f}_t$. Specifically, a spatiotemporal model is presented that judiciously captures the dynamics over space and time. Based on this model the KRR criterion over time and space is formulated, and an online algorithm is derived with affordable computational complexity, when the kernels are preselected. To bypass the need for selecting an appropriate kernel, this section discusses a data-adaptive multikernel learning extension of the KRR estimator that learns the optimal kernel "on-the-fly."

#### *8.3.2.1 Spatiotemporal Models*

Consider modeling the dynamics of $\mathbf{f}_t$ separately over time and space as $f(v_n,t) = f^{(\nu)}(v_n,t) + f^{(\chi)}(v_n,t)$, or in vector form

$$\mathbf{f}_t = \mathbf{f}_t^{(\nu)} + \mathbf{f}_t^{(\chi)}, \tag{8.42}$$

where $\mathbf{f}_t^{(\nu)} := [f^{(\nu)}(v_1,t),\ldots,f^{(\nu)}(v_N,t)]^T$ and $\mathbf{f}_t^{(\chi)} := [f^{(\chi)}(v_1,t),\ldots,f^{(\chi)}(v_N,t)]^T$. The first term $\{\mathbf{f}_t^{(\nu)}\}_t$ captures only spatial dependencies and can be thought of as exogenous input to the graph that does not affect the evolution of the function in time.

The second term $\{\mathbf{f}_t^{(\chi)}\}_t$ accounts for spatiotemporal dynamics. A popular approach [54, Chapter 3] models $\mathbf{f}_t^{(\chi)}$ with the state equation

$$\mathbf{f}_t^{(\chi)} = \mathbf{A}_{t,t-1}\mathbf{f}_{t-1}^{(\chi)} + \boldsymbol{\eta}_t, \quad t = 1,2,\ldots, \tag{8.43}$$

where $\mathbf{A}_{t,t-1}$ is a generic transition matrix that can be chosen, *e.g.*, as the $N \times N$ adjacency of a possibly directed "transition graph," with $\mathbf{f}_0^{(\chi)} = \mathbf{0}$ and $\boldsymbol{\eta}_t \in \mathbb{R}^N$ capturing the state error. The state transition matrix $\mathbf{A}_{t,t-1}$ can be selected in accordance with the prior information available. Simplicity in estimation is a major advantage of the random walk model [55], where $\mathbf{A}_{t,t-1} = \alpha\mathbf{I}_N$ with $\alpha > 0$. On the other hand, adherence to the graph prompts the selection $\mathbf{A}_{t,t-1} = \alpha\mathbf{A}$, in which case (8.43) amounts to a diffusion process on the time-invariant graph $\mathcal{G}$. The recursion in (8.43) is a vector autoregressive model (VARM) of order one and offers flexibility in tracking multiple forms of temporal dynamics [54, Chapter 3]. The model in (8.43) captures the dependence between $f^{(\chi)}(v_n,t)$ and its time lagged versions $\{f^{(\chi)}(v_n,t-1)\}_{n=1}^N$.

Next, a model with increased flexibility is presented to account for instantaneous spatial dependencies as well. We have

$$\mathbf{f}_t^{(\chi)} = \mathbf{A}_{t,t}\mathbf{f}_t^{(\chi)} + \mathbf{A}_{t,t-1}\mathbf{f}_{t-1}^{(\chi)} + \boldsymbol{\eta}_t, \quad t = 1,2,\ldots, \tag{8.44}$$

where $\mathbf{A}_{t,t}$ encodes the instantaneous relation between $f^{(\chi)}(v_n,t)$ and $\{f^{(\chi)}(v_{n'},t)\}_{n' \neq n}$. The recursion in (8.44) amounts to a structural vector autoregressive model (SVARM) [44]. Interestingly, (8.44) can be rewritten as

$$\mathbf{f}_t^{(\chi)} = (\mathbf{I}_N - \mathbf{A}_{t,t})^{-1}\mathbf{A}_{t,t-1}\mathbf{f}_{t-1}^{(\chi)} + (\mathbf{I}_N - \mathbf{A}_{t,t})^{-1}\boldsymbol{\eta}_t, \tag{8.45}$$

where $\mathbf{I}_N - \mathbf{A}_{t,t}$ is assumed invertible. After defining $\tilde{\boldsymbol{\eta}}_t := (\mathbf{I}_N - \mathbf{A}_{t,t})^{-1}\boldsymbol{\eta}_t$ and $\tilde{\mathbf{A}}_{t,t-1} := (\mathbf{I}_N - \mathbf{A}_{t,t})^{-1}\mathbf{A}_{t,t-1}$, (8.44) boils down to

$$\mathbf{f}_t^{(\chi)} = \tilde{\mathbf{A}}_{t,t-1}\mathbf{f}_{t-1}^{(\chi)} + \tilde{\boldsymbol{\eta}}_t, \tag{8.46}$$

which is equivalent to (8.43). This section will focus on deriving estimators based on (8.43), but can also accommodate (8.44) using the aforementioned reformulation.

Modeling $\mathbf{f}_t$ as the superposition of a term $\mathbf{f}_t^{(\chi)}$, capturing the slow dynamics over time with a state space equation, and a term $\mathbf{f}_t^{(\nu)}$ accounting for fast dynamics is motivated by the application at hand [56,55,57]. In the kriging terminology [56], $\mathbf{f}_t^{(\nu)}$ is said to model small-scale *spatial fluctuations*, whereas $\mathbf{f}_t^{(\chi)}$ captures the so-called *trend*. The decomposition (8.42) is often dictated by the sampling interval; while $\mathbf{f}_t^{(\chi)}$ captures slow dynamics relative to the sampling interval, fast variations are modeled with $\mathbf{f}_t^{(\nu)}$. Such a modeling approach is advised in the prediction of network delays [55], where $\mathbf{f}_t^{(\chi)}$ represents the queuing delay while $\mathbf{f}_t^{(\nu)}$ represents the propagation, transmission and processing delays. Likewise, when predicting prices across different stocks, $\mathbf{f}_t^{(\chi)}$ captures the daily evolution of the stock market, which is correlated across stocks and time samples, while $\mathbf{f}_t^{(\nu)}$ describes unexpected changes, such as the daily drop of the stock market due to political statements, which are assumed uncorrelated over time.

### 8.3.2.2 Kernel Kriged Kalman Filter

The spatiotemporal model in (8.42), (8.43) can represent multiple forms of spatiotemporal dynamics by judicious selection of the associated parameters. The *batch* KRR estimator over time yields

$$\begin{aligned}
&\underset{\{\mathbf{f}_\tau^{(\chi)},\boldsymbol{\eta}_\tau,\mathbf{f}_\tau^{(\nu)},\mathbf{f}_\tau\}_{\tau=1}^t}{\arg\min} \quad \sum_{\tau=1}^{t} \frac{1}{S(\tau)}\|\mathbf{y}_\tau - \mathbf{S}_\tau\mathbf{f}_\tau\|^2 + \mu_1 \sum_{\tau=1}^{t} \|\boldsymbol{\eta}_\tau\|^2_{\mathbf{K}_\tau^{(\eta)}} + \mu_2 \sum_{\tau=1}^{t} \|\mathbf{f}_\tau^{(\nu)}\|^2_{\mathbf{K}_\tau^{(\nu)}} \\
&\text{s.t.} \quad \boldsymbol{\eta}_\tau = \mathbf{f}_\tau^{(\chi)} - \mathbf{A}_{\tau,\tau-1}\mathbf{f}_{\tau-1}^{(\chi)}, \ \mathbf{f}_\tau = \mathbf{f}_\tau^{(\nu)} + \mathbf{f}_\tau^{(\chi)}, \quad \tau = 1,\ldots,t.
\end{aligned} \tag{8.47}$$

The first term in (8.47) penalizes the fitting error in accordance with (8.1). The scalars $\mu_1, \mu_2 \geq 0$ are regularization parameters controlling the effect of the kernel regularizers, while prior information about $\{\mathbf{f}_\tau^{(\nu)}, \boldsymbol{\eta}_\tau\}_{\tau=1}^t$ may guide the selection of the appropriate kernel matrices. The constraints in (8.47) imply adherence to (8.43) and (8.42). Since the $\mathbf{f}_\tau^{(\nu)}, \boldsymbol{\eta}_\tau$ are defined over the time evolving $\mathcal{G}(\tau)$, a potential approach is to select Laplacian kernels as $\mathbf{K}_\tau^{(\nu)}, \mathbf{K}_\tau^{(\eta)}$; see §8.2.2. Next, we rewrite (8.47) in a form amenable to online solvers, namely

$$\begin{aligned}
\underset{\{\mathbf{f}_\tau^{(\chi)},\mathbf{f}_\tau^{(\nu)}\}_{\tau=1}^t}{\arg\min} \quad & \sum_{\tau=1}^{t} \frac{1}{S(\tau)}\|\mathbf{y}_\tau - \mathbf{S}_\tau\mathbf{f}_\tau^{(\chi)} - \mathbf{S}_\tau\mathbf{f}_\tau^{(\nu)}\|^2 + \\
& + \mu_1 \sum_{\tau=1}^{t} \|\mathbf{f}_\tau^{(\chi)} - \mathbf{A}_{\tau,\tau-1}\mathbf{f}_{\tau-1}^{(\chi)}\|^2_{\mathbf{K}_\tau^{(\eta)}} + \\
& + \mu_2 \sum_{\tau=1}^{t} \|\mathbf{f}_\tau^{(\nu)}\|^2_{\mathbf{K}_\tau^{(\nu)}}.
\end{aligned} \tag{8.48}$$

**Algorithm 3:** Kernel Kriged Kalman filter (KeKriKF).

**Input:** $\mathbf{K}_t^{(\eta)}; \mathbf{K}_t^{(\nu)} \in \mathbb{S}_+^N; \mathbf{A}_{t,t-1} \in \mathbb{R}^{N \times N}; \mathbf{y}_t \in \mathbb{R}^{S(t)}; \mathbf{S}_t \in \{0,1\}^{S(t) \times N}$.

$\bar{\mathbf{K}}_t^{(\chi)} = \frac{1}{\mu_2}\mathbf{S}_t\mathbf{K}_t^{(\nu)}\mathbf{S}_t^T + S(t)\mathbf{I}_{S(t)}$

$\hat{\mathbf{f}}_{t|t-1}^{(\chi)} = \mathbf{A}_{t,t-1}\hat{\mathbf{f}}_{t-1|t-1}^{(\chi)}$

$\mathbf{M}_{t|t-1} = \mathbf{A}_{t,t-1}\mathbf{M}_{t-1|t-1}\mathbf{A}_{t,t-1}^T + \frac{1}{\mu_1}\mathbf{K}_t^{(\eta)}$

$\mathbf{G}_t = \mathbf{M}_{t|t-1}\mathbf{S}_t^T(\bar{\mathbf{K}}_t^{(\chi)} + \mathbf{S}_t\mathbf{M}_{t|t-1}\mathbf{S}_t^T)^{-1}$

$\mathbf{M}_{t|t} = (\mathbf{I} - \mathbf{G}_t\mathbf{S}_t)\mathbf{M}_{t|t-1}$

$\hat{\mathbf{f}}_{t|t}^{(\chi)} = \hat{\mathbf{f}}_{t|t-1}^{(\chi)} + \mathbf{G}_t(\mathbf{y}_t - \mathbf{S}_t\hat{\mathbf{f}}_{t|t-1}^{(\chi)})$

$\hat{\mathbf{f}}_{t|t}^{(\nu)} = \mathbf{K}_t^{(\nu)}\mathbf{S}_t^T{\bar{\mathbf{K}}_t^{(\chi)}}^{-1}(\mathbf{y}_t - \mathbf{S}_t\hat{\mathbf{f}}_{t|t}^{(\chi)})$

**Output:** $\hat{\mathbf{f}}_{t|t}^{(\chi)}; \hat{\mathbf{f}}_{t|t}^{(\nu)}; \mathbf{M}_{t|t}$.

In a batch form the optimization in (8.48) yields $\{\hat{\mathbf{f}}_{\tau|t}^{(\nu)}$ and $\hat{\mathbf{f}}_{\tau|t}^{(\chi)}\}_{\tau=1}^t$ per slot $t$ with complexity that grows with $t$. Fortunately, the *filtered* solutions $\{\hat{\mathbf{f}}_{\tau|\tau}^{(\nu)}, \hat{\mathbf{f}}_{\tau|\tau}^{(\chi)}\}_{\tau=1}^t$ of (8.48) are attained by the kernel kriged Kalman filter (KeKriKF) in an *online* fashion. For the proof the reader is referred to [58]. One iteration of the proposed KeKriKF is summarized as Algorithm 3. This online estimator—with computational complexity $\mathcal{O}(N^3)$ per $t$—tracks the temporal variations of the signal of interest through (8.43) and promotes desired properties such as smoothness over the graph, using $\mathbf{K}_t^{(\nu)}$ and $\mathbf{K}_t^{(\eta)}$. Different from existing KriKF approaches over graphs [55], the KeKriKF takes into account the underlying graph structure in estimating $\mathbf{f}_t^{(\nu)}$ as well as $\mathbf{f}_t^{(\chi)}$. Furthermore, by using $\mathbf{L}_t$ in (8.16), it can also accommodate dynamic graph topologies. Finally, it should be noted that KeKriKF encompasses as a special case the KriKF, which relies on knowing the statistical properties of the function [55–57,60].

Lack of prior information prompts the development of data-driven approaches that efficiently learn the appropriate kernel matrix. Section 8.3.2.3 proposes an online MKL approach for achieving this goal.

#### 8.3.2.3 Online Multikernel Learning

To cope with a lack of prior information about the pertinent kernel, the following dictionaries of kernels will be considered: $\mathcal{D}_\nu := \{\mathbf{K}^{(\nu)}(m) \in \mathbb{S}_+^N\}_{m=1}^{M_\nu}$ and $\mathcal{D}_\eta := \{\mathbf{K}^{(\eta)}(m) \in \mathbb{S}_+^N\}_{m=1}^{M_\eta}$. For the following assume that $\mathbf{K}_\tau^{(\nu)} = \mathbf{K}^{(\nu)}$, $\mathbf{K}_\tau^{(\eta)} = \mathbf{K}^{(\eta)}$ and $\mathbf{S}_\tau = \mathbf{S}$, $\forall \tau$. Moreover, we postulate that the kernel matrices are of the form $\mathbf{K}^{(\nu)} = \mathbf{K}^{(\nu)}(\boldsymbol{\theta}^{(\nu)}) = \sum_{m=1}^{M_\nu} \theta^{(\nu)}(m)\mathbf{K}^{(\nu)}(m)$ and $\mathbf{K}^{(\eta)} = \mathbf{K}^{(\eta)}(\boldsymbol{\theta}^{(\eta)}) = \sum_{m=1}^{M_\eta} \theta^{(\eta)}(m)\mathbf{K}^{(\eta)}(m)$, where $\theta^{(\eta)}(m), \theta^{(\nu)}(m) \geq 0$, $\forall m$.

Next, in accordance with §8.2.3 the coefficients $\boldsymbol{\theta}^{(\nu)} = [\theta^{(\nu)}(1), \ldots, \theta^{(\nu)}(M)]^T$ and $\boldsymbol{\theta}^{(\eta)} = [\theta^{(\eta)}(1), \ldots, \theta^{(\eta)}(M)]^T$ can be found by jointly minimizing (8.48) with respect to $\{\mathbf{f}_\tau^{(\chi)}, \mathbf{f}_\tau^{(\nu)}\}_{\tau=1}^t, \boldsymbol{\theta}^{(\nu)}$ and $\boldsymbol{\theta}^{(\eta)}$, which yields

$$\underset{\substack{\{\mathbf{f}_\tau^{(\chi)},\mathbf{f}_\tau^{(\nu)}\}_{\tau=1}^t,\\ \theta^{(\nu)}\geq\mathbf{0},\theta^{(\eta)}\geq\mathbf{0}}}{\arg\min} \sum_{\tau=1}^{t} \frac{1}{S}\|\mathbf{y}_\tau - \mathbf{S}\mathbf{f}_\tau^{(\chi)} - \mathbf{S}\mathbf{f}_\tau^{(\nu)}\|^2 + \mu_1 \sum_{\tau=1}^{t} \|\mathbf{f}_\tau^{(\chi)} - \mathbf{A}_{\tau,\tau-1}\mathbf{f}_{\tau-1}^{(\chi)}\|^2_{\mathbf{K}^{(\eta)}(\theta^{(\eta)})}$$

$$+\mu_2 \sum_{\tau=1}^{t} \|\mathbf{f}_\tau^{(\nu)}\|^2_{\mathbf{K}^{(\nu)}(\theta^{(\nu)})} + t\rho_\nu\|\theta^{(\nu)}\|_2^2 + t\rho_\eta\|\theta^{(\eta)}\|_2^2, \tag{8.49}$$

where $\rho_\nu, \rho_\eta \geq 0$ are regularization parameters that effect a ball constraint on $\theta^{(\nu)}$ and $\theta^{(\eta)}$, weighted by $t$ to account for the first three terms that are growing with $t$. Observe that the optimization problem in (8.49) gives time varying estimates $\theta_t^{(\nu)}$ and $\theta_t^{(\eta)}$ allowing to track the optimal $\mathbf{K}^{(\nu)}$ and $\mathbf{K}^{(\eta)}$ that change over time respectively.

The optimization problem in (8.49) is not jointly convex in $\{\mathbf{f}_\tau^{(\chi)}, \mathbf{f}_\tau^{(\nu)}\}_{\tau=1}^t, \theta^{(\nu)}, \theta^{(\eta)}$, but it is separately convex in these variables. To solve (8.49) alternating minimization strategies will be employed that suggest optimizing with respect to one variable, while keeping the other variables fixed [61]. If $\theta^{(\nu)}, \theta^{(\eta)}$ are considered fixed, (8.49) reduces to (8.48), which can be solved by Algorithm 3 for the estimates $\hat{\mathbf{f}}_{t|t}^{(\chi)}, \hat{\mathbf{f}}_{t|t}^{(\nu)}$ at each $t$. For $\{\mathbf{f}_\tau^{(\chi)}, \mathbf{f}_\tau^{(\nu)}\}_{\tau=1}^t$ fixed and replaced by $\{\hat{\mathbf{f}}_{\tau|\tau}^{(\chi)}, \hat{\mathbf{f}}_{\tau|\tau}^{(\nu)}\}_{\tau=1}^t$ in (8.48) the time varying estimates of $\theta^{(\nu)}, \theta^{(\eta)}$ are found by

$$\hat{\theta}_t^{(\eta)} = \underset{\theta^{(\eta)}\geq\mathbf{0}}{\arg\min}\frac{1}{t}\sum_{\tau=1}^{t} \|\hat{\mathbf{f}}_{\tau|\tau}^{(\chi)} - \mathbf{A}_{\tau,\tau-1}\hat{\mathbf{f}}_{\tau-1|\tau-1}^{(\chi)}\|^2_{\mathbf{K}^{(\eta)}(\theta^{(\eta)})} + \frac{\rho_\eta}{\mu_1}\|\theta^{(\eta)}\|_2^2, \tag{8.50a}$$

$$\hat{\theta}_t^{(\nu)} = \underset{\theta^{(\nu)}\geq\mathbf{0}}{\arg\min}\frac{1}{t}\sum_{\tau=1}^{t} \|\hat{\mathbf{f}}_{\tau|\tau}^{(\nu)}\|^2_{\mathbf{K}^{(\nu)}(\theta^{(\nu)})} + +\frac{\rho_\nu}{\mu_2}\|\theta^{(\nu)}\|_2^2. \tag{8.50b}$$

The optimization problems (8.50a) and (8.50b) are strongly convex and iterative algorithms are available based on projected gradient descent (PGD) [62], or the Frank–Wolfe algorithm [63]. When the kernel matrices belong to the Laplacian family (8.16), efficient algorithms that exploit the common eigenspace of the kernels in the dictionary have been developed in [58]. The proposed method reduces the per step computational complexity of PGD from a prohibitive $\mathcal{O}(N^3M)$ for general kernels to a more affordable $\mathcal{O}(NM)$ for Laplacian kernels. The proposed algorithm, termed multikernel KriKF (MKriKF), alternates between computing $\hat{\mathbf{f}}_{t|t}^{(\chi)}$ and $\hat{\mathbf{f}}_{t|t}^{(\nu)}$ utilizing the KKriKF and estimating $\hat{\theta}_t^{(\nu)}$ and $\hat{\theta}_t^{(\eta)}$ from solving (8.50b) and (8.50a).

### 8.3.3 NUMERICAL TESTS

This section compares the performance of the methods we discussed in §8.3.1 and §8.3.2 with state-of-the-art alternatives and illustrates some of the trade-offs inherent to time varying function reconstruction through real-data experiments. The source code for the simulations is available at the authors' websites.

Unless otherwise stated, the compared estimators include distributed least squares reconstruction (DLSR) [49] with step size $\mu_{\text{DLSR}}$ and parameter $\beta_{\text{DLSR}}$; the least mean squares (LMS) algorithm in [50] with step size $\mu_{\text{LMS}}$; the BL instantaneous estimator (BL-IE), which results after applying [11,

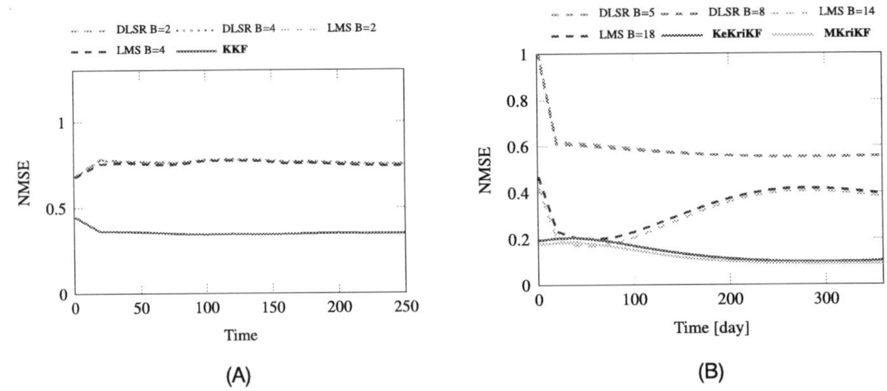


**FIGURE 8.3**

NMSE for real data simulations. (A) NMSE for the ECoG data set ($\sigma = 1.2$, $\mu = 10^{-4}$, $\mu_{\text{DLSR}} = 1.2$, $b^{(\mathcal{T})} = 0.01$, $\beta_{\text{DLSR}} = 0.5$, $\mu_{\text{LMS}} = 0.6$). (B) NMSE of temperature estimates ($\mu_1 = 1$, $\mu_2 = 1$, $\mu_{\text{DLSR}} = 1.6$, $\beta_{\text{DLSR}} = 0.5$, $\mu_{\text{LMS}} = 0.6$, $\alpha = 10^{-3}$, $\mu_\eta = 10^{-5}$, $r_\eta = 10^{-6}$, $\mu_\nu = 2$, $r_\nu = 0.5$, $M_\nu = 40$, $M_\eta = 40$).

13,15] separately per $t$; and the KRR instantaneous estimator (KRR-IE) in (8.36) with a diffusion kernel with parameter $\sigma$. DLSR, LMS and BL-IE also use a bandwidth parameter $B$.

**Reconstruction via extended graphs.** For our first experiment we use a dataset obtained from an epilepsy study [65], which is used to showcase an example analysis of ECoG data (analysis of ECoG data is a standard tool in diagnosing epilepsy). Our next experiment utilizes the ECoG time series in [65] from $N = 76$ electrodes implanted in a patient's brain before and after the onset of a seizure. A symmetric time-invariant adjacency matrix $\mathbf{A}$ was obtained using the method in [44] with ECoG data before the onset of the seizure. Function $f(v_n, t)$ comprises the electrical signal at the $n$th electrode and $t$th sampling instant after the onset of the seizure, for a period of $T = 250$ samples. The values of $f(v_n, t)$ were normalized by subtracting the temporal mean of each time series before the onset of the seizure. The goal of the experiment is to illustrate the reconstruction performance of KKF in capturing the complex spatiotemporal dynamics of brain signals.

Fig. 8.3A depicts the NMSE$(t, \{\mathcal{S}_\tau\}_{\tau=1}^t)$, averaged over all sets $\mathcal{S}_t = \mathcal{S},\ \forall t$, of size $S = 53$. For the KKF, a space–time kernel was created (see [51]) with $\mathbf{K}_t$ a time-invariant covariance kernel $\mathbf{K}_t = \hat{\boldsymbol{\Sigma}}$, where $\hat{\boldsymbol{\Sigma}}$ was set to the sample covariance matrix of the time series before the onset of the seizure, with a time-invariant $\mathbf{B}^{(\mathcal{T})} = b^{(\mathcal{T})}\mathbf{I}$. The results clearly show the superior reconstruction performance of the KKF, which successfully exploits the statistics of the signal when available, among competing approaches, even with a small number of samples. This result suggests that the ECoG diagnosis technique could be efficiently conducted even with a smaller number of intracranial electrodes, which may have a positive impact on the patient's experience.

**Reconstruction via KeKriKF.** The second dataset is provided by the National Climatic Data Center [66] and comprises hourly temperature measurements at $N = 109$ measuring stations across the continental United States in 2010. A time-invariant graph was constructed as in [51], based on geographical distances. The value $f(v_n, t)$ represents the temperature recorded at the $n$th station and $t$th day.

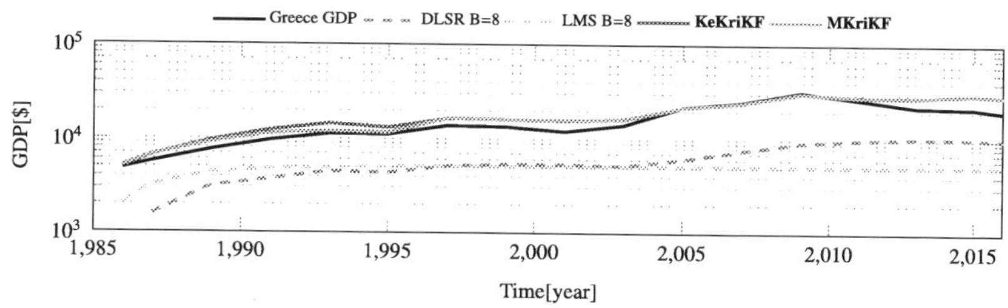


**FIGURE 8.4**

NMSE for GDP data. Tracking of GDP ($\mu_{\text{DLSR}} = 1.6$, $\beta_{\text{DLSR}} = 0.4$, $\mu_{\text{LMS}} = 1.6$, $\rho_\nu = 10^5$, $\rho_\eta = 10^5$).

Fig. 8.3B reports the performance of different reconstruction algorithms in terms of NMSE, for $S = 40$. The KeKriKF Algorithm 3 adopts a diffusion kernel for $\mathbf{K}^{(\nu)}$ with $\sigma = 1.8$ and for $\mathbf{K}^{(\eta)} = s_\eta \mathbf{I}_N$ with $s_\eta = 10^{-5}$. The MKriKF is configured with: $\mathcal{D}_\nu$ that contains $M_\nu$ diffusion kernels with parameters $\{\sigma(m)\}_{m=1}^{M_\nu}$ drawn from a Gaussian distribution with mean $\mu_\nu$ and variance $r_\nu$; $\mathcal{D}_\eta$ that contains $M_\eta$ $s_\eta \mathbf{I}_N$ with parameters $\{s_\eta(m)\}_{m=1}^{M_\eta}$ drawn from a Gaussian distribution with mean $\mu_\eta$ and variance $r_\eta$. The specific kernel selection for KeKriKF leads to the smallest NMSE error and was selected using cross validation. Observe that MKriKF captures the spatiotemporal dynamics, successfully explores the pool of available kernels and achieves superior performance.

The third dataset is provided by the World Bank Group [67] and is comprised of the gross domestic product (GDP) per capita values for $N = 127$ countries for the years 1960–2016. A time-invariant graph was constructed using the correlation between the GDP values for the first 25 years of different countries. The graph function $f(v_n, t)$ denotes the GDP value reported for the $n$th country and $t$th year for $t = 1985, \ldots, 2016$. The graph Fourier transform of the GDP values shows that the graph frequencies $\check{f}_k$, $4 < k < 120$, take small values and large values otherwise. Motivated by the aforementioned observation, the KKriKF is configured with a band-reject kernel $\mathbf{K}^{(\nu)}$ that results after applying $r(\lambda_n) = \beta$ for $k \leq n \leq N - l$ and $r(\lambda_n) = 1/\beta$ otherwise in (8.16) with $k = 3, l = 6, \beta = 15$ and for $\mathbf{K}^{(\eta)} = s_\eta \mathbf{I}_N$ with $s_\eta = 10^{-4}$. The MKriKF adopts a $\mathcal{D}_\nu$ that contains band-reject kernels with $k \in [2, 4]$, $l \in [3, 6]$, $\beta = 15$ that result in $M_\nu = 12$ kernels and a $\mathcal{D}_\eta$ that contains $\{s_\eta(m)\mathbf{I}_N\}_{m=1}^{40}$ with $s_\eta(m)$ drawn from a Gaussian distribution with mean $\mu_\eta = 10^{-5}$ and variance $r_\eta = 10^{-6}$. Next, the performance of different algorithms in tracking the GDP value is evaluated after sampling $S = 38$ countries.

Fig. 8.4 illustrates the actual GDP as well as GDP estimates for Greece, which is not contained in the sampled countries. Clearly, MKriKF, which learns the pertinent kernels from the data, achieves roughly the same performance as KKriKF, which is configured manually to obtain the smallest possible NMSE.

## 8.3.4 SUMMARY

The task of reconstructing functions defined on graphs arises naturally in a plethora of applications. The kernel-based approach offers a clear, principled and intuitive way for tackling this problem. In this chapter, we gave a contemporary treatment of this framework focusing on both time-invariant and time evolving domains. The methods presented herein offer the potential of providing an expressive way to tackle interesting real-world problems. Besides illustrating the effectiveness of the discussed approaches, our tests were also chosen to showcase interesting application areas as well as reasonable modeling approaches for the interested readers to build upon. For further details about the models discussed here and their theoretical properties, the reader is referred to [23,39,59,58,68,69] and the references therein.

## ACKNOWLEDGMENTS

The research was supported by NSF grants 1442686, 1500713, 1508993, 1509040, and 1739397.

## REFERENCES

[1] E.D. Kolaczyk, Statistical Analysis of Network Data: Methods and Models, Springer, New York, 2009.

[2] R.I. Kondor, J. Lafferty, Diffusion kernels on graphs and other discrete structures, in: Proc. Int. Conf. Mach. Learn., Sydney, Australia, 2002, pp. 315–322.

[3] A.J. Smola, R.I. Kondor, Kernels and regularization on graphs, in: Learning Theory and Kernel Machines, Springer, 2003, pp. 144–158.

[4] D.I. Shuman, S.K. Narang, P. Frossard, A. Ortega, P. Vandergheynst, The emerging field of signal processing on graphs: extending high-dimensional data analysis to networks and other irregular domains, IEEE Signal Processing Magazine 30 (3) (2013) 83–98.

[5] A. Sandryhaila, J.M.F. Moura, Discrete signal processing on graphs, IEEE Transactions on Signal Processing 61 (7) (2013) 1644–1656.

[6] O. Chapelle, B. Schölkopf, A. Zien, et al., Semi-Supervised Learning, MIT Press, Cambridge, 2006.

[7] O. Chapelle, V. Vapnik, J. Weston, Transductive inference for estimating values of functions, in: Proc. Advances Neural Inf. Process. Syst., vol. 12, Denver, Colorado, 1999, pp. 421–427.

[8] C. Cortes, M. Mohri, On transductive regression, in: Proc. Advances Neural Inf. Process. Syst., Vancouver, Canada, 2007, pp. 305–312.

[9] M. Belkin, P. Niyogi, V. Sindhwani, Manifold regularization: a geometric framework for learning from labeled and unlabeled examples, Journal of Machine Learning Research 7 (2006) 2399–2434.

[10] S.K. Narang, A. Gadde, A. Ortega, Signal processing techniques for interpolation in graph structured data, in: Proc. IEEE Int. Conf. Acoust., Speech, Sig. Process., IEEE, Vancouver, Canada, 2013, pp. 5445–5449.

[11] S.K. Narang, A. Gadde, E. Sanou, A. Ortega, Localized iterative methods for interpolation in graph structured data, in: Global Conf. Sig. Inf. Process., IEEE, Austin, Texas, 2013, pp. 491–494.

[12] A. Gadde, A. Ortega, A probabilistic interpretation of sampling theory of graph signals, in: Proc. IEEE Int. Conf. Acoust., Speech, Sig. Process., Brisbane, Australia, 2015, pp. 3257–3261.

[13] M. Tsitsvero, S. Barbarossa, P. Di Lorenzo, Signals on graphs: uncertainty principle and sampling, IEEE Transactions on Signal Processing 64 (18) (2016) 4845–4860.

[14] S. Chen, R. Varma, A. Sandryhaila, J. Kovacevic, Discrete signal processing on graphs: sampling theory, IEEE Transactions on Signal Processing 63 (24) (2015) 6510–6523.

[15] A. Anis, A. Gadde, A. Ortega, Efficient sampling set selection for bandlimited graph signals using graph spectral proxies, IEEE Transactions on Signal Processing 64 (14) (2016) 3775–3789.

[16] X. Wang, P. Liu, Y. Gu, Local-set-based graph signal reconstruction, IEEE Transactions on Signal Processing 63 (9) (2015) 2432–2444, https://doi.org/10.1109/TSP.2015.2411217.

[17] A.G. Marques, S. Segarra, G. Leus, A. Ribeiro, Sampling of graph signals with successive local aggregations, IEEE Transactions on Signal Processing 64 (7) (2016) 1832–1843.

[18] B. Schölkopf, A.J. Smola, Learning with Kernels: Support Vector Machines, Regularization, Optimization, and Beyond, MIT Press, 2002.

[19] B. Schölkopf, R. Herbrich, A.J. Smola, A generalized representer theorem, in: Computational Learning Theory, Springer, 2001, pp. 416–426.
[20] V.N. Vapnik, Statistical Learning Theory, vol. 1, Wiley, New York, 1998.
[21] G. Kimeldorf, G. Wahba, Some results on Tchebycheffian spline functions, Journal of Mathematical Analysis and Applications 33 (1) (1971) 82–95.
[22] D. Zhou, B. Schölkopf, A regularization framework for learning from graph data, in: ICML Workshop Stat. Relational Learn. Connections Other Fields, vol. 15, Banff, Canada, 2004, pp. 67–68.
[23] D. Romero, M. Ma, G.B. Giannakis, Kernel-based reconstruction of graph signals, IEEE Transactions on Signal Processing 65 (3) (2017) 764–778.
[24] L. Page, S. Brin, R. Motwani, T. Winograd, The PageRank Citation Ranking: Bringing Order to the Web, tech. rep., Stanford InfoLab, 1999.
[25] A.N. Nikolakopoulos, A. Korba, J.D. Garofalakis, Random surfing on multipartite graphs, in: 2016 IEEE International Conference on Big Data (Big Data), 2016, pp. 736–745.
[26] A.N. Nikolakopoulos, J.D. Garofalakis, Random surfing without teleportation, in: Algorithms, Probability, Networks, and Games, Springer International Publishing, 2015, pp. 344–357.
[27] J.A. Bazerque, G.B. Giannakis, Nonparametric basis pursuit via kernel-based learning, IEEE Signal Processing Magazine 28 (30) (2013) 112–125.
[28] V. Sindhwani, P. Niyogi, M. Belkin, Beyond the point cloud: from transductive to semi-supervised learning, in: Proc. Int. Conf. Mach. Learn., ACM, 2005, pp. 824–831.
[29] M. Gönen, E. Alpaydın, Multiple kernel learning algorithms, Journal of Machine Learning Research 12 (Jul) (2011) 2211–2268.
[30] J.A. Bazerque, G. Mateos, G.B. Giannakis, Group-lasso on splines for spectrum cartography, IEEE Transactions on Signal Processing 59 (10) (2011) 4648–4663, https://doi.org/10.1109/TSP.2011.2160858.
[31] G.B. Giannakis, Q. Ling, G. Mateos, I.D. Schizas, H. Zhu, Decentralized learning for wireless communications and networking, arXiv preprint, arXiv:1503.08855, 2016.
[32] C.A. Micchelli, M. Pontil, Learning the kernel function via regularization, Journal of Machine Learning Research (2005) 1099–1125.
[33] S. Segarra, A.G. Marques, G. Leus, A. Ribeiro, Reconstruction of graph signals through percolation from seeding nodes, IEEE Transactions on Signal Processing 64 (16) (2016) 4363–4378.
[34] A.S. Zamzam, V.N. Ioannidis, N.D. Sidiropoulos, Coupled graph tensor factorization, in: Proc. Asilomar Conf. Sig., Syst., Comput., Pacific Grove, CA, 2016, pp. 1755–1759.
[35] A.N. Nikolakopoulos, J.D. Garofalakis, NCDawareRank: a novel ranking method that exploits the decomposable structure of the web, in: Proceedings of the Sixth ACM International Conference on Web Search and Data Mining, ACM, 2013, pp. 143–152.
[36] A.N. Nikolakopoulos, J.D. Garofalakis, Top-n recommendations in the presence of sparsity: an NCD-based approach, in: Web Intelligence, vol. 13, IOS Press, 2015, pp. 247–265.
[37] A.N. Nikolakopoulos, M.A. Kouneli, J.D. Garofalakis, Hierarchical itemspace rank: exploiting hierarchy to alleviate sparsity in ranking-based recommendation, Neurocomputing 163 (2015) 126–136.
[38] A.N. Nikolakopoulos, J.D. Garofalakis, NCDREC: a decomposability inspired framework for top-n recommendation, in: 2014 IEEE/WIC/ACM International Joint Conferences on Web Intelligence (WI) and Intelligent Agent Technologies (IAT), vol. 1, 2014, pp. 183–190.
[39] V.N. Ioannidis, A.N. Nikolakopoulos, G.B. Giannakis, Semi-parametric graph kernel-based reconstruction, in: IEEE Global Conf. Sig. Inf. Process., Montreal, Canada, 2017.
[40] V. Vapnik, The Nature of Statistical Learning Theory, Springer, 2013.
[41] Bureau of Transportation, United States [online], available: http://www.transtats.bts.gov/, 2016.
[42] S. Chen, R. Varma, A. Singh, J. Kovačević, Signal representations on graphs: tools and applications, arXiv preprint, arXiv:1512.05406, 2015 [online].
[43] U. Von Luxburg, A tutorial on spectral clustering, Statistics and Computing 17 (4) (2007) 395–416.
[44] Y. Shen, B. Baingana, G.B. Giannakis, Nonlinear structural vector autoregressive models for inferring effective brain network connectivity, arXiv preprint, arXiv:1610.06551, 2016.
[45] B. Baingana, G.B. Giannakis, Tracking switched dynamic network topologies from information cascades, IEEE Transactions on Signal Processing 65 (4) (2017) 985–997.

[46] F.R. Bach, M.I. Jordan, Learning graphical models for stationary time series, IEEE Transactions on Signal Processing 52 (8) (2004) 2189–2199.

[47] J. Mei, J.M.F. Moura, Signal processing on graphs: causal modeling of big data, arXiv preprint, arXiv:1503.00173v3, 2016.

[48] P.A. Forero, K. Rajawat, G.B. Giannakis, Prediction of partially observed dynamical processes over networks via dictionary learning, IEEE Transactions on Signal Processing 62 (13) (2014) 3305–3320.

[49] X. Wang, M. Wang, Y. Gu, A distributed tracking algorithm for reconstruction of graph signals, IEEE Journal of Selected Topics in Signal Processing 9 (4) (2015) 728–740.

[50] P.D. Lorenzo, S. Barbarossa, P. Banelli, S. Sardellitti, Adaptive least mean squares estimation of graph signals, IEEE Transactions on Signal and Information Processing over Networks 2 (4) (2017) 555–568.

[51] D. Romero, V.N. Ioannidis, G.B. Giannakis, Kernel-based reconstruction of space-time functions on dynamic graphs, IEEE Journal of Selected Topics in Signal Processing 11 (6) (2017) 1–14.

[52] G. Strang, K. Borre, Linear Algebra, Geodesy, and GPS, Siam, 1997.

[53] I.D. Schizas, G.B. Giannakis, S.I. Roumeliotis, A. Ribeiro, Consensus in ad hoc WSNs with noisy links—part ii: distributed estimation and smoothing of random signals, IEEE Transactions on Signal Processing 56 (4) (2008) 1650–1666.

[54] T.W. Anderson, An Introduction to Multivariate Statistical Analysis, vol. 2, Wiley, New York, 1958.

[55] K. Rajawat, E. Dall'Anese, G.B. Giannakis, Dynamic network delay cartography, IEEE Transactions on Information Theory 60 (5) (2014) 2910–2920.

[56] C.K. Wikle, N. Cressie, A dimension-reduced approach to space-time Kalman filtering, Biometrika (1999) 815–829.

[57] S.J. Kim, E. Dall'Anese, G.B. Giannakis, Cooperative spectrum sensing for cognitive radios using kriged Kalman filtering, IEEE Journal of Selected Topics in Signal Processing 5 (1) (2011) 24–36, https://doi.org/10.1109/JSTSP.2010.2053016.

[58] V.N. Ioannidis, D. Romero, G.B. Giannakis, Inference of spatio-temporal functions over graphs via multi-kernel kriged Kalman filtering, arXiv preprint, arXiv:1711.09306, 2017.

[59] V.N. Ioannidis, D. Romero, G.B. Giannakis, Inference of spatiotemporal processes over graphs via kernel kriged Kalman filtering, in: Proc. European Sig. Process. Conf., Kos, Greece, 2017.

[60] K.V. Mardia, C. Goodall, E.J. Redfern, F.J. Alonso, The kriged Kalman filter, Test 7 (2) (1998) 217–282.

[61] I. Csiszár, G. Tusnády, Information geometry and alternating minimization procedures, Statistics and Decisions (1984) 205–237.

[62] L. Zhang, D. Romero, G.B. Giannakis, Fast convergent algorithms for multi-kernel regression, in: Proc. Workshop Stat. Sig. Process., Palma de Mallorca, Spain, 2016.

[63] L. Zhang, G. Wang, D. Romero, G.B. Giannakis, Randomized block Frank–Wolfe for convergent large-scale learning, IEEE Transactions on Signal Processing 65 (24) (2017) 6448–6461.

[64] A.N. Nikolakopoulos, V. Kalantzis, E. Gallopoulos, J.D. Garofalakis, Factored proximity models for top-N recommendations, in: Proc. 2017 IEEE International Conference on Big Knowledge, ICBK, Hefei, China, 2017, pp. 80–87.

[65] M.A. Kramer, E.D. Kolaczyk, H.E. Kirsch, Emergent network topology at seizure onset in humans, Epilepsy Research 79 (2) (2008) 173–186.

[66] 1981–2010 U.S. climate normals [online], available: https://www.ncdc.noaa.gov.

[67] GDP per capita (current US) [online], available: https://data.worldbank.org/indicator/NY.GDP.PCAP.CD.

[68] V.N. Ioannidis, D. Romero, G.B. Giannakis, Kernel-based reconstruction of space-time functions via extended graphs, in: Proc. Asilomar Conf. Sig., Syst., Comput., Pacific Grove, CA, 2016, pp. 1829–1833.

[69] D. Romero, M. Ma, G.B. Giannakis, Estimating signals over graphs via multi-kernel learning, in: Proc. Workshop Stat. Sig. Process., Palma de Mallorca, Spain, 2016.

PART

# 3

# NONLINEAR MODELING WITH MULTIPLE LEARNING MACHINES

CHAPTER

# ONLINE NONLINEAR MODELING VIA SELF-ORGANIZING TREES

# 9

**Nuri Denizcan Vanli***, **Suleyman Serdar Kozat**†

**Massachusetts Institute of Technology, Laboratory for Information and Decision Systems, Cambridge, MA, United States* †*Bilkent University, Ankara, Turkey*

## CONTENTS



## CHAPTER POINTS

- We present a nonlinear modeling method for online supervised learning problems.
- Nonlinear modeling is introduced via SOTs, which adaptively partition the feature space to minimize the loss of the algorithm.
- Experimental validation shows huge empirical performance improvements with respect to the state-of-the-art methods.

Adaptive Learning Methods for Nonlinear System Modeling. DOI: 10.1016/B978-0-12-812976-0.00012-9

## 9.1 INTRODUCTION

Nonlinear adaptive learning is extensively investigated in the signal processing [1–4] and machine learning literature [5–7], especially for applications where linear modeling is inadequate and hence does not provide satisfactory results due to the structural constraint on linearity. Although nonlinear approaches can be more powerful than linear methods in modeling, they usually suffer from overfitting and stability and convergence issues [8], which considerably limit their application to signal processing and machine learning problems. These issues are especially exacerbated in adaptive filtering due to the presence of feedback, which is hard to control even for linear models [9]. Furthermore, for applications involving big data, which require the processing of input vectors with considerably large dimensions, nonlinear models are usually avoided due to unmanageable computational complexity increase [10]. To overcome these difficulties, tree-based nonlinear adaptive filters or regressors are introduced as elegant alternatives to linear models since these highly efficient methods retain the breadth of nonlinear models while mitigating the overfitting and convergence issues [11–13].

In its most basic form, a tree defines a hierarchical or nested partitioning of the feature space [12]. As an example, consider the binary tree in Fig. 9.1, which partitions a two-dimensional feature space. On this tree, each node is constructed by a bisection of the feature space (where we use hyperplanes for separation), which results in a complete nested and disjoint partitioning of the feature space. After the partitions are defined, the local learners in each region can be chosen as desired. As an example, to solve a regression problem, one can train a linear regressor in each region, which yields an overall piecewise linear regressor. In this sense, tree-based modeling is a natural nonlinear extension of linear models via a tractable nested structure.

Although nonlinear modeling using trees is a powerful and efficient method, there exist several algorithmic parameters and design choices that affect their performance in many applications [11]. Tuning these parameters is a difficult task for applications involving nonstationary data exhibiting saturation effects, threshold phenomena or chaotic behavior [14]. In particular, the performance of tree-based models heavily depends on a *careful partitioning* of the feature space. Selection of a good partition is essential to balance the bias and variance of the regressor [12]. As an example, even for a uniform binary tree, while increasing the depth of the tree improves the modeling power, such an increase usually results in overfitting [15]. To address this issue, there exist nonlinear modeling algorithms that avoid such a direct commitment to a particular partition but instead construct a weighted average of all possible partitions (or equivalently, piece-wise models) defined on a tree [6,7,16,17]. Note that a full binary tree of depth $d$ defines a doubly exponential number of different partitions of the feature space [18]; for an example, see Fig. 9.2. Each of these partitions can be represented by a certain collection of the nodes of the tree, where each node represents a particular region of the feature space. Any of these partitions can be used to construct a nonlinear model, *e.g.*, by training a linear model in each region, we can obtain a piece-wise linear model. Instead of selecting one of these partitions and fixing it as the nonlinear model, one can run all partitions in parallel and combine their outputs using a mixture-of-experts approach. Such methods are shown to mitigate the bias–variance tradeoff in a deterministic framework [6,7,16,19]. However, these methods are naturally constrained to work on a fixed partitioning structure, *i.e.*, the partitions are fixed and cannot be adapted to data.

Although there exist numerous methods to partition the feature space, many of these split criteria are typically chosen *a priori* and fixed such as the dyadic partitioning [20] and a specific loss (*e.g.*, the Gini index [21]) is minimized separately for each node. For instance, multivariate trees are extended

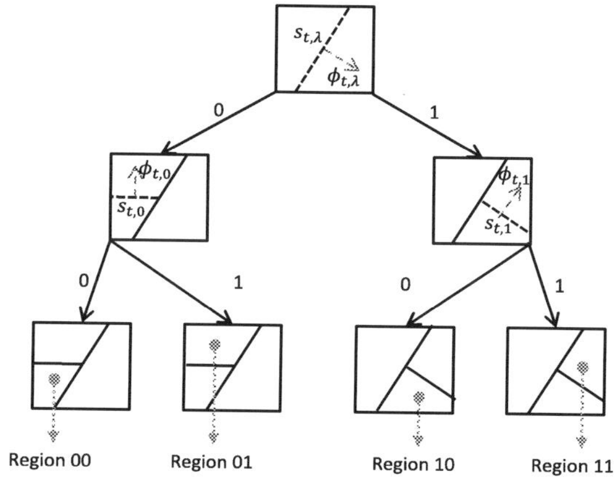


**FIGURE 9.1**

Feature space partitioning using a binary tree. The partitioning of a two-dimensional feature space using a complete tree of depth-2 with hyperplanes for separation. The feature space is first bisected by $s_{t,\lambda}$, which is defined by the hyperplane $\boldsymbol{\phi}_{t,\lambda}$, where the region on the direction of the $\boldsymbol{\phi}_{t,\lambda}$ vector corresponds to the child with label "1". We then continue to bisect children regions using $s_{t,0}$ and $s_{t,1}$, defined by $\boldsymbol{\phi}_{t,0}$ and $\boldsymbol{\phi}_{t,1}$, respectively.

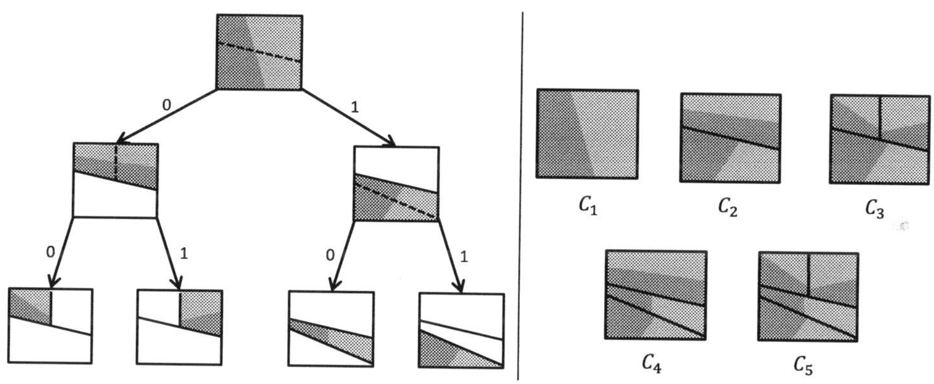


**FIGURE 9.2**

Example partitioning for a binary classification problem. The left figure shows an example partitioning of a two-dimensional feature space using a depth-2 tree. The active region corresponding to a node is shown colored, where the dashed line represents the separating hyperplane at that node, and the two different colored subregions in a node represent the local classifier trained in that region. The right figure shows all different partitions (and consequently classifiers) defined by the tree on the left.

to allow the simultaneous use of functional inner and leaf nodes to draw a decision in [13]. Similarly, the node-specific individual decisions are combined in [22] via the context tree weighting method [23] and a piece-wise linear model for sequential classification is obtained. Since the partitions in these methods are fixed and chosen even before the processing starts, the nonlinear modeling capability of such methods is very limited and significantly deteriorates in cases of high dimensionality [24].

To resolve this issue, we introduce self-organizing trees (SOTs) that jointly learn the optimal feature space partitioning to minimize the loss of the algorithm. In particular, we consider a binary tree where a separator (*e.g.*, a hyperplane) is used to bisect the feature space in a nested manner, and an online linear predictor is assigned to each node. The sequential losses of these node predictors are combined (with their corresponding weights that are sequentially learned) into a global loss that is parameterized via the separator functions and the parameters of the node predictors. We minimize this global loss using online gradient descent, *i.e.*, by updating the complete set of SOT parameters, *i.e.*, the separators, the node predictors and the combination weights, at each time instance. The resulting predictor is a highly dynamical SOT structure that jointly (and in an online and adaptive manner) learns the region classifiers and the optimal feature space partitioning and hence provides an efficient nonlinear modeling with multiple learning machines. In this respect, the proposed method is remarkably robust to drifting source statistics, *i.e.*, nonstationarity. Since our approach is essentially based on a finite combination of linear models, it generalizes well and does not overfit or limitedly overfits (as also shown by an extensive set of experiments).

## 9.2 SELF-ORGANIZING TREES FOR REGRESSION PROBLEMS

In this section, we consider the sequential nonlinear regression problem, where we observe a desired signal $\{d_t\}_{t\geq 1}$, $d_t \in \mathbb{R}$, and regression vectors $\{\boldsymbol{x}_t\}_{t\geq 1}$, $\boldsymbol{x}_t \in \mathbb{R}^p$, such that we sequentially estimate $d_t$ by

$$\hat{d}_t = f_t(\boldsymbol{x}_t),$$

where $f_t(\cdot)$ is the adaptive nonlinear regression function defined by the SOT. At each time $t$, the regression error of the algorithm is given by

$$e_t = d_t - \hat{d}_t$$

and the objective of the algorithm is to minimize the square error loss $\sum_{t=1}^{T} e_t^2$, where $T$ is the number of observed samples.

### 9.2.1 NOTATION

We first introduce a labeling for the tree nodes following [23]. The root node is labeled with an empty binary string $\lambda$ and assuming that a node has a label $n$, where $n$ is a binary string, we label its upper and lower children as $n1$ and $n0$, respectively. Here we emphasize that a string can only take its letters from the binary alphabet $\{0, 1\}$, where 0 refers to the lower child and 1 refers to the upper child of a node. We also introduce another concept, *i.e.*, the definition of the prefix of a string. We say that a string $n' = q'_1 \ldots q'_{l'}$ is a prefix to string $n = q_1 \ldots q_l$ if $l' \leq l$ and $q'_i = q_i$ for all $i = 1, \ldots, l'$ and the empty string $\lambda$ is a prefix to all strings. Let $\mathcal{P}(n)$ represent all prefixes to the string $n$, *i.e.*, $\mathcal{P}(n) \triangleq \{n_0, \ldots, n_l\}$, where $l \triangleq l(n)$ is the length of the string $n$, $n_i$ is the string with $l(n_i) = i$ and $n_0 = \lambda$ is the empty string, such that the first $i$ letters of the string $n$ form the string $n_i$ for $i = 0, \ldots, l$.

For a given SOT of depth $D$, we let $\mathcal{N}_D$ denote all nodes defined on this SOT and $\mathcal{L}_D$ denote all leaf nodes defined on this SOT. We also let $\beta_D$ denote the number of partitions defined on this SOT. This yields the recursion $\beta_{j+1} = \beta_j^2 + 1$ for all $j \geq 1$, with the base case $\beta_0 = 1$. For a given partition $k$, we let $\mathcal{M}_k$ denote the set of all nodes in this partition.

For a node $n \in \mathcal{N}_D$ (defined on the SOT of depth $D$), we define $\mathcal{S}_D(n) \triangleq \{\acute{n} \in \mathcal{N}_D \mid \mathcal{P}(\acute{n}) = n\}$ as the set of all nodes of the SOT of depth $D$, whose set of prefixes includes node $n$.

For a node $n \in \mathcal{N}_D$ (defined on the SOT of depth $D$) with length $l(n) \geq 1$, the total number of partitions that contain $n$ can be found by the following recursion:

$$\gamma_d\big(l(n)\big) \triangleq \prod_{j=1}^{l(n)} \beta_{d-j}.$$

For the case where $l(n) = 0$ (*i.e.*, for $n = \lambda$), one can clearly observe that there exists only one partition containing $\lambda$, therefore $\gamma_d(0) = 1$.

For two nodes $n, \acute{n} \in \mathcal{N}_D$ (defined on the SOT of depth $D$), we let $\rho(n, \acute{n})$ denote the number of partitions that contain both $n$ and $\acute{n}$. Trivially, if $\acute{n} = n$, then $\rho(n, \acute{n}) = \gamma_d(l(n))$. If $n \neq \acute{n}$, then letting $\bar{n}$ denote the longest prefix to both $n$ and $\acute{n}$, *i.e.*, the longest string in $\mathcal{P}(n) \cap \mathcal{P}(\acute{n})$, we obtain

$$\rho(n, \acute{n}) \triangleq \begin{cases} \gamma_d(l(n)), & \text{if } n = \acute{n}, \\ \frac{\gamma_d(l(n))\gamma_{d-l(\bar{n})-1}(l(\acute{n})-l(\bar{n})-1)}{\beta_{d-l(\bar{n})-1}}, & \text{if } n \notin \mathcal{P}(\acute{n}) \cup \mathcal{S}_D(\acute{n}), \\ 0, & \text{otherwise.} \end{cases} \tag{9.1}$$

Since $l(\bar{n}) + 1 \leq l(n)$ and $l(\bar{n}) + 1 \leq l(\acute{n})$ from the definition of the SOT, we naturally have $\rho(n, \acute{n}) = \rho(\acute{n}, n)$.

## 9.2.2 CONSTRUCTION OF THE ALGORITHM

For each node $n$ on the SOT, we define a node predictor

$$\hat{d}_{t,n} = \boldsymbol{v}_{t,n}^T \boldsymbol{x}_t, \tag{9.2}$$

whose parameter $\boldsymbol{v}_{t,n}$ is updated using the online gradient descent algorithm. We also define a separator function for each node $p$ on the SOT except the leaf nodes (note that leaf nodes do not have any children) using the sigmoid function

$$s_{t,n} = \frac{1}{1 + \exp(\boldsymbol{\phi}_{t,n}^T \boldsymbol{x}_t)}, \tag{9.3}$$

where $\boldsymbol{\phi}_{t,n}$ is the normal to the separation plane. We then define the prediction of any partition according to the hierarchical structure of the SOT as the weighted sum of the prediction of the nodes in that partition, where the weighting is determined by the separator functions of the nodes between the leaf node and the root node. In particular, the prediction of the $k$th partition at time $t$ is defined as follows:

$$\hat{d}_t^{(k)} = \sum_{n \in \mathcal{M}_k} \left( \hat{d}_{t,n} \prod_{i=0}^{l(n)-1} s_{t,n_i}^{q_i} \right), \tag{9.4}$$

where $n_i \in \mathcal{P}(n)$ is the prefix to string $n$ with length $i-1$, $q_i$ is the $i$th letter of the string $n$, *i.e.*, $n_{i+1} = n_i q_i$, and finally $s_{t,n_i}^{q_i}$ denotes the value of the separator function at node $n_i$ such that

$$s_{t,n_i}^{q_i} \triangleq \begin{cases} s_{t,n_i}, & \text{if } q_i = 0, \\ 1 - s_{t,n_i}, & \text{otherwise,} \end{cases} \tag{9.5}$$

with $s_{t,n_i}$ defined as in (9.3). We emphasize that we dropped the $n$ dependency of $q_i$ and $n_i$ to simplify notation. Using these definitions, we can construct the final estimate of our algorithm as

$$\hat{d}_t = \sum_{k \in \beta_D} w_t^{(k)} \hat{d}_t^{(k)}, \tag{9.6}$$

where $w_t^{(k)}$ represents the weight of partition $k$ at time $t$.

Having found a method to combine the predictions of all partitions to generate the final prediction of the algorithm, we next aim to obtain a low-complexity representation since there are $\mathcal{O}(1.5^{2^D})$ different partitions defined on the SOT and (9.6) requires a storage and computational complexity of $\mathcal{O}(1.5^{2^D})$. To this end, we denote the product terms in (9.4) as follows:

$$\hat{\delta}_{t,n} \triangleq \hat{d}_{t,n} \prod_{i=0}^{l(n)-1} s_{t,n_i}^{q_i}, \tag{9.7}$$

where $\hat{\delta}_{t,n}$ can be viewed as the estimate of the node $n$ at time $t$. Then (9.4) can be rewritten as follows:

$$\hat{d}_t^{(k)} = \sum_{p \in \mathcal{M}_k} \hat{\delta}_{t,n}.$$

Since we now have a compact form to represent the tree and the outputs of each partition, we next introduce a method to calculate the combination weights of $\mathcal{O}(1.5^{2^D})$ partitions in a simplified manner. For this, we assign a particular linear weight to each node. We denote the weight of node $n$ at time $t$ as $w_{t,n}$ and then define the weight of the $k$th partition as the sum of the weights of its nodes, *i.e.*,

$$w_t^{(k)} = \sum_{n \in \mathcal{M}_k} w_{t,n},$$

for all $k \in \{1, \ldots, \beta_D\}$. Since we use online gradient descent to update the weight of each partition, the weight of partition $k$ is recursively updated as

$$w_{t+1}^{(k)} = w_t^{(k)} + \mu_t e_t \hat{d}_t^{(k)}.$$

This yields the following recursive update on the node weights:

$$w_{t+1,n} = w_{t,n} + \mu_t e_t \hat{\delta}_{t,n}, \tag{9.8}$$

where $\hat{\delta}_{t,n}$ is defined as in (9.7). This result implies that instead of managing $\mathcal{O}(1.5^{2^D})$ memory locations and making $\mathcal{O}(1.5^{2^D})$ calculations, only keeping track of the weights of every node is sufficient and the number of nodes in a depth-$D$ model is $|\mathcal{N}_D| = 2^{D+1} - 1$. Therefore, we can reduce the storage and computational complexity from $\mathcal{O}(1.5^{2^D})$ to $\mathcal{O}(2^D)$ by performing the update in (9.8) for all $n \in \mathcal{N}_D$.

Using these node predictors and weights, we construct the final estimate of our algorithm as follows:

$$\hat{d}_t = \sum_{k=1}^{\beta_d} \left\{ \left( \sum_{n \in \mathcal{M}_k} w_{t,n} \right) \left( \sum_{n \in \mathcal{M}_k} \hat{\delta}_{t,n} \right) \right\}.$$

Here, we observe that for arbitrary two nodes $n, \acute{n} \in \mathcal{N}_d$, the product $w_{t,n}\hat{\delta}_{t,\acute{n}}$ appears $\rho(n, \acute{n})$ times in $\hat{d}_t$ (*cf.* (9.1)). Hence, the combination weight of the estimate of the node $n$ at time $t$ can be calculated as follows:

$$\kappa_{t,n} = \sum_{\acute{n} \in \mathcal{N}_d} \rho(n, \acute{n}) w_{t,\acute{n}}. \tag{9.9}$$

Using the combination weight (9.9), we obtain the final estimate of our algorithm as follows:

$$\hat{d}_t = \sum_{n \in \mathcal{N}_D} \kappa_{t,n}\, \hat{\delta}_{t,n}. \tag{9.10}$$

Note that (9.10) is equal to (9.6) with a storage and computational complexity of $\mathcal{O}(4^D)$ instead of $\mathcal{O}(1.5^{2^D})$.

As we derived all the update rules for the node weights and the parameters of the individual node predictors, what remains is to provide an update scheme for the separator functions. To this end, we use the online gradient descent update

$$\boldsymbol{\phi}_{t+1,n} = \boldsymbol{\phi}_{t,n} - \frac{1}{2}\eta_t \nabla e_t^2(\boldsymbol{\phi}_{t,n}), \tag{9.11}$$

for all nodes $n \in \mathcal{N}_D \setminus \mathcal{L}_D$, where $\eta_t$ is the learning rate of the algorithm and $\nabla e_t^2(\boldsymbol{\phi}_{t,n})$ is the derivative of $e_t^2(\boldsymbol{\phi}_{t,n})$ with respect to $\boldsymbol{\phi}_{t,n}$. After some algebra, we obtain

$$\begin{aligned}
\boldsymbol{\phi}_{t+1,n} &= \boldsymbol{\phi}_{t,n} + \eta_t e_t \frac{\partial \hat{d}_t}{\partial s_{t,n}} \frac{\partial s_{t,n}}{\partial \boldsymbol{\phi}_{t,n}} \\
&= \boldsymbol{\phi}_{t,n} + \eta_t e_t \left\{ \sum_{\acute{n} \in \mathcal{N}_D} \kappa_{t,\acute{n}} \frac{\partial \hat{\delta}_{t,\acute{n}}}{\partial s_{t,n}} \right\} \frac{\partial s_{t,n}}{\partial \boldsymbol{\phi}_{t,n}} \\
&= \boldsymbol{\phi}_{t,n} + \eta_t e_t \left\{ \sum_{q=0}^{1} \sum_{\acute{n} \in \mathcal{S}_D(nq)} (-1)^q \kappa_{t,\acute{n}} \frac{\hat{\delta}_{t,\acute{n}}}{s_{t,n}^q} \right\} \frac{\partial s_{t,n}}{\partial \boldsymbol{\phi}_{t,n}},
\end{aligned} \tag{9.12}$$

**Algorithm 1** Self-Organizing Tree Regressor (SOTR).

**for** $t = 1$ **to** $n$ **do**
  Calculate separator functions $s_{t,p}$, for all $p \in \mathcal{N}_D \setminus \mathcal{L}_D$ using (9.14).
  Calculate node predictors $\hat{d}_{t,p}$, for all $p \in \mathcal{L}_D$ using (9.2).
  Define $\alpha_{t,p} = \prod_{i=1}^{l(p)} s_{t,v_i}^{q_i}$ and calculate $\hat{\delta}_{t,p}$, for all $p \in \mathcal{L}_D$ using (9.7).
  Calculate combination weights $\kappa_{t,p}$, for all $p \in \mathcal{L}_D$ using (9.9).
  Construct the final estimate $\hat{d}_t$ using (9.10).
  Observe the error $e_t = d_t - \hat{d}_t$.
  Update the node predictors $\boldsymbol{v}_{t+1,p} = \boldsymbol{v}_{t,p} + \mu_t e_t \alpha_{t,p} \boldsymbol{x}_t$ for all $p \in \mathcal{L}_D$.
  Update the node weights $w_{t+1,p} = w_{t,p} + \mu_t e_t \hat{\delta}_{t,p}$ for all $p \in \mathcal{L}_D$.
  Update the separator functions $\boldsymbol{\phi}_{t,p}$ for all $p \in \mathcal{N}_D \setminus \mathcal{L}_D$ using (9.12).
**end for**

where we use the logistic regression classifier as our separator function, *i.e.*, $s_{t,n} = \left(1+\exp(\boldsymbol{x}_t^T \boldsymbol{\phi}_{t,n})\right)^{-1}$. Therefore, we have

$$\begin{aligned}\frac{\partial s_{t,n}}{\partial \boldsymbol{\phi}_{t,n}} &= -\left(1+\exp(\boldsymbol{x}_t^T \boldsymbol{\phi}_{t,n})\right)^{-2} \exp(\boldsymbol{x}_t^T \boldsymbol{\phi}_{t,n}) \boldsymbol{x}_t \\ &= -s_{t,n}(1-s_{t,n})\boldsymbol{x}_t. \end{aligned} \tag{9.13}$$

We emphasize that other separator functions can also be used in a similar way by simply calculating the gradient with respect to the extended direction vector and plugging in (9.12) and (9.13). From (9.13), we observe that $\nabla e_t^2(\boldsymbol{\phi}_{t,n})$ includes the product of $s_{t,n}$ and $1 - s_{t,n}$ terms; hence, in order not to slow down the learning rate of our algorithm, we restrict $s^+ \leq s_t \leq 1 - s^+$ for some $0 < s^+ < 0.5$. In accordance with this restriction, we define the separator functions as follows:

$$s_t = s^+ + \frac{1-2s^+}{1+e^{\boldsymbol{x}_t^T \boldsymbol{\phi}_t}}. \tag{9.14}$$

According to the update rule in (9.12), the computational complexity of the introduced algorithm results in $\mathcal{O}(p4^d)$. This concludes the construction of the algorithm and a pseudocode is given in Algorithm 1.

### 9.2.3 CONVERGENCE OF THE ALGORITHM

For Algorithm 1, we have the following convergence guarantee, which implies that our regressor (given in Algorithm 1) asymptotically achieves the performance of the best linear combination of the $\mathcal{O}(1.5^{2^D})$ different adaptive models that can be represented using a depth-$D$ tree with a computational complexity $\mathcal{O}(p4^D)$. While constructing the algorithm, we refrain from any statistical assumptions on the underlying data, and our algorithm works for any sequence of $\{d_t\}_{t\geq 1}$ with an arbitrary length of $n$. Furthermore, one can use this algorithm to learn the region boundaries and then feed this information to the first algorithm to reduce computational complexity.

**Theorem 1.** *Let $\{d_t\}_{t\geq 1}$ and $\{\boldsymbol{x}_t\}_{t\geq 1}$ be arbitrary, bounded and real-valued sequences. The predictor $\hat{d}_t$ given in Algorithm 1 when applied to these sequences yields*

$$\sum_{t=1}^{T}(d_t-\hat{d}_t)^2-\min_{\boldsymbol{w}\in\mathbb{R}^{\beta_d}}\sum_{t=1}^{T}(d_t-\boldsymbol{w}^T\hat{\boldsymbol{d}}_t)^2\leq\mathcal{O}(\log(T)), \tag{9.15}$$

*for all $T$, when $e_t^2(\boldsymbol{w})$ is strongly convex $\forall t$, where $\hat{\boldsymbol{d}}_t=[\hat{d}_t^{(1)},\ldots,\hat{d}_t^{(\beta_d)}]^T$ and $\hat{d}_t^{(k)}$ represents the estimate of $d_t$ at time $t$ for the adaptive model $k=1,\ldots,\beta_d$.*

Proof of this theorem can be found in Appendix 9.A.1.

## 9.3 SELF-ORGANIZING TREES FOR BINARY CLASSIFICATION PROBLEMS

In this section, we study online binary classification, where we observe feature vectors $\{\boldsymbol{x}_t\}_{t\geq 1}$ and determine their labels $\{y_t\}_{t\geq 1}$ in an online manner. In particular, the aim is to learn a classification function $f_t(\boldsymbol{x}_t)$ with $\boldsymbol{x}_t\in\mathbb{R}^p$ and $y_t\in\{-1,1\}$ such that, when applied in an online manner to any streaming data, the empirical loss of the classifier $f_t(\cdot)$, *i.e.*,

$$L_T(f_t)\triangleq\sum_{t=1}^{T}\mathbb{1}_{\{f_t(\boldsymbol{x}_t)\neq d_t\}}, \tag{9.16}$$

is asymptotically as small (after averaging over $T$) as the empirical loss of the best partition classifier defined over the SOT of depth $D$. To be more precise, we measure the relative performance of $f_t$ with respect to the performance of a partition classifier $f_t^{(k)}$, where $k\in\{1,\ldots,\beta_D\}$, using the following regret:

$$R_T(f_t;f_t^{(k)})\triangleq\frac{L_T(f_t)-L_T(f_t^{(k)})}{T}, \tag{9.17}$$

for any arbitrary length $T$. Our aim is then to construct an online algorithm with guaranteed upper bounds on this regret for any partition classifier defined over the SOT.

### 9.3.1 CONSTRUCTION OF THE ALGORITHM

Using the notations described in Section 9.2.1, the output of a partition classifier $k\in\{1,\ldots,\beta_D\}$ is constructed as follows. Without loss of generality, suppose that the feature $\boldsymbol{x}_t$ has fallen into the region represented by the leaf node $n\in\mathcal{L}_D$. Then $\boldsymbol{x}_t$ is contained in the nodes $n_0,\ldots,n_D$, where $n_d$ is the $i$ letter prefix of $n$, *i.e.*, $n_D=n$ and $n_0=\lambda$. For example, if node $n_d$ is contained in partition $k$, then one can simply set $f_t^{(k)}(\boldsymbol{x}_t)=f_{t,n_d}(\boldsymbol{x}_t)$. Instead of making a hard selection, we allow an error margin for the classification output $f_{t,n_d}(\boldsymbol{x}_t)$ in order to be able to update the region boundaries later in the proof. To achieve this, for each node contained in partition $k$, we define a parameter called *path probability* to measure the contribution of each leaf node to the classification task at time $t$. This parameter is equal to the multiplication of the separator functions of the nodes from the respective node to the root node, which represents the probability that $\boldsymbol{x}_t$ should be classified using the region classifier of node $n_d$. This path probability (similar to the node predictor definition in (9.7)) is defined as

$$P_{t,n_d}(\boldsymbol{x}_t) \triangleq \prod_{i=0}^{d-1} s_{t,n_i}^{q_{i+1}}(\boldsymbol{x}_t), \tag{9.18}$$

where $p_{t,n_i}^{q_{i+1}}(\cdot)$ represents the value of the partitioning function corresponding to node $n_i$ towards the $q_{i+1}$ direction as in (9.5). We consider that the classification output of node $n_d$ can be trusted with a probability of $P_{t,n_d}(\boldsymbol{x}_t)$. This and the other probabilities in our development are independently defined for ease of exposition and gaining intuition, *i.e.*, these probabilities are not related to the unknown data statistics in any way and they definitely cannot be regarded as certain assumptions on the data. Indeed, we do not take any assumptions about the data source.

Intuitively, the path probability is low when the feature vector is close to the region boundaries; hence we may consider to classify that feature vector by another node classifier (*e.g.*, the classifier of the sibling node). Using these path probabilities, we aim to update the region boundaries by learning whether an efficient node classifier is used to classify $\boldsymbol{x}_t$, instead of directly assigning $\boldsymbol{x}_t$ to node $n_d$ and lose a significant degree of freedom. To this end, we define the final output of each node classifier according to a Bernoulli random variable with outcomes $\{-f_{t,n_d}(\boldsymbol{x}_t), f_{t,n_d}(\boldsymbol{x}_t)\}$, where the probability of the latter outcome is $P_{t,n_d}(\boldsymbol{x}_t)$. Although the final classification output of node $n_d$ is generated according to this Bernoulli random variable, we continue to call $f_{t,n_d}(\boldsymbol{x}_t)$ the final classification output of node $n_d$, with an abuse of notation. Then the classification output of the partition classifier is set to $f_t^{(k)}(\boldsymbol{x}_t) = f_{t,n_d}(\boldsymbol{x}_t)$.

Before constructing the SOT classifier, we first introduce certain definitions. Let the instantaneous empirical loss of the proposed classifier $f_t$ at time $t$ be denoted by $\ell_t(f_t) \triangleq \mathbb{1}_{\{f_t(\boldsymbol{x}_t)\neq y_t\}}$. Then the expected empirical loss of this classifier over a sequence of length $T$ can be found by

$$L_T(f_t) = E\left[\sum_{t=1}^{T} \ell_t(f_t)\right], \tag{9.19}$$

with the expectation taken with respect to the randomization parameters of the classifier $f_t$. We also define the effective region of each node $n_d$ at time $t$ as follows: $\mathcal{R}_{t,n_d} \triangleq \left\{\boldsymbol{x} : P_{t,n_d}(\boldsymbol{x}) \geq (0.5)^d\right\}$. According to the aforementioned structure of partition classifiers, the node $n_d$ classifies an instance $\boldsymbol{x}_t$ only if $\boldsymbol{x}_t \in \mathcal{R}_{t,n_d}$. Therefore, the time accumulated empirical loss of any node $n$ during the data stream is given by

$$L_{T,n} \triangleq \sum_{t\leq T:\{\boldsymbol{x}_t\}_{t\geq 1}\in\mathcal{R}_{t,n}} \ell_t(f_{t,n}). \tag{9.20}$$

Similarly, the time accumulated empirical loss of a partition classifier $k$ is $L_T^{(k)} \triangleq \sum_{n\in\mathcal{M}_k} L_{T,n}$.

We then use a mixture-of-experts approach to achieve the performance of the best partition classifier that minimizes the accumulated classification error. To this end, we set the final classification output of our algorithm as $f_t(\boldsymbol{x}_t) = f_t^{(k)}$ with probability $w_t^{(k)}$, where

$$w_t^{(k)} = \frac{1}{Z_{t-1}} 2^{-J(k)} \exp\left(-b\, L_{t-1}^{(k)}\right),$$

$b \geq 0$ is a constant controlling the learning rate of the algorithm, $J(k) \leq 2|\mathcal{L}(k)| - 1$ represents the number of bits required to code the partition $k$ (which satisfies $\sum_{k=1}^{\beta_D} J(k) = 1$) and $Z_t = \sum_{k=1}^{\beta_D} 2^{-J(k)} \exp\left(-b\, L_t^{(k)}\right)$ is the normalization factor.

Although this randomized method can be used as the SOT classifier, in its current form, it requires a computational complexity $\mathcal{O}(1.5^{2^D} p)$ since the randomization $w_t^{(k)}$ is performed over the set $\{1, \ldots, \beta_D\}$ and $\beta_D \approx 1.5^{2^D}$. However, the set of all possible classification outputs of these partitions has a cardinality as small as $D+1$ since $\boldsymbol{x}_t \in \mathcal{R}_{t,n_D}$ for the corresponding leaf node $n_D$ (in which $\boldsymbol{x}_t$ is included) and $f_t^{(k)} = f_{t,n_d}$ for some $d = 0, \ldots, D$, $\forall k \in \{1, \ldots, \beta_D\}$. Hence, evaluating all the partition classifiers in $k$ at the instance $\boldsymbol{x}_t$ to produce $f_t(\boldsymbol{x}_t)$ is unnecessary. In fact, the computational complexity for producing $f_t(\boldsymbol{x}_t)$ can be reduced from $\mathcal{O}(1.5^{2^D} p)$ to $\mathcal{O}(Dp)$ by performing the exact same randomization over $f_{t,n_d}$'s using the new set of weights $w_{t,n_d}$, which can be straightforwardly derived as follows:

$$w_{t,n_d} = \sum_{k=1}^{\beta_D} w_t^{(k)} \mathbb{1}_{f_t^{(k)}(\boldsymbol{x}_t) = f_{t,n_d}(\boldsymbol{x}_t)}. \tag{9.21}$$

To efficiently calculate (9.21) with complexity $\mathcal{O}(Dp)$, we consider the universal coding scheme and let

$$M_{t,n} \triangleq \begin{cases} \exp\left(-bL_{t,n}\right), & \text{if } n \text{ has depth } D, \\ \frac{1}{2}\left[M_{t,n0} M_{t,n1} + \exp\left(-bL_{t,n}\right)\right], & \text{otherwise} \end{cases} \tag{9.22}$$

for any node $n$ and observe that we have $M_{t,\lambda} = Z_t$ [23]. Therefore, we can use the recursion (9.22) to obtain the denominator of the randomization probabilities $w_t^{(k)}$. To efficiently calculate the numerator of (9.21), we introduce another intermediate parameter as follows. Letting $n_d'$ denote the sibling of node $n_d$, we recursively define

$$\kappa_{t,n_d} \triangleq \begin{cases} \frac{1}{2}, & \text{if } d = 0 \\ \frac{1}{2} M_{t-1,n_d'} \kappa_{t,n_{d-1}}, & \text{if } 0 < d < D, \\ M_{t-1,n_d'} \kappa_{t,n_{d-1}}, & \text{if } d = D \end{cases} \tag{9.23}$$

$\forall d \in \{0, \ldots, D\}$, where $\boldsymbol{x}_t \in \mathcal{R}_{t,n_D}$. Using the intermediate parameters in (9.22) and (9.23), it can be shown that we have

$$w_{t,n_d} = \frac{\kappa_{t,n_d} \exp\left(-b\, L_{t,n_d}\right)}{M_{t,\lambda}}. \tag{9.24}$$

Hence, we obtain the final output of the algorithm as $f_t(\boldsymbol{x}_t) = f_{t,n_d}(\boldsymbol{x}_t)$ with probability $w_{t,n_d}$, where $d \in \{0, \ldots, D\}$ (*i.e.*, with a computational complexity $\mathcal{O}(D)$).

We then use the final output of the introduced algorithm and update the region boundaries of the tree (*i.e.*, organize the tree) to minimize the final classification error. To this end, we minimize the loss $E\left[\ell_t(f_t)\right] = E[\mathbb{1}_{\{f_t(\boldsymbol{x}_t) \neq y_t\}}] = \frac{1}{4} E\left[(y_t - f_t(\boldsymbol{x}_t))^2\right]$ with respect to the region boundary parameters, *i.e.*, we use the stochastic gradient descent method as follows:

**Algorithm 2** Self-Organizing Tree Classifier (SOTC).

```
for t = 1 to T do
    Propagate {x_t}_{t≥1} from the root to the leaf and obtain the visited nodes n_0, ..., n_D.
    Calculate P_{t,n_d}(x_t) for all d ∈ 0, ..., D using (9.18).
    Calculate w_{t,n_d}(x_t) for all d ∈ 0, ..., D using (9.24).
    Draw a node among n_0, ..., n_D with probabilities w_{t,n_0}, ..., w_{t,n_D}, respectively; suppose that
    n_d is drawn.
    Draw a classification output {1, −1} with probabilities P_{t,n_d}(x_t) and 1 − P_{t,n_d}(x_t), respectively;
    f_t(x_t) is equated to the selected output.
    Update the region classifiers (perceptron) at the visited nodes [25].
    ℓ_t(f_t) ← 1_{f_t(x_t)≠y_t}
    Update L_{t,n_d} for all d ∈ 0, ..., D using (9.20).
    Apply the recursion in (9.22) to update M_{t+1,n_d} for all d ∈ 0, ..., D.
    Update the separator parameters φ using (9.27).
end for
```

$$\begin{aligned}\boldsymbol{\phi}_{t+1,n_d} &= \boldsymbol{\phi}_{t,n_d} - \eta \nabla E\left[\ell_t(f_t)\right] \\ &= \boldsymbol{\phi}_{t,n_d} - (-1)^{q_{d+1}} \eta \left(y_t - f_t(\boldsymbol{x}_t)\right) s_{t,n_d}^{q'_{d+1}}(\boldsymbol{x}_t) \left[\sum_{i=d+1}^{D} f_{t,n_i}(\boldsymbol{x}_t)\right] \boldsymbol{x}_t,\end{aligned} \tag{9.25}$$

$\forall d \in \{0, \ldots, D-1\}$, where $\eta$ denotes the learning rate of the algorithm and $q'_{d+1}$ represents the complementary letter to $q_{d+1}$ from the binary alphabet $\{0, 1\}$. Defining a new intermediate variable

$$\pi_{t,n_d} \triangleq \begin{cases} f_{t,n_d}(\boldsymbol{x}_t), & \text{if } d = D-1, \\ \pi_{t,n_{d+1}} + f_{t,n_d}(\boldsymbol{x}_t), & \text{if } d < D-1, \end{cases} \tag{9.26}$$

one can perform the update in (9.25) with a computational complexity $\mathcal{O}(p)$ for each node $n_d$, where $d \in \{0, \ldots, D-1\}$, resulting in an overall computational complexity of $\mathcal{O}(Dp)$ as follows:

$$\boldsymbol{\phi}_{t+1,n_d} = \boldsymbol{\phi}_{t,n_d} - (-1)^{m_{d+1}} \eta \left(y_t - f_t(\boldsymbol{x}_t)\right) \pi_{t,n_d} s_{t,n_d}^{q'_{d+1}}(\boldsymbol{x}_t)\, \boldsymbol{x}_t. \tag{9.27}$$

This concludes the construction of the algorithm and the pseudocode of the SOT classifier can be found in Algorithm 2.

### 9.3.2 CONVERGENCE OF THE ALGORITHM

In this section, we illustrate that the performance of Algorithm 2 is asymptotically as good as the best partition classifier such that, as $T \to \infty$, we have $R_T(f_t; f_t^{(k)}) \to 0$. Hence, Algorithm 2 asymptotically achieves the performance of the best partition classifier among $\mathcal{O}(1.5^{2^D})$ different classifiers that can be represented using the SOT of depth $D$ with a significantly reduced computational complexity of $\mathcal{O}(Dp)$ without any statistical assumptions on data.

**Theorem 2.** *Let $\{x_t\}_{t\geq 1}$ and $\{y_t\}_{t\geq 1}$ be arbitrary and real-valued sequences of feature vectors and their labels, respectively. Then Algorithm 2, when applied to these data sequences, sequentially yields*

$$\max_{k\in\{1,\ldots,\beta_D\}} E\left[R_T(f_t; f_t^{(k)})\right] \leq O\left(\sqrt{\frac{2^D}{T}}\right), \tag{9.28}$$

*for all T with a computational complexity $\mathcal{O}(Dp)$, where p represents the dimensionality of the feature vectors and the expectation is with respect to the randomization parameters.*

Proof of this theorem can be found in Appendix 9.A.2.

## 9.4 NUMERICAL RESULTS

In this section, we illustrate the performance of SOTs under different scenarios with respect to state-of-the-art methods. The proposed method has a wide variety of application areas, such as channel equalization [26], underwater communications [27], nonlinear modeling in big data [28], speech and texture analysis [29, Chapter 7] and health monitoring [30]. Yet, in this section, we consider nonlinear modeling for fundamental regression and classification problems.

### 9.4.1 NUMERICAL RESULTS FOR REGRESSION PROBLEMS

Throughout this section, "SOTR" represents the self-organizing tree regressor defined in Algorithm 1, "CTW" represents the context tree weighting algorithm of [16], "OBR" represents the optimal batch regressor, "VF" represents the truncated Volterra filter [1], "LF" represents the simple linear filter, "B-SAF" and "CR-SAF" represent the Beizer and the Catmull–Rom spline adaptive filter of [2], respectively, and "FNF" and "EMFNF" represent the Fourier and even mirror Fourier nonlinear filter of [3], respectively. Finally, "GKR" represents the Gaussian-kernel regressor and it is constructed using $n$ node regressors, say $\hat{d}_{t,1}, \ldots, \hat{d}_{t,n}$, and a fixed Gaussian mixture weighting (that is selected according to the underlying sequence in hindsight), giving

$$\hat{d}_t = \sum_{i=1}^{n} f\left(x_t; \mu_i, \Sigma_i\right) \hat{d}_{t,i},$$

where $\hat{d}_{t,i} = v_{t,i}^T x_t$ and

$$f\left(x_t; \mu_i, \Sigma_i\right) \triangleq \frac{1}{2\pi\sqrt{|\Sigma_i|}} e^{-\frac{1}{2}(x_t-\mu_i)^T \Sigma_i^{-1}(x_t-\mu_i)},$$

for all $i = 1, \ldots, n$.

For a fair performance comparison, in the corresponding experiments in Subsection 9.4.1.2, the desired data and the regressor vectors are normalized between $[-1, 1]$ since the satisfactory performance of several algorithms requires the knowledge on the upper bounds (such as the B-SAF and the CR-SAF) and some require these upper bounds to be between $[-1, 1]$ (such as the FNF and the EMFNF).

Moreover, in the corresponding experiments in Subsection 9.4.1.1, the desired data and the regressor vectors are normalized between $[-1, 1]$ for the VF, the FNF and the EMFNF algorithms due to the aforementioned reason. The regression errors of these algorithms are then scaled back to their original values for a fair comparison.

Considering the illustrated examples in the respective papers [2,3,16], the orders of the FNF and the EMFNF are set to 3 for the experiments in Subsection 9.4.1.1 and 2 for the experiments in Subsection 9.4.1.2. The order of the VF is set to 2 for all experiments. Similarly, the depth of the trees for the SOTR and CTW algorithms is set to 2 for all experiments. For these tree-based algorithms, the feature space is initially partitioned by the direction vectors $\boldsymbol{\phi}_{t,n} = [\phi_{t,n}^{(1)}, \ldots, \phi_{t,n}^{(p)}]^T$ for all nodes $n \in \mathcal{N}_D \setminus \mathcal{L}_D$, where $\phi_{t,n}^{(i)} = -1$ if $i \equiv l(n) \pmod{D}$, *e.g.*, when $D = p = 2$, we have the four quadrants as the four leaf nodes of the tree. Finally, we use cubic B-SAF and CR-SAF algorithms, whose number of knots are set to 21 for all experiments. We emphasize that both these parameters and the learning rates of these algorithms are selected to give equal rates of performance and convergence.

### 9.4.1.1 Mismatched Partitions

In this subsection, we consider the case where the desired data is generated by a piece-wise linear model that mismatches with the initial partitioning of the tree-based algorithms. Specifically, the desired signal is generated by the following piece-wise linear model:

$$d_t = \begin{cases} \boldsymbol{w}^T \boldsymbol{x}_t + \pi_t, & \text{if } \boldsymbol{\phi}_0^T \boldsymbol{x}_t \geq 0.5 \text{ and } \boldsymbol{\phi}_1^T \boldsymbol{x}_t \geq 1, \\ -\boldsymbol{w}^T \boldsymbol{x}_t + \pi_t, & \text{if } \boldsymbol{\phi}_0^T \boldsymbol{x}_t \geq 0.5 \text{ and } \boldsymbol{\phi}_1^T \boldsymbol{x}_t < 1, \\ -\boldsymbol{w}^T \boldsymbol{x}_t + \pi_t, & \text{if } \boldsymbol{\phi}_0^T \boldsymbol{x}_t < 0.5 \text{ and } \boldsymbol{\phi}_2^T \boldsymbol{x}_t \geq -1, \\ \boldsymbol{w}^T \boldsymbol{x}_t + \pi_t, & \text{if } \boldsymbol{\phi}_0^T \boldsymbol{x}_t < 0.5 \text{ and } \boldsymbol{\phi}_2^T \boldsymbol{x}_t < -1, \end{cases} \tag{9.29}$$

where $\boldsymbol{w} = [1, 1]^T$, $\boldsymbol{\phi}_0 = [4, -1]^T$, $\boldsymbol{\phi}_1 = [1, 1]^T$, $\boldsymbol{\phi}_2 = [1, 2]^T$, $\boldsymbol{x}_t = [x_{1,t}, x_{2,t}]^T$, $\pi_t$ is a sample function from a zero mean white Gaussian process with variance 0.1 and $x_{1,t}$ and $x_{2,t}$ are sample functions of a jointly Gaussian process of mean $[0, 0]^T$ and variance $\boldsymbol{I}_2$. The learning rates are set to 0.005 for SOTR and CTW, 0.1 for FNF, 0.025 for B-SAF and CR-SAF, 0.05 for EMFNF and VF. Moreover, in order to match the underlying partition, the mass points of GKR are set to $\boldsymbol{\mu}_1 = [1.4565, 1.0203]^T$, $\boldsymbol{\mu}_2 = [0.6203, -0.4565]^T$, $\boldsymbol{\mu}_3 = [-0.5013, 0.5903]^T$ and $\boldsymbol{\mu}_4 = [-1.0903, -1.0013]^T$ with the same covariance matrix as in the previous example.

Fig. 9.3 shows the normalized time accumulated regression error of the proposed algorithms. We emphasize that the SOTR algorithm achieves a better error performance compared to its competitors. Comparing the performances of the SOTR and CTW algorithms, we observe that the CTW algorithm fails to accurately predict the desired data, whereas the SOTR algorithm learns the underlying partitioning of the data, which significantly improves the performance of the SOTR. This illustrates the importance of the initial partitioning of the regressor space for tree-based algorithms to yield a satisfactory performance.

In particular, the CTW algorithm converges to the best batch regressor having the predetermined leaf nodes (*i.e.*, the best regressor having the four quadrants of two-dimensional space as its leaf nodes). However that regressor is suboptimal since the underlying data is generated using another constellation; hence its time accumulated regression error is always lower bounded by $\mathcal{O}(1)$ compared to the global optimal regressor. The SOTR algorithm, on the other hand, adapts its region boundaries and captures

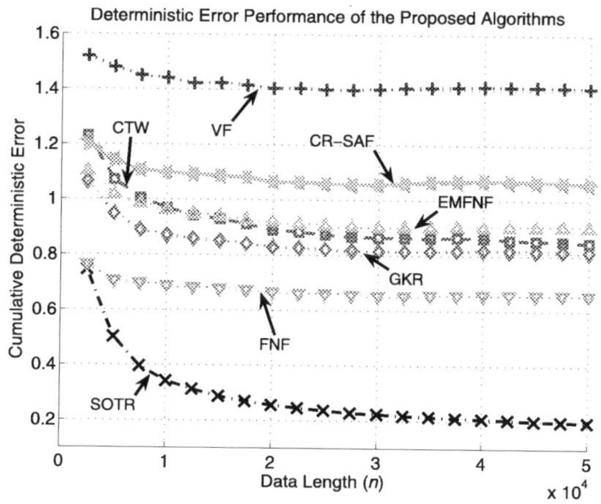


**FIGURE 9.3**

Regression error performances for the second-order piece-wise linear model in (9.29).

the underlying unevenly rotated and shifted regressor space partitioning perfectly. Fig. 9.4 shows how our algorithm updates its separator functions and illustrates the nonlinear modeling power of SOTs.

### *9.4.1.2 Chaotic Signals*

In this subsection, we illustrate the performance of the SOTR algorithm when estimating chaotic data generated by the Henon map and the Lorenz attractor [31].

First, we consider a zero mean sequence generated by the Henon map, a chaotic process given by

$$d_t = 1 - \zeta\, d_{t-1}^2 + \eta\, d_{t-2}, \tag{9.30}$$

known to exhibit chaotic behavior for the values of $\zeta = 1.4$ and $\eta = 0.3$. The desired data at time $t$ is denoted as $d_t$ whereas the extended regressor vector is $\boldsymbol{x}_t = [d_{t-1}, d_{t-2}, 1]^T$, *i.e.*, we consider a prediction framework. The learning rate is set to 0.025 for B-SAF and CR-SAF, whereas it is set to 0.05 for the rest.

Fig. 9.5 (left plot) shows the normalized regression error performance of the proposed algorithms. One can observe that the algorithms whose basis functions do not include the necessary quadratic terms and the algorithms that rely on a fixed regressor space partitioning yield unsatisfactory performance. On the other hand, VF can capture the salient characteristics of this chaotic process since its order is set to 2. Similarly, FNF can also learn the desired data since its basis functions can well approximate the chaotic process. The SOTR algorithm, however, uses a piece-wise linear modeling and still achieves asymptotically the same performance as the VF algorithm, while outperforming the FNF algorithm.

Second, we consider the chaotic signal set generated using the Lorenz attractor [31] that is defined by the following three discrete-time equations:

$$\begin{aligned} x_t &= x_{t-1} + (\sigma(y - x))dt, \\ y_t &= y_{t-1} + (x_{t-1}(\rho - z_{t-1}) - y_{t-1})dt, \end{aligned} \tag{9.31}$$

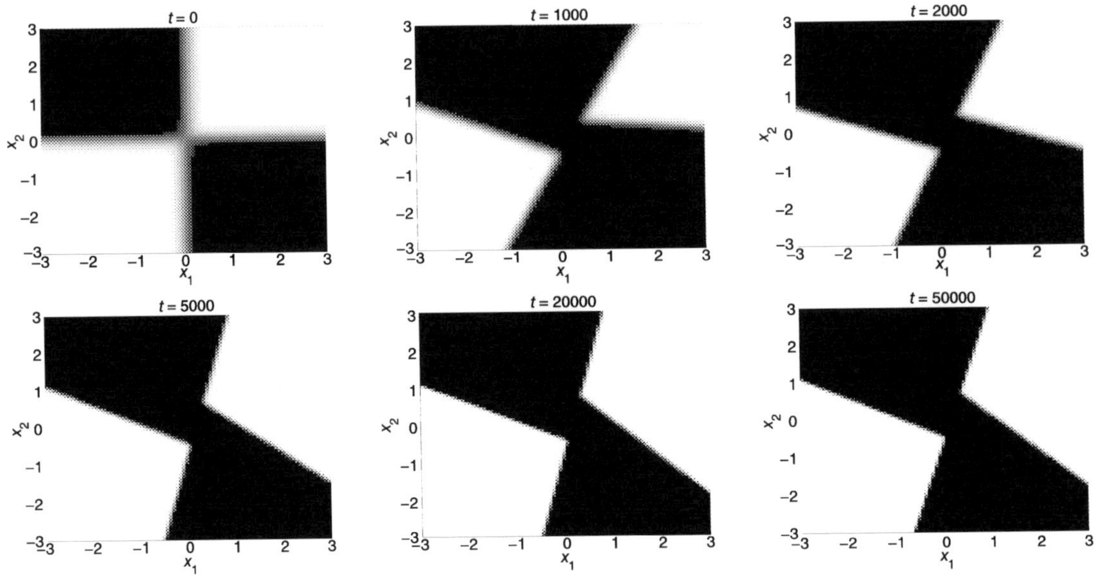


**FIGURE 9.4**

Changes in the boundaries of the leaf nodes of the SOT of depth 2 generated by the SOTR algorithm at time instances $t = 0, 1000, 2000, 5000, 20000, 50000$. The separator functions adaptively learn the boundaries of the piece-wise linear model in (9.29).

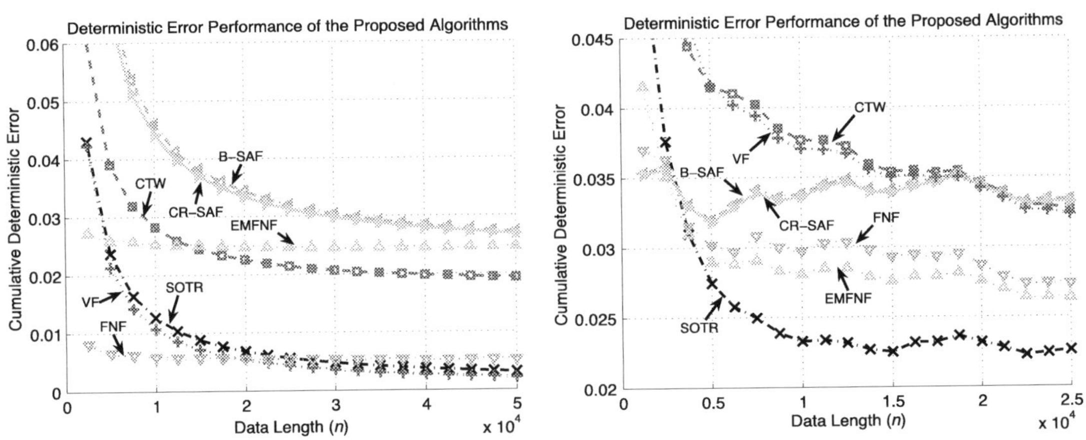


**FIGURE 9.5**

Regression error performances of the proposed algorithms for the signal generated by the Henon map in (9.30) (left figure) and for the Lorenz attractor in (9.31) with parameters $dt = 0.01$, $\rho = 28$, $\sigma = 10$ and $\beta = 8/3$ (right figure).

$$z_t = z_{t-1} + (x_{t-1}y_{t-1} - \beta z_{t-1})dt,$$

where we set $dt = 0.01$, $\rho = 28$, $\sigma = 10$ and $\beta = 8/3$ to generate the well-known chaotic solution of the Lorenz attractor. In the experiment, $x_t$ is selected as the desired data and the two-dimensional region represented by $y_t, z_t$ is set as the regressor space, that is, we try to estimate $x_t$ with respect to $y_t$ and $z_t$. The learning rates are set to 0.01 for all algorithms.

Fig. 9.5 (right plot) illustrates the nonlinear modeling power of the SOTR algorithm even when estimating a highly nonlinear chaotic signal set. As can be observed from Fig. 9.5, the SOTR algorithm significantly outperforms its competitors and achieves a superior error performance since it tunes its region boundaries to the optimal partitioning of the regressor space, whereas the performances of the other algorithms directly rely on the initial selection of the basis functions and/or tree structures and partitioning.

## 9.4.2 NUMERICAL RESULTS FOR CLASSIFICATION PROBLEMS

### 9.4.2.1 Stationary Data

In this section, we consider stationary classification problems and compare the SOTC algorithm with the following methods: *Perceptron,* "PER" [25]; *Online AdaBoost,* "OZAB" [32]; *Online Gradient-Boost,* "OGB" [33]; *Online SmoothBoost,* "OSB" [34]; and *Online Tree–Based Nonadaptive Competitive Classification,* "TNC" [22]. The parameters for all of these compared methods are set as in their original proposals. For "OGB" [33], which uses $K$ weak learners per $M$ selectors, essentially resulting in $MK$ weak learners in total, we use $K = 1$, as in [34], for a fair comparison along with the logit loss that has been shown to consistently outperform other choices in [33]. The TNC algorithm is nonadaptive, *i.e.*, not self-organizing, in terms of the space partitioning, which we use in our comparisons to illustrate the gain due to the proposed self-organizing structure. We use the Perceptron algorithm as the weak learners and node classifiers in all algorithms. We set the learning rate of the SOTC algorithm to $\eta = 0.05$ in all of our stationary as well as nonstationary data experiments. We use $N = 100$ weak learners for the boosting methods, whereas we use a depth-4 tree in SOTC and TNC algorithms, which corresponds to $31 = 2^5 - 1$ local node classifiers. The SOTC algorithm has linear complexity in the depth of the tree, whereas the compared methods have linear complexity in the number of weak learners.

As can be observed in Table 9.1, the SOTC algorithm consistently outperforms the compared methods. In particular, the compared methods essentially fail to classify Banana and BMC datasets, which indicates that these methods are not able to extend to complex nonlinear classification problems. On the contrary, the SOTC algorithm successfully models these complex nonlinear relations with piece-wise linear curves and provides a superior performance. In general, the SOTC algorithm has significantly better transient characteristics and the TNC algorithm occasionally performs poorly (such as on BMC and Banana data sets) depending on the mismatch between the initial partitions defined on the tree and the underlying optimal separation of the data. This illustrates the importance of learning the region boundaries in piece-wise linear models.

### 9.4.2.2 Nonstationary Data: Concept Change/Drift

In this section, we apply the SOTC algorithm to nonstationary data, where there might be continuous or sudden/abrupt changes in the source statistics, *i.e.*, concept change. Since the SOTC algorithm processes data in a sequential manner, we choose the Dynamically Weighted Majority (DWM) algorithm

**Table 9.1 Average classification errors (in percentage) of algorithms on benchmark datasets**

| Data Set | PER | OZAB | OGB | OSB | TNC | SOTC |
|---|---|---|---|---|---|---|
| Heart | 24.66 | 23.96 | 23.28 | 23.63 | 21.75 | **20.09** |
| Breast cancer | 5.77 | 5.44 | 5.71 | 5.23 | 4.84 | **4.65** |
| Australian | 20.82 | 20.26 | 19.70 | 20.01 | 15.92 | **14.86** |
| Diabetes | 32.25 | 32.43 | 33.49 | 31.33 | 26.89 | **25.75** |
| German | 32.45 | 31.86 | 32.72 | 31.86 | 28.13 | **26.74** |
| BMC | 47.09 | 45.72 | 46.92 | 46.37 | 25.37 | **17.03** |
| Splice | 33.42 | 32.59 | 32.79 | 32.81 | 18.88 | **18.56** |
| Banana | 48.91 | 47.96 | 48.00 | 48.84 | 27.98 | **17.60** |

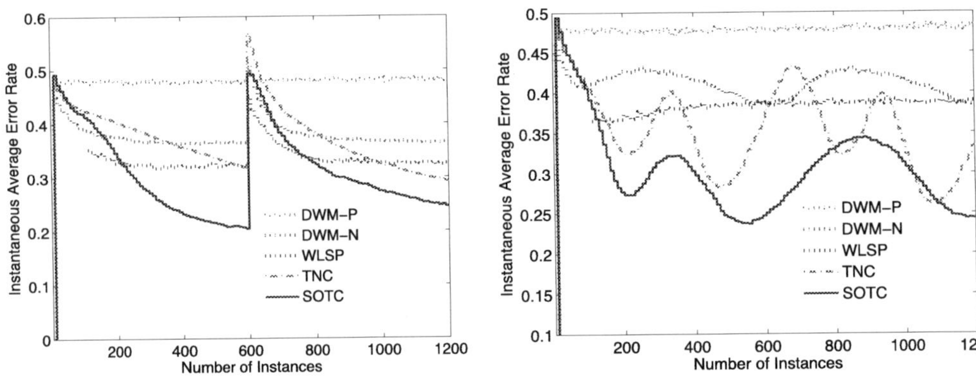


**FIGURE 9.6**

Performances of the algorithms in case of the abrupt and continuous concept changes in the BMC dataset. On the left figure, at the 600th instance, there is a 180° clock-wise rotation around the origin that is effectively a label flip. On the right figure, at each instance, there is a 180°/1200 clock-wise rotation around the origin.

(DWM) [35] with Perceptron (DWM-P) or naive Bayes (DWM-N) experts for the comparison, since the DWM algorithm is also an online algorithm. Although the batch algorithms do not truly fit into our framework, we still devise an online version of the tree-based local space partitioning algorithm [24] (which also learns the space partitioning and the classifier using the coordinate ascent approach) using a sliding-window approach and abbreviate it as the WLSP algorithm. For the DWM method, which allows the addition and removal of experts during the stream, we set the initial number of experts to 1, where the maximum number of experts is bounded by 100. For the WLSP method, we use a window size of 100. The parameters for these compared methods are set as in their original proposals.

We run these methods on the BMC dataset (1200 instances, Fig. 9.6), where a sudden/abrupt concept change is obtained by rotating the feature vectors (clock-wise around the origin) 180° after the 600th instance. This is effectively equivalent to flipping the label of the feature vectors; hence the resulting dataset is denoted as BMC-F. For a continuous concept drift, we rotate each feature vector $180°/1200 = 0.15°$ starting from the beginning; the resulting dataset is denoted as BMC-C. In Fig. 9.6, we present the classification errors for the compared methods averaged over 1000 trials. At each 10th

**Table 9.2 Running times (in seconds) of the compared methods when processing the BMC data set on a daily-use machine (Intel(R) Core(TM) i5-3317U CPU @ 1.70 GHz with 4 GB memory)**

| PER | OZAB | OGB | OSB | TNC | DWM-P | DWM-N | WLSP | SOTC |
|---|---|---|---|---|---|---|---|---|
| 0.06 | 12.90 | 3.57 | 3.91 | 0.43 | 2.06 | 6.91 | 68.40 | 0.62 |

instance, we test the algorithms with 1200 instances drawn from the active set of statistics (active concept).

Since the BMC dataset is non-Gaussian with strongly nonlinear class separations, the DWM method does not perform well on the BMC-F data. For instance, DWM-P operates with an error rate fluctuating around 0.48–0.49 (random guess). This results since the performance of the DWM method is directly dependent on the expert success and we observe that both base learners (Perceptron or the naive Bayes) fail due to the high separation complexity in the BMC-F data. On the other hand, the WLSP method quickly converges to steady state, however it is also asymptotically outperformed by the SOTC algorithm in both experiments. Increasing the window size is clearly expected to boost the performance of WLSP, however at the expense of an increased computational complexity. It is already significantly slower than the SOTC method even when the window size is 100 (for a more detailed comparison, see Table 9.2). The performance of the WLSP method is significantly worse on the BMC-C data set compared to the BMC-F data set, since in the former scenario, WLSP is trained with batch data of a continuous mixture of concepts in the sliding windows. Under this continuous concept drift, the SOTC method always (not only asymptotically as in the case of the BMC-F data set) performs better than the WLSP method. Hence, the sliding-window approach is sensitive to the continuous drift. Our discussion about the DWM method on the concept change data (BMC-F) remains valid in the case of the concept drift (BMC-C) as well. In these experiments, the power of self-organizing trees is obvious as the SOTC algorithm almost always outperforms the TNC algorithm. We also observe from Table 9.2 that the SOTC algorithm is computationally very efficient and the cost of region updates (compared to the TNC algorithm) does not increase the computational complexity of the algorithm significantly.

# APPENDIX 9.A

## 9.A.1 PROOF OF THEOREM 1

For the SOT of depth $D$, suppose $\hat{d}_t^{(k)}$, $k = 1, \ldots, \beta_d$, are obtained as described in Section 9.2.2. To achieve the upper bound in (9.15), we use the online gradient descent method and update the combination weights as

$$\boldsymbol{w}_{t+1} = \boldsymbol{w}_t - \frac{1}{2}\eta_t \nabla e_t^2(\boldsymbol{w}_t) = \boldsymbol{w}_t + \eta_t e_t \hat{\boldsymbol{d}}_t, \tag{9.32}$$

where $\eta_t$ is the learning rate of the online gradient descent algorithm. We derive an upper bound on the sequential learning regret $R_n$, which is defined as

$$R_T \triangleq \sum_{t=1}^{T} e_t^2(\boldsymbol{w}_t) - \sum_{t=1}^{T} e_t^2(\boldsymbol{w}_n^*),$$

where $\boldsymbol{w}_T^*$ is the optimal weight vector over $T$, *i.e.*,

$$\boldsymbol{w}_T^* \triangleq \arg\min_{\boldsymbol{w}\in\mathbb{R}^{\beta d}} \sum_{t=1}^{T} e_t^2(\boldsymbol{w}).$$

Following [36], using Taylor series approximation, for some point $z_t$ on the line segment connecting $\boldsymbol{w}_t$ to $\boldsymbol{w}_T^*$, we have

$$e_t^2(\boldsymbol{w}_T^*) = e_t^2(\boldsymbol{w}_t) + \left(\nabla e_t^2(\boldsymbol{w}_t)\right)^T (\boldsymbol{w}_T^* - \boldsymbol{w}_t) + \frac{1}{2}(\boldsymbol{w}_T^* - \boldsymbol{w}_t)^T \nabla^2 e_t^2(z_t)(\boldsymbol{w}_T^* - \boldsymbol{w}_t). \tag{9.33}$$

According to the update rule in (9.32), at each iteration the update on weights are performed as $\boldsymbol{w}_{t+1} = \boldsymbol{w}_t - \frac{\eta_t}{2}\nabla e_t^2(\boldsymbol{w}_t)$. Hence, we have

$$\begin{aligned}
\left|\left|\boldsymbol{w}_{t+1} - \boldsymbol{w}_T^*\right|\right|^2 &= \left|\left|\boldsymbol{w}_t - \frac{\eta_t}{2}\nabla e_t^2(\boldsymbol{w}_t) - \boldsymbol{w}_T^*\right|\right|^2 \\
&= \left|\left|\boldsymbol{w}_t - \boldsymbol{w}_T^*\right|\right|^2 - \eta_t \left(\nabla e_t^2(\boldsymbol{w}_t)\right)^T (\boldsymbol{w}_t - \boldsymbol{w}_T^*) + \frac{\eta_t^2}{4}\left|\left|\nabla e_t^2(\boldsymbol{w}_t)\right|\right|^2.
\end{aligned}$$

This yields

$$\left(\nabla e_t^2(\boldsymbol{w}_t)\right)^T (\boldsymbol{w}_t - \boldsymbol{w}_T^*) = \frac{\left|\left|\boldsymbol{w}_t - \boldsymbol{w}_T^*\right|\right|^2 - \left|\left|\boldsymbol{w}_{t+1} - \boldsymbol{w}_T^*\right|\right|^2}{\eta_t} + \eta_t \frac{\left|\left|\nabla e_t^2(\boldsymbol{w}_t)\right|\right|^2}{4}.$$

Under the mild assumptions that $\left|\left|\nabla e_t^2(\boldsymbol{w}_t)\right|\right|^2 \leq A^2$ for some $A > 0$ and $e_t^2(\boldsymbol{w}_T^*)$ is $\lambda$-strong convex for some $\lambda > 0$ [36], we achieve the following upper bound:

$$e_t^2(\boldsymbol{w}_t) - e_t^2(\boldsymbol{w}_T^*) \leq \frac{\left|\left|\boldsymbol{w}_t - \boldsymbol{w}_T^*\right|\right|^2 - \left|\left|\boldsymbol{w}_{t+1} - \boldsymbol{w}_T^*\right|\right|^2}{\eta_t} - \frac{\lambda}{2}\left|\left|\boldsymbol{w}_t - \boldsymbol{w}_T^*\right|\right|^2 + \eta_t \frac{A^2}{4}. \tag{9.34}$$

By selecting $\eta_t = \frac{2}{\lambda t}$ and summing up the regret terms in (9.34), we get

$$\begin{aligned}
R_n &= \sum_{t=1}^{n} \left\{e_t^2(\boldsymbol{w}_t) - e_t^2(\boldsymbol{w}_T^*)\right\} \\
&\leq \sum_{t=1}^{n} \left|\left|\boldsymbol{w}_t - \boldsymbol{w}_T^*\right|\right|^2 \left(\frac{1}{\eta_t} - \frac{1}{\eta_{t-1}} - \frac{\lambda}{2}\right) + \frac{A^2}{4}\sum_{t=1}^{n} \eta_t \\
&= \frac{A^2}{4}\sum_{t=1}^{n} \frac{2}{\lambda t} \leq \frac{A^2}{2\lambda}(1 + \log(n)).
\end{aligned}$$

### 9.A.2 PROOF OF THEOREM 2

Since $Z_t$ is a summation of terms that are all positive, we have $Z_t \geq 2^{-J(k)} \exp\left(-b\, L_t^{(k)}\right)$ and after taking the logarithm of both sides and rearranging the terms, we get

$$-\frac{1}{b} \log Z_T \leq L_T^{(k)} + \frac{J(k) \log 2}{b} \tag{9.35}$$

for all $k \in \{1, \dots, \beta_D\}$ at the (last) iteration at time $T$. We then make the following observation:

$$\begin{aligned} Z_T &= \prod_{t=1}^{T} \frac{Z_t}{Z_{t-1}} = \prod_{t=1}^{T} \sum_{k=1}^{\beta_D} \frac{2^{-J(k)} \exp\left(-b\, L_{t-1}^{(k)}\right)}{Z_{t-1}} \exp\left(-b\, \ell_t(f_t^{(k)})\right) \\ &\leq \exp\left(-b\, L_T(f_t) + \frac{T b^2}{8}\right), \end{aligned} \tag{9.36}$$

where the second line follows from the definition of $Z_t$ and the last line follows from the Hoeffding inequality by treating the $w_t^{(k)} = 2^{-J(k)} \exp\left(-b\, L_{t-1}^{(k)}\right) / Z_{t-1}$ terms as the randomization probabilities. Note that $L_T(f_t)$ represents the expected loss of the final algorithm; *cf.* (9.19). Combining (9.35) and (9.36), we obtain

$$\frac{L_T(f_t)}{T} \leq \frac{L_T^{(k)}}{T} + \frac{J(k) \log 2}{T b} + \frac{b}{8}$$

and choosing $b = \sqrt{2^D / T}$, we find the desired upper bound in (9.28) since $J(k) \leq 2^{D+1} - 1$, for all $k \in \{1, \dots, \beta_D\}$.

## ACKNOWLEDGMENTS

The authors would like to thank Huseyin Ozkan for his contributions in this work.

## REFERENCES

[1] M. Schetzen, The Volterra and Wiener Theories of Nonlinear Systems, John Wiley & Sons, NJ, 1980.
[2] M. Scarpiniti, D. Comminiello, R. Parisi, A. Uncini, Nonlinear spline adaptive filtering, Signal Processing 93 (4) (2013) 772–783.
[3] A. Carini, G.L. Sicuranza, Fourier nonlinear filters, Signal Processing 94 (0) (2014) 183–194.
[4] N.D. Vanli, M.O. Sayin, I. Delibalta, S.S. Kozat, Sequential nonlinear learning for distributed multiagent systems via extreme learning machines, IEEE Transactions on Neural Networks and Learning Systems 28 (3) (March 2017) 546–558.
[5] D.P. Helmbold, R.E. Schapire, Predicting nearly as well as the best pruning of a decision tree, Machine Learning 27 (1) (1997) 51–68.
[6] E. Takimoto, A. Maruoka, V. Vovk, Predicting nearly as well as the best pruning of a decision tree through dynamic programming scheme, Theoretical Computer Science 261 (2001) 179–209.
[7] E. Takimoto, M.K. Warmuth, Predicting nearly as well as the best pruning of a planar decision graph, Theoretical Computer Science 288 (2002) 217–235.
[8] D.P. Bertsekas, Nonlinear Programming, Athena Scientific, 1999.

[9] A.H. Sayed, Fundamentals of Adaptive Filtering, John Wiley & Sons, NJ, 2003.
[10] S. Dasgupta, Y. Freund, Random projection trees for vector quantization, IEEE Transactions on Information Theory 55 (7) (2009) 3229–3242.
[11] W.-Y. Loh, Classification and regression trees, Wiley Interdisciplinary Reviews: Data Mining and Knowledge Discovery 1 (1) (2011) 14–23.
[12] L. Brieman, J. Friedman, C.J. Stone, R.A. Olsen, Classification and Regression Trees, Chapman & Hall, 1984.
[13] J. Gama, Functional trees, Machine Learning 55 (3) (2004) 219–250.
[14] O.J.J. Michel, A.O. Hero, A.-E. Badel, Tree-structured nonlinear signal modeling and prediction, IEEE Transactions on Signal Processing 47 (11) (Nov 1999) 3027–3041.
[15] H. Ozkan, N.D. Vanli, S.S. Kozat, Online classification via self-organizing space partitioning, IEEE Transactions on Signal Processing 64 (15) (Aug 2016) 3895–3908.
[16] S.S. Kozat, A.C. Singer, G.C. Zeitler, Universal piecewise linear prediction via context trees, IEEE Transactions on Signal Processing 55 (7) (2007) 3730–3745.
[17] N.D. Vanli, M.O. Sayin, S.S. Kozat, Predicting nearly as well as the optimal twice differentiable regressor, CoRR, arXiv:1401.6413, 2014.
[18] A.V. Aho, N.J.A. Sloane, Some doubly exponential sequences, Fibonacci Quarterly 11 (1970) 429–437.
[19] N.D. Vanli, K. Gokcesu, M.O. Sayin, H. Yildiz, S.S. Kozat, Sequential prediction over hierarchical structures, IEEE Transactions on Signal Processing 64 (23) (Dec 2016) 6284–6298.
[20] C. Scott, R.D. Nowak, Minimax-optimal classification with dyadic decision trees, IEEE Transactions on Information Theory 52 (4) (2006) 1335–1353.
[21] C. Strobl, A.-L. Boulesteix, T. Augustin, Unbiased split selection for classification trees based on the Gini index, Computational Statistics & Data Analysis 52 (1) (2007) 483–501.
[22] H. Ozkan, M.A. Donmez, O.S. Pelvan, A. Akman, S.S. Kozat, Competitive and online piecewise linear classification, in: IEEE International Conference on Acoustics, Speech and Signal Processing (ICASSP), May 2013, pp. 3452–3456.
[23] F.M.J. Willems, Y.M. Shtarkov, T.J. Tjalkens, The context-tree weighting method: basic properties, IEEE Transactions on Information Theory 41 (3) (1995) 653–664.
[24] J. Wang, V. Saligrama, Local supervised learning through space partitioning, in: Advances in Neural Information Processing Systems (NIPS), 2012, pp. 91–99.
[25] Y. Freund, R.E. Schapire, Large margin classification using the perceptron algorithm, Machine Learning 37 (3) (1999) 277–296.
[26] K. Kim, N. Kalantarova, S.S. Kozat, A.C. Singer, Linear MMSE-optimal turbo equalization using context trees, IEEE Transactions on Neural Networks and Learning Systems 61 (12) (April 2013) 3041–3055.
[27] D. Kari, N.D. Vanli, S.S. Kozat, Adaptive and efficient nonlinear channel equalization for underwater acoustic communication, Physical Communication 24 (2017) 83–93.
[28] F. Khan, D. Kari, I.A. Karatepe, S.S. Kozat, Universal nonlinear regression on high dimensional data using adaptive hierarchical trees, IEEE Transactions on Big Data 2 (2) (2016) 175–188.
[29] T. Kohonen, Self-Organizing Maps, 3rd edition, Springer-Verlag, Inc., New York, 2001.
[30] H.G. Basara, M. Yuan, Community health assessment using self-organizing maps and geographic information systems, International Journal of Health Geographics 7 (1) (Dec 2008).
[31] E.N. Lorenz, Deterministic nonperiodic flow, Journal of the Atmospheric Sciences 20 (2) (1963) 130–141.
[32] N.C. Oza, S. Russell, Online bagging and boosting, in: Artificial Intelligence and Statistics, 2001, pp. 105–112.
[33] C. Leistner, A. Saffari, P.M. Roth, H. Bischof, On robustness of on-line boosting-a competitive study, in: IEEE 12th International Conference on Computer Vision Workshops, 2009, pp. 1362–1369.
[34] S.-T. Chen, H.-T. Lin, C.-J. Lu, An online boosting algorithm with theoretical justifications, in: International Conference on Machine Learning, 2012.
[35] J. Zico Kolter, Marcus Maloof, Dynamic weighted majority – an ensemble method for drifting concepts, Journal of Machine Learning Research 8 (2007) 2755–2790.
[36] E. Hazan, A. Agarwal, S. Kale, Logarithmic regret algorithms for online convex optimization, Machine Learning 69 (2–3) (2007) 169–192.

CHAPTER

# ADAPTATION AND LEARNING OVER NETWORKS FOR NONLINEAR SYSTEM MODELING

# 10

**Simone Scardapane***, **Jie Chen**†, **Cédric Richard**‡

*Sapienza University of Rome, Rome, Italy †Northwestern Polytechnical University, Xi'an, School of Marine Science and Technology, Xi'an, China ‡Université Côte d'Azur, Laboratoire Lagrange (UMR CNRS 7293), Nice, France

## CONTENTS



## CHAPTER POINTS

- We introduce the problem of distributed filtering over networks of communicating agents, such as sensors in a wireless network.
- We summarize the state-of-the-art research on nonlinear distributed filtering.
- We introduce a novel kernel approach to perform nonlinear distributed filtering in multitask scenarios.
- We validate our proposal on an experimental benchmark problem.

Adaptive Learning Methods for Nonlinear System Modeling. DOI: 10.1016/B978-0-12-812976-0.00013-0

## 10.1 INTRODUCTION

Adaptive filters have been at the heart of digital signal processing over the last century, thanks to their capability of rapidly adapting to streams of incoming data. At the same time, classical filtering approaches have not been satisfactory to handle the challenges posed by large-scale, unstructured big data scenarios that are common today. This has fostered the recent development of techniques to deal with such problems, allowing signal processing to scale to truly massive datasets [1] and to work with less structured data types, such as graphs [2,3]. In this chapter, we look at one peculiar aspect unifying several big data problems, namely, their *distributed* nature, where the data are naturally generated and aggregated at different locations with possibly poor or expensive network connectivity. Examples of problems in this category abound and include (but are not limited to) wireless sensor networks (WSNs) [4], distributed databases, robotic swarms and fog computing platforms. In all these scenarios, the agents in the network can be severely limited in their capabilities, in terms of either energy constraints (*e.g.*, low-power devices in WSNs), connectivity, privacy, or other aspects. As a consequence, any solution devised for learning and inference over networks needs to be aware of these constraints, making this a challenging problem with wide applications.

Distributed learning can be cast as a decentralized optimization problem, which has a long history in the optimization field [5] and in artificial intelligence. In recent years, this problem has gained a renewed interest from the machine learning community, with the development of a number of learning protocols for a variety of models, including boosting [6], support vector machines [7,8], kernel regression [4] and sparse linear models [9,10], to cite a few. Several of these were also applied in signal processing problems, most notably in order to provide distributed inference capabilities in WSNs [4]. A large majority of them, however, is only applicable in *batch* situations, where each agent is allowed to manipulate its entire (local) dataset at each iteration. Distributed filtering algorithms, on the contrary, require the development of online solutions, where the data are received and processed, sequentially, by the agents.

In the filtering literature, a recent series of works was initiated by the development of the so-called 'diffusion filtering' (DF) algorithms, starting from the diffusion LMS [11] and the diffusion RLS [12], up to their more general formulation in terms of generic convex cost functions [13–15], which are considered here. DF algorithms are characterized by interleaving local updates in parallel (mimicking classical filters), with communication steps, during which the agents exchange information on their current estimate with their neighbors. Following the development of the main theory, in the subsequent years, several authors have extended classical linear filters to the distributed case (*e.g.*, sparse LMS [10] and group LMS [16]), while others have focused on nonlinear filters, as we review further on in the chapter. For the interested reader, we refer to the guest editorial in [17] and references therein for a recent overview of the literature.

All the works mentioned up to now assume that the agents are identifying a *common* model. This is a reasonable assumption in several contexts, particularly when the statistics of the data do not depend on the spatial locations of the agents, making it very common in distributed machine learning problems [7,15]. In general, however, the agents could be interested in different identification problems, which are similar in some quantifiable sense. As an example, consider the setup illustrated in Fig. 10.1. If the agents are sensors deployed over an environment, trying to predict some quantity of interest (*e.g.*, some pollutant concentration), the model might be different among groups (clusters) of agents, possibly due to the spatial conformation of the ground (*e.g.*, sensors deployed over a mountain vs. sensors

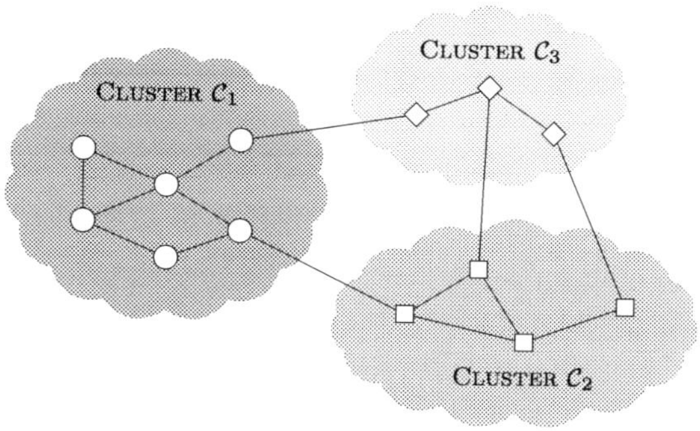


**FIGURE 10.1**

An example of a network with 12 agents subdivided in three different clusters.

deployed in a valley). Nonetheless, since they are all trying to predict the same quantity of interest, communication between agents belonging to different clusters can be beneficial.

The first DF solution to this problem, which is termed *multitask* network, was analyzed in [18]. Following this, a range of algorithms was proposed in the linear case. Most notably, Chen *et al.* [19] and Zhao and Sayed [20] investigated the possibility of unsupervised learning of the clusters structure when it is not available *a priori*. Additional developments include the extension to asynchronous networks where, *e.g.*, links may fail or agents might disconnect [21]; proximal updates for nondifferentiable regularization terms [22]; total least squares approaches [23]; and, finally, multitask learning over (linear) latent subspaces [24,25]. Almost no work, however, has addressed the problem of learning in a multitask network with *nonlinear* models.

Based on the previous discussion, this chapter has three separate aims. First, we introduce the fundamental concept of DF in Section 10.2, which serves as a very general introduction to the topic. Next, we summarize recent works on nonlinear DF algorithms in Section 10.3, with an emphasis on three classes of solutions. In order to motivate research on multitask learning with nonlinear models, in Section 10.4 we propose a multitask kernel algorithm based on a functional formulation of DF. Finally, we validate the algorithm on an experimental benchmark in Section 10.5, before making some final remarks in Section 10.6.

## 10.2 MATHEMATICAL FORMULATION OF THE PROBLEM

This section is intended to familiarize the reader with some basic theoretical elements underlying most distributed learning scenarios. We start by providing a setup for the problem in Section 10.2.1. Next, we describe a general class of algorithms based on diffusion protocols in Section 10.2.2. In Section 10.2.3, we show how these algorithms can be customized to address multitask scenarios. For conciseness, we only focus on a selection of key items, without providing a comprehensive treatment of these thematics. We refer the interested reader to [14,15] for introductory references.

### 10.2.1 PROBLEM SETUP

Let us consider a generic network of $N$ agents (*e.g.*, sensors in a WSN) as the one depicted in Fig. 10.1. We assume that time is slotted and, at every time instant $n$, each agent receives a new observation $\left(\mathbf{u}_{k,n}, d_k(n)\right)$, where $\mathbf{u}_{k,n} \in \mathbb{R}^M$ is the model input vector at agent $k$ (*e.g.*, a buffer of the last $M$ samples) and $d_k(n)$ the corresponding desired response. For simplicity, we assume that $d_k(n)$ is a scalar. For the rest of the chapter, we shall use the subscript $k$ to denote a quantity specific to one of the agents.

Following the standard supervised learning approach, the desired input/output relation can be modeled by choosing a function $f$ in some hypothesis space $\mathcal{H}$. For simplicity, in this section we suppose that each function is parameterized by a vector of tunable parameters $\mathbf{w} \in \mathbb{R}^q$, *e.g.*, a linear predictor.[1] With this setting, each agent is interested in finding a set of parameters $\mathbf{w}_k^*$ which minimizes some local cost function $J_k(\cdot)$ defined over the hypothesis space from streaming data. Specifically, for most estimation problems in practice, these local cost functions are defined as the expectation of some *error function* $L(\cdot,\cdot)$ with respect to the statistics of the local stream of data,

$$J_k(\mathbf{w}_k) = \mathbb{E}\left\{L\left(d_k, f_k(\mathbf{u}_k)\right)\right\}, \tag{10.1}$$

where we use the shorthand $f_k(\mathbf{u}_k) = f(\mathbf{u}_k; \mathbf{w}_k)$. The global optimization problem to be solved at the network level is then given by the sum of the local cost functions,

$$\min_{\mathbf{w}_1,\ldots,\mathbf{w}_N \in \mathbb{R}^q} \left\{J_{\text{glob}}(\mathbf{w}_1,\ldots,\mathbf{w}_N)\right\} = \sum_{k=1}^{N} J_k(\mathbf{w}_k), \tag{10.2}$$

where $\mathbf{w}_k$ is the estimate at the $k$th agent. If we assume that no relation holds between the local cost functions, then (10.2) reduces to a set of $N$ optimization problems that can be solved in parallel by every agent, independently of all the others. A more interesting formulation arises by assuming some form of relation among the cost functions (detailed below). In this case, the information gathered by one agent during its optimization process can potentially be used by the other agents to speed up their convergence, or even converge to a better solution using some shared information.

The difficulty arises from the fact that each agent has direct access to its local cost function, but it has no access to the local cost functions of the other agents for all the reasons mentioned in the introduction. Depending on the relation between cost functions, we can distinguish between three classes of distributed problems:

- **Single-task problems**: in this case, $\mathbf{w}_k^* = \mathbf{w}^*$, $k = 1,\ldots,N$, *i.e.*, all the cost functions have the same minimizer which must be attained by all agents. This is the scenario which has drawn most attention in the literature, being particularly useful in distributed machine learning problems [7], where it is common to assume that the data of interest are generated by a single underlying distribution.
- **Multitask problems**: in this scenario, each local cost function has possibly a different minimizer $\mathbf{w}_k$.[2] In order to make the problem interesting, we assume that these minimizers are 'similar' (in

[1] Distributed kernel filters (Section 10.3.2) are an example of a nonparametric formulation, which is recast as a parametric problem thanks to the representer's theorem.

[2] Some readers might recognize that in the machine learning literature, the term 'multitask learning' is employed in a slightly different meaning. It refers to the problem of solving several learning tasks defined on the same (or similar) input domain(s)

some sense to be properly defined) among pairs of neighboring agents, so that communicating can increase their speed of convergence and possibly counter noisy environments.

- **Clustered multitask problems**: in this intermediate case, each agent belongs to one of $T$ different groups (clusters), such that all agents belonging to the same cluster have the same minimizer and *vice versa*, as shown in Fig. 10.1. Clearly, both single-task and multitask problems can be derived as extreme cases of this class of problems, by setting $T = 1$ and $T = N$.

Multitask problems can be further subdivided, depending on whether the similarity between tasks is known *a priori*, or whether it must be inferred from the data. An example of the former case is the algorithm in [18], while examples of the latter case can be found in [19,20]. Inferring knowledge about the groups might require the inclusion of some decentralized clustering procedure in the learning process, which is an interesting problem in its own right [29]. For simplicity, in this chapter we will focus on the multitask formulation, but we underline that almost all multitask algorithms can be generalized to handle the clustered multitask case; *e.g.*, see [18].

### 10.2.2 DIFFUSION-BASED ALGORITHMS

In order to describe a family of algorithms to solve the previous problem, we first need to define a model of communication between the different agents. Most of the literature focuses on the case where the communication links form a static, undirected, connected graph $\mathcal{G}$. At each time step, the $k$th agent is allowed to communicate with its set of direct neighbors $\mathcal{N}_k$, while it cannot send messages to agents to which it is not directly linked.[3] Nonetheless, the overall connectedness of the graph ensures that information can flow throughout the entire network. In this scenario, connectivity can be described by a symmetric, real-valued matrix $\mathbf{A} \in \mathbb{R}^{N \times N}$ such that $A_{kl} \neq 0$ only if agents $k$ and $l$ are connected and

$$\sum_{k=1}^{N} A_{kl} = 1 \,, \quad A_{kl} \geq 0 \;\; \text{for any } k, l = 1, \dots, N. \tag{10.3}$$

These weights are used by the agents to scale and combine information received by their neighbors. The previous condition ensures that each row of the matrix defines a convex combination, so that the range of the information to be combined is always preserved (formally, the condition requires that $\mathbf{A}$ be left stochastic). There are several strategies allowing agents to build such matrices, as we show later. This formulation can also be extended in several ways, most notably with the use of asynchronous formulations [30] (thus avoiding the need for a common clock throughout the network) and mixing matrices that do not respect double stochasticity [31]. Nonetheless, since this chapter is only intended as an introduction, we will focus on the simpler case detailed before.

In the case of a single agent, (10.2) could be solved by a simple gradient descent algorithm. In order to counteract the lack of global information, the basic idea of diffusion algorithms is to interleave local

---

[26,27]. While the setup and the objectives in the two cases do not perfectly overlap, we speculate that exploring the connections between them is of particular significance, particularly due to the increasing interest given by deep neural networks [28].

[3] The graph describes all *feasible* communication links. This is a relatively general formulation, since every multihop network can be described with an equivalent single-hop network by considering all possible paths as a direct link in the corresponding graph.

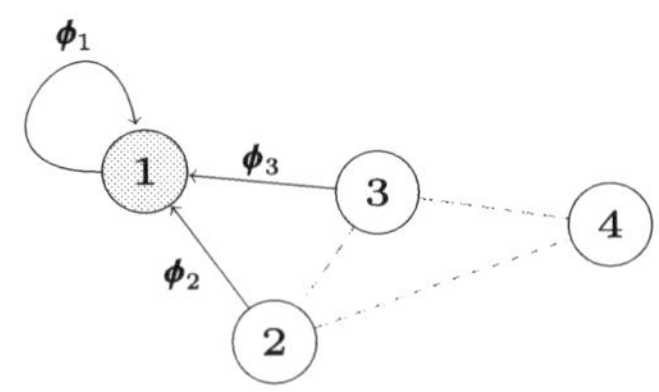


**FIGURE 10.2**

An example of a diffusion step relative to agent 1 in a simple network with four agents. Gray links are inactive.

optimization steps with communication steps, where each agent combines its own estimate with those of its neighbors. In the filtering literature, this strategy is generally denoted as adapt-then-combine (ATC):

$$\boldsymbol{\phi}_{k,n} = \mathbf{w}_{k,n-1} - \mu_k \nabla J_k(\mathbf{w}_{k,n-1}) \,, \tag{10.4}$$

$$\mathbf{w}_{k,n} = \sum_{l=1}^{N} A_{lk} \boldsymbol{\phi}_{l,n} \,, \tag{10.5}$$

where $\mu_k$ is a (possibly time-dependent) step size and, similarly to before, we use a double subscript $(k, n)$ to denote the estimate of agent $k$ at time $n$. Practically, the gradient term in (10.4) can be replaced with a noisy version, for example using an instantaneous approximation computed from the current data sample. A diffusion step example is given in Fig. 10.2.

These algorithms are particularly suitable in single-task scenarios, where their convergence properties in the convex case have been analyzed extensively [14,15]. Interestingly, they can still have good convergence properties in the multitask case, as shown in [19]. However, they provide a building block for all other formulations, as we show in Section 10.2.3. Before that, we describe an example of diffusion algorithm for nonlinear learning with a distributed logistic regression model.

#### 10.2.2.1 An Example: Logistic Regression Over Networks

Consider a binary classification problem where $d_k(n) \in \{0, 1\}$, *i.e.*, each output is a single bit representing whether the corresponding input belongs to a certain class or not. We approximate the underlying relation using a logistic predictor $f(\mathbf{u}) = \sigma(\mathbf{w}^T \mathbf{u})$, where $\sigma(\cdot)$ is the sigmoid function, as follows:

$$\sigma(s) = \frac{1}{1 + \exp\{-s\}} \,, \tag{10.6}$$

ensuring that the outputs of the model are properly scaled as valid probabilities. Each agent wishes to minimize the (regularized) expected cross-entropy over its stream of data:

$$J_k(\mathbf{w}) = \mathbb{E}\Big\{-d_k \cdot \log\big(f_k(\mathbf{u}_k)\big) - (1 - d_k) \cdot \log\big(1 - f_k(\mathbf{u}_k)\big)\Big\} + \frac{\lambda}{2N} \|\mathbf{w}_k\|_2^2 \,, \tag{10.7}$$

where the factor $1/N$ in the regularization term ensures that the total penalization in (10.2), when summed over all agents, is equal to $\frac{\lambda}{2}$. By taking instantaneous approximations to the gradient, simple

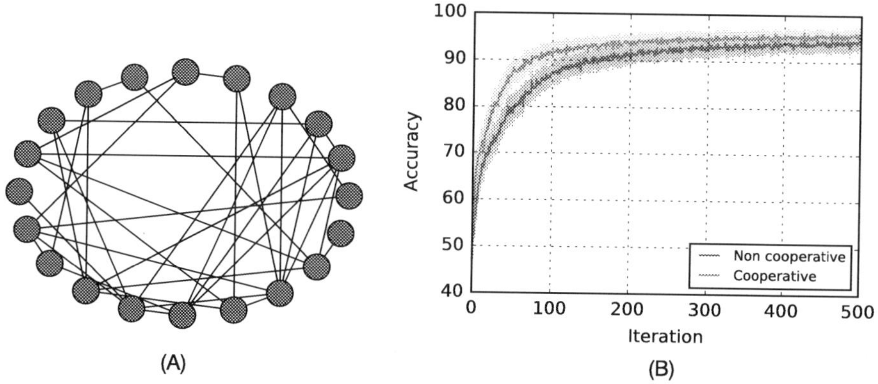


**FIGURE 10.3**

An example of distributed (online) logistic regression on the WBCD dataset. (A) The network of 20 agents used in this experiment. (B) Average accuracy for the diffusion algorithm, in blue (light gray in print version), and for a network of $N$ noncooperating agents (which is equivalent to set $\mathbf{A} = \mathbf{I}$), in red (dark gray in print version). We use shaded contours to plot standard deviation.

algebra manipulations show that the update steps in (10.4) are given by

$$\boldsymbol{\phi}_{k,n} = \mathbf{w}_{k,n-1} + \mu_k \big(d_k(n) - f_{k,n-1}(\mathbf{u}_{k,n})\big)\mathbf{u}_{k,n} - \frac{1}{N}\mu_k \lambda \mathbf{w}_{k,n-1}\,. \tag{10.8}$$

In order to show the speedup obtained by such procedure, in Fig. 10.3 we plot the average accuracy (see below) obtained with a network of 20 agents, whose connectivity is generated randomly, where at every iteration each agent receives a randomly chosen example taken from the well-known Wisconsin Breast Cancer Database (WBCD).[4] The accuracy is defined as 1 if the agent makes a correct prediction (*i.e.*, the sign of $f(\mathbf{u})$ agrees with $d$), 0 otherwise. It is computed before the adaptation step, and it is averaged with respect to the different agents and the different simulations.

### 10.2.3 EXTENSION TO MULTITASK LEARNING

The general ideas exposed in Section 10.2.2 can be easily extended in order to be more efficient in the multitask scenario. Here, we detail a simple extension originally proposed in [18] to show one example of such extensions. We refer to the large number of works cited in the introduction for more recent proposals.

Suppose that, given two agents $k$ and $l$, we have a way to quantify the similarity among their respective minimizers. In order to leverage this information, we can augment the original cost function in (10.2) with a regularization term forcing the similarity among minimizers, in terms of their Euclidean

[4] https://archive.ics.uci.edu/ml/datasets/Breast+Cancer+Wisconsin+(Diagnostic).

distance, as follows:

$$J^{\text{glob}}(\mathbf{w}_1, \ldots, \mathbf{w}_N) = \sum_{k=1}^{N} J_k(\mathbf{w}_k) + \eta \sum_{k=1}^{N} \sum_{l \neq k, l \in \mathcal{N}_k} \rho_{kl} \left\| \mathbf{w}_k - \mathbf{w}_l \right\|_2^2 , \tag{10.9}$$

where $\eta$ is a regularization factor and the nonnegative coefficients $\rho_{kl} \geq 0$ quantify our *a priori* knowledge about the similarities. We assume that, for each agent, the weights are positive and sum to one. We have

$$\sum_{l=1}^{N} \rho_{kl} = 1 \quad \text{and} \quad \rho_{kl} = 0 \text{ if } l \notin \mathcal{N}_k, \ \forall k \in \{1, \ldots, N\} . \tag{10.10}$$

As a consequence of their definition, these regularization factors mirror the communication network of the agents. Due to this, taking a local optimization step with respect to the estimate of the $k$th agent immediately requires a diffusion step. We have

$$\mathbf{w}_{k,n} = \mathbf{w}_{k,n-1} - \mu_k \nabla J_k(\mathbf{w}_{k,n-1}) - \mu_k \eta \sum_{l \neq k, l \in \mathcal{N}_k} \frac{(\rho_{kl} + \rho_{lk})}{2} \left( \mathbf{w}_{k,n-1} - \mathbf{w}_{l,n-1} \right) . \tag{10.11}$$

Since the estimates are already exchanged, the previous update step can be implemented without the need for additional combination steps like in Section 10.2.2. As an example, by considering a simple linear predictor $f_k(\mathbf{u}) = \mathbf{w}_k^T \mathbf{u}$ and instantaneous approximations to the gradient, we obtain the following update rule for the multitask diffusion LMS, presented in [18]:

$$\mathbf{w}_{k,n} = \mathbf{w}_{k,n-1} + \mu_k \left( d_k(n) - \mathbf{w}_{k,n-1}^T \mathbf{u}_{k,n} \right) \mathbf{u}_{k,n} - \mu_k \eta \sum_{l \neq k, l \in \mathcal{N}_k} \frac{(\rho_{kl} + \rho_{lk})}{2} \left( \mathbf{w}_{k,n-1} - \mathbf{w}_{l,n-1} \right) . \tag{10.12}$$

If we assume that the mixing weights are symmetric, $\frac{(\rho_{kl} + \rho_{lk})}{2}$ simplifies to $\rho_{kl}$. It is possible to obtain asymmetric regularization terms by considering a game-theoretical formulation of the optimization problem; see the discussion in [18].

## 10.3 EXISTING APPROACHES TO NONLINEAR DISTRIBUTED FILTERING

In this section we describe three approaches to extend the previous formulation to nonlinear models in an efficient way. We underline that all these algorithms have been devised mostly for the case of single-task networks. Further extensions to the multitask scenario are the topic of Section 10.4.

### 10.3.1 EXPANSION OVER RANDOM BASES

One immediate idea is to project the original input vector $\mathbf{u}$ to a high-dimensional space via some fixed function $\mathbf{h}(\mathbf{x}) : \mathbb{R}^M \rightarrow \mathbb{R}^B$ before using it with a linear predictor, where $d$ is the dimensionality of the input vector $\mathbf{u}$ and $B$ is a parameter which (in general) can be chosen by the user. In the distributed case, this requires only a small communication overhead in the beginning for the agents to agree on a specific projection function. Any distributed linear algorithm, such as the diffusion LMS or the diffusion RLS, can then be used.

Generally speaking, deterministic mappings (such as those mentioned in Chapters 2 and 4) are not efficient, because their size might grow exponentially with respect to $M$. A different idea is to use basis functions whose parameters are assigned stochastically, *e.g.*, a parameterized sigmoid,

$$h_i(\mathbf{u}) = \frac{1}{1 + \exp\left\{-\mathbf{a}_i^T \mathbf{u} - b_i\right\}},$$

where $\mathbf{a}_i$ and $b_i$ might be extracted randomly from some uniform distribution (whose range is generally chosen in order to provide a good accuracy; see the discussion in [32]). Interestingly, it is possible to show that the resulting estimator (called a random vector functional-link network) is a universal approximator over compact functions provided that $B$ is chosen large enough [33]. It is also possible to interpret it as a degenerate case of the echo state network described in Chapter 12, where connections between different nodes have been removed. The idea of using RVFL networks in a distributed context was proposed in [34] for the batch case and in [35] for the online case with DF algorithms.

A different approach is to design a feature mapping $\mathbf{h}(\cdot)$ approximating a specific kernel $\mathcal{K}(\cdot,\cdot)$ function[5] chosen by the user:

$$\mathcal{K}(\mathbf{u}_1, \mathbf{u}_2) \approx \langle \mathbf{h}(\mathbf{u}_1), \mathbf{h}(\mathbf{u}_2) \rangle. \tag{10.13}$$

This idea was popularized by [36] for approximating shift-invariant kernels (*e.g.*, the Gaussian kernel) in large-scale applications of kernel methods. In particular, it is possible to show that this class of kernels can be easily approximated with very simple stochastic mappings. The work [37] was the first to apply this idea explicitly to kernel filters, and similar algorithms were independently reintroduced in [38]. Since Chapter 7 is entirely devoted to this idea, we will not go further into it. We refer the interested reader to [32] for a recent overview on random feature methods.

### 10.3.2 DISTRIBUTED KERNEL FILTERS

An alternative line of research is devoted to distributed strategies for kernel filters, working directly on some reproducing kernel Hilbert space (RKHS), instead of approximating the kernel function as in Section 10.3.1. As we stated in the introduction, several distributed algorithms for kernel ridge regression were devised in the context of WSNs [4], followed by algorithms for the distributed optimization of SVMs [39,7]. Any approach to dealing with kernels faces the challenge of working with a kernel-based model that depends explicitly on all the data in the training set. A naive distributed implementation would thus require to exchange all the local datasets between the agents, which can become infeasible.

In an online context, this is made worse by the growing nature of the kernel model [40]. This is a fundamental drawback underlying any kernel filter algorithm [41–43]. An initial investigation in developing a fully distributed version of the kernel LMS (KLMS) was made in [44], where the basic idea is to consider diffusion algorithms directly in a functional form. In particular, let us assume that

[5] We refer to Chapters 5–7 for introductory material on kernel filters.

the data received from the $k$th agent satisfies a model of the following form:

$$d_k(n) = \psi_k^o(\mathbf{u}_{k,n}) + \nu_k(n), \tag{10.14}$$

where $\psi_k^o$ belongs to an RKHS $\mathcal{H}$, while $\nu_k(n)$ is a zero mean white noise with variance $\sigma_k^2$. Restricting our attention to a generic single-task network, we have

$$\psi_k^o = \psi^o \;\; \forall k \in \{1, \ldots, N\}.$$

Considering the classical squared error function, the gradient of the local cost functions can now be computed in terms of their Fréchet derivatives as[6]

$$\nabla J_k(\psi_k) = -2\mathbb{E}\left\{\big(d_k - \psi_k(\mathbf{u}_k)\big)\kappa(\cdot, \mathbf{u}_k)\right\}, \tag{10.15}$$

where $\kappa$ is the kernel function associated to the RKHS. Considering instantaneous approximations for the expectation as was done earlier, we arrive at a functional equivalent of the ATC diffusion framework,

$$\delta_{k,n} = \psi_{k,n-1} + \mu_k \left(d_k(n) - \psi_{k,n-1}(\mathbf{u}_{k,n})\right) \kappa(\cdot, \mathbf{u}_{k,n}), \tag{10.16}$$

$$\psi_{k,n} = \sum_{l=1}^{N} A_{lk}\delta_{l,n}. \tag{10.17}$$

Although this formulation is extremely general, one has still to solve the problem of the growing structure of the kernel functions. The idea pursued in [44] is to assume some shared dictionary $\mathcal{D}$ among nodes, whose selection is (at the moment) an open research question. Using this approximation, we can rewrite the desired function as

$$\psi_{k,n} = \boldsymbol{\beta}_{k,n}^T \mathbf{k}_{k,n}, \tag{10.18}$$

where $\mathbf{k}_{k,n}$ is the vector of kernel values computed between the current input vector $\mathbf{u}_{k,n}$ and the shared dictionary $\mathcal{D}$. Each function $\psi_{k,n}$ is now parameterized by the set of linear coefficients $\boldsymbol{\beta}_{k,n}$. The previous algorithm can be rewritten as[7]

$$\boldsymbol{\delta}_{k,n} = \boldsymbol{\beta}_{k,n-1} + \mu_k \left(d_k(n) - \boldsymbol{\beta}_{k,n-1}^T \mathbf{k}_{k,n}\right) \mathbf{k}_{k,n}, \tag{10.19}$$

$$\boldsymbol{\beta}_{k,n} = \sum_{l=1}^{N} A_{lk}\boldsymbol{\delta}_{l,n}. \tag{10.20}$$

[6]A functional derivative is needed because the dimensionality of $\mathcal{H}$ can be infinite. See [45] for an introduction to Fréchet derivatives in the context of kernel methods and [46] for an introductory textbook on functional analysis.

[7]Note that we consider a simplified formulation with respect to [44], where two combination steps are used. Also, we use the same symbol $\boldsymbol{\delta}_k$ for the result of the adaptation step as in (10.16), but in boldface to underline that it is now a vector-valued quantity.

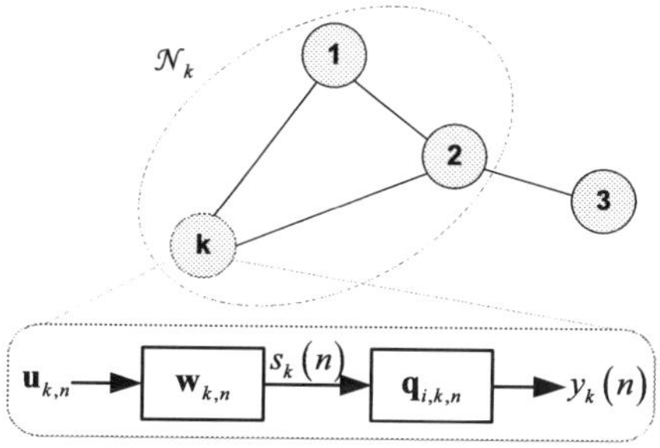


**FIGURE 10.4**

Illustration of SAF interpolation performed over a network of agents (adapted and reprinted with permission from [50]).

The idea of preselecting a dictionary is not new in the kernel literature. In fact, one of the earliest algorithms for distributed SVMs [39] exploited a similar idea, which is termed semiparametric SVM. In [47], a fixed dictionary is used to analyze the convergence behavior of the KLMS algorithm. A derivation of the functional diffusion KLMS algorithm when removing the fixed dictionary constraint is given in [48]. Another extension is presented in [49], where a set of consensus constraints is included in the problem to ensure convergence and speed up the algorithm.

### 10.3.3 DIFFUSION SPLINE FILTERS

Another possibility for nonlinear learning over networks is given by considering spline adaptive filters (SAFs).[8] The idea of using SAFs in a distributed environment was recently introduced in [50]. In the following we briefly describe the distributed algorithm. The interested readers can refer to [51,52] for introductory material on the SAF model.

Let us assume that the data are generated according to a restricted Wiener model given by

$$d_k(n) = f_k^o\left(\mathbf{w}_k^T \mathbf{u}_{k,n}\right) + \nu_k(n), \tag{10.21}$$

where $f_k^o$ is any smooth nonlinear function and $\nu_k(n)$ is a noise term. A SAF mimics this architecture, where the nonlinear term is approximated via spline interpolation over a set of $Q$ fixed control points that are adapted during learning. Hence, in a distributed scenario each agent has estimates of the local part of the filter, $\mathbf{w}_{k,n}$, and of the aforementioned control points, $\mathbf{q}_{k,n}$, as shown schematically in Fig. 10.4.

Consider a single-task scenario, where each agent tries to minimize the expected squared loss. Following [50], we consider a combine-then-adapt (CTA) scheme, where the combination is performed before the adaptation. The two steps are applied to both sets of parameters $\mathbf{w}_{k,n}$ and $\mathbf{q}_{k,n}$ simultaneously. However, by exploiting the SAF structure we can avoid exchanging the full weight vector $\mathbf{q}_{k,n}$, as described in the following.

[8] SAFs are the topic of Chapter 3.

The combination step starts with the agents exchanging their current estimates of the linear weights as

$$\boldsymbol{\psi}_{k,n-1} = \sum_{l \in \mathcal{N}_k} A_{lk} \mathbf{w}_{l,n-1}. \tag{10.22}$$

The new weights are used to compute the output of the linear part of the filter, denoted as $s_k(n) = \boldsymbol{\psi}_{k,n-1}^T \mathbf{u}_{k,n}$. We use $i$ to denote the index of the closest control point to $s_k(n)$ in our set of fixed control points. As described in Chapter 3, the final output of a Wiener SAF depends only on the $i$th control point and its $P$ right neighbors, with $P$ being the order of interpolation. Let us denote by $\mathbf{q}_{i,k,n-1}$ the set of such 'active' control points for agent $k$, which are called the 'span' of the filter (see Chapter 3 for more details). We use a third subscript to denote dependence with respect to the span. The second combination step is performed only with respect to the current span:

$$\boldsymbol{\xi}_{k,n-1} = \sum_{l \in \mathcal{N}_k} A_{lk} \mathbf{q}_{i,l,n-1}. \tag{10.23}$$

In the case of cubic interpolation, each $\boldsymbol{\xi}_{k,n-1}$ has dimensionality 4, making its exchange extremely efficient, with only a fixed overhead with respect to a classical diffusion LMS. Practically, every agent sends its current span index $i$ to its neighbors and receives back the vectors $\mathbf{q}_{i,l,n-1}$. For simplicity, the mixing weights $A_{lk}$ in the two diffusion steps are assumed identical.

Next, we proceed to the adaptation step. The complete SAF output given the new span is obtained (again following the general rules of Chapter 5) as

$$y_k(n) = \mathbf{u}^T \mathbf{B} \boldsymbol{\xi}_{k,n-1}, \tag{10.24}$$

where the vector $\mathbf{u}$ is constructed by taking powers up to a fixed order of the normalized value $\frac{s_k(n)}{\Delta x} - \left\lfloor \frac{s_k(n)}{\Delta x} \right\rfloor$, where $\Delta x$ is the sampling precision of the spline, and $\mathbf{B}$ is the spline matrix, *e.g.*, the Catmull-Rom spline given by

$$\mathbf{B} = \frac{1}{2} \begin{bmatrix} -1 & 3 & -3 & 1 \\ 2 & -5 & 4 & -1 \\ -1 & 0 & 1 & 0 \\ 0 & 2 & 0 & 0 \end{bmatrix}. \tag{10.25}$$

Adaptation is carried out by performing two parallel gradient descent steps:

$$\mathbf{w}_{k,n} = \boldsymbol{\psi}_{k,n-1} + \mu_k e_{k,n} \varphi'(s_k(n)) \mathbf{u}_{k,n}, \tag{10.26}$$

$$\mathbf{q}_{i,k,n} = \boldsymbol{\xi}_{k,n-1} + \mu_k e_{k,n} \mathbf{B}^T \mathbf{u}, \tag{10.27}$$

where $e_{k,n}$ is the instantaneous local error and $\varphi'(s_k(n))$ is the spline derivative with respect to the linear weights. Note that the diffusion LMS can be obtained as a special case, where each node initializes its nonlinearity as the identity, and the step size of the nonlinear part is set to zero.

## 10.4 A DISTRIBUTED KERNEL FILTER FOR MULTITASK PROBLEMS

As we saw in Section 10.2.3, several ideas have been proposed to model nonlinear systems in a distributed fashion, but almost none is framed for the multitask scenario. As a first step towards this line of research, in this section we briefly combine some of the previous ideas to devise an efficient kernel-based diffusion algorithm for multitask networks. In a nutshell, we combine the multitask diffusion LMS presented in Section 10.2.3 with the functional diffusion KLMS of Section 10.3.2. To this end, consider again the data model in (10.14), where we assumed that all the minimizers are the same across the agents. More generally, we can consider the case where two functions $\psi_k^o$ and $\psi_l^o$ are assumed to be 'close' in the sense of the norm $\|\cdot\|_{\mathcal{H}}$ of the RKHS, whenever the corresponding agents are spatial neighbors. We have

$$\psi_k^o \sim \psi_l^o \ \text{ if } l \in \mathcal{N}_k, \tag{10.28}$$

where $\mathcal{N}_k$ denotes the set of neighbors of $k$ and $\sim$ denotes similarity. To recover the unknown functions and leveraging over the basic idea described in Section 10.2.3, we aim at minimizing the following global cost function in a decentralized fashion:

$$J^{\text{glob}}(\psi_1,\dots,\psi_N) = \sum_{k=1}^{N} \mathbb{E}\left\{\left|d_k(n) - \psi_k(\mathbf{u}_{k,n})\right|^2\right\} + \eta \sum_{k=1}^{N} \sum_{l \neq k, l \in \mathcal{N}_k} \rho_{kl} \left\|\psi_k - \psi_l\right\|_{\mathcal{H}}^2, \tag{10.29}$$

where $\eta > 0$ is a regularization factor and the nonnegative coefficients $\rho_{kl} \geq 0$ weight the similarity between different functions. Once again, we assume that, for each agent, the weights are positive and sum to one, so we have

$$\sum_{l=1}^{N} \rho_{kl} = 1 \ \text{ and } \ \rho_{kl} = 0 \text{ if } l \notin \mathcal{N}_k, \ \forall k \in \{1,\dots,N\}. \tag{10.30}$$

Thus, each agent is interested in minimizing the local expected mean squared error, under suitable proximity constraints on its function and the functions of its neighbors. The previous problem decomposes as a sum of local cost functions defined as

$$J_k^{\text{loc}}(\psi_1,\dots,\psi_N) = \mathbb{E}\left\{\left|d_k(n) - \psi_k(\mathbf{u}_{k,n})\right|^2\right\} + \eta \sum_{l \neq k, l \in \mathcal{N}_k} \rho_{kl} \left\|\psi_k - \psi_l\right\|_{\mathcal{H}}^2. \tag{10.31}$$

Each local cost function is independent of the estimate of agents which are not in its immediate neighborhood. Taking the Fréchet derivative of (10.31) gives us

$$\nabla J_k^{\text{loc}}(\cdot) = -2\mathbb{E}\left\{\left(d_k(n) - \psi_k(\mathbf{u}_{k,n})\right)\kappa(\cdot,\mathbf{u}_{k,n})\right\} + 2\eta \sum_{l \neq k, l \in \mathcal{N}_k} \rho_{kl}\left(\psi_{k,n-1} - \psi_{l,n-1}\right), \tag{10.32}$$

where $\kappa(\cdot,\cdot)$ is the reproducing kernel associated with $\mathcal{H}$. For simplicity, we assume that the mixing weights $\rho_{kl}$ are symmetrical (see the discussion at the end of Section 10.2.3). Making an instantaneous approximation for the expectation gives us the following local update rule in functional form at time instant $n$:

$$\psi_{k,n} = \psi_{k,n-1} + \mu_k \left[ d_k(n) - \psi_{k,n-1}(\mathbf{u}_{k,n}) \right] \kappa(\cdot, \mathbf{u}_{k,n}) \\ - \mu_k \eta \sum_{l \neq k, l \in \mathcal{N}_k} \rho_{kl} \left( \psi_{k,n-1} - \psi_{l,n-1} \right), \tag{10.33}$$

where the factor 2 has been included in the step size $\mu_k$. Considering $\mathcal{H}$ as the space of linear predictors over $\mathbf{u}_{k,n}$, (10.33) reduces to the diffusion LMS for multitask networks presented earlier. In order to have a feasible implementation, once again we assume a shared dictionary $\mathcal{D}$ among agents. Eq. (10.33) reduces to

$$\boldsymbol{\beta}_{k,n} = \boldsymbol{\beta}_{k,n-1} + \mu_k \left[ d_k(n) - \boldsymbol{\beta}_{k,n-1}^T \mathbf{k}_{k,n} \right] \mathbf{k}_{k,n} \\ - \mu_k \eta \sum_{l \neq k, l \in \mathcal{N}_k} \rho_{kl} \left( \boldsymbol{\beta}_{k,n-1} - \boldsymbol{\beta}_{l,n-1} \right). \tag{10.34}$$

## 10.5 EXPERIMENTAL EVALUATION

### 10.5.1 EXPERIMENT SETUP

In this section, we evaluate the proposed method on a simulated multitask nonlinear problem. The output at each agent is given by the following equation:

$$d_k(n) = f(\mathbf{u}_{k,n}) + (\mathbf{w}_0 + \mathbf{w}_k)^T \mathbf{u}_{k,n} + \nu_k(n), \tag{10.35}$$

which is composed of a common nonlinear part $f(\cdot)$, a common linear term with weight vector $\mathbf{w}_0$, a local linear part $\mathbf{w}_k^T \mathbf{u}_{k,n}$ and a local noise of variance $\sigma_k^2$. In particular, we considered a three dimensional input vector $\mathbf{u} = [u_1, u_2, u_3]^T$, with the following nonlinearity:

$$f(\mathbf{u}) = a u_1^2 + b u_2 u_3,$$

where $a$ and $b$ were generated from a normal distribution, similarly to the local and shared coefficient vectors $\mathbf{w}_k$. Noise variances were generated uniformly for each agent in the interval $[0, 0.3]$. We added an additional level of diversity over the network by randomly assigning the learning rates to the agents from the uniform distribution over the interval $[0, 0.1]$. We considered a network of 9 agents, whose connectivity was randomly assigned such that each agent is connected in average with one-fifth of the other agents, with the requirement that the overall graph is connected. The resulting network connectivity, an example of desired output and a plot of the noise variances and learning rates are all shown in Fig. 10.5.

We trained the network over a sequence of 1000 time instants, with white Gaussian inputs with zero mean and unitary variance. The mixing matrix was chosen according to the max-degree heuristic:

$$A_{lk} = \begin{cases} \frac{1}{\text{deg}_{\max}+1}, & \text{if } l \text{ is connected to } k, \\ 1 - \frac{\text{deg}_k}{\text{deg}_{\max}+1}, & \text{if } k = l, \\ 0, & \text{otherwise}, \end{cases} \tag{10.36}$$

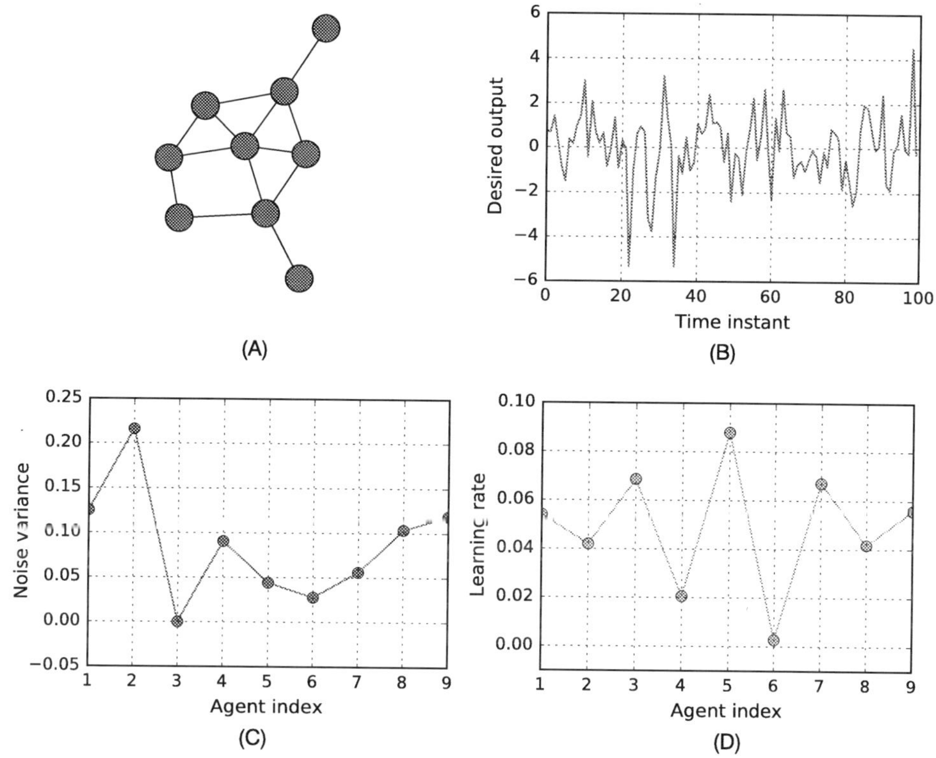


**FIGURE 10.5**

General setup for the experimental section. (A) The network of 9 agents used in all experiments. (B) The first 100 samples of the desired output for the first agent. (C) Noise variance for each agent. (D) Learning rate for each agent. See the text for a description of how (B)–(D) were generated.

where $\deg_k$ is the degree of node $k$ and $\deg_{\max}$ is the maximum degree of the network.[9] Each experiment was averaged over 500 different runs, by keeping fixed the assignments shown in Fig. 10.5.

### 10.5.2 RESULTS AND DISCUSSION

We compared the performance of a standard diffusion LMS (D-LMS), a multitask D-LMS as described in Section 10.2.3 (D-MT-LMS), the diffusion KLMS described in Section 10.3.2 and the proposed multitask D-KLMS introduced in Section 10.4 (D-MT-KLMS). For the kernel algorithms, we used a Gaussian kernel,

$$\mathcal{K}(\mathbf{u}_1, \mathbf{u}_2) = \exp\left\{-\gamma \,\|\mathbf{u}_1 - \mathbf{u}_2\|_2^2\right\},$$

[9] The degree of a node is the cardinality of the set of its direct neighbors.

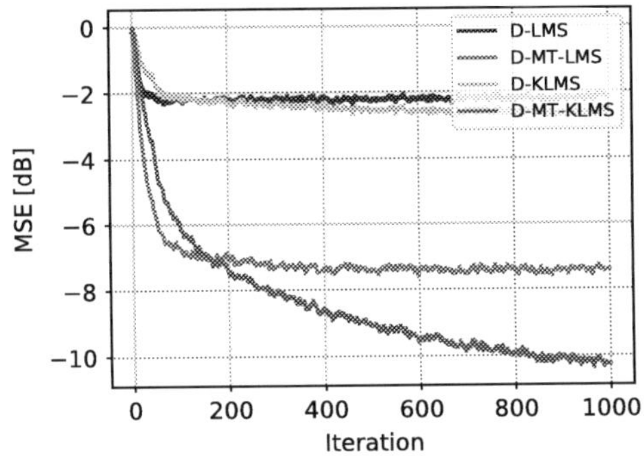


**FIGURE 10.6**

Average MSE for the algorithms under consideration, averaged both over agents and over 500 independent runs.

where $\gamma$ was chosen as the inverse of the dimensionality of $\mathbf{u}$, which was found to provide a good accuracy. For the multitask algorithms, the regularization coefficients were selected uniformly as

$$\rho_{kl} = \begin{cases} \frac{1}{\deg_k}, & \text{if } k \text{ is connected to } l, \\ 0, & \text{otherwise,} \end{cases} \tag{10.37}$$

while the regularization factor was set to $\eta = 0.01$. For the kernel algorithms, we fixed *a priori* a dataset of size 100 with randomly extracted elements. The average MSE in dB across all runs is shown in Fig. 10.6.

As expected, the D-LMS was the poorest performing algorithm, due to the doubly incorrect assumptions that the agents share the same minimizer and that the underlying function is linear. By relaxing one of the two assumptions, D-MT-LMS performed better, with an accuracy that is slightly inferior to D-KLMS. Clearly, D-MT-KLMS was the best algorithm in this case, showing that it can be an effective solution for nonlinear multitask problems.

In Fig. 10.7 we show the MSE evolution for three representative agents. As expected, their performance is different depending on the selected learning rate and amount of noise, but the multitask algorithm is able to effectively combine the learning curves to obtain the average behavior as in the purple (darkest gray in print version) line of Fig. 10.6. Finally, in Fig. 10.8 we show the average MSE evolution when varying the size of the dictionary. Increasing the size improves the accuracy (up to a given upper bound), at the cost of a larger computational burden.

## 10.6 DISCUSSION AND OPEN PROBLEMS

Distributed inference is a fundamental tool from the perspective of today's technological trends. In the adaptive filtering community, many classical algorithms can be readily extended to the distributed scenario by exploiting diffusion principles, where local adaptation steps are interleaved with communication steps between neighbors. The resulting algorithms are both computationally efficient and

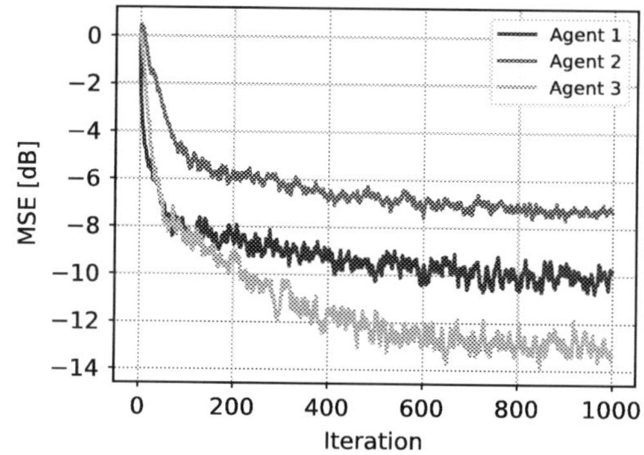


**FIGURE 10.7**

Local MSE averaged over 100 runs for three representative agents over the network.

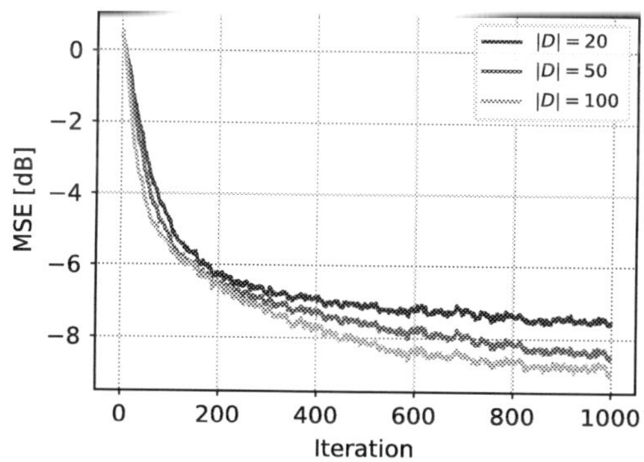


**FIGURE 10.8**

Average MSE for three different dictionary sizes.

deployable over a large set of scenarios. In this chapter, we reviewed the basic tools of this field, and we briefly surveyed some of the nonlinear extensions that have been proposed.

An important distinction can be made between single-task problems, where all agents share the same minimizer, and multitask problems, where the minimizers can be different, but it is known that they share some similarities. We underlined how little work has been done on the nonlinear multitask case, and we proposed a simple kernel-based diffusion algorithm to this end. Many extensions to the basic setup of this chapter are possible, most notably a way to remove the assumption of a shared dictionary, an adaptive way to build the regularization coefficients, a theoretical analysis of the algorithm, or additional extensions towards asynchronous networks. Finally, we can consider mixing multitask networks with multiobjective algorithms [53], such that each agent is interested in minimizing multiple objectives simultaneously.

## ACKNOWLEDGMENTS

The work of Simone Scardapane was supported in part by Italian MIUR, "*Progetti di Ricerca di Rilevante Interesse Nazionale*", GAUChO project, under Grant 2015YPXH4W_004. The work of Jie Chen was supported in part by the National Natural Science Foundation of China (NSFC grant 61671382).

## REFERENCES

[1] V. Cevher, S. Becker, M. Schmidt, Convex optimization for big data: scalable, randomized, and parallel algorithms for big data analytics, IEEE Signal Processing Magazine 31 (5) (2014) 32–43.

[2] A. Sandryhaila, J.M.F. Moura, Big data analysis with signal processing on graphs: representation and processing of massive data sets with irregular structure, IEEE Signal Processing Magazine 31 (5) (2014) 80–90.

[3] P. Di Lorenzo, S. Barbarossa, P. Banelli, S. Sardellitti, Adaptive least mean squares estimation of graph signals, IEEE Transactions on Signal and Information Processing over Networks 2 (4) (2016) 555–568.

[4] J.B. Predd, S.B. Kulkarni, H.V. Poor, Distributed learning in wireless sensor networks, IEEE Signal Processing Magazine 23 (4) (2006) 56–69.

[5] J. Tsitsiklis, D. Bertsekas, M. Athans, Distributed asynchronous deterministic and stochastic gradient optimization algorithms, IEEE Transactions on Automatic Control 31 (9) (1986) 803–812.

[6] A. Lazarevic, Z. Obradovic, The distributed boosting algorithm, in: Seventh ACM SIGKDD International Conference on Knowledge Discovery and Data Mining, ACM, 2001, pp. 311–316.

[7] P.A. Forero, A. Cano, G.B. Giannakis, Consensus-based distributed support vector machines, Journal of Machine Learning Research 11 (May) (2010) 1663–1707.

[8] S. Scardapane, R. Fierimonte, P. Di Lorenzo, M. Panella, A. Uncini, Distributed semi-supervised support vector machines, Neural Networks 80 (2016) 43–52.

[9] G. Mateos, J.A. Bazerque, G.B. Giannakis, Distributed sparse linear regression, IEEE Transactions on Signal Processing 58 (10) (2010) 5262–5276.

[10] P. Di Lorenzo, A.H. Sayed, Sparse distributed learning based on diffusion adaptation, IEEE Transactions on Signal Processing 61 (6) (2013) 1419–1433.

[11] C.G. Lopes, A.H. Sayed, Diffusion least-mean squares over adaptive networks: formulation and performance analysis, IEEE Transactions on Signal Processing 56 (7) (2008) 3122–3136.

[12] F.S. Cattivelli, C.G. Lopes, A.H. Sayed, Diffusion recursive least-squares for distributed estimation over adaptive networks, IEEE Transactions on Signal Processing 56 (5) (2008) 1865–1877.

[13] J. Chen, A.H. Sayed, Diffusion adaptation strategies for distributed optimization and learning over networks, IEEE Transactions on Signal Processing 60 (8) (2012) 4289–4305.

[14] A.H. Sayed, Adaptive networks, Proceedings of the IEEE 102 (4) (2014) 460–497.

[15] A.H. Sayed, Adaptation, learning, and optimization over networks, Foundations and Trends® in Machine Learning 7 (4–5) (2014) 311–801.

[16] J. Chen, S.K. Ting, C. Richard, A.H. Sayed, Group diffusion LMS, in: 2016 IEEE International Conference on Acoustics, Speech and Signal Processing, ICASSP, IEEE, 2016, pp. 4925–4929.

[17] V. Matta, C. Richard, V. Saligrama, A.H. Sayed, Guest editorial inference and learning over networks, IEEE Transactions on Signal and Information Processing over Networks 2 (4) (2016) 423–425.

[18] J. Chen, C. Richard, A.H. Sayed, Multitask diffusion adaptation over networks, IEEE Transactions on Signal Processing 62 (16) (2014) 4129–4144.

[19] J. Chen, C. Richard, A.H. Sayed, Diffusion LMS over multitask networks, IEEE Transactions on Signal Processing 63 (11) (2015) 2733–2748.

[20] X. Zhao, A.H. Sayed, Distributed clustering and learning over networks, IEEE Transactions on Signal Processing 63 (13) (2015) 3285–3300.

[21] R. Nassif, C. Richard, A. Ferrari, A.H. Sayed, Multitask diffusion adaptation over asynchronous networks, IEEE Transactions on Signal Processing 64 (11) (2016) 2835–2850.

[22] R. Nassif, C. Richard, A. Ferrari, A.H. Sayed, Proximal multitask learning over networks with sparsity-inducing coregularization, IEEE Transactions on Signal Processing 64 (23) (2016) 6329–6344.

[23] C. Li, S. Huang, Y. Liu, Y. Liu, Distributed TLS over multitask networks with adaptive intertask cooperation, IEEE Transactions on Aerospace and Electronic Systems 52 (6) (2016) 3036–3052.
[24] J. Chen, C. Richard, A.O. Hero, A.H. Sayed, Diffusion LMS for multitask problems with overlapping hypothesis subspaces, in: 2014 IEEE International Workshop on Machine Learning for Signal Processing, MLSP, IEEE, 2014, pp. 1–6.
[25] J. Chen, C. Richard, A. Sayed, Multitask diffusion adaptation over networks with common latent representations, IEEE Journal of Selected Topics in Signal Processing 11 (3) (2017) 563–579.
[26] T. Evgeniou, M. Pontil, Regularized multi–task learning, in: Tenth ACM SIGKDD International Conference on Knowledge Discovery and Data Mining, ACM, 2004, pp. 109–117.
[27] A. Argyriou, T. Evgeniou, M. Pontil, Multi-task feature learning, Advances in Neural Information Processing Systems 19 (2007) 41.
[28] A.A. Rusu, N.C. Rabinowitz, G. Desjardins, H. Soyer, J. Kirkpatrick, K. Kavukcuoglu, R. Pascanu, R. Hadsell, Progressive neural networks, arXiv preprint, arXiv:1606.04671, 2016.
[29] R. Jin, A. Goswami, G. Agrawal, Fast and exact out-of-core and distributed k-means clustering, Knowledge and Information Systems 10 (1) (2006) 17–40.
[30] X. Zhao, A.H. Sayed, Asynchronous adaptation and learning over networks—part i: modeling and stability analysis, IEEE Transactions on Signal Processing 63 (4) (2015) 811–826.
[31] K. Yuan, B. Ying, X. Zhao, A.H. Sayed, Exact diffusion for distributed optimization and learning—part i: algorithm development, arXiv preprint, arXiv:1702.05122, 2017.
[32] S. Scardapane, D. Wang, Randomness in neural networks: an overview, Wiley Interdisciplinary Reviews: Data Mining and Knowledge Discovery 7 (2) (2017).
[33] B. Igelnik, Y.-H. Pao, Stochastic choice of basis functions in adaptive function approximation and the functional-link net, IEEE Transactions on Neural Networks 6 (6) (1995) 1320–1329.
[34] S. Scardapane, D. Wang, M. Panella, A. Uncini, Distributed learning for random vector functional-link networks, Information Sciences 301 (2015) 271–284.
[35] S. Huang, C. Li, Distributed extreme learning machine for nonlinear learning over network, Entropy 17 (2) (2015) 818–840.
[36] A. Rahimi, B. Recht, Random features for large-scale kernel machines, in: Advances in Neural Information Processing Systems, vol. 3, 2007, pp. 1–5.
[37] A. Singh, N. Ahuja, P. Moulin, Online learning with kernels: overcoming the growing sum problem, in: 2012 IEEE International Workshop on Machine Learning for Signal Processing, MLSP, IEEE, 2012, pp. 1–6.
[38] P. Bouboulis, S. Pougkakiotis, S. Theodoridis, Efficient KLMS and KRLS algorithms: a random Fourier feature perspective, in: 2016 IEEE Statistical Signal Processing Workshop, SSP, IEEE, 2016, pp. 1–5.
[39] A. Navia-Vazquez, D. Gutierrez-Gonzalez, E. Parrado-Hernández, J.J. Navarro-Abellan, Distributed support vector machines, IEEE Transactions on Neural Networks 17 (4) (2006) 1091–1097.
[40] P. Honeine, M. Essoloh, C. Richard, H. Snoussi, Distributed regression in sensor networks with a reduced-order kernel model, in: 2008 IEEE Global Telecommunications Conference, GLOBECOM, IEEE, 2008, pp. 1–5.
[41] C. Richard, J.-C.M. Bermudez, P. Honeine, Online prediction of time series data with kernels, IEEE Transactions on Signal Processing 57 (3) (2009) 1058–1067.
[42] W.D. Parreira, J.-C.M. Bermudez, C. Richard, J.-Y. Tourneret, Stochastic behavior analysis of the Gaussian kernel least-mean-square algorithm, IEEE Transactions on Signal Processing 60 (5) (2012) 2208–2222.
[43] P. Honeine, C. Richard, J.-C.M. Bermudez, On-line nonlinear sparse approximation of functions, in: 2007 IEEE International Symposium on Information Theory, ISIT, IEEE, 2007, pp. 956–960.
[44] W. Gao, J. Chen, C. Richard, J. Huang, Diffusion adaptation over networks with kernel least-mean-square, in: 2015 IEEE 6th International Workshop on Computational Advances in Multi-Sensor Adaptive Processing, CAMSAP, IEEE, 2015, pp. 217–220.
[45] P. Bouboulis, S. Theodoridis, Extension of Wirtinger's calculus to reproducing kernel Hilbert spaces and the complex kernel LMS, IEEE Transactions on Signal Processing 59 (3) (2011) 964–978.
[46] A.V. Balakrishnan, Applied Functional Analysis, Springer Science & Business Media, 2012.
[47] J. Chen, W. Gao, C. Richard, J.-C.M. Bermudez, Convergence analysis of kernel LMS algorithm with pre-tuned dictionary, in: 2014 IEEE International Conference on Acoustics, Speech and Signal Processing, ICASSP, IEEE, 2014, pp. 7243–7247.
[48] B.-S. Shin, H. Paul, A. Dekorsy, Distributed kernel least squares for nonlinear regression applied to sensor networks, in: 2016 24th European Signal Processing Conference, EUSIPCO, IEEE, 2016, pp. 1588–1592.
[49] S. Chouvardas, M. Draief, A diffusion kernel LMS algorithm for nonlinear adaptive networks, in: 2016 IEEE International Conference on Acoustics, Speech and Signal Processing, ICASSP, IEEE, 2016, pp. 4164–4168.

[50] S. Scardapane, M. Scarpiniti, D. Comminiello, A. Uncini, Diffusion spline adaptive filtering, in: 2016 24th European Signal Processing Conference, EUSIPCO, IEEE, 2016, pp. 1498–1502.

[51] M. Scarpiniti, D. Comminiello, R. Parisi, A. Uncini, Nonlinear spline adaptive filtering, Signal Processing 93 (4) (2013) 772–783.

[52] M. Scarpiniti, D. Comminiello, G. Scarano, R. Parisi, A. Uncini, Steady-state performance of spline adaptive filters, IEEE Transactions on Signal Processing 64 (4) (2016) 816–828.

[53] J. Chen, A.H. Sayed, Distributed Pareto optimization via diffusion strategies, IEEE Journal of Selected Topics in Signal Processing 7 (2) (2013) 205–220.

CHAPTER

# COMBINED FILTERING ARCHITECTURES FOR COMPLEX NONLINEAR SYSTEMS

# 11

**Luis A. Azpicueta-Ruiz**[*,#], **Jerónimo Arenas-García**[*,##], **Magno T.M. Silva**[†,###], **Renato Candido**[†]

[*]*Universidad Carlos III de Madrid, Madrid, Spain* [†]*Universidade de São Paulo, São Paulo, SP, Brasil*

## CONTENTS



## CHAPTER POINTS

- Cooperative strategies can be exploited in nonlinear adaptive filtering to improve the performance of most available schemes and to facilitate the selection of their parameters.

[#]The work of Azpicueta-Ruiz has been partly supported by MINECO Project TEC2014-52289-R and TEC2017-83838-R.
[##]The work of Arenas-García has been partly supported by MINECO Projects TEC2014-52289-R, TEC2017-83838-R and TEC2016-81900-REDT/AEI and by Comunidad de Madrid Project PRICAM S2013/ICE-2933.
[###]The work of Silva has been partly supported by CNPq under Grant 304715/2017-4 and FAPESP under Grant 2017/20378-9.

**Adaptive Learning Methods for Nonlinear System Modeling. DOI: 10.1016/B978-0-12-812976-0.00014-2**

- Another issue with nonlinear filters is structure selection. To this end, combination approaches can also be used to select the useful components and virtually remove the gradient noise that would be introduced by unnecessary units.

## 11.1 INTRODUCTION

Recently, combined schemes are increasingly receiving attention as a method to improve the performance of adaptive filters, mainly due to their versatility and robustness against the lack of knowledge about the filtering scenario. One of the most popular of such cooperative strategies is based on the adaptive combination of adaptive filters [1]. This approach can be considered as both competitive and cooperative, in the sense that individual filters compete to provide the best model for the unknown system, but the combination strategy allows the cooperative use of all solutions and can even provide a global solution of improved accuracy with respect to the models of individual filter components. Although these schemes usually entail an increment in terms of computational cost with respect to other adaptive solutions, they have been successfully employed in a wide range of signal processing applications, including acoustic echo cancellation [2], adaptive array beamforming [3], active noise control [4], line enhancement [5] and adaptive blind equalization [6].

The combination of adaptive filters approach was initially proposed to combine linear adaptive filters [7,8]. However, because of the versatility, robustness and efficacy of the idea, it has also been employed in several applications of nonlinear signal processing, such as nonlinear acoustic echo cancellation [9,10], signal modality characterization [11], adaptive nonlinear equalization for communications [12], nonlinear adaptive noise cancellation [13,14] and nonlinear audio equalization [15]. Focusing on the types of nonlinear filters used within combination schemes, we can find Volterra Filters (VFs) [16], Functional Link Adaptive Filters (FLAFs) [17], Power Filters [18] and Kernel Adaptive Filters (KAFs) [19], among others.

In this chapter, we will review some of these works, focusing on how combination schemes can be exploited to improve the performance of nonlinear adaptive filters. In particular, we will study two different applications in this context:

- Introduction of diversity among the adaptive filter parameters in order to simplify their selection and to improve the overall performance of the filter.
- Adjustment of the nonlinear filter structure, deactivating unnecessary units that do not contribute to reduce the filter error, but whose adaptation introduces additional gradient noise.

To illustrate the general principles underlying these applications of combined schemes, we will consider FLAFs, VFs and KAFs as the combination components. In all these cases, an optimal adjustment of the filter parameters and structure would require some *a priori* knowledge about the filtering scenario that is rarely available, and the proposed schemes increase the robustness of the nonlinear adaptive filters to this lack of knowledge at a moderate cost.

In passing, we note that combined schemes can also be used for other objectives. For instance, [13] proposes to use a combination of VFs and FLAFs to keep the best performing characteristics of both families of filters. Since in this chapter we attempt to present some general principles which are not dependent on particular filter technologies, these combinations with heterogeneous filter components will not be analyzed any further in spite of their interest.

The rest of the chapter is organized as follows. Section 11.2 summarizes the main features of the nonlinear filters employed in this chapter. Section 11.3 reviews different possibilities for implementing the combination of adaptive filters, explaining how the mixing parameters can be adapted. Sections 11.4 and 11.5 are dedicated to each of the two application scenarios we have already mentioned, respectively, to simplify the selection of the setting parameters and the structure of the nonlinear filter. Although the proposed schemes are not tied to any particular application, we will illustrate their advantages in acoustic echo cancellation, system identification and signal prediction. The chapter finishes drawing our conclusions and suggesting future lines of research.

## 11.2 NONLINEAR ADAPTIVE FILTERS

In this section, we will explain the working mechanisms of mainstream implementations of VFs, FLAFs and KAFs, discussing some of their most interesting properties, as well as limitations that can motivate the use of combination strategies.

### 11.2.1 ADAPTIVE VOLTERRA FILTERS

VFs are able to model a large class of nonlinearities with memory [16]. The output $y(n)$ of a VF can be expressed as the superposition of partial outputs $y_p(n),\ p = 1, 2, \cdots, P$, coming from kernels of different orders that represent different degrees of nonlinearity, *i.e.*,

$$y(n) = \sum_{p=1}^{P} y_p(n) = \sum_{p=1}^{P} \sum_{i_1=0}^{N_p-1} \cdots \sum_{i_p=i_{p-1}}^{N_p-1} h_{p,i_1,\ldots,i_p}(n) \prod_{q=1}^{p} x(n-i_q), \tag{11.1}$$

where $P$ is the number of kernels, $N_p$ is the length of each kernel and $h_{p,i_1,\ldots,i_p}(n)$ stands for the adaptive coefficients of the $p$th kernel. The adaptation of the filter coefficients $h_{p,i_1,\ldots,i_p}(n)$ is performed with the goal of minimizing the second power of the error signal, $e(n) = d(n) - y(n)$, with $d(n)$ being the desired filter output.

In view of Eq. (11.1), an important advantage of VFs is their simplicity for modeling nonlinearities with memory, assigning just an adaptive coefficient to a special input sample implemented as a cross-product of input samples with different time indexes. However, related to this interesting feature, the main drawback of VFs lies in the high computational cost associated with the operation of high-order kernels.

A second disadvantage of VFs, which is also intrinsically related to their ability to model nonlinearities of different orders with memory, is that the adaptation of each coefficient is typically subject to a certain amount of gradient noise. This is compensated when the reduction of the mean square error (MSE) which can be achieved is larger than the introduced noise. In any case, it would be very useful to identify which units of the VF contribute to a reduction of the MSE and remove the gradient noise associated to unnecessary units. In general, this is not an easy task, since the particular kind of nonlinearities that should be implemented are not known *a priori*, and they can even change during the filter operation. As we will explain in Section 11.5, combination approaches can be used precisely

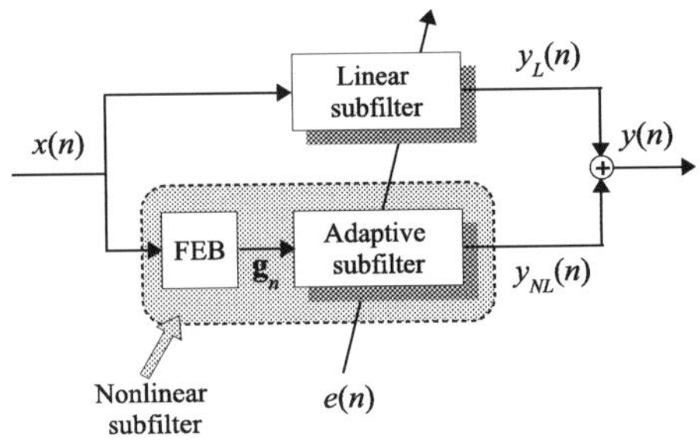


**FIGURE 11.1**

Structure of a Split Functional Link Adaptive Filter. The filter consists of a linear and a nonlinear branch, where the nonlinear subfilter consists of a functional expansion block followed by standard "linear-in-the-parameters" adaptive filter.

with the goal of activating and deactivating units in an online manner, so that the most convenient filter structure is identified at each time.

## 11.2.2 SPLIT FUNCTIONAL LINK ADAPTIVE FILTERS

Although different versions of FLAFs can be employed to design combination schemes, in this chapter we focus on the Split FLAF (SFLAF), proposed in [10] and depicted in Fig. 11.1. The SFLAF is a parallel filter with a linear and a nonlinear branch. Both subfilters adapt in order to minimize the power of a common error, *i.e.*, $e(n) = d(n) - y(n)$, where $y(n)$ is the output of the SFLAF, which can be calculated as the addition of the linear and nonlinear outputs, $y_L(n) + y_{FL}(n)$. However, the split architecture allows each subfilter to utilize different settings, for instance different lengths or step size parameters [10]. In this way, the linear subfilter is only focused on the modeling of the linear part and the nonlinear subfilter tries to model exclusively the nonlinearities.

As Fig. 11.1 illustrates, the linear branch is composed of a linear filter, which we will characterize by a length-$N$ vector $\mathbf{w}_{L,n}$. The nonlinear branch is based on two elements: a nonlinear functional expansion block (FEB) by means of a series of functions (functional links) that satisfy universal approximation constraints and an adaptive filter $\mathbf{w}_{\mathrm{FL},n}$ of length $N_e$ that receives the expanded input $\mathbf{g}_n$ provided by the FEB.

One of the key characteristics of the FLAF is how the FEB maps $x(n)$ into $\mathbf{g}_n$. Different choices can be considered to generate the expansion. In this chapter, we restrict ourselves to the memoryless nonlinear trigonometric expansion that expands each input signal sample $x(n-i)$ as follows:

$$\varphi_j\left[x(n-i)\right] = \begin{cases} \sin\left[p\pi x(n-i)\right], & j = 2p-2, \\ \cos\left[p\pi x(n-i)\right], & j = 2p-1, \end{cases} \tag{11.2}$$

where $p = 1, \ldots, P$ is the expansion index, with $P$ being the expansion order, and $j = 0, \ldots, 2P-1$ is the functional link index. Therefore, considering a nonlinear buffer of $N_i$ input samples, both the length of $\mathbf{g}_n$ and of $\mathbf{w}_{\mathrm{FL},n}$ would be $N_e = 2PN_i$.

In spite of being a powerful strategy for modeling nonlinearities of any kind, the performance of an SFLAF filter is subject to the selection of certain parameters:

- Parameters associated with the adaptation of the filter coefficients. Both VF and SFLAF involve subfilters that can be adapted using strategies similar to those of linear adaptive filters [20,21].
- Parameters related to the nonlinear filter structure. In particular, for the case of FLAFs, the expansion order $P$ and the length of both adaptive filters, $N$ and $N_e$, play an important role in the performance of the adaptive filter (AF).

As for the VF case, the selection of both the structure of an SFLAF and the parameters for its adaptation would require *a priori* knowledge of the filtering scenario. Alternatively, this problem can be simplified recurring to combination schemes.

### 11.2.3 KERNEL ADAPTIVE FILTERS

Kernel adaptive filters are able to solve nonlinear problems by implicitly projecting the input samples to a high dimensional space, where a standard linear adaptive filter is applied (see, *e.g.*, [19,22,23] and Chapters 5 and 6 of this book). KAFs have been mainly applied in system identification [24], echo cancellation [25], time series prediction [19,26–28] and channel equalization [19,29].

The kernel version of the least mean square (LMS) algorithm, denoted as KLMS [29], is the most popular among the different kernel adaptive filters in the literature. KLMS maps the input column vector $\mathbf{x}(n) \in \mathbb{R}^N$ into a high-dimensional feature space $\mathcal{F}$ as $\varphi(\mathbf{x}(n))$, using a Mercer kernel. Mercer's kernel is a continuous, symmetric and positive definite function $\kappa : \mathbb{R}^N \times \mathbb{R}^N \to \mathbb{R}$, such that

$$\kappa(\mathbf{x}, \mathbf{x}') = \varphi(\mathbf{x})^T \varphi(\mathbf{x}'). \tag{11.3}$$

In the space $\mathcal{F}$, the LMS algorithm is used to update the filter weight column vector $\mathbf{\Omega}(n-1)$ in order to estimate the desired signal $d(n)$. The Representer's Theorem states that $\mathbf{\Omega}(n-1)$ can be expressed as a linear combination of past input samples, so that the filter output can be obtained as

$$y(n) = \varphi(\mathbf{x}(n))^T \mathbf{\Omega}(n-1) = \mu \sum_{i=1}^{n-1} e(i)\kappa(\mathbf{x}(n), \mathbf{x}(i)), \tag{11.4}$$

where $\mu$ is a step size.

KAFs fall in the area of interest of kernel methods and offer a powerful alternative to VFs and SFLAFs. However, from a practical point of view their main drawbacks are the selection of an appropriate kernel and the high computational burden and memory, since the dictionary size (expansion in (11.4)) grows linearly with the incoming samples. Throughout this chapter, we focus on the Gaussian kernel, which is the most commonly used kernel in the literature and is defined as

$$\kappa(\mathbf{x}, \mathbf{x}') = \exp\left(-\frac{\|\mathbf{x} - \mathbf{x}'\|^2}{2\sigma^2}\right), \tag{11.5}$$

where $\sigma > 0$ is the kernel width [19]. Selection of an appropriate width remains an important issue to achieve good performance and can indeed benefit from the application of combination strategies. With respect to the problem of unbounded complexity, we review next the quantized KLMS algorithm, one of the possible solutions described in the literature to avoid the linear growth of the dictionary.

#### 11.2.3.1 The Quantized KLMS Algorithm

One of the main challenges of KAFs is to reduce the dictionary size in order to minimize the computational burden and the memory. The dictionary, also known as codebook, is defined as the set of support vectors $\mathcal{C}(n) = \{\mathbf{x}(c_j)\}_{j=1}^{N_c(n)}$, where $\mathbf{x}(c_j)$ is the $j$th element and $N_c(n)$ is its cardinality, which can vary from a time instant to another. The index $c_j \in \{1, 2, \cdots, n-1\}$ is used to distinguish the dictionary elements $\mathbf{x}(c_1), \cdots, \mathbf{x}(c_{N_c(n)})$ from the input vector $\mathbf{x}(n)$. With no dictionary restriction, we can observe from (11.4) that to estimate $d(n)$, we should use all the input vectors since the algorithm initialization until the previous time instant, *i.e.*, $\mathcal{C}(n) = \{\mathbf{x}(i)\}_{i=1}^{n-1}$ and, in this case, the dictionary size $N_c(n) = n - 1$, grows linearly with the incoming samples. To avoid this linear growth, different sparsification techniques have been proposed in the literature to update the dictionary by including in it only informative data (see, *e.g.*, [30–34]).

The quantized KLMS (QKLMS) algorithm was proposed in [23] as an alternative to sparsification techniques. It is obtained by quantizing the feature vector $\varphi(\mathbf{x}(n))$, which leads to an algorithm similar to the sparsified KLMS with novelty criterion [23]. The main difference is that QKLMS uses the "redundant" data to locally update the coefficient of the closest vector. This coefficient update can enhance the efficient use of that vector. Therefore, better accuracy and a more compact dictionary can be achieved [23]. Due to its inherent advantages, we consider QKLMS in all schemes and simulations of this chapter.

## 11.3 DIFFERENT APPROACHES TO COMBINE NONLINEAR ADAPTIVE FILTERS

Once we have reviewed some of the most important types of nonlinear adaptive filters, in this section we will present different structures that can be used to combine them. We will focus just on general structures for combining nonlinear adaptive filters, postponing to Section 11.4 the presentation of particular combinations of interest. Among the different combination strategies that have been proposed in the literature, in this chapter we will focus on the following two:

- A first possibility would be to implement a basic combination of filters using the algorithm in [7]. This general approach can be used to combine the outputs of any two filters, with the same or a different structure.
- The other approach is specially suited for nonlinear filters composed by different subfilters or branches that operate in parallel. In this case, instead of combining the final outputs of the nonlinear filters, we can incorporate the combination into the filter structure, replacing a subfilter in a branch by a combination of subfilters of the same nature. This combination strategy was firstly proposed for VFs [9], and it has also been successfully employed with FLAFs [10,35].

Fig. 11.2A depicts a combination of two complete nonlinear filters where each filter has two subfilters or branches. In this case, the combined output is calculated as

$$y_c(n) = \eta(n)y_1(n) + [1 - \eta(n)]y_2(n),$$

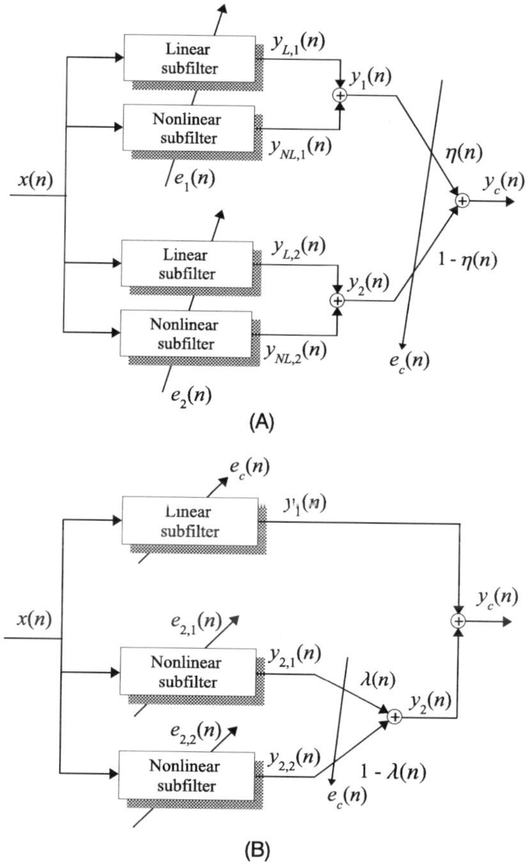


**FIGURE 11.2**

Two different combination strategies for nonlinear adaptive filters. This figure illustrates the two combination structures that will be used in the paper. (A) Combination of the outputs of two filters. (B) Combination of two subfilters at the nonlinear branch.

where $\eta(n)$ represents the mixing parameter employed to adaptively combine the outputs of each nonlinear filter, $y_1(n)$ and $y_2(n)$. Each adaptive nonlinear filter adapts independently of each other according to its own error, $e_i(n) = d(n) - y_i(n)$ with $i = 1, 2$, while the combination parameter $\eta(n)$ is adapted to minimize the square of the combined error $e_c(n) = d(n) - y_c(n)$. If the combination parameter is properly adapted, the combined filter behaves at least as the best of the filter components and, under certain circumstances, even better than both components.

Fig. 11.2B depicts a modified nonlinear filter with two branches, where the nonlinear subfilter has been replaced by a combination of two subfilters of the same nature but with different settings. The output of this scheme can be calculated as

$$y_c(n) = y_1(n) + y_2(n) = y_1(n) + \lambda(n)y_{2,1}(n) + [1 - \lambda(n)]y_{2,2}(n), \tag{11.6}$$

where $y_1(n)$, $y_{2,1}(n)$ and $y_{2,2}(n)$ are the outputs of each subfilter and $\lambda(n)$ is the mixing parameter that combines both subfilters of the same branch. An important particularity of this scheme is related to the error signal employed to update each subfilter [9]. The subfilter of the first branch adapts trying to minimize the power of the combined error, *i.e.*, $e_c(n) = d(n) - y_c(n)$. However, both subfilters of the second branch adapt in order to minimize an error signal that ignores the presence of the other subfilter of the same branch, *i.e.*, $e_{2,i}(n) = d(n) - [y_1(n) + y_{2,i}(n)]$. Finally, the mixing parameter $\lambda(n)$ is adapted with the goal of minimizing the square of the combined error.

Apart from the particular combination structure, adaptation of the mixing parameter can also follow different rules. A first issue is to decide whether constraints must be placed in the range for the mixing parameter. In this chapter, we consider just convex combinations in which the mixing parameters are constrained to lie in the interval [0, 1]. For a more comprehensive study about the impact of this constraint and a comparison between convex and affine combinations of adaptive filters, please refer to [1].

When a convex combination of filters is selected (first configuration in Fig. 11.2), an easy way to ensure that $\eta(n) \in [0, 1]$ is to define an auxiliary mixing parameter $a(n)$ univocally related to the mixing parameter by a sigmoidal activation function. We have

$$\eta(n) = \text{sgm}[a_\eta(n)] = \frac{1}{1 + e^{-a_\eta(n)}}. \tag{11.7}$$

Thus, instead of directly adjusting the mixing parameter, we adapt the auxiliary parameter $a_\eta(n)$. A popular scheme for the adaptation of $a_\eta(n)$ follows a power normalized stochastic gradient descent method to minimize $e_c^2(n)$,

$$a_\eta(n+1) = a_\eta(n) + \frac{\mu_a}{p(n)}\eta(n)[1 - \eta(n)]\big[y_1(n) - y_2(n)\big]e_c(n), \tag{11.8}$$

where parameter $p(n) = \beta p(n-1) + (1-\beta)[y_1(n) - y_2(n)]^2$ represents a low-pass filtered version of $[y_1(n) - y_2(n)]^2$, $0 << \beta < 1$ being a smoothing factor.

The presence of the factor $\eta(n)[1 - \eta(n)]$ in (11.8) explains another advantage of the sigmoidal activation: when $\eta(n)$ is close to its limiting values 0 or 1, the effective step size in (11.8) gets very small and the gradient noise in the adaptation of $a_\eta(n)$ is reduced. It should be noted that $\eta(n) \approx 1$ and $\eta(n) \approx 0$ represent two important cases since the combined filter in these cases is behaving as one of its components. Another practical issue to be considered in the adaptation of $a_\eta(n)$ comes from the observation that (11.8) could stop if $\eta(n) = 0$ or $\eta(n) = 1$. In order to avoid this, the values of $a_\eta(n)$ are normally restricted to the interval $[-4, 4]$.[1]

Eqs. (11.7) and (11.8) were initially proposed to adapt the mixing parameter of a combination of linear filters [7]. However, they can be generalized to the subfilter combination of Fig. 11.2B, defining $\lambda(n) = \text{sgm}[a_\lambda(n)]$ and following [9] to update the combination as follows:

$$a_\lambda(n+1) = a_\lambda(n) + \frac{\mu_a}{p(n)}\lambda(n)[1 - \lambda(n)][y_{2,1}(n) - y_{2,2}(n)]e_c(n), \tag{11.9}$$

with $p(n) = \beta p(n-1) + [1-\beta][y_{2,1}(n) - y_{2,2}(n)]^2$.

[1] As a consequence of this restriction, when Eq. (11.7) is used, the mixing parameter is not able to reach the limit values 0 and 1. To exactly consider the whole range [0, 1] for $\eta(n)$, a modified version of (11.7) has to be employed [1]. We have not included it for the sake of simplicity in the discourse.

## 11.4 COMBINED NONLINEAR FILTERS WITH DIVERSITY IN THE PARAMETERS

One of the most immediate uses of combination schemes is to facilitate the selection of the parameters intrinsic to any adaptive filter to enhance their operation and increase their robustness with respect to the lack of knowledge about the characteristics of the filtering scenario. In this section, we enumerate some examples of combination schemes using VFs, FLAFs and KAFs that follow this approach.

One of the first schemes to combine VFs was proposed to simplify the selection of the step sizes of second-order VFs adapted using the separate NLMS (SNLMS) algorithm [9]. It is a well-known fact that the selection of the step size of LMS-type schemes imposes a tradeoff between convergence speed and residual error after convergence. In [9], two strategies were followed to alleviate this compromise both in the linear and the quadratic kernels of a VF.

A first possibility consists of a hierarchical combination of four VFs that only differ in the step sizes employed to adapt both kernels, *i.e.*, $[\mu_{l,1}, \mu_{q,1}]$, $[\mu_{l,1}, \mu_{q,2}]$, $[\mu_{l,2}, \mu_{q,1}]$ and $[\mu_{l,2}, \mu_{q,2}]$, with $\mu_{l,1} >> \mu_{l,2}$ and $\mu_{q,1} >> \mu_{q,2}$, for each VF, respectively. The experimental evaluation of this scheme in [9] showed a better tradeoff between convergence speed and final error than every filter component and evidenced its robustness with respect to abrupt changes that occur just in the linear or nonlinear parts of the unknown system.

A second strategy in [9] was based on the combination of subfilters and required the combination of just two linear (with step sizes $\mu_{l,1}$ and $\mu_{l,2}$) and two quadratic kernels (with step sizes $\mu_{q,1}$ and $\mu_{q,2}$). This scheme permits to reach a similar operation as the hierarchical combination of filters, being the preferred scheme due to its reduced computational cost (approximately twice the complexity of a standard VF, but just half the complexity of the hierarchical structure).

Combined schemes have also been designed to improve the adaptation of FLAFs. For instance, a modified FLAF is proposed in [35], where three nonlinear branches are combined following the approach of Fig. 11.2B. All the subfilters of the combined filter (the linear and the three nonlinear) employ the Affine Projection Algorithm (APA) for their adaptation, and they differ in their projection orders.

Recently, combined schemes based on FLAFs that adapt by means of the improved proportionate NLMS (IPNLMS) algorithm have been proposed [36,37]. A modified FLAF with a convex combination of two nonlinear branches only differing in the proportionate factor was designed in [36] as a method to improve the nonlinear modeling capability when the nonlinear degree of the filtering scenario is *a priori* unknown or time varying. A deeper analysis about the sparse nature of the nonlinear functional expansion of FLAFs has permitted to design improved combined schemes based on block-wise combinations of the FEB units [37].

Combined schemes have also been employed to improve the adaptation of other nonlinear models, such as radial basis functions networks [38] and adaptive kernels [22]. An interesting approach was proposed in [22], where a kernel recursive least squares (KRLS) and a KLMS were combined. The combined scheme was tested in a nonlinear channel equalization scenario showing advantages with respect to both filter components.

In view of the works that have been listed in this section, we can conclude that combination strategies provide a powerful and versatile strategy to improve the performance of nonlinear adaptive filters. We have seen that combination strategies have been used with different schemes and to simplify the selection of several kinds of parameters (step size, projection order, asymmetry factor, etc.) Given the large amount of work in this direction, in this section we have preferred to briefly describe some of the most relevant works. The interested reader is referred to the corresponding papers for details on the particular implementations and experimental work.

## 11.5 COMBINATION SCHEMES TO SIMPLIFY THE SELECTION OF THE FILTER STRUCTURE

Apart from selecting the parameters of the adaptive filter, one needs to propose a structure which is suitable for the particular application at hand. Normally, a linear branch is implemented in parallel with the nonlinear subfilter, which generally admits sufficient structure flexibility independently of the selected technology:

- In VFs we need to select the memory of the filter and the number of diagonals. We could also consider to incorporate just a quadratic kernel or terms originating from higher-order nonlinearities.
- In FLAFs, we can select the type or the number of units of the FEB.
- In KAFs, we need to select the type of kernel, dictionary elements, etc.

The correct adjustment of these elements is critical for obtaining a good performance, since inability to correctly model the nonlinear part will result in increased MSE. One approach to ensure sufficient expressive capability would be to expand the nonlinear part and have multiple nonlinear AFs working in parallel. In this way, sufficient expressive power would be ensured. However, this solution does not work well, since every element is subject to adaptation gradient noise, and therefore unnecessary units will contaminate the filter output.

Note that the tradeoff we have just mentioned between expressive power and introduced gradient noise is relevant even from the point of view of whether a nonlinear subfilter should be included at all, since when the presence of nonlinearities in the desired filter output has low power (*i.e.*, the linear-to-nonlinear ratio, LNLR, in $d(n)$ is large) it may actually be better not to identify the nonlinear distortions if this implies a large amount of gradient noise.

In this section, we propose a general strategy that can be used to minimize the degrading effects of unnecessary units: to use multiplicative factors that can either activate or suppress different elements of the nonlinear filter depending on whether the filter performance improvement they achieve exceeds the amount of gradient noise they introduce or not.

This strategy is based on the work for biasing the outputs of AFs [39] and has already been successfully applied to VFs and FLAFs. In this section, we study also the behavior of this idea when working with multikernel adaptive filters.

### 11.5.1 ROBUST VFS FOR NONLINEAR ACOUSTIC ECHO CANCELLATION

In the last years, due to the widespread utilization of hands-free communication devices, acoustic echo cancelers (AECs) have been forced to incorporate nonlinear models to compensate the nonlinear distortions mainly caused by low-cost loudspeakers and amplifiers driven at high power levels. In this scenario, linear AECs operate inefficiently, which has made nonlinear acoustic echo cancellation

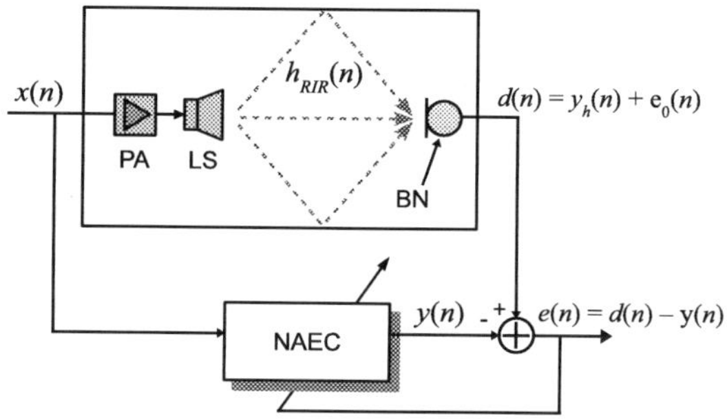


**FIGURE 11.3**

Block diagram of nonlinear echo cancellation. The nonlinear adaptive filter tries to generate a replica of the echo received at the microphone, so that the echo can be suppressed before sending the microphone signal to the far-end speaker.

become a demanding field of application of nonlinear models, including VFs [40,9], FLAFs [10] and KAFs [41,25].

Fig. 11.3 shows a block diagram of a nonlinear AEC (NAEC). The NAEC has access to both the input signal $x(n)$ that is going to be emitted by the loudspeaker and the desired signal that is transducted by the microphone $d(n) = y_h(n) + e_0(n)$, where $y_h(n)$ is related to the transduction and propagation of $x(n)$ and $e_0(n)$ depends on the background noise at the microphone location. It is important to remark the relation between $x(n)$ and $y_h(n)$. Signal $x(n)$ represents the electric signal before the power amplifier and $y_h(n)$ is the electric signal that would be transduced by the microphone if the background noise at the microphone location was non-existent, *i.e.*, $e_0(n) = 0$. Therefore, $x(n)$ has to be amplified by the power amplifier, transducted by the loudspeaker, acoustically propagated through the room with room impulse response $h_{RIR}(n)$ and transduced by the microphone in reception. In this process, the main sources of nonlinear distortions are the power amplifier and the loudspeaker operation.

The objective of the NAEC is to obtain a copy as faithfully as possible of $y_h(n)$, denoted as $y(n)$, using $x(n)$ and seeking to minimize the power of the cancellation error $e^2(n) = [d(n) - y(n)]^2$. If the cancellation is perfect, $e(n) = e_0(n)$. For this task, the NAEC includes adaptive complex nonlinear models to try to compensate both linear and nonlinear distortions, normally with and without memory.

Recently, combination schemes have been proposed to improve the performance of NAECs. An important combined scheme, to which we will refer as robust NAEC (RNAEC), was firstly proposed in [9] considering VFs and has also been employed with FLAFs [10]. The objective of RNAEC is to improve the robustness of NAECs with respect to the presence or absence of nonlinear distortions in the filtering scenario. In other words, RNAEC aims to mitigate a disadvantage of NAECs: if the nonlinear distortion in the cancellation scenario is negligible and only linear distortions have to be compensated, a purely linear model would outperform a complex NAEC, since the latter introduces gradient noise related to the nonlinear adaptive subfilters. In addition, it is quite difficult to know in advance which canceler (linear or nonlinear) to employ in each situation, since the LNLR of signal powers, which relates the power of the linear and nonlinear distortions in the scenario, is normally *a priori* unknown and can even be time varying.

Fig. 11.4 depicts a block diagram of the RNAEC, which is composed of two branches:

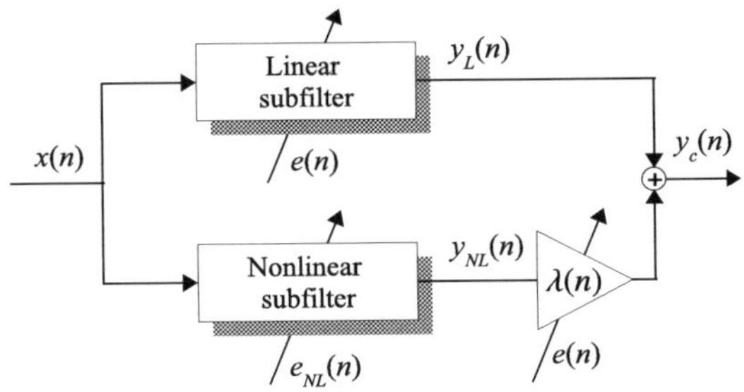


**FIGURE 11.4**

Block diagram of the RNAEC. In order to make the NAEC robust to the possible absence of nonlinear distortions, the output of the nonlinear filter branch is multiplied by a biasing factor $\lambda(n) \in [0, 1]$ that is adapted according to the overall filter performance.

- The linear branch: Its output $y_L(n)$ will be always considered in the model since the canceler at least has to model the room impulse response, which is linear in nature.
- The nonlinear branch: It incorporates a purely nonlinear model, for which we can recur to VFs, FLAFs or KAFs, among other alternatives [25].

To make the scheme robust to unknown or time varying LNLR, the objective is to achieve that the output of the nonlinear branch, $y_{NL}(n)$, is incorporated into the output of the RNAEC only if its operation implies more benefits than disadvantages, *i.e.*, depending on whether the improvement of the quality of the cancellation achieved exceeds the amount of gradient noise introduced or not.

As can be seen in Fig. 11.4, the output of the RNAEC can be calculated as follows:

$$y(n) = y_L(n) + \lambda(n) y_{NL}(n), \tag{11.10}$$

where $\lambda(n) \in [0, 1]$ represents a biasing parameter that permits to weight the nonlinear output. Eq. (11.10) allows to reinterpret the scheme of Fig. 11.4 as a combined scheme where the nonlinear branch is composed of a convex combination of the existing subfilter and a fixed filter whose output is always null, *i.e.*,

$$y(n) = y_L(n) + \lambda(n) y_{NL}(n) + [1 - \lambda(n)] \cdot 0.$$

In this way, adaptation of $\lambda(n)$ can be performed with (11.9). Thus, when only linear distortions are present in the scenario (*i.e.*, LNLR $= \infty$), $\lambda(n) \to 0$. However, when the nonlinear distortions are nonnegligible, $\lambda(n) \to 1$ and $y(n) \approx y_L(n) + y_{NL}(n)$. The sigmoidal activation function minimizes the gradient noise near $\lambda(n) = 0$ and $\lambda(n) = 1$, allowing this scheme to behave as a purely linear or nonlinear AEC.

In [42], a combined scheme was proposed based on the RNAEC with a second-order VF where, instead of using only a parameter $\lambda(n)$ to multiply the output of the quadratic kernel, different biasing parameters are employed, each weighting the output of a group of adjacent diagonals. In this way, the combined scheme reaches an improved performance, enhancing its versatility and robustness in scenarios where the memory of the nonlinear distortions is unknown. Following the approach of [42], in

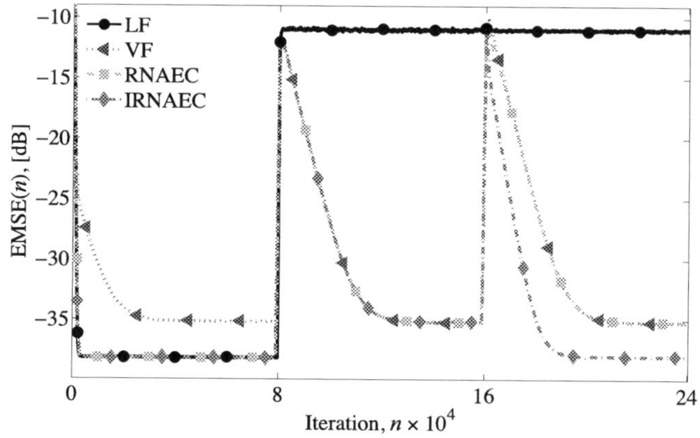


**FIGURE 11.5**

Performance of the improved RNAEC scheme. Performance of the four AECs is shown as a function of time for different echo paths. Initially, the echo is purely linear (LNLR $=\infty$). After $n=8\cdot 10^4$, LNLR decreases to 10 dB, with a memory (memoryless) echo path before (after) $n=16\cdot 10^4$.

this chapter we propose a modified RNAEC based on second-order VFs, denoted as improved RNAEC (IRNAEC). The idea of the IRNAEC is to increase the robustness of RNAEC with respect to the type of nonlinear distortions: with memory (mainly related to loudspeaker operation due to its electromechanical nature) or without memory (mainly caused by power amplifiers). To do this, the output of the IRNAEC can be calculated as

$$y(n)=y_L(n)+\lambda_M(n)y_{2,M}(n)+\lambda_R(n)y_{2,R}(n), \tag{11.11}$$

where $y_{2,M}(n)$ is the output of the main diagonal of a quadratic Volterra kernel, *i.e.*, involving only products of input signals with the same time index, $x^2(n-i)$, and $y_{2,R}(n)$ includes the output of the rest of diagonals, *i.e.*, those that consider quadratic input signals of different time instants $x(n-i)x(n-j)$ with $i\neq j$. Both partial outputs are respectively weighted by their corresponding biasing parameters.

The proposed IRNAEC has been evaluated in a nonlinear acoustic echo cancellation scenario where the nature of the nonlinear distortion is changing over time, comparing its performance with that of three different schemes: a linear filter, a second-order VF and the RNAEC. The input signal is a white Gaussian noise with zero mean and unit variance and the signal-to-noise (SNR) ratio at the microphone location is adjusted to SNR $=30$, considering $e_0(n)$ as an additive Gaussian independent and identically distributed noise.

Fig. 11.5 shows the algorithms behavior in terms of the excess MSE, EMSE$(n)=\mathrm{E}\{[e(n)-e_0(n)]^2\}$, estimated averaging the results of over 600 independent realizations of the input and noise signals. The linear filter adapts its coefficients by means of the NLMS algorithm with $\mu=1$, whereas the VF, the RNAEC and the IRNAEC update the coefficients of their kernels following the SNLMS algorithm with $\mu_l=1$ and $\mu_q=0.3$ for the linear and the quadratic kernels, respectively. We have selected $\mu_a=0.1$ to adapt the mixing parameters of all schemes.

The experiment starts considering a scenario where only linear distortions are present in the echo path, *i.e.*, LNLR $=\infty$ dB. In this case, the VF provides the worst model, due to the large amount of gradient noise introduced by the adaptation of the coefficients of the quadratic kernel, which are unnecessary in this situation. In this case, both the RNAEC and the IRNAEC perform satisfactory, behaving as the linear filter, since their biasing parameters converge to zero deactivating completely the quadratic kernel. At $n = 8 \cdot 10^4$, the echo path presents a nonlinear distortion with memory that makes LNLR $= 10$ dB. In this case, the best model is provided by the VF, the linear filter obtaining a very poor performance. As expected in this situation, both the IRNAEC and the RNAEC reach a similar cancellation as the VF, since $\lambda(n) \approx \lambda_M(n) \approx \lambda_R(n) \approx 1$. However, the performances of the IRNAEC and the RNAEC differ after $n = 16 \cdot 10^4$, where the nonlinearity included in the echo path is memoryless. In this case, only the coefficients of the main diagonal of the quadratic kernel are necessary to model the nonlinear distortion. For this reason, both the VF and the RNAEC perform similarly, since the latter only is able to activate or deactivate the whole quadratic kernel. However, the IRNAEC obtains the best cancellation, since its biasing parameters permit to maintain the coefficients of the main diagonal (when $\lambda_M(n) \approx 1$) and to eliminate the gradient noise of unnecessary units of the quadratic kernel out of such diagonal (when $\lambda_R(n) \approx 0$).

## 11.5.2 IMPROVED MULTIKERNEL ADAPTIVE FILTERS WITH SELECTIVE BIAS

In order to simplify the selection of an appropriate kernel in KAFs, several authors have already proposed approaches based on combination schemes. Essentially, these schemes incorporate several filters with different kernels and/or kernel parameters and adjust the combination to optimize overall filter performance. In this section, we explain how the performance of these combined filters can be improved even further by introducing selective shrinkage factors similarly to what we explained in Section 11.5.1.

### *11.5.2.1 Multiple-Kernel Approaches*

Many adaptive multiple-kernel approaches have been proposed in the literature, as the single-input multikernel LMS (SI-KLMS) [43], the convex combination of two KLMS filters with different kernels [26,22] and the multiple-input multikernel LMS (MI-KLMS) [27,28], among others.

The SI-KLMS scheme was introduced in [43] for classification and has been widely studied since then. In this approach, the kernel function $\kappa$ is a convex combination of kernels, *i.e.*,

$$\kappa(\mathbf{x}, \mathbf{x}') = \sum_{\ell=1}^{L} \beta_\ell \kappa_\ell(\mathbf{x}, \mathbf{x}'), \tag{11.12}$$

where $\beta_\ell$, $\ell = 1, \cdots, L$ are coefficients fixed beforehand with $\beta_\ell > 0$ and $\sum_{\ell=1}^{L} \beta_\ell = 1$. The component kernels $\kappa_\ell(\cdot, \cdot)$ can simply be usual kernels, such as Gaussian kernels, with different parameter settings. Since the sum of the Mercer kernels $\sum_{\ell=1}^{L} \beta_\ell \kappa_\ell(\cdot, \cdot)$ is also a valid Mercer kernel, this approach is equivalent to the monokernel KLMS but considers a different feature space.

A convex combination of two KLMS filters was also proposed to select a proper kernel function [26,22]. These works recur to the standard combination of two filters that is represented in Fig. 11.2A and adapted according to (11.7) and (11.8). Depending on the value of $\eta(n)$, the combination may show a preference for one of the kernels. If the mixing parameter takes an intermediate value, both filters are used simultaneously, and the combination may perform better than each individual filter.

A final approach we consider is known as multiple-input KLMS (MI-KLMS). It works by augmenting the dimensionality of the feature space by mapping input samples to multiple feature elements, which in turn are linearly combined to yield the output estimate. This has been proven advantageous in some scenarios when compared with the standard kernel regression paradigm, allowing the learning of multiple nonlinear features presented in the data [28]. Considering the multidimensional mapping of $\mathbf{x}(n)$ to a feature space $\mathbb{M}$, given by

$$\mathbf{\Phi}(\mathbf{x}(n)) = [\varphi_1(\mathbf{x}(n)) \cdots \varphi_L(\mathbf{x}(n))]^T, \tag{11.13}$$

with $\varphi_l \in \mathcal{F}_l$ for $l = 1, \cdots, L$ and the inner product in $\mathbb{M}$ defined as

$$\langle \mathbf{\Phi}, \mathbf{\Phi}' \rangle_{\mathbb{M}} = \sum_{l=1}^{L} \langle \varphi_l, \varphi_l' \rangle, \tag{11.14}$$

it is possible to show that $\mathbb{M}$ is a Hilbert space and presents the reproducing properties [28]. Therefore, there is a kernel $\mathcal{K}$ associated with the inner product in $\mathbb{M}$, $\mathcal{K}(\mathbf{x}, \mathbf{x}') = \langle \mathbf{\Phi}(\mathbf{x}), \mathbf{\Phi}(\mathbf{x}') \rangle_{\mathbb{M}}$.

Thus, the filter output can be calculated by generalizing (11.4) to the new kernel,

$$y(n) = \sum_{i=1}^{n-1} \mu e(i) \sum_{l=1}^{L} \kappa_l(\mathbf{x}(i), \mathbf{x}(n)). \tag{11.15}$$

This approach is equivalent to considering $L$ KLMS filters in parallel and adapting them using a single error signal given by $e(n) = d(n) - y(n)$.

In terms of performance, simulation results show that MI-KLMS generally tends to present a lower MSE in comparison with SI-KLMS as a tradeoff to its higher computational cost. In some applications, MI-KLMS may present better performance in terms of MSE in comparison to the convex combination of two KLMS filters. However, in certain situations, if the parameters of one kernel component are poorly adjusted, then the convex combination is able to select the best component filter (making $\eta(n) \approx 0$ or $\eta(n) \approx 1$) and may outperform SI-KLMS and MI-KLMS, which have their output degraded by the poorly adjusted kernel. In the next section, we propose a scheme to circumvent this problem, minimizing the degrading effects of poorly adjusted kernel filters in the MI-KLMS scheme.

#### *11.5.2.2 Improving the MI-KLMS Scheme*

Based on [39], we propose a scheme to minimize the degrading effects of unnecessary kernel filters in the MI-KLMS structure by using multiplicative factors that can either activate or suppress the output of useless kernels in the global filter output. This scheme, referred to here as robust MI-KLMS (R-MI-KLMS), is depicted in Fig. 11.6 for an MI-KLMS scheme with two kernels.

The overall output of the scheme of Fig. 11.6 is calculated as

$$y(n) = \lambda_1(n) y_1(n) + \lambda_2(n) y_2(n), \tag{11.16}$$

where $\lambda_i(n) \in [0, 1]$, $i = 1, 2$, are multiplicative factors (shrinkage or biasing parameters) that can be used to activate or remove each of the kernels from the output. Adapting these shrinkage factors will result in $\lambda_i(n) \to 1$ when the kernel contributes to a reduction of the MSE, whereas $\lambda_i(n) \to 0$

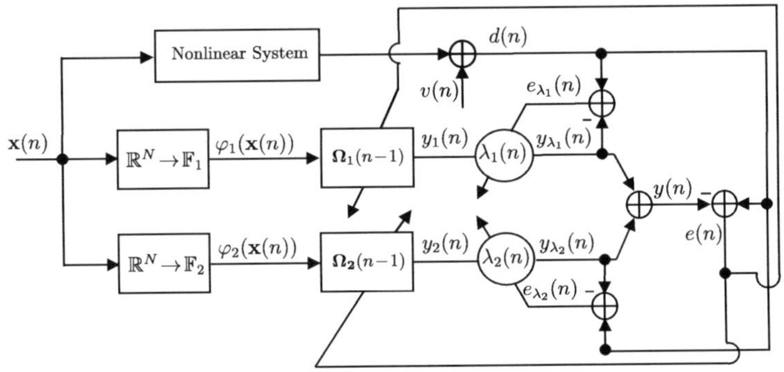


**FIGURE 11.6**

R-MI-KLMS applied to nonlinear system identification. The output of each kernel is multiplied by a shrinkage factor that can suppress the contribution of bad kernels, together with the gradient noise generated by its adaptation.

otherwise. In the latter case, not only the output of the corresponding kernel will be suppressed, but also the gradient noise incurred by its adaptation (please refer to [39] for a further explanation of the working principles of this approach).

In Table 11.1 we provide implementation details for the proposed scheme when combining an arbitrary number of kernel AFs, using the quantized version of KLMS to implement the filter components.

### *11.5.2.3 Experiments*

In the first set of simulations, we consider the problem of nonlinear prediction, where the desired sequence is computed as [27]

$$\begin{aligned} d(n) = & \left[0.8 - 0.5\exp(-d^2(n-1))\right] d(n-1) \\ & - \left[0.3 + 0.9\exp(-d^2(n-1))\right] d(n-2) \\ & + 0.1\sin(d(n-1)\pi), \qquad \text{for } 0 \le n < N_{\text{it}}/2, \end{aligned} \tag{11.17}$$

and

$$\begin{aligned} d(n) = & \left[0.2 - 0.7\exp(-d^2(n-1))\right] d(n-1) \\ & - \left[0.8 + 0.8\exp(-d^2(n-1))\right] d(n-2) \\ & + 0.2\sin(d(n-1)\pi), \qquad \text{for } N_{\text{it}}/2 \le n < N_{\text{it}}, \end{aligned} \tag{11.18}$$

with $N_{\text{it}}$ being the number of iterations and $d(n) = 0.1$ for $n < 0$.

Considering $M = 2$, $N_{\text{it}} = 10^4$, combinations of two QKLMS filters in all multiple kernel schemes and adjusting the parameters as in Table 11.2, Fig. 11.7 shows the learning curves, evolution of the mixing parameter $\eta(n)$ for the convex combination and evolution of parameters $\lambda_1(n)$ and $\lambda_2(n)$ for the

**Table 11.1 Summary of the R-MI-QKLMS algorithm considering $L$ QKLMS filters**

**Input:** $\{\mathbf{x}(n) \in \mathbb{R}^N, d(n)\}, \quad n = 1, 2, \cdots$

**Initialization:** Choose step sizes $\mu_\ell > 0$ and $\mu_{\alpha_\ell}$, $\ell = 1 \cdots, L$,
kernel widths $\sigma_\ell > 0$, thresholds $\varepsilon_\ell \geq 0$, $\ell = 1, \cdots, L$, forgetting factor $\beta$
and initialize the dictionaries $\mathcal{C}_\ell(1) = \{\mathbf{x}(1)\}$, $\ell = 1, \cdots, L$,
the parameters $\alpha_\ell(1) = \alpha^+$, $p_\ell(1) = 1$, $\ell = 1, \cdots, L$,
and coefficient vectors $\mathbf{a}_\ell(1) = [\mu d(1)]$, $\ell = 1, \cdots, L$.

**For** $n = 2, 3, \cdots$, do:

  **For** $\ell = 1, \cdots, L$, do:

$$\lambda_\ell(n) = \psi[\alpha_\ell(n-1)] = \frac{\mathrm{sgm}[\alpha_\ell(n-1)] - \mathrm{sgm}[-\alpha^+]}{\mathrm{sgm}[\alpha^+] - \mathrm{sgm}[-\alpha^+]}$$

$$y_\ell(n) = \sum_{j=1}^{N_{c,\ell}(n)} a_{j,\ell}(n-1)\kappa_\ell(\mathbf{x}(n), \mathbf{x}(c_{j,\ell}))$$

$$y_{\lambda_\ell}(n) = \lambda_\ell(n) y_\ell(n)$$

$$e_{\lambda_\ell}(n) = d(n) - y_{\lambda_\ell}(n)$$

$$\psi'[\alpha_\ell(n-1)] = \frac{\mathrm{sgm}[\alpha_\ell(n-1)]\{1 - \mathrm{sgm}[\alpha_\ell(n-1)]\}}{\mathrm{sgm}[\alpha^+] - \mathrm{sgm}[-\alpha^+]}$$

$$p_\ell(n) = \beta\, p_\ell(n-1) + (1-\beta)\, y_\ell^2(n)$$

$$\alpha_\ell(n) = \alpha_\ell(n-1) + \frac{\mu_{\alpha_\ell}}{p_\ell(n)} e_{\lambda_\ell}(n) y_\ell(n) \psi'[\alpha_\ell(n-1)]$$

  **if** $|\alpha_\ell(n)| > \alpha^+$

   $\alpha_\ell(n) \leftarrow \mathrm{sign}[\alpha_\ell(n)]\alpha^+$

  **end**

  **end**

$$y(n) = \sum_{\ell=1}^{L} y_{\lambda_\ell}(n)$$

$$e(n) = d(n) - y(n)$$

  **For** $\ell = 1, \cdots, L$, do:

$$\mathrm{dis}(\mathbf{x}(n), \mathcal{C}_\ell(n)) = \min_{1 \leq j \leq N_{c,\ell}(n)} \|\mathbf{x}(n) - \mathbf{x}(c_{j,\ell})\|$$

  If $\mathrm{dis}(\mathbf{x}(n),\ \mathcal{C}_\ell(n)) \leq \varepsilon_\ell$, keep the dictionary unchanged, *i.e.*, $\mathcal{C}_\ell(n+1) = \mathcal{C}_\ell(n)$, and quantize $\mathbf{x}(n)$ to the closest vector through updating the coefficient of that vector, *i.e.*,

$$a_{j^*,\ell}(n) = a_{j^*,\ell}(n-1) + \mu e(n),$$

  where

$$j^* = \arg \min_{1 \leq j \leq N_{c,\ell}(n)} \|\mathbf{x}(n) - \mathbf{x}(c_{j,\ell})\|$$

  Otherwise, assign a new center and corresponding new coefficient:

$$\mathcal{C}_\ell(n+1) = \{\mathcal{C}_\ell(n), \mathbf{x}(n)\},$$

$$\mathbf{a}_\ell(n) = [\mathbf{a}_\ell(n-1), \mu e(n)].$$

  **end**

**end**

**Table 11.2 Parameters used in the simulations of Figs. 11.7 and 11.8**

| Algorithm | Parameters |
|---|---|
| $\text{QKLMS}_2$ | $\mu_1 = 0.05, \sigma_1, \varepsilon_1 = 0.05$ |
| $\text{QKLMS}_2$ | $\mu_2 = 0.5, \sigma_2, \varepsilon_2 = 0.5$ |
| Convex Comb. | $\mu_a = 1, \beta = 0.9$ |
| $\text{SI-QKLMS}_1$ | $\mu = 0.05, \beta_1 = \beta_2 = 0.5, \sigma_1, \sigma_2, \varepsilon = 0.05$ |
| $\text{SI-QKLMS}_2$ | $\mu = 0.5, \beta_1 = \beta_2 = 0.5, \sigma_1, \sigma_2, \varepsilon = 0.5$ |
| MI-QKLMS | $\mu_1 = 0.05, \mu_2 = 0.5, \sigma_1, \sigma_2, \varepsilon_1 = 0.05, \varepsilon_2 = 0.5$ |
| R-MI-QKLMS | $\mu_1 = 0.05, \mu_2 = 0.5, \sigma_1, \sigma_2, \varepsilon_1 = 0.05, \varepsilon_2 = 0.5,$ $\mu_{a_1} = \mu_{a_2} = 1, \beta = 0.9$ |

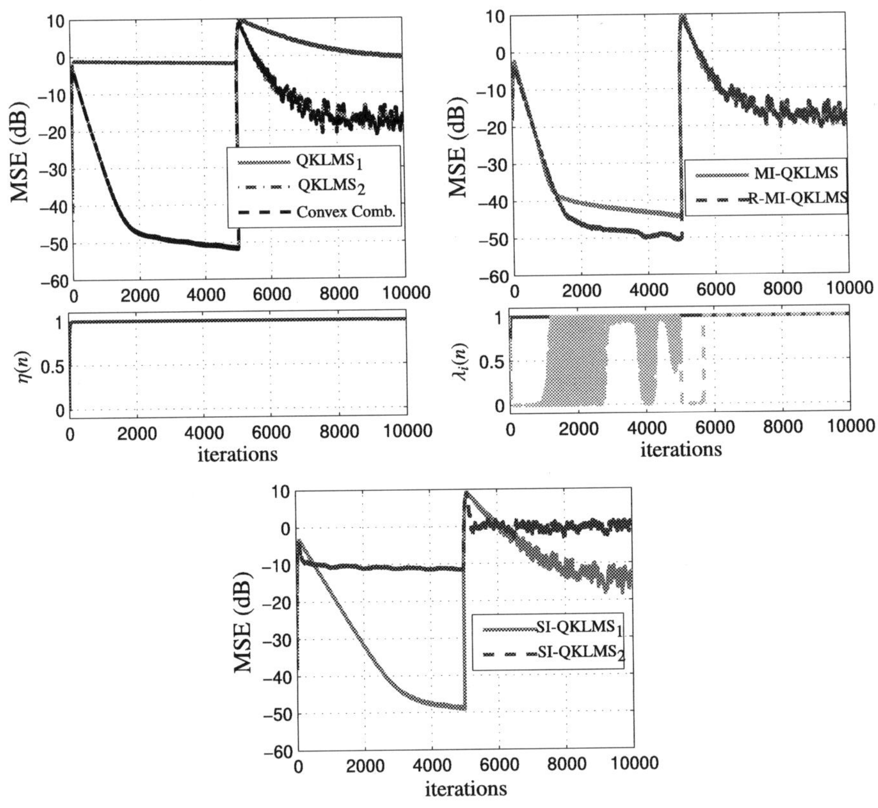


**FIGURE 11.7**

Performance of the multi-kernel Adaptive Filters in a series prediction task. Mean-square error considering the prediction of the sequence $d(n)$ defined in (11.17) and (11.18) for convex combination of two QKLMS, MI-MQKLMS, and SI-QKLMS. Kernel widths $\sigma_1 = 0.2$ and $\sigma_2 = 100$ and other parameters as in Table 11.2; to facilitate the visualization, the MSE curves were filtered by a moving average filter with 64 taps.

R-MI-QKLMS scheme, considering kernel widths as $\sigma_1 = 0.2$ and $\sigma_2 = 100$. We can observe that the kernel width $\sigma_2 = 100$ does not lead to good results for a monokernel filter as shown by the learning curve of the QKLMS$_2$ filter. The convex combination follows QKLMS$_1$ since the mixing parameter converges to one. By adjusting adequately the step size and dictionary threshold, the SI-QKLMS$_1$ scheme is able to achieve a good performance, but presents a smaller convergence rate than the monokernel QKLMS$_1$. Finally, the kernel width $\sigma_2 = 100$ degrades the performance of MI-QKLMS for $1200 < n < 5000$, which is avoided by the R-MI-QKLMS scheme. After iteration, $n = 6000$, the kernel width $\sigma_2 = 100$ is considered in the overall output, since it does not degrade the performance of MI-QKLMS anymore.

In the second set of simulations, we consider a nonlinear system identification application with two different systems modified from [24]. For the first one, the desired sequence was given by

$$d_1(n) = s_1(n) + z_1(n), \tag{11.19}$$

with

$$s_1(n) = \frac{s_1(n-1)}{1 + s_1^2(n-1)} + x^3(n-1), \tag{11.20}$$

where $x(n)$ is the input signal and $z_1(n)$ is a zero mean Gaussian noise with variance $\sigma_{z_1}^2 = 10^{-4}$. For the second nonlinear system, the desired sequence was computed as

$$d_2(n) = g(s_2(n)) + z_2(n), \tag{11.21}$$

with $z_2(n)$ being a zero mean Gaussian noise with variance $\sigma_{z_2}^2 = 10^{-6}$,

$$s_2(n) = x(n) + 0.5x(n-1) - 0.2s_2(n-1) + 0.35s_2(n-2) \tag{11.22}$$

and

$$g(s) = \begin{cases} \dfrac{s}{3\sqrt{0.1 + 0.9s^2}}, & \text{for } s \geq 0, \\ \dfrac{-s^2[1 - \exp(0.7s)]}{3}, & \text{for } s < 0. \end{cases} \tag{11.23}$$

Fig. 11.8 shows the learning curves, evolution of the mixing parameter $\eta(n)$ for the convex combination and evolution of the biased parameters $\lambda_1(n)$ and $\lambda_2(n)$ for the R-MI-QKLMS scheme, considering kernel widths as $\sigma_1 = 1$ and $\sigma_2 = 0.1$. The identification of the first system was considered, with the desired signal given by $d(n) = d_1(n)$ from $n = 0$ until $n = 5 \times 10^4$ and an abrupt change to the identification of the second system, with $d(n) = d_2(n)$ from $n = 50001$ until $n = 10^5$. The input signal $x(n)$ was independently and identically distributed randomly generated from a zero mean Gaussian distribution with standard deviation $\sigma_x = 0.15$; we considered an average of 1000 runs. We can observe that the convex combination of two QKLMS filters performs as the best component filter. The performance of SI-QKLMS$_2$ is better than that of SI-QKLMS$_1$ in the identification of the first system but for the second system, SI-QKLMS$_1$ outperforms SI-QKLMS$_2$, emphasizing the importance of adjusting the step size and dictionary threshold adequately for this scheme. This result also shows that, even considering a smaller number of elements in the dictionary (by using a lower value for the dictionary

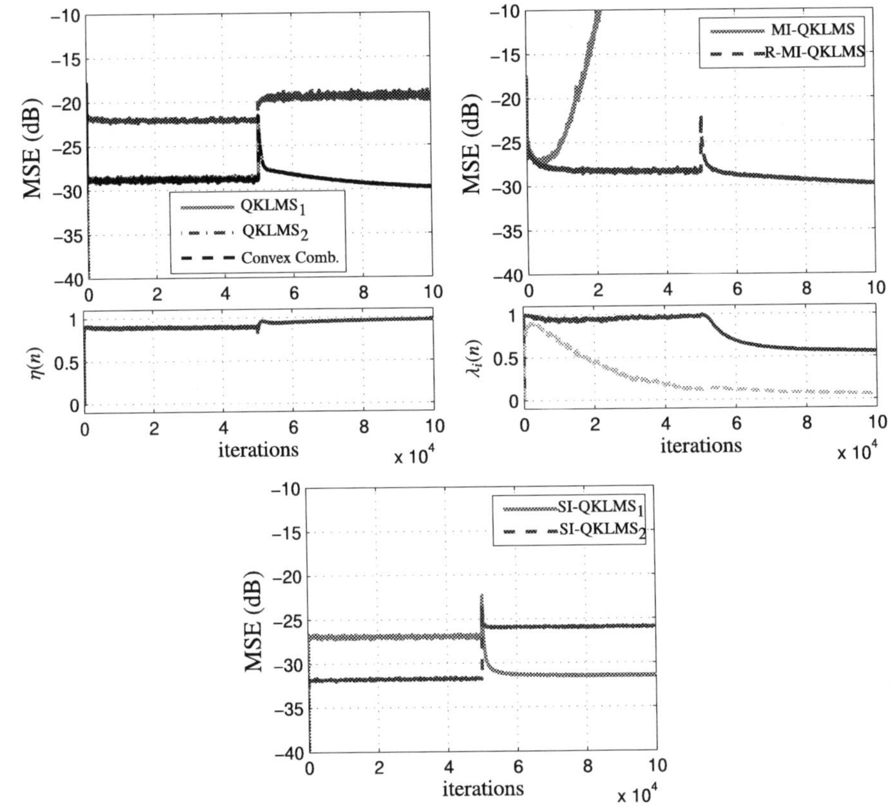


**FIGURE 11.8**

Performance of the multikernel Adaptive Filters in system identification task. Mean square error considering the identification of the nonlinear system described by Eqs. (11.19)–(11.23) for convex combination of two QKLMS, MI-MQKLMS and SI-QKLMS. Kernel widths are $\sigma_1 = 1$ and $\sigma_2 = 0.1$ and other parameters as in Table 11.2; average of 1000 independent runs; to facilitate the visualization, the MSE curves were filtered by a moving average filter with 64 taps.

threshold), it is possible to obtain an improved performance, depending on the other parameters. In this scenario, MI-QKLMS presents a poor performance and it is possible to observe the importance of using the improved version R-MI-QKLMS. This robust version is able to minimize the degrading effects of unnecessary kernels, achieving a performance similar to that of the convex combination.

## 11.6 CONCLUSIONS

Combination approaches constitute an easy and versatile approach to improve the performance of adaptive filters and in particular to optimize the design of nonlinear adaptive filters. We have reviewed different situations in which the use of combined schemes is of interest and contribute to improve the final performance of the filter. The versatility of the approach has been demonstrated by considering a number of particular systems involving VFs, FLAFs and KAFs and in the context of nonlinear echo cancellation, signal prediction and nonlinear system identification.

## REFERENCES

[1] J. Arenas-García, L.A. Azpicueta-Ruiz, M.T.M. Silva, V.H. Nascimento, A.H. Sayed, Combinations of adaptive filters: performance and convergence properties, IEEE Signal Processing Magazine 33 (1) (2016) 120–140.

[2] J. Arenas-García, A.R. Figueiras-Vidal, Adaptive combination of proportionate filters for sparse echo cancelation, IEEE Transactions on Audio, Speech and Language Processing 17 (6) (2009) 1087–1098.

[3] S. Lu, J. Sun, G. Wang, J. Tian, A novel GSC beamformer using a combination of two adaptive filters for smart antenna array, IEEE Antennas Wireless Propagation Letters 11 (2012) 377–380.

[4] M. Ferrer, A. Gonzalez, M. de Diego, G. Piñero, Convex combination filtered-X algorithms for active noise control systems, IEEE Transactions on Audio, Speech and Language Processing 21 (1) (2013) 156–167.

[5] T. Trump, A combination of two NLMS filters in an adaptive line enhancer, in: Proceedings of the International Conference on Digital Signal Processing, Corfu, Greece, July 6–8, 2011, pp. 1–6.

[6] M.T.M. Silva, J. Arenas-García, A soft-switching blind equalization scheme via convex combination of adaptive filters, IEEE Transactions on Signal Processing 61 (5) (2013) 1171–1182.

[7] J. Arenas-García, A.R. Figueiras-Vidal, A.H. Sayed, Mean-square performance of a convex combination of two adaptive filters, IEEE Transactions on Signal Processing 54 (3) (2006) 1078–1090.

[8] S.S. Kozat, A.C. Singer, A.T. Erdogan, A.H. Sayed, Unbiased model combinations for adaptive filtering, IEEE Transactions on Signal Processing 58 (3) (2010) 4421–4427.

[9] L.A. Azpicueta-Ruiz, M. Zeller, A.R. Figueiras-Vidal, J. Arenas-García, W. Kellermann, Adaptive combination of Volterra kernels and its application to nonlinear acoustic echo cancelation, IEEE Transactions on Audio, Speech and Language Processing 19 (1) (2011) 97–110.

[10] D. Comminiello, M. Scarpiniti, L.A. Azpicueta-Ruiz, J. Arenas-García, A. Uncini, Functional link adaptive filters for nonlinear acoustic echo cancellation, IEEE Transactions on Audio, Speech and Language Processing 21 (7) (2013) 1502–1512.

[11] B. Jefs, S. Javidi, P. Vayanos, D.P. Mandic, Characterization of signal modality: exploiting signal nonlinearity in machine learning and signal processing, IEEE Journal of Signal Processing Systems 61 (1) (2010) 105–115.

[12] H. Zhao, J. Zhang, Adaptively combined FIR and functional link artificial neural network equalizer for nonlinear communication channel, IEEE Transactions on Neural Networks 20 (4) (2009) 665–674.

[13] N.V. George, A. Gonzalez, Convex combination of nonlinear adaptive filters for active noise control, Applied Acoustics 76 (2014) 157–161.

[14] M. Scarpiniti, D. Comminiello, R. Parisi, A. Uncini, A collaborative filter approach to adaptive noise cancellation, in: Proceedings of the 22nd Italian Workshop on Neural Nets, WIRN12, Vietri sul Mare, Italy, May 17–19, 2012, pp. 101–109.

[15] L. Fuster, M. de Diego, M. Ferrer, A. Gonzalez, Combination of filtered-x adaptive filters for nonlinear listening-room compensation, in: Proceedings of the European Signal Processing Conference, EUSIPCO, Budapest, Hungary, August 28–September 2, 2016, pp. 1773–1777.

[16] V.J. Mathews, Adaptive polynomial filters, IEEE Signal Processing Magazine 8 (3) (1991) 10–26.

[17] G.L. Sicuranza, A. Carini, A generalized FLANN filter for nonlinear active noise control, IEEE Transactions on Audio, Speech and Language Processing 19 (8) (2011) 2412–2417.

[18] F. Küch, W. Kellermann, Orthogonalized power filters for nonlinear acoustic echo cancellation, Signal Processing 86 (6) (2006) 1168–1181.

[19] W. Liu, J.C. Príncipe, S. Haykin, Kernel Adaptive Filtering: A Comprehensive Introduction, John Wiley & Sons, New York, 2010.

[20] S. Haykin, Adaptive Filter Theory, Prentice Hall, Upper Saddle River, 2001.

[21] A.H. Sayed, Fundamentals of Adaptive Filtering, John Wiley & Sons, NJ, 2003.

[22] W. Gao, C. Richard, J.C.M. Bermudez, J. Huang, Convex combinations of kernel adaptive filters, in: Proceedings of the IEEE International Workshop on Machine Learning for Signal Processing, MLSP, Reims, France, September 21–24, 2014, pp. 1–5.

[23] B. Chen, S. Zhao, P. Zhu, J.C. Príncipe, Quantized kernel least mean square algorithm, IEEE Transactions on Neural Networks and Learning Systems 23 (1) (2012) 22–32.

[24] W.D. Parreira, J.C.M. Bermudez, C. Richard, J.Y. Tourneret, Stochastic behavior analysis of the Gaussian kernel least-mean-square algorithm, IEEE Transactions on Signal Processing 60 (5) (2012) 2208–2222.

[25] S. Van Vaerenbergh, L.A. Azpicueta-Ruiz, D. Comminiello, A split kernel adaptive filtering architecture for nonlinear acoustic echo cancellation, in: Proceedings of the European Signal Processing Conference, EUSIPCO, Budapest, Hungary, August 28–September 2, 2016, pp. 1768–1772.

[26] M. Scarpiniti, D. Comminiello, R. Parisi, A. Uncini, A collaborative approach to time-series prediction, in: Proceedings of the 21st Italian Workshop on Neural Nets, WIRN11, Vietri sul Mare, Italy, June 3–5, 2011, pp. 178–185.

[27] M. Yukawa, Multikernel adaptive filtering, IEEE Transactions on Signal Processing 60 (9) (2012) 4672–4682.

[28] F.A. Tobar, S.Y. Kung, D.P. Mandic, Multikernel least mean square algorithm, IEEE Transactions on Neural Networks and Learning Systems 24 (2) (2014) 265–277.

[29] W. Liu, P.P. Pokharel, J.C. Príncipe, The kernel least-mean-square algorithm, IEEE Transactions on Signal Processing 56 (2) (2008) 543–554.

[30] J. Platt, A resource-allocating network for function interpolation, Neural Computation 2 (1991) 213–225.

[31] Y. Engel, S. Mannor, R. Meir, The kernel recursive least-squares algorithm, IEEE Transactions on Signal Processing 52 (8) (2004) 2275–2285.

[32] W. Liu, I. Park, J.C. Príncipe, An information theoretic approach of designing sparse kernel adaptive filters, IEEE Transactions on Neural Networks 20 (12) (2009) 1950–1961.

[33] C. Richard, J.C.M. Bermudez, P. Honeine, Online prediction of time series data with kernels, IEEE Transactions on Signal Processing 57 (3) (2009) 1058–1067.

[34] K. Slavakis, P. Bouboulis, S. Theodoridis, Online learning in reproducing kernel Hilbert spaces, in: Academic Press Library in Signal Processing: Signal Processing Theory and Machine Learning, vol. 1, chapter 17, Academic Press, Chennai, 2014, pp. 883–987.

[35] D. Comminiello, M. Scarpiniti, S. Scardapane, R. Parisi, A. Uncini, Improving nonlinear modeling capabilities of functional link adaptive filters, Neural Networks 69 (2015) 51–59.

[36] D. Comminiello, M. Scarpiniti, L.A. Azpicueta-Ruiz, J. Arenas-García, A. Uncini, A nonlinear architecture involving a combination of proportionate functional link adaptive filters, in: Proceedings of the European Signal Processing Conference, EUSIPCO, Nice, France, August 31–September 4, 2015, pp. 2919–2923.

[37] D. Comminiello, M. Scarpiniti, L.A. Azpicueta-Ruiz, J. Arenas-García, A. Uncini, Combined nonlinear filtering architectures involving sparse functional link adaptive filters, Signal Processing 35 (7) (2017) 168–178.

[38] X. Zeng, H. Zhao, W. Jin, Z. He, T. Li, Identification of nonlinear dynamic systems using convex combinations of multiple adaptive radius basis function networks, Measurement 46 (2013) 628–638.

[39] M. Lázaro-Gredilla, L.A. Azpicueta-Ruiz, A.R. Figueiras-Vidal, J. Arenas-García, Adaptively biasing the weights of adaptive filters, Transactions on Signal Processing 58 (7) (2010) 3890–3895.

[40] F. Küch, W. Kellermann, Nonlinear Acoustic Echo Cancellation, Springer, Heidelberg, Germany, 2006.

[41] J.M. Gil-Cacho, M. Signoretto, T. Van Waterschoot, M. Moonen, S.H. Jesen, Nonlinear acoustic echo cancellation based on a sliding-window leaky kernel affine projection algorithm, IEEE Transactions on Audio, Speech and Language Processing 21 (9) (2013) 1867–1878.

[42] L.A. Azpicueta-Ruiz, M. Zeller, A.R. Figueiras-Vidal, W. Kellermann, J. Arenas-García, Enhanced adaptive Volterra filtering by automatic attenuation of memory regions and its application to acoustic echo cancellation, IEEE Transactions on Signal Processing 61 (11) (2013) 2745–2750.

[43] G.R.G. Lanckriet, N. Cristianini, P. Bartlett, L. El Ghaoui, M.I. Jordan, Learning the kernel matrix with semidefinite programming, Journal of Machine Learning Research 5 (2004) 27–72.

PART

# 4

# NONLINEAR MODELING BY NEURAL SYSTEMS

CHAPTER

# ECHO STATE NETWORKS FOR MULTIDIMENSIONAL DATA: EXPLOITING NONCIRCULARITY AND WIDELY LINEAR MODELS

12

**Yili Xia*, Min Xiang†, Zhe Li*, Danilo P. Mandic†**

**Southeast University, School of Information Science and Engineering, Nanjing, P. R. China †Imperial College London, London, United Kingdom*

## CONTENTS



## CHAPTER POINTS

- Quaternion-valued echo state networks (QESNs) are introduced for nonlinear adaptive filtering of 3D and 4D signals.
- The exploitation of full quaternion nonlinear activation functions with local analyticity allows for first-order differentiability at the operating point to design gradient descent–based quaternion nonlinear learning algorithms.
- The widely linear model is incorporated into QESNs for optimal performance for second-order noncircular quaternion signals.

Adaptive Learning Methods for Nonlinear System Modeling. DOI: 10.1016/B978-0-12-812976-0.00016-6

## 12.1 INTRODUCTION

Recurrent neural networks (RNNs) represent a large and varied class of computational models that are designed by more or less detailed analogy with biological brain modules. In RNNs, numerous abstract neurons are interconnected by abstracted synaptic connections, which enable activations to propagate through the network. The characteristic feature of RNNs that distinguishes them from the feedforward neural networks is that the connection topology possesses feedback cycles. The computational power of RNNs stems from their ability to act as universal approximators for any continuous function on a compact domain [1,2], and owing to their rich inherent memory through feedback, RNNs have found wide applications in the modeling of highly nonlinear and nonstationary dynamic systems and the associated attractor dynamics, including system identification [3], time series prediction [4] and adaptive noise cancellation [5], where, for the nonstationary and nonlinear nature of the signals and typically long impulse responses, using the class of static feedforward networks or transversal filters would result in undermodeling [2,6].

Recent progress in sensing technology has made possible the recording from multidimensional data sources which are typically 3D and 4D, such as measurements from inertial sensors in body sensor networks, three-axis ultrasonic anemometers in wind measurement and three-axis seismometers in oil exploration. Real-valued multichannel signal processing considers these measurements as vectors in $\mathbb{R}^3$ and $\mathbb{R}^4$ and cater for the "coupled" nature of the available information across data channels, but the information in the correlation matrices is not easy to use, due to their multiple and scattered block structures. In practice, when 4D data are modeled using real-valued vectors, they produce e.g., 10 covariance matrices. Also, accuracy may be compromised due to the deficiencies of the nondivision vector algebra (gimbal lock). These problems are largely mitigated when modeling 3D and 4D signals by quaternions, a number system first introduced by William Rowan Hamilton in 1843, since its 4D division algebra naturally accounts for the coupling between data channels and also provides parsimonious representations, e.g., with only four covariance matrices needed. For this reason, quaternions, which were traditionally used in aerospace engineering, because of their advantage in modeling 3D rotations and orientations, are rapidly gaining popularity in various applications, including computer graphics [7], molecular modeling [8], color image processing [9,10] and 3D polarized signal representation for vector sensor array processing [11,12].

In the context of learning systems, quaternion approaches include both Kalman filtering [13–15] and stochastic gradient algorithms, such as the quaternion least mean square [16]. However, in the context of nonlinear learning systems like RNNs, quaternion-valued processing is still relatively underexplored, mainly because of problems arising from the lack of analytic nonlinearities in the quaternion domain $\mathbb{H}$. Although quaternion nonlinear functions have been implemented, for example, in the quaternion independent component analysis [17], the analyticity of these functions has not been rigorously examined, mainly because the very stringent Cauchy–Riemann–Fueter (CRF) conditions admit only linear functions and constants as globally analytic quaternion-valued functions [18]. This is a serious obstacle that prevents us from choosing the standard nonlinear activation functions (tanh, logistic) as the nonlinearities in the design of quaternion-valued nonlinear learning algorithms.

To partially overcome this issue, some approaches use a "split" quaternion function that treats each quaternion component separately (as a real channel), passed through a real-valued smooth nonlinearity [19,20]. Although this may yield enhanced performance over vector-based processing, the noncommutativity aspect of the quaternion algebra is overlooked, thus prohibiting rigorous treatment of the cross-information and not exploiting the full potential of quaternions. Recognizing that gradient-based learning algorithms, such as nonlinear gradient descent (NGD) and real-time recurrent learning (RTRL) [2], require gradient evaluation at a point, it is possible to adopt a local alternative, such as the local analyticity condition (LAC) [21], to relax the global analyticity requirement for quaternion nonlinearities by the CRF conditions. The work in [22] uses LAC to establish a class of neural networks in $\mathbb{H}$, which are a generic extension of those in $\mathbb{R}$ and $\mathbb{C}$, allowing us to use standard activation functions, such as tanh. The learning algorithms introduced in this way include quaternion nonlinear gradient decent for nonlinear FIR filters [22] and quaternion real-time recurrent learning for fully connected neural networks [23]. However, another obvious obstacle which hinders the further development of quaternion RNNs is the high computational complexity associated with their training, as a quaternion addition requires four real-valued additions whereas a quaternion multiplication requires 16 real-valued multiplications and 12 real-valued additions.

Recently, a class of discrete-time RNNs, called echo state networks (ESNs), has been introduced, with the aim to reduce the complexity of computation encountered by standard RNNs [24,25]. The principle behind ESNs is to separate the RNN architecture into two constituent components: a recurrent architecture, called the "dynamical reservoir" or "hidden layer" and a memoryless output layer, called the "readout neuron." The recurrent architecture consists of a randomly generated group of hidden neurons with a specified degree of recurrent connections and should satisfy the so-called "echo state property" to maintain stability [24,26]. This way, the high computational complexity of RNNs is significantly reduced due to the sparse connections among the hidden neurons; in addition, the learning requirements are reduced to only the weights connecting the hidden layer and the readout neuron, and the emergence of ESNs enables the possibility to design computationally efficient quaternion-valued RNNs with moderate complexity.

As quaternions can be conveniently constructed as a pair of complex numbers through the Cayley–Dickson representation, it is natural to ask whether some recent breakthroughs in statistical signal processing of complex variables can be generalized to quaternions. One such breakthrough in complex-valued signal processing has been due to the widely linear model and augmented complex statistics [27,28]. This model, together with the corresponding augmented complex statistics, has been successfully used to design enhanced algorithms in communications and adaptive filters [29,30]. A number of studies have shown that widely linear modeling offers theoretical and practical advantages over the standard strictly linear model, and it is applicable to the generality of complex-valued random signals, both second-order circular (proper) and noncircular (improper) [28,31–33]. The concept of $\mathbb{H}$ properness was introduced in [34] as the invariance of the probability density function (pdf) of a quaternion-valued variable $q$ under some specific rotations around the angle of $\pi/2$, and this restriction was later relaxed to an arbitrary axis and angle of rotation $\phi$ [35]. In this sense, a variable $q$ is said to be proper, if $\text{pdf}(q) = \text{pdf}(e^{p\phi}q)$ for any pure unit quaternion $p$. These discoveries were followed by the quaternion widely linear model, together with the augmented quaternion statistics, to account for the generality of quaternion signals, both proper and improper [36–39]. Such models have been shown to outperform the traditional strictly linear quaternion processing, which relies only on the standard covariance matrix, and a rigorous quantitative analysis of statistical advantages of widely linear over strictly linear processing in $\mathbb{H}$ has been performed in [40].

Therefore, in this chapter, we introduce quaternion-valued echo state networks (QESNs), as a generic extension of real-valued and complex-valued ESNs [24,25,41], to enable the processing of hypercomplex 3D and 4D signals. To further equip the QESNs with enhanced modeling capability for noncircular quaternion signals (rotation-dependent distribution), we incorporate the widely linear model into QESNs, making them second-order optimal for the generality of quaternion signals (both circular and noncircular) [42]. The so introduced widely linear model for state space estimation is not limited to QESNs but is also applicable to general quaternion-valued RNN architectures. Simulations and experimental results in the prediction setting on both benchmark circular and noncircular signals and on real-world noncircular, nonlinear and nonstationary 3D body motion and wind data support the analysis.

## 12.2 MATHEMATICAL BACKGROUND

The necessary mathematical background on the design of QESNs and the associated nonlinear learning algorithms will be briefly discussed in this section.

### 12.2.1 QUATERNION ALGEBRA

As a noncommutative extension of the complex domain $\mathbb{C}$, the quaternion domain $\mathbb{H}$ provides a natural framework for processing 3D and 4D signals. A quaternion variable $q \in \mathbb{H}$ is a skew field over $\mathbb{R}$ and comprises four real-valued components ($q_r$, $q_\imath$, $q_\jmath$ and $q_\kappa$) and three imaginary units ($\imath$, $\jmath$ and $\kappa$), to give

$$q = q_r + \imath q_\imath + \jmath q_\jmath + \kappa q_\kappa = Sq + Vq. \tag{12.1}$$

It can also be represented by a real part (scalar), denoted by $Sq = \Re(q) = q_r$, and a vector part $Vq$ (also known as a pure quaternion $\Im(q)$), consisting of the three imaginary components, $Vq = \Im(q) = \imath q_\imath + \jmath q_\jmath + \kappa q_\kappa$. The pure quaternions $\Im(q)$ are usually used for 3D signal modeling. The imaginary units $\imath$, $\jmath$ and $\kappa$ obey the following rules:

$$\imath\jmath = \kappa,\ \jmath\kappa = i,\ \kappa\imath = \jmath,\ \imath^2 = \jmath^2 = \kappa^2 = \imath\jmath\kappa = -1. \tag{12.2}$$

Note that the quaternion multiplication is not commutative, that is, $\imath\jmath \neq \jmath\imath = -\kappa$. For any two quaternions, $q_1$ and $q_2$, their product is given by

$$\begin{aligned} q_1 q_2 &= (Sq_1 + Vq_1)(Sq_2 + Vq_2) \\ &= Sq_1 Sq_2 - Vq_1 \cdot Vq_2 + Sq_1 Vq_2 + Sq_2 Vq_1 + Vq_1 \times Vq_2, \end{aligned} \tag{12.3}$$

where the symbol "·" denotes the dot-product and "×" the cross-product. The cross-product makes the quaternion multiplication noncommutative. The norm of a quaternion variable $q$ is given by

$$\|q\| = \sqrt{qq^*} = \sqrt{q_r^2 + q_\imath^2 + q_\jmath^2 + q_\kappa^2} \tag{12.4}$$

while the quaternion conjugate of $q$, denoted by $q^*$, is given by

$$q^* = Sq - Vq = q_r - \imath q_\imath - \jmath q_\jmath - \kappa q_\kappa. \tag{12.5}$$

Another important notion in the quaternion domain is the so-called "quaternion involution," which defines a self-inverse mapping, analogous to the complex conjugate. The general involution of a quaternion variable $q$ is defined as $\phi_\alpha(q) \equiv q^\alpha = \alpha q \alpha^{-1} = -\alpha q \alpha$, where $\alpha \in \mathbb{H}$, while the special cases of involutions about the $\imath$, $\jmath$ and $\kappa$ imaginary axes are given by [9]

$$
\begin{aligned}
q^\imath &= -\imath q \imath = q_r + \imath q_\imath - \jmath q_\jmath - \kappa q_\kappa, \\
q^\jmath &= -\jmath q \jmath = q_r - \imath q_\imath + \jmath q_\jmath - \kappa q_\kappa, \\
q^\kappa &= -\kappa q \kappa = q_r - \imath q_\imath - \jmath q_\jmath + \kappa q_\kappa,
\end{aligned}
\tag{12.6}
$$

with the following properties ($\forall \eta \in \{\imath, \jmath, \kappa\}$):

$$
(q^\eta)^\eta = q, \ (q^\eta)^* = (q^*)^\eta, \ (q_1 q_2)^\eta = q_1^\eta q_2^\eta, \ (q_1 + q_2)^\eta = q_1^\eta + q_2^\eta. \tag{12.7}
$$

Similar to the complex conjugate, involutions allow for the components of a quaternion variable $q$ to be expressed in terms of the actual variable $q$ and its "partial conjugate," $q^\imath$, $q^\jmath$ and $q^\kappa$, that is,

$$
\begin{aligned}
q_r &= \frac{1}{4}(q + q^\imath + q^\jmath + q^\kappa), \ q_\imath = \frac{1}{4\imath}(q + q^\imath - q^\jmath - q^\kappa), \\
q_\jmath &= \frac{1}{4\jmath}(q - q^\imath + q^\jmath - q^\kappa), \ q_\kappa = \frac{1}{4\kappa}(q - q^\imath - q^\jmath + q^\kappa).
\end{aligned}
\tag{12.8}
$$

In this way, the relationship between the involutions and the quaternion variable $q$ and its conjugate is specified by

$$
q = \frac{1}{2}(q^{\imath *} + q^{\jmath *} + q^{\kappa *} - q^*), \ q^* = \frac{1}{2}(q^\imath + q^\jmath + q^\kappa - q). \tag{12.9}
$$

### 12.2.2 QUATERNION WIDELY LINEAR MODEL AND QUATERNION AUGMENTED STATISTICS

Now, consider a quaternion-valued mean square error (MSE) estimator given by

$$
\hat{y} = E[y|\mathbf{q}], \tag{12.10}
$$

where $\hat{y}$ is the estimated process, $\mathbf{q}$ is the observed variable and $E[\cdot]$ is the statistical expectation operator. For zero mean jointly Gaussian $\mathbf{q}$ and $y$, the strictly linear estimation solution, similar to those in $\mathbb{R}$ and $\mathbb{C}$, is given by [13,16]

$$
\hat{y} = \mathbf{w}^{\mathrm{T}}\mathbf{q}, \tag{12.11}
$$

where $\mathbf{w}$ and $\mathbf{q}$ are, respectively, the coefficient and regressor vector. Observe, however, that for all the components $\{y_r, y_\imath, y_\jmath, y_\kappa\}$, we have

$$
\hat{y}_\eta = E[y_\eta|\mathbf{q}_r, \mathbf{q}_\imath, \mathbf{q}_\jmath, \mathbf{q}_\kappa], \quad \eta \in \{r, \imath, \jmath, \kappa\}, \tag{12.12}
$$

so that by using the involutions in (12.6), we can express each element of a quaternion variable as in (12.8). This gives, for instance, the real component of a quaternion variable $\mathbf{q}_r = (\mathbf{q} + \mathbf{q}^\imath + \mathbf{q}^\jmath + \mathbf{q}^\kappa)/4$, leading to the general expression for all the components

$$\hat{y}_\eta = E[y_\eta|\mathbf{q}, \mathbf{q}^\imath, \mathbf{q}^\jmath, \mathbf{q}^\kappa] \text{ and } \hat{y} = E[y|\mathbf{q}, \mathbf{q}^\imath, \mathbf{q}^\jmath, \mathbf{q}^\kappa] \tag{12.13}$$

and allowing us to arrive at the widely linear model [36–38]

$$y = \mathbf{w}^{a\mathrm{T}}\mathbf{q}^a = \mathbf{a}^\mathrm{T}\mathbf{q} + \mathbf{b}^\mathrm{T}\mathbf{q}^\imath + \mathbf{c}^\mathrm{T}\mathbf{q}^\jmath + \mathbf{d}^\mathrm{T}\mathbf{q}^\kappa, \tag{12.14}$$

where $\mathbf{w}^a = [\mathbf{a}^\mathrm{T}, \mathbf{b}^\mathrm{T}, \mathbf{c}^\mathrm{T}, \mathbf{d}^\mathrm{T}]^\mathrm{T}$ is the augmented weight vector.

Current statistical signal processing in $\mathbb{H}$ by and large employs the strictly linear model in (12.11), drawing upon the covariance matrix $\mathbf{R} = E[\mathbf{q}\mathbf{q}^\mathrm{H}]$, which is sufficient to describe only second-order circular (proper) signals with equal power data components. However, based on (12.14), the modeling of both the second-order circular (proper) and noncircular (improper) signals is only possible using the augmented covariance matrix, given by [36–39],

$$\mathbf{R}^a = E[\mathbf{q}^a\mathbf{q}^{a\mathrm{H}}] = \begin{bmatrix} \mathbf{R} & \mathbf{P} & \mathbf{S} & \mathbf{T} \\ \mathbf{P}^\imath & \mathbf{R}^\imath & \mathbf{T}^\imath & \mathbf{S}^\imath \\ \mathbf{S}^\jmath & \mathbf{T}^\jmath & \mathbf{R}^\jmath & \mathbf{P}^\jmath \\ \mathbf{T}^\kappa & \mathbf{S}^\kappa & \mathbf{P}^\kappa & \mathbf{R}^\kappa \end{bmatrix}, \tag{12.15}$$

where $\mathbf{R} = E[\mathbf{q}\mathbf{q}^H]$ and the three pseudocovariances are termed the $\imath$-covariance $\mathbf{P}$, $\jmath$-covariance $\mathbf{S}$ and $\kappa$-covariance $\mathbf{T}$, given by

$$\mathbf{P} = E[\mathbf{q}\mathbf{q}^{\imath\mathrm{H}}],\ \ \mathbf{S} = E[\mathbf{q}\mathbf{q}^{\jmath\mathrm{H}}],\ \ \mathbf{T} = E[\mathbf{q}\mathbf{q}^{\kappa\mathrm{T}}]. \tag{12.16}$$

Second-order circular (proper) quaternion-valued signals have probability distributions that are rotation-invariant with respect to all the six possible pairs of axes (combinations of $\imath$, $\jmath$ and $\kappa$) [38] and thus equal powers in all the components, so that the three pseudocovariance matrices **P**, **S** and **T** vanish. However, in most real-world applications, probability density functions are rotation-dependent and hence require the use of the augmented quaternion statistics and the quaternion widely linear model. Also note that real-valued multichannel processing in $\mathbb{R}^4$ requires ten covariance matrices, as opposed to four in the quaternion domain, since only **R**, **P**, **S** and **T** are needed to fully describe $\mathbf{R}^a$ in (12.15).

### 12.2.3 $\mathbb{HR}$-CALCULUS AND QUATERNION GRADIENT OPERATIONS

In the design of quaternion-valued adaptive learning systems, a common optimization objective is to minimize a positive real function of quaternion variables. However, gradient-based optimization in $\mathbb{H}$ is quite restrictive, as the standard CRF conditions do not admit derivatives of nonanalytic real-valued cost functions [18]. It is only recently that the $\mathbb{HR}$-calculus has made possible the differentiation of both analytic and nonanalytic functions in $\mathbb{H}$ [43], through exploiting the duality with their isomorphic quadrivariate real functions, which is built upon the relationships between quaternion involutions and their four real-valued components, as indicated by (12.6). Compared with other existing quaternion gradients, which typically consider the real part and the vector part of a quaternion separately [21, 44], the $\mathbb{HR}$-calculus enables the possibility to treat a quaternion as an entry, giving more physically intuitive gradient calculation results with compact expressions.

The $\mathbb{HR}$- and $\mathbb{HR}^*$-derivatives are respectively given by [43]

$$\frac{\partial f(q,q^{\imath},q^{\jmath},q^{\kappa})}{\partial q}=\frac{1}{4}\left(\frac{\partial f}{\partial q_r}-\imath\frac{\partial f}{\partial q_\imath}-\jmath\frac{\partial f}{\partial q_\jmath}-\kappa\frac{\partial f}{\partial q_\kappa}\right) \tag{12.17}$$

and

$$\frac{\partial f(q^*,q^{\imath *},q^{\jmath *},q^{\kappa *})}{\partial q^*}=\frac{1}{4}\left(\frac{\partial f}{\partial q_r}+\imath\frac{\partial f}{\partial q_\imath}+\jmath\frac{\partial f}{\partial q_\jmath}+\kappa\frac{\partial f}{\partial q_\kappa}\right). \tag{12.18}$$

We shall now consider some examples on how to use the $\mathbb{HR}$-calculus for derivatives typically encountered in the design of quaternion-valued RNNs.

- *Derivative of the analytic function* $f(q)=q$. We first express it in terms of the involutions $\{q,q^{\imath},q^{\jmath},q^{\kappa}\}$ using (12.9) and then differentiate it with respect to $q$, giving $\partial f(q)/\partial q=1$. This is equivalent to the standard CRF derivative, which gives $f'(q)=1$.
- *Derivative of the nonanalytic function* $f(q,q^*)=qq^*$. Such a cost function is dependent on both $q$ and its conjugate $q^*$ and is nonanalytic in the CRF sense. However, the $\mathbb{HR}$-calculus circumvents this problem through (12.17) and (12.18). For example, consider $\partial f(q,q^*)/\partial q^*=\partial(qq^*)/\partial q^*=\partial q/\partial q^*\cdot q^*+q\cdot\partial q^*/\partial q^*$. The $\mathbb{HR}^*$-derivative gives $\partial q/\partial q^*=-1/2$ and the $\mathbb{HR}$-derivative $\partial q^*/\partial q^*=\partial q/\partial q=1$, and hence $\partial f(q,q^*)/\partial q^*=\partial(qq^*)/\partial q^*=\partial q/\partial q^*\cdot q^*+q\cdot\partial q^*/\partial q^*=-q^*/2+q$.
- *Derivatives of involutions* $f(q)=q^{\eta}$, $\eta=\{\imath,\jmath,\kappa\}$. As desired, the $\mathbb{HR}$-derivatives give

$$\frac{\partial q^{\eta}}{\partial q^{\theta}}=\begin{cases}1, & \eta=\theta,\quad \eta,\theta\in\{\imath,\jmath,\kappa\},\\ 0, & \eta\neq\theta,\quad \eta,\theta\in\{\imath,\jmath,\kappa\}.\end{cases} \tag{12.19}$$

### 12.2.4 QUATERNION NONLINEAR ACTIVATION FUNCTIONS

Hypercomplex nonlinear learning systems, like quaternion-valued RNNs, are more difficult to design than their real-valued counterparts, owing to the lack of analytic nonlinear activation functions. This is due to the stringent CRF conditions [18]. For instance, a CRF differentiable quaternion function $f(q)$ should satisfy

$$\frac{\partial f}{\partial q_r}+\imath\frac{\partial f}{\partial q_\imath}+\jmath\frac{\partial f}{\partial q_\jmath}+\kappa\frac{\partial f}{\partial q_\kappa}=0\Leftrightarrow\frac{\partial f}{\partial q^*}=0. \tag{12.20}$$

It is important to note that only linear quaternion functions and constants fulfill these conditions, yet nonlinear learning algorithms in $\mathbb{H}$ require differentiable nonlinear functions. To circumvent the analyticity problem, recent work in [22] adopted the local analyticity condition (LAC) [21], based on a complex-valued representation of a quaternion, to give

$$\frac{\partial f}{\partial q_r}=-\zeta\frac{\partial f}{\partial \alpha}, \tag{12.21}$$

where $\zeta$ and $\alpha$ are respectively given by

$$\zeta = \frac{\imath q_\imath + \jmath q_\jmath + \kappa q_\kappa}{\alpha}, \quad \alpha = \sqrt{q_\imath^2 + q_\jmath^2 + q_\kappa^2}. \tag{12.22}$$

In this way, an "imaginary" unit $\zeta$ comprises the vector part of quaternions. Although the local analyticity condition only guarantees first-order differentiability at the current operating point, this is a perfect match for gradient descent–based quaternion-valued learning algorithms, which require gradient evaluation at a point only.

**Proposition 1.** *The quaternion exponential* $e^q = e^{q_r + \imath q_\imath + \jmath q_\jmath + \kappa q_\kappa}$ *satisfies the LAC in* (12.21).

*Proof.* We can expand $e^q$ using the Euler formula as

$$\begin{aligned} e^q &= e^{q_r}\left(\cos(\alpha) + \zeta \sin(\alpha)\right) \\ &= e^{q_r}\left(\cos(\alpha) + \frac{\imath q_\imath \sin(\alpha)}{\alpha} + \frac{\jmath q_\jmath \sin(\alpha)}{\alpha} + \frac{\kappa q_\kappa \sin(\alpha)}{\alpha}\right), \end{aligned} \tag{12.23}$$

where $\zeta$ and $\alpha$ are defined in (12.22), to give

$$\frac{\partial e^q}{\partial q_r} = e^q = -\zeta \frac{e^q}{\alpha}. \tag{12.24}$$

Note that the quaternion exponential $e^{-q} = e^{-(q_r + \imath q_\imath + \jmath q_\jmath + \kappa q_\kappa)}$ also satisfies the LAC in (12.21). This is straightforward to show using the same approach as in **Proposition 1**. In this way, quaternion transcendental nonlinear functions, constructed on the basis of quaternion exponentials $e^q$ and $e^{-q}$, are a generic extension of those in $\mathbb{R}$ and $\mathbb{C}$, and they also satisfy the LAC [22]. In this chapter, we employ a fully quaternion hyperbolic tangent function to design the quaternion ESNs, defined as [22]

$$\tanh(q) = \frac{e^q - e^{-q}}{e^q + e^{-q}} = \frac{e^{2q} - 1}{e^{2q} + 1}, \tag{12.25}$$

for which the first derivative is given by

$$\frac{\partial \tanh(q)}{\partial q} = \mathrm{sech}^2(q), \quad \mathrm{sech}(q) = \frac{2}{e^q + e^{-q}}. \tag{12.26}$$

## 12.3 QUATERNION ESNS

The existence of fully quaternion nonlinear activation functions enables the design of RNNs in $\mathbb{H}$ [22, 23]. In this section, the quaternion ESNs (QESNs) and its augmented (widely linear) variant (AQESNs) are introduced.

### 12.3.1 STANDARD QESNS

The architecture of a standard ESN, which is a recurrent discrete-time neural network composed of $K$ external inputs, $N$ internal neurons (also referred to as dynamical reservoir) and $L$ readout neurons, is

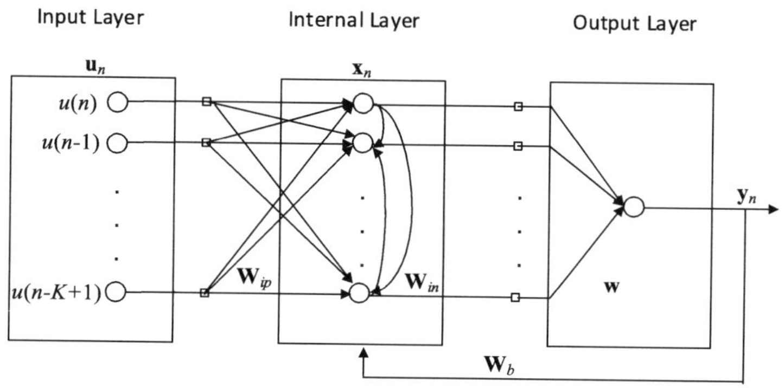


**FIGURE 12.1**

The architecture of an Echo State Network (ESN).

shown in Fig. 12.1. The $N \times K$ weight matrix $\mathbf{W}_{ip}$ and the $N \times N$ weight matrix $\mathbf{W}_{in}$ respectively contain the connections between the input units and the internal units and those between the internal units themselves, whereas the feedback connections between the internal neurons and the readout neurons are stored in the $N \times L$ feedback weight matrix $\mathbf{W}_b$. The $N \times 1$ internal state vector $\mathbf{x}_n$, $K \times 1$ input vector $\mathbf{u}_n$ and $L \times 1$ output vector $\mathbf{y}_n$ are all at time instant $n$. By concatenating the input $\mathbf{u}_n$, the internal state $\mathbf{x}_n$ and the delayed output $\mathbf{y}_{n-1}$, the overall network state, denoted by $\mathbf{s}_n$, is defined as

$$\mathbf{s}_n = [u(n), \ldots, u(n-K+1), x_1(n), \ldots, x_N(n), y(n-1), \ldots, y(n-L)]^{\mathrm{T}}, \tag{12.27}$$

while the internal unit dynamics are updated according to

$$\mathbf{x}_n = \Phi\big(\mathbf{W}_{ip}\mathbf{u}_n + \mathbf{W}_{in}\mathbf{x}_{n-1} + \mathbf{W}_b\mathbf{y}_{n-1}\big), \tag{12.28}$$

where $\Phi(\cdot)$ here is a quaternion-valued nonlinear activation of the neurons within the reservoir. The echo state property plays a key role in ensuring adequate operation of the dynamical reservoir of ESNs, which means that the internal neurons $\mathbf{x}_n$ within the dynamical reservoir of ESNs should display systematic variants of the exciting external signal $\mathbf{u}_n$, working as its echo functions. A practical procedure to ensure well-prepared RNNs with echo state property adopts a two-step operation on $\mathbf{W}_{in}$: (1) an internal weight matrix $\mathbf{W}_{in}$ is randomly drawn from a uniform distribution over a symmetric interval, (2) $\mathbf{W}_{in}$ is rescaled by $\mathbf{W}_{in} \leftarrow \mathbf{W}_{in}/|\lambda_{\max}|$, where $|\lambda_{\max}|$ is the largest absolute eigenvalue of $\mathbf{W}_{in}$ (spectral radius). In this way, the initialization of input and feedback connections stored in $\mathbf{W}_{ip}$ and $\mathbf{W}_b$ can be arbitrary [24,25]. To find recent advances in echo state property analysis, readers are advised to refer to [45–47]. The output of a nonlinear output mapping of ESNs is given by

$$y(n) = \Phi\big(\mathbf{w}_n^{\mathrm{T}}\mathbf{s}_n\big), \tag{12.29}$$

where $\mathbf{w}_n$ is the weight vector associated with the output layer, for which the value is updated by minimizing suitable cost functions. The cost function chosen here is the instantaneous squared error, defined as

$$E(n)=\frac{1}{2}|e(n)|^2=\frac{1}{2}e(n)e^*(n), \tag{12.30}$$

where $d(n)$ is the desired (teaching) signal and $e(n)$ is the instantaneous output error, given by $e(n)=d(n)-y(n)$. Note that $E(n)$ in (12.30) is real-valued, dependent on both quaternion-valued $e(n)$ and $e^*(n)$, and for such functions, it has been shown by $\mathbb{HR}$ calculus that the maximum change of gradient lies in the direction of the conjugate gradient, which is in accordance with the corresponding solution in $\mathbb{C}$ [43]. In this way, the minimization of $E(n)$ through gradient descent is given by

$$\mathbf{w}_{n+1}=\mathbf{w}_n-\mu\nabla_{\mathbf{w}^*}E(n), \tag{12.31}$$

where $\mu$ is the step size, a small positive constant. Using the chain rule, the gradient $\nabla_{\mathbf{w}^*}E(n)$ in (12.31) can be derived as

$$\begin{aligned}\nabla_{\mathbf{w}^*}E(n)&=\frac{1}{2}\frac{\partial|e(n)|^2}{\partial\mathbf{w}_n^*}\\&=\frac{1}{2}\left(e(n)\frac{\partial e^*(n)}{\partial\mathbf{w}_n^*}+\frac{\partial e(n)}{\partial\mathbf{w}_n^*}e^*(n)\right),\end{aligned} \tag{12.32}$$

where

$$e(n)=d(n)-y(n)=d(n)-\Phi\big(\mathbf{w}_n^{\mathrm{T}}\mathbf{s}_n\big). \tag{12.33}$$

Note that due to the noncommutativity of the quaternion product, the partial derivatives in (12.32) cannot be swapped with $e(n)$ and $e^*(n)$. Using the fact that $\Phi^*(q)=\Phi(q^*)$ and $(q_1q_2)^*=q_2^*q_1^*$, the output error conjugate $e^*(n)$ can be derived as

$$\begin{aligned}e^*(n)&=d^*(n)-\Phi^*\big(\mathbf{w}_n^{\mathrm{T}}\mathbf{s}_n\big)\\&=d^*(n)-\Phi\big((\mathbf{w}_n^{\mathrm{T}}\mathbf{s}_n)^*\big)\\&=d^*(n)-\Phi\big(\mathbf{s}_n^{\mathrm{H}}\mathbf{w}_n^*\big).\end{aligned} \tag{12.34}$$

Substituting it into the first gradient on the right-hand side of (12.32), we have

$$\begin{aligned}\frac{\partial e^*(n)}{\partial\mathbf{w}_n^*}&=\frac{\partial\big(d^*(n)-\Phi(\mathbf{s}_n^{\mathrm{H}}\mathbf{w}_n^*)\big)}{\partial\mathbf{w}_n^*}\\&=-\frac{\partial\Phi\big(\mathbf{s}_n^{\mathrm{H}}\mathbf{w}_n^*\big)}{\partial\mathbf{w}_n^*}\\&=-\Phi'\big(\mathbf{s}_n^{\mathrm{H}}\mathbf{w}_n^*\big)\frac{\partial\big(\mathbf{s}_n^{\mathrm{H}}\mathbf{w}_n^*\big)}{\partial\mathbf{w}_n^*}\\&=-\Phi'^*\big(\mathbf{w}_n^{\mathrm{T}}\mathbf{s}_n\big)\mathbf{s}_n^*,\end{aligned} \tag{12.35}$$

where the last step is achieved by using the $\mathbb{HR}$-derivative introduced in (12.17) in Section 12.2.3. On the other hand,

$$\frac{\partial e(n)}{\partial \mathbf{w}_n^*} = \frac{\partial\left(d(n) - \Phi(\mathbf{w}_n^{\mathrm{T}}\mathbf{s}_n)\right)}{\partial \mathbf{w}_n^*} = -\frac{\partial \Phi(\mathbf{w}_n^{\mathrm{T}}\mathbf{s}_n)}{\partial \mathbf{w}_n^*} = -\Phi'(\mathbf{w}_n^{\mathrm{T}}\mathbf{s}_n)\frac{\partial\left(\mathbf{w}_n^{\mathrm{T}}\mathbf{s}_n\right)}{\partial \mathbf{w}_n^*}. \tag{12.36}$$

Using the $\mathbb{HR}^*$-derivative given in (12.18), we obtain

$$\frac{\partial\left(\mathbf{w}_n^{\mathrm{T}}\mathbf{s}_n\right)}{\partial \mathbf{w}_n^*} = -\frac{\mathbf{s}_n}{2} \tag{12.37}$$

and hence

$$\frac{\partial e(n)}{\partial \mathbf{w}_n^*} = \frac{\Phi'\left(\mathbf{w}_n^{\mathrm{T}}\mathbf{s}_n\right)\mathbf{s}_n}{2}. \tag{12.38}$$

Finally, we arrive at the update of the weight vector $\mathbf{w}_n$ in the output layer, given by

$$\mathbf{w}_{n+1} = \mathbf{w}_n - \mu\nabla_{\mathbf{w}^*}E(n) = \mathbf{w}_n + \mu\left(e(n)\Phi'^*\left(\mathbf{w}_n^{\mathrm{T}}\mathbf{s}_n\right)\mathbf{s}_n^* - \frac{1}{2}\Phi'\left(\mathbf{w}_n^{\mathrm{T}}\mathbf{s}_n\right)\mathbf{s}_n e^*(n)\right), \tag{12.39}$$

where the factor $1/2$ in the cost function (12.30) is absorbed into the step size $\mu$.

### 12.3.2 AUGMENTED QESNS

In order to make QESNs optimal for general quaternion signals (both second-order circular and noncircular), we use recent developments in augmented quaternion statistics, as described in Section 12.2.2, to incorporate the widely linear model into the QESN architecture, which gives the augmented version of QESNs (AQESNs). This means that the network state should be "augmented" in such a way as

$$\mathbf{s}_n^a = \left[\mathbf{u}_n, \mathbf{u}_n^{\imath}, \mathbf{u}_n^{\jmath}, \mathbf{u}_n^{\kappa}, \mathbf{x}_n^a, \mathbf{y}_n\right]^{\mathrm{T}}. \tag{12.40}$$

This augmented network state design is generally applicable for not only the architecture of QESNs, but also quaternion-valued extensions of other types of RNNs. Since the input weights of QESNs contained in the matrix $\mathbf{W}_{ip}$ are randomly chosen prior to training, we can use three other matrices $\mathbf{W}_{ip1}$, $\mathbf{W}_{ip2}$ and $\mathbf{W}_{ip3}$ to initialize the weights associated with the input involutions $\mathbf{u}_n^{\imath}, \mathbf{u}_n^{\jmath}, \mathbf{u}_n^{\kappa}$. To this point, the update of the internal state dynamics within the AQESNs is given by

$$\mathbf{x}_n^a = \Phi\left(\mathbf{W}_{ip}\mathbf{u}_n + \mathbf{W}_{ip1}\mathbf{u}_n^{\imath} + \mathbf{W}_{ip2}\mathbf{u}_n^{\jmath} + \mathbf{W}_{ip3}\mathbf{u}_n^{\kappa} + \mathbf{W}_{in}\mathbf{x}_{n-1} + \mathbf{W}_b\mathbf{y}_{n-1}\right). \tag{12.41}$$

Due to the specific properties of the ESN output layer, the output $y(n)$ is now governed by an asymmetric version of the quaternion widely linear model in (12.14) to yield

$$\begin{aligned} net(n) &= \mathbf{a}_n^{\mathrm{T}}\mathbf{v}_n + \mathbf{b}_n^{\mathrm{T}}\mathbf{u}_n^{\imath} + \mathbf{c}_n^{\mathrm{T}}\mathbf{u}_n^{\jmath} + \mathbf{d}_n^{\mathrm{T}}\mathbf{u}_n^{\kappa}, \\ y(n) &= \Phi\big(net(n)\big), \end{aligned} \tag{12.42}$$

where $\mathbf{v}_n = [u(n), \ldots, u(n-K+1), x_1^a(n), \ldots x_N^a(n), y(n-1), \ldots, y(n-L)]^{\mathrm{T}}$ is a subset of the augmented network state $\mathbf{s}_n^a$ and has the same dimension $(K+N+L)$ as the state vector $\mathbf{s}_n$ within standard ESNs; however, the internal state dynamics are updated using (12.41). The recursion of the output weight vectors $\{\mathbf{a}_n, \mathbf{b}_n, \mathbf{c}_n, \mathbf{d}_n\}$ are made gradient adaptive according to

$$\begin{aligned} \mathbf{a}_{n+1} &= \mathbf{a}_n - \mu\nabla_{\mathbf{a}^*}E(n), \\ \mathbf{b}_{n+1} &= \mathbf{b}_n - \mu\nabla_{\mathbf{b}^*}E(n), \\ \mathbf{c}_{n+1} &= \mathbf{c}_n - \mu\nabla_{\mathbf{c}^*}E(n), \\ \mathbf{d}_{n+1} &= \mathbf{d}_n - \mu\nabla_{\mathbf{d}^*}E(n). \end{aligned} \tag{12.43}$$

We notice that the gradient $\nabla_{\mathbf{a}^*}E(n)$ in (12.43) and its counterpart $\nabla_{\mathbf{s}^*}E(n)$ in (12.31) are equivalent, so that they have the same dimension, and hence

$$\mathbf{a}_{n+1} = \mathbf{a}_n + \mu\big(e(n)\Phi'^*(net(n))\mathbf{v}_n^* - \frac{1}{2}\Phi'(net(n))\mathbf{v}_n\big). \tag{12.44}$$

The error gradient $\nabla_{\mathbf{b}^*}E(n)$ for the weight vector $\mathbf{b}_n$ corresponding to the $\imath$-involution $\mathbf{u}_n^{\imath}$ of the input $\mathbf{u}$ is given by

$$\nabla_{\mathbf{b}^*}E(n) = \frac{1}{2}\left(e(n)\frac{\partial e^*(n)}{\partial \mathbf{b}_n^*} + \frac{\partial e(n)}{\partial \mathbf{b}_n^*}e^*(n)\right), \tag{12.45}$$

where

$$\begin{aligned} \frac{\partial e^*(n)}{\partial \mathbf{b}_n^*} &= -\frac{\partial\Phi\big(net^*(n)\big)}{\partial \mathbf{b}_n^*} \\ &= -\Phi'^*\big(net(n)\big)\frac{\partial net^*(n)}{\partial \mathbf{b}_n^*} \\ &= -\Phi'^*\big(net(n)\big)\mathbf{u}_n^{\imath *} \end{aligned} \tag{12.46}$$

and

$$\begin{aligned} \frac{\partial e(n)}{\partial \mathbf{b}_n^*} &= -\frac{\partial\Phi\big(net(n)\big)}{\partial \mathbf{b}_n^*} \\ &= -\Phi'\big(net(n)\big)\frac{\partial net(n)}{\partial \mathbf{b}_n^*} \\ &= \frac{1}{2}\Phi'\big(net(n)\big)\mathbf{u}_n^{\imath}. \end{aligned} \tag{12.47}$$

Substituting the partial derivatives in (12.46) and (12.47) into the error gradient $\nabla_{\mathbf{b}^*} E(n)$ given in (12.45) yields

$$\mathbf{b}_{n+1} = \mathbf{b}_n + \mu\Big(e(n)\Phi'^*\big(net(n)\big)\mathbf{u}_n^{\imath *} - \frac{1}{2}\Phi'\big(net(n)\big)\mathbf{u}_n^{\imath}e^*(n)\Big).$$

Proceeding in a similar manner, the weight updates for $\mathbf{c}_n$ and $\mathbf{d}_n$ are found to be

$$\mathbf{c}_{n+1}=\mathbf{c}_n + \mu\Big(e(n)\Phi'^*\big(net(n)\big)\mathbf{u}_n^{\jmath *} - \frac{1}{2}\Phi'\big(net(n)\big)\mathbf{u}_n^{\jmath}e^*(n)\Big) \tag{12.48}$$

and

$$\mathbf{d}_{n+1}=\mathbf{d}_n + \mu\Big(e(n)\Phi'^*\big(net(n)\big)\mathbf{u}_n^{\kappa *} - \frac{1}{2}\Phi'\big(net(n)\big)\mathbf{u}_n^{\kappa}e^*(n)\Big). \tag{12.49}$$

For convenience, the final weight update of the gradient decent algorithm used to train the output layer of the AQESNs can be written in an augmented form using (12.40) as

$$\mathbf{w}_{n+1}^a = \mathbf{w}_n^a + \mu\Big(e(n)\Phi'^*\big(net(n)\big)\mathbf{s}_n^{a*} - \frac{1}{2}\Phi'\big(net(n)\big)\mathbf{s}_n^{a}e^*(n)\Big), \tag{12.50}$$

where $\mathbf{w}_n^a = [\mathbf{a}_n^{\mathrm{T}}, \mathbf{b}_n^{\mathrm{T}}, \mathbf{c}_n^{\mathrm{T}}, \mathbf{d}_n^{\mathrm{T}}]^{\mathrm{T}}$ is the augmented weight vector and $net(n) = \mathbf{w}_n^{a\mathrm{T}}\mathbf{s}_n^a$.

### 12.3.3 STABILITY ANALYSIS OF QESNS AND AQESNS

In order to ensure satisfactory performances of both QESNs and AQESNs, we now address the set bounds on the values of the step size $\mu$ to guarantee their stability. Following the approach from [2], we employ the mean square convergence criterion, given by

$$E\Big[|\bar{e}(n)|^2\Big] < E\Big[|\tilde{e}(n)|^2\Big], \tag{12.51}$$

where $\bar{e}(n)$ and $\tilde{e}(n)$ are respectively the *a posteriori* and the *a priori* output error, defined as

$$\begin{aligned}\bar{e}(n) &= d(n) - \Phi\big(\mathbf{w}_{n+1}^{\mathrm{T}}\mathbf{s}_n\big),\\ \tilde{e}(n) &= d(n) - \Phi\big(\mathbf{w}_n^{\mathrm{T}}\mathbf{s}_n\big).\end{aligned} \tag{12.52}$$

The error terms $\bar{e}(n)$ and $\tilde{e}(n)$ in (12.52) can be related by the first-order Taylor series expansion (TSE) as [48]

$$\begin{aligned}|\bar{e}(n)|^2 &= |\tilde{e}(n)|^2 + \frac{\partial|\tilde{e}(n)|^2}{\partial \mathbf{w}_n^{\mathrm{T}}}\Delta\mathbf{w}_n + \frac{\partial|\tilde{e}(n)|^2}{\partial \mathbf{w}_n^{\imath\mathrm{T}}}\Delta\mathbf{w}_n^{\imath} + \frac{\partial|\tilde{e}(n)|^2}{\partial \mathbf{w}_n^{\jmath\mathrm{T}}}\Delta\mathbf{w}_n^{\jmath} + \frac{\partial|\tilde{e}(n)|^2}{\partial \mathbf{w}_n^{\kappa\mathrm{T}}}\Delta\mathbf{w}_n^{\kappa}\\ &= |\tilde{e}(n)|^2 + 4\Re\left(\frac{\partial|\tilde{e}(n)|^2}{\partial \mathbf{w}_n^{\mathrm{T}}}\Delta\mathbf{w}_n\right).\end{aligned} \tag{12.53}$$

To simplify the derivation of (12.53), notice that $\Re(q_1 q_2) = \Re((q_1 q_2)^*) = \Re(q_2^* q_1^*)$ for any pair $\{q_1, q_2\} \in \mathbb{H}$ to give

$$|\bar{e}(n)|^2 = |\tilde{e}(n)|^2 + 4\Re\left(\Delta\mathbf{w}_n^{\mathrm{H}} \frac{\partial |\tilde{e}(n)|^2}{\partial \mathbf{w}_n^*}\right), \tag{12.54}$$

where the partial derivative $\partial|\tilde{e}(n)|^2/\partial\mathbf{w}_n^*$ is effectively the gradient of the cost function with respect to $\mathbf{w}_n^*$, given in (12.32). The term $\Delta\mathbf{w}_n^{\mathrm{H}}$ can be expressed by using (12.39) as

$$\begin{aligned}\Delta\mathbf{w}_n^{\mathrm{H}} &= \left(\mathbf{w}_{n+1} - \mathbf{w}_n\right)^{\mathrm{H}}\\ &= \mu\left(\tilde{e}(n)\Phi'^*(\mathbf{w}_n^{\mathrm{T}}\mathbf{s}_n)\mathbf{s}_n^* - \frac{1}{2}\Phi'(\mathbf{w}_n^{\mathrm{T}}\mathbf{s}_n)\mathbf{s}_n\tilde{e}^*(n)\right)^{\mathrm{H}}\\ &= \mu\left(\mathbf{s}_n^{\mathrm{T}}\Phi'(\mathbf{w}_n^{\mathrm{T}}\mathbf{s}_n)\tilde{e}^*(n) - \frac{1}{2}\tilde{e}(n)\mathbf{s}_n^{\mathrm{H}}\Phi'^*(\mathbf{w}_n^{\mathrm{T}}\mathbf{s}_n)\right).\end{aligned} \tag{12.55}$$

Substituting (12.32) and (12.55) into the TSE in (12.54) yields

$$\begin{aligned}|\bar{e}(n)|^2 &= |\tilde{e}(n)|^2 - 4\mu\Re\Big(\big(\mathbf{s}_n^{\mathrm{T}}\Phi'(\mathbf{w}_n^{\mathrm{T}}\mathbf{s}_n)\tilde{e}^*(n) - \frac{1}{2}\tilde{e}(k)\mathbf{s}_n^{\mathrm{H}}\Phi'^*(\mathbf{w}_n^{\mathrm{T}}\mathbf{s}_n)\big)\\ &\quad\cdot\big(\tilde{e}(n)\Phi'^*(\mathbf{w}_n^{\mathrm{T}}\mathbf{s}_n)\mathbf{s}_n^* - \frac{1}{2}\Phi'(\mathbf{w}_n^{\mathrm{T}}\mathbf{s}_n)\mathbf{s}_n\tilde{e}^*(n)\big)\Big)\\ &= |\tilde{e}(n)|^2\left(1 - 5\mu|\Phi'(\mathbf{w}_n^{\mathrm{T}}\mathbf{s}_n)|^2\|\mathbf{s}_n\|_2^2|\right)\\ &\quad + 4\mu\Re\left(\mathbf{s}_n^{\mathrm{T}}\Phi'(\mathbf{w}_n^{\mathrm{T}}\mathbf{s}_n)\tilde{e}^*(n)\Phi'(\mathbf{w}_n^{\mathrm{T}}\mathbf{s}_n)\mathbf{s}_n\tilde{e}^*(n)\right).\end{aligned} \tag{12.56}$$

Given that the term $\Re(\cdot)$ is negligible, upon applying the statistical expectation operator on both sides of (12.56), we have

$$E\left[|\bar{e}(n)|^2\right] = E\left[|\tilde{e}(n)|^2(1 - 5\mu|\Phi'(\mathbf{w}_n^{\mathrm{T}}\mathbf{s}_n)|^2\|\mathbf{s}_n\|_2^2|)\right]. \tag{12.57}$$

By using the standard statistical independence assumption between $\tilde{e}(n)$ and $\mathbf{s}_n$, we have

$$E\left[|\bar{e}(n)|^2\right] = E\left[|\tilde{e}(n)|^2\right]E\left[1 - 5\mu|\Phi'(\mathbf{w}_n^{\mathrm{T}}\mathbf{s}_n)|^2\|\mathbf{s}_n\|_2^2\right]. \tag{12.58}$$

Therefore, the convergence condition in (12.51) is satisfied for

$$0 < E\left[5\mu|\Phi'(\mathbf{w}_n^{\mathrm{T}}\mathbf{s}_n)|^2\|\mathbf{s}_n\|_2^2\right] < 1. \tag{12.59}$$

Solving for $\mu$, we obtain the range of the step size for QESNs to converge as

$$0 < \mu < \frac{1}{E\left[5|\Phi'(\mathbf{w}_n^{\mathrm{T}}\mathbf{s}_n)|^2\|\mathbf{s}_n\|_2^2\right]}. \tag{12.60}$$

Similarly, the bounds on $\mu$ to ensure the convergence of AQESNs are given by

$$0 < \mu < \frac{1}{E\left[5|\Phi'(\mathbf{w}_n^{a\mathrm{T}}\mathbf{s}_n^a)|^2\|\mathbf{s}_n^a\|_2^2\right]}. \tag{12.61}$$

The upper bounds of both (12.60) and (12.61) are governed by the expected value of the network state vector and the gradient of the fully quaternion nonlinearity. However, we should note that the upper bound of $\mu$ for the AQESNs in (12.61) is smaller than that of the standard QESNs in (12.60), due to the larger size of the augmented network state vector $\mathbf{s}_n^a$.

## 12.4 SIMULATIONS

To verify the suitability of QESNs to cater for 3D and 4D nonlinear processes and our idea of exploiting the noncircularity of multidimensional data through the widely linear model, simulations in a prediction setting were performed for both benchmark synthetic second-order circular (proper) and noncircular (improper) 3D and 4D signals and real-world noncircular, nonlinear and nonstationary 3D body motion and wind data. In order to implement the adequate buildup of QESNs, which are guaranteed to maintain the echo state property, the randomly selected input, internal and feedback weights $\mathbf{W}_{ip}$, $\mathbf{W}_{in}$ and $\mathbf{W}_b$ were generated from a uniform distribution in range $[-1, 1]$, and the spectral radius of $\mathbf{W}_{in}$ was 0.8. The internal weights connections contained in $\mathbf{W}_{in}$ were sparse with 5% interconnectivity, so that the reservoir contained many loosely coupled subsystems, and a rich variety of dynamics of different internal units can be achieved [24,25]. The fully quaternion $\tanh(q)$ function was employed as the activation function of QESNs. The step size $\mu$ of the gradient descent algorithms used to train the output layers of QESNs and augmented (widely linear) QESNs (AQESNs) was set at $\mu = 0.02$. To assess the quantitative performance of learning algorithms, it is convenient to use the prediction gain $\mathrm{R_p}$, defined as

$$\mathrm{R_p} = 10\log_{10}\frac{\hat{\sigma}_q^2}{\hat{\sigma}_e^2}, \tag{12.62}$$

which is a logarithmic ratio between the estimated signal variance $\hat{\sigma}_q^2$ and estimated prediction error variance $\hat{\sigma}_e^2$. Five input neurons and five output neurons were used and the different dynamical reservoir sizes of ESNs were investigated. Due to their random network preparation, all the simulation results were obtained by averaging 100 independent simulation trials.

In the first set of simulations, we considered a benchmark synthetic second-order circular (proper) 4D autoregressive AR(4) process, given by [2]

$$u(n) = 1.79u(n-1) - 1.85u(n-2) + 1.27u(n-3) - 0.41u(n-4) + q(n), \tag{12.63}$$

driven by a circular quaternion white Gaussian noise (QWGN) $q(n) = q_r(n) + \imath q_\imath(n) + \jmath q_\jmath(n) + \kappa q_\kappa(n)$, where $q_r(n)$, $q_\imath(n)$, $q_\jmath(n)$ and $q_\kappa(n)$ are independent realizations of real-valued WGN $\sim \mathcal{N}(0, 1)$.

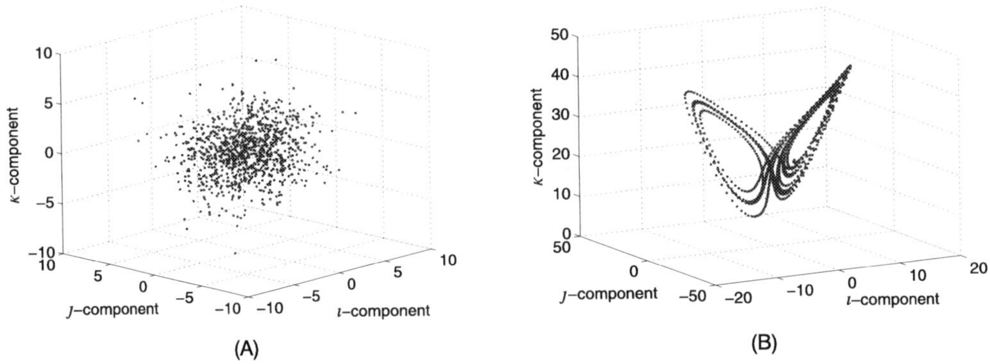


**FIGURE 12.2**

Geometric view of circularity via a 3D scatter diagram. (A) Circular 4D AR(4) signal. (B) Noncircular and nonlinear 3D chaotic Lorenz signal. For illustration purpose, only the ı-, ȷ- and κ-components of the quaternion AR(4) signal are provided.

**Table 12.1 Comparison of prediction gains $R_p$ of standard QESNs and AQESNs, as well as strictly linear QLMS and its augmented (widely linear) version AQLMS adaptive learning algorithms, for the benchmark signals**

| $R_p$ [dB] | 4D AR(4) | 3D Lorenz |
|---|---|---|
| QLMS | 3.42 | 5.24 |
| AQLMS | 3.38 | 10.02 |
| QESNs | 3.39 | 14.54 |
| AQESNs | 3.35 | 17.96 |

The benchmark noncircular and nonlinear signal considered was the chaotic 3D Lorenz attractor, governed by coupled partial differential equations [49]

$$\frac{\partial x}{\partial t} = \alpha(y - x), \frac{\partial y}{\partial t} = x(\rho - z) - y, \frac{\partial z}{\partial t} = xy - \beta z, \tag{12.64}$$

where $\alpha = 10$, $\rho = 28$ and $\beta = 8/3$, and it was constructed as a pure quaternion by using the three imaginary components.

Fig. 12.2 shows the 3D scatter plots of the considered benchmark signals, in which the 4D AR(4) process had a rotation invariant distribution (circular), while the 3D Lorenz signal possessed an obvious noncircular nature. Table 12.1 compares the prediction gains $R_p$ of QESNs and AQESNs with 100 internal neurons in the dynamical reservoir, as well as strictly linear quaternion least mean square (QLMS) and its widely linear version augmented QLMS (AQLMS) adaptive learning algorithms [16, 37,43], for the benchmark signals considered. For the strictly linear and circular 4D AR(4) signal,

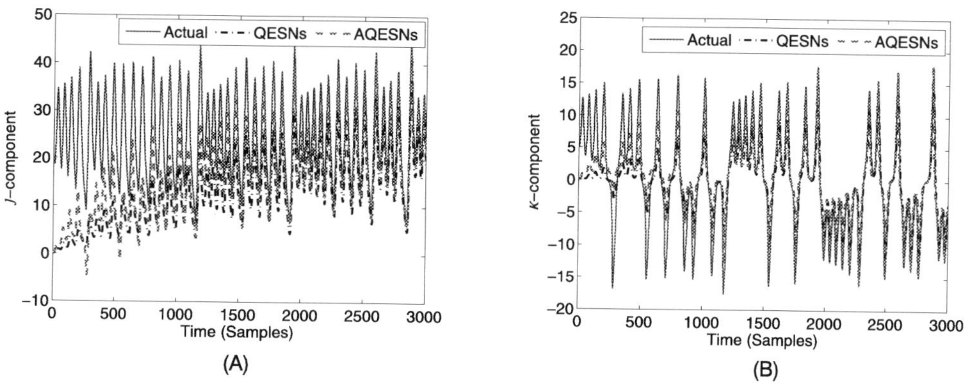


**FIGURE 12.3**

One-step ahead prediction of the noncircular and nonlinear 3D Lorenz signal using QESNs and AQESNs. Amplitudes of the actual and predicted signals are plotted. (A) $\jmath$-component. (B) $\kappa$-component.

the performance of AQESNs was comparable but slightly worse than that of QESNs. This is because the widely linear model–based input layer of AQESNs simplifies into the strictly linear one used in QESNs, as the weights associated with the involutions of the input vector, that is, $\{\mathbf{b}, \mathbf{c}, \mathbf{d}\}$ in (12.41), vanish theoretically; however, in practice, small weight deviations from zero arise from the employed learning algorithm. For the noncircular 3D Lorenz signal, there was a significant overall performance improvement in terms of the prediction gain $R_p$ when the widely linear model–based AQESNs were employed as compared with the strictly linear QESNs. This is due to their faster and more accurate tracking capability of the signal dynamics arising from their natural exploitation of data noncircularity, as indicated by Fig. 12.3. The main motivation to design quaternion-valued nonlinear learning algorithms with enhanced modeling capacities for nonlinear dynamic systems, as compared with the strictly linear learning algorithms, has been also justified by observing the higher prediction gains achieved by QESNs and AQESNs than those of QLMS and AQLMS algorithms, on the nonlinear and noncircular 3D Lorenz signal. As expected, for strictly linear and circular 4D AR(4) process, all the considered learning algorithms yielded comparable prediction results.

In the next set of simulations, we investigated the usefulness of the QESNs and AQENs for a real-world 3D noncircular and nonstationary body motion tracking experiment. The 3D motion data were recorded using the XSense MTx 3-DOF orientation tracker, placed on the left and right hands, the left and right arms and the waist of an athlete performing Tai Chi movements. The movement of the left arm was used as a pure quaternion input, and QESNs and AQESNs with 200 internal neurons were employed for this experiment. Fig. 12.4A–C show that along all the three dimensions, AQESNs were able to track the left arm dynamics of the athlete more quickly and more accurately than their strictly linear counterparts, which is further supported by the evolution of their mean square errors (MSEs), shown in Fig. 12.4D.

Wind forecasting at short scales plays an important role in renewable energy, smart grids, air pollution modeling and aviation safety [50]. To further demonstrate the modeling ability of the designed quaternion-valued nonlinear learning systems in the fusion of heterogeneous data sources, a 4D wind model was considered in such a way that the 3D noncircular, nonlinear and nonstationary wind speed

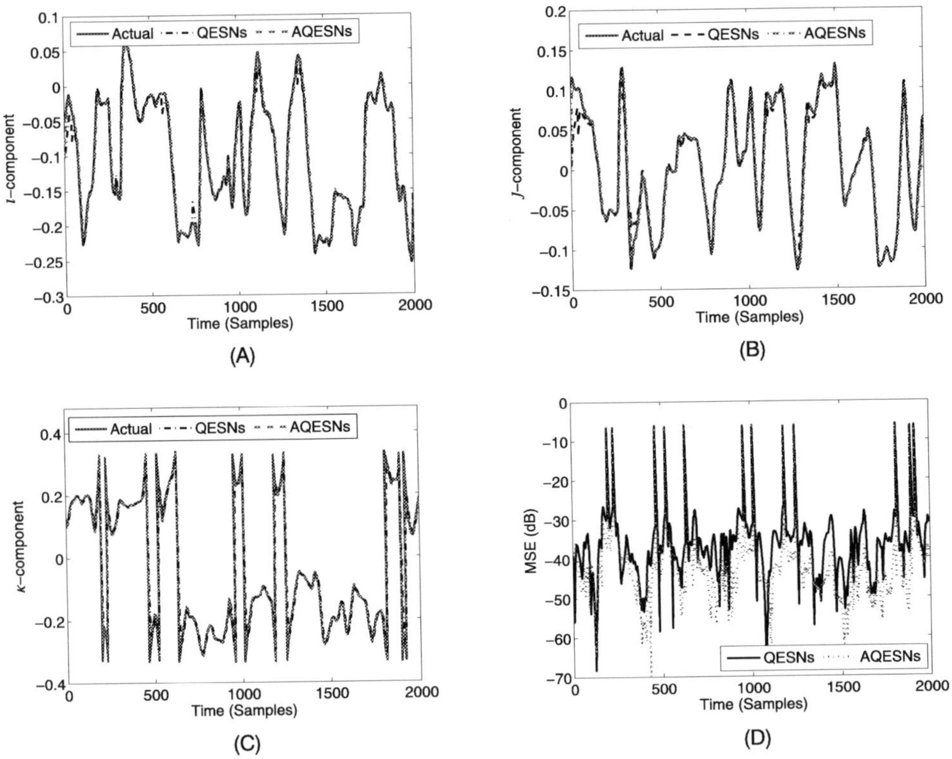


**FIGURE 12.4**

Tracking capabilities and MSE performances, evaluated over all the three dimensions, of QESNs and AQESNs on a one-step ahead prediction of the 3D left arm motion of an athlete performing Tai Chi movements. (A) $\imath$-component. (B) $\jmath$-component. (C) $\kappa$-component. (D) MSE (dB).

data in east, north and vertical directions constituted the vector part of the quaternion, and the air temperature was used as a scalar (real) part. Note that the validity of incorporating air temperature in wind forecasting has been justified by the fact that the air temperature has a high degree of correlation with the wind speed [51]. Fig. 12.5A shows the overall prediction performance over the three dimensions of the wind speed data for QESNs with 100 internal neurons, where an improved accuracy was achieved by incorporating the widely linear model into QESNs. It is interesting to note that the widely linear model enabled AQESNs to more closely follow the highly nonstationary trend of the air temperature, as indicated by Fig. 12.5B. From Fig. 12.5A and Fig. 12.5B, we also observe that the strictly linear QLMS learning algorithm, and even its augmented version AQLMS, were not able to track the highly nonlinear dynamics of the heterogeneous 4D wind data. To further illustrate the robustness of the performance advantage of using widely linear model within QESNs, we compared the performances of both augmented and standard QESNs against the size of the dynamical reservoir, an important parameter that influences the performance of ESNs, as it reflects their universal approximation capability. Generally speaking, an ESN with a larger reservoir can learn the signal dynamics with higher accu-

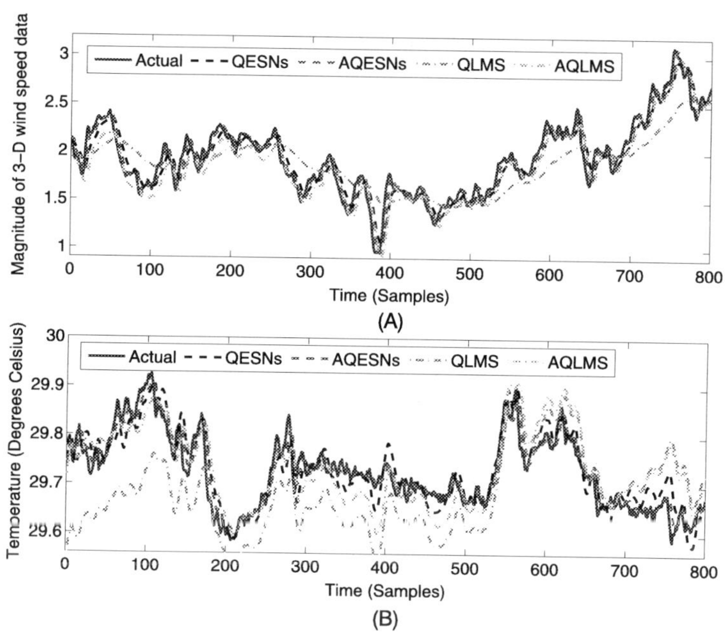


**FIGURE 12.5**

One-step ahead prediction of the 4D wind data using QESNs and AQESNs with 100 internal neurons, and strictly linear QLMS and AQLMS learning algorithms. (A) Magnitude of the 3D wind speed. (B) Air temperature.

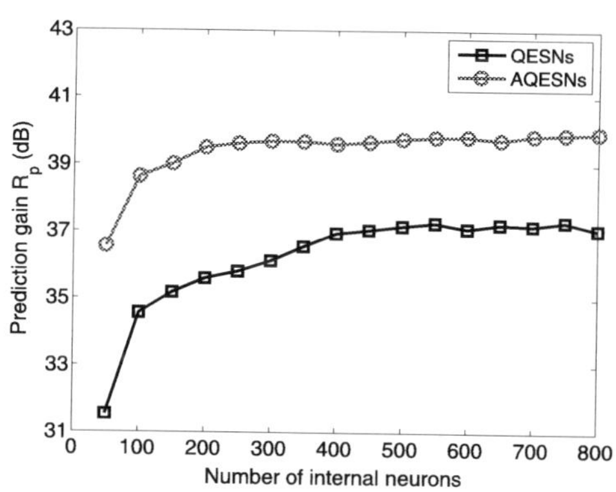


**FIGURE 12.6**

Performance comparison of QESNs and AQESNs over a range of reservoir sizes on one-step ahead prediction of the 4D wind data.

racy. This is confirmed in Fig. 12.6, where the performances of QESNs and AQESNs are investigated against the reservoir size. In all cases, the AQESNs outperformed their standard counterparts.

## 12.5 DISCUSSION AND CONCLUSION

We have introduced QESNs for nonlinear adaptive filtering of 3D and 4D processes. This is achieved by employing full quaternion nonlinear activation functions with local analyticity, allowing for first-order differentiability at the operating point to design quaternion-valued gradient descent algorithms. Unlike currently existing quaternion-valued RNNs, such as the fully connected quaternion RNNs and the quaternion Hopfield neural networks [52], with well-prepared sparse connected dynamical reservoir prior to training, QESNs require training only for the weights connecting the hidden layer and the readout neurons, significantly reduce computational complexity and, hence, enable the possibility to build up large-scale quaternion RNNs with acceptable computational costs. To further make QESNs optimal for the generality of quaternion-valued signals (both second-order circular and noncircular), the widely linear model has been incorporated into QESNs to introduce AQESNs. The performance advantage of exploiting data noncircularity through the widely linear model has been illustrated by simulations over a range of noncircular synthetic 3D and 4D signals and for real-world noncircular, nonlinear and nonstationary 3D body motion recordings and 4D wind data via a fusion of heterogeneous data sources.

## ACKNOWLEDGMENTS

This work was partially supported by the National Natural Science Foundation of China under Grants 61401094 and 61771124, the Natural Science Foundation of Jiangsu Province under Grant BK20140645 and the Fundamental Research Funds for the Central Universities under Grant 2242016K41050.

## REFERENCES

[1] G. Cybenko, Approximation by superpositions of a sigmoidal function, Mathematics of Control, Signals, and Systems 2 (4) (1989) 303–314.

[2] D.P. Mandic, J. Chambers, Recurrent Neural Networks for Prediction: Learning Algorithms, Architectures and Stability, Wiley, New York, NY, USA, 2001.

[3] K.S. Narendra, K. Parthasarathy, Identification and control of dynamical systems using neural networks, IEEE Transactions on Neural Networks 1 (1) (1990) 4–27.

[4] S. Haykin, L. Li, Nonlinear adaptive prediction of nonstationary signals, IEEE Transactions on Signal Processing 43 (2) (1995) 526–535.

[5] S.A. Billings, C.F. Fung, Recurrent radial basis function networks for adaptive noise cancellation, Neural Networks 8 (2) (1995) 273–290.

[6] J.L. Elman, Finding structure in time, Cognitive Science 14 (2) (1990) 179–211.

[7] S.B. Choe, J.J. Faraway, Modeling head and hand orientation during motion using quaternions, Journal of Aerospace Engineering 113 (1) (2004) 186–192.

[8] C.F. Karney, Quaternions in molecular modeling, Journal of Molecular Graphics & Modelling 25 (5) (2007) 595–604.

[9] T.A. Ell, S.J. Sangwine, Hypercomplex Fourier transforms of color images, IEEE Transactions on Image Processing 16 (1) (2007) 22–35.

[10] A. Kolaman, O. Yadid-Pecht, Quaternion structural similarity: a new quality index for color images, IEEE Transactions on Image Processing 21 (4) (2012) 1526–1536.

[11] N. Le Bihan, J. Mars, Singular value decomposition of quaternion matrices: a new tool for vector-sensor signal processing, Signal Processing 84 (7) (2004) 1177–1199.

[12] N. Le Bihan, S. Miron, J. Mars, MUSIC algorithm for vector-sensors array using biquaternions, IEEE Transactions on Signal Processing 55 (9) (2007) 4523–4533.

[13] A.M. Sabatini, Quaternion-based extended Kalman filter for determining orientation by inertial and magnetic sensing, IEEE Transactions on Aerospace and Electronic Systems 53 (7) (2006) 1346–1356.
[14] D. Choukkroun, I.Y. Bar-Itzhack, Y. Ohsman, Novel quaternion Kalman filter, IEEE Transactions on Aerospace and Electronic Systems 42 (1) (2006) 174–190.
[15] C. Jahanchahi, D.P. Mandic, A class of quaternion Kalman filters, IEEE Transactions on Neural Networks and Learning Systems 25 (3) (2014) 533–544.
[16] C. Cheong-Took, D.P. Mandic, The quaternion LMS algorithm for adaptive filtering of hypercomplex processes, IEEE Transactions on Signal Processing 57 (4) (2009) 1316–1327.
[17] J. Via, D.P. Palomar, L. Viela, I. Santamaria, Quaternion ICA from second-order statistics, IEEE Transactions on Signal Processing 59 (4) (2011) 1586–1600.
[18] A. Sudbery, Quaternionic analysis, Mathematical Proceedings of the Cambridge Philosophical Society 85 (2) (1979) 199–225.
[19] T. Isokawa, H. Nishuimura, N. Kamiura, N. Matsui, Associative memory in quaternionic Hopfield neural network, International Journal of Neural Systems 18 (2) (2008) 135–145.
[20] B.C. Ujang, C. Cheong-Took, D.P. Mandic, Split quaternion nonlinear adaptive filtering, Neural Networks 23 (3) (2010) 426–434.
[21] S. De Leo, P. Rotelli, Quaternion analyticity, Applied Mathematics Letters 16 (7) (2003) 1077–1081.
[22] B.C. Ujang, C. Cheong-Took, D.P. Mandic, Quaternion-valued nonlinear adaptive filtering, IEEE Transactions on Neural Networks 22 (8) (2011) 1193–1206.
[23] B.C. Ujang, C. Cheong-Took, D.P. Mandic, On quaternion analyticity: enabling quaternion-valued nonlinear adaptive filtering in: Proceedings of the IEEE International Conference on Acoustics, Speech, and Signal Processing, Kyoto, Japan, March 25–March 30, 2012, pp. 2117–2120.
[24] H. Jaeger, H. Haas, Harnessing nonlinearity: predicting chaotic systems and saving energy in wireless communication, Science 304 (5667) (2004) 78–80.
[25] M. Lukosevicius, H. Jaeger, Reservoir computing approaches to recurrent neural network training, Computer Science Review 3 (3) (2009) 127–149.
[26] I.B. Yildiz, H. Jaeger, S.J. Kiebel, Re-visiting the echo state property, Neural Networks 35 (2012) 1–9.
[27] B. Picinbono, On noncircularity, IEEE Transactions on Signal Processing 42 (12) (1994) 3473–3483.
[28] B. Picinbono, P. Chevailer, Widely linear estimation with complex data, IEEE Transactions on Signal Processing 43 (8) (1995) 2030–2033.
[29] D.P. Mandic, S.L. Goh, Complex Valued Nonlinear Adaptive Filters: Noncircularity, Widely Linear and Neural Models, Wiley, New York, NY, USA, 2009.
[30] P.J. Schreier, L.L. Scharf, Statistical Signal Processing of Complex-Valued Data: The Theory of Improper and Noncircular Signals, Cambridge University Press, Cambridge, UK, 2010.
[31] L. Anttila, M. Valkama, M. Renfors, Circularity-based I/Q imbalance compensation in wideband direct-conversion receivers, IEEE Transactions on Vehicular Technology 57 (4) (2008) 2099–2113.
[32] Y. Xia, S.C. Douglas, D.P. Mandic, Adaptive frequency estimation in smart grid applications: exploiting noncircularity and widely linear adaptive estimators, IEEE Signal Processing Magazine 29 (5) (2012) 44–54.
[33] Y. Xia, D.P. Mandic, Complementary mean square analysis of augmented CLMS for second-order noncircular Gaussian signals, IEEE Signal Processing Letters 24 (9) (2017) 1413–1417.
[34] N.N. Vakhania, Random vectors with values in quaternion Hilbert spaces, Theory of Prabability & Its Application 43 (1) (1999) 99–115.
[35] P.O. Amblard, N. Le Bihan, On properness of quaternion valued random variables, in: Proceedings of the Sixth IMA International Conference on Mathematics in Signal Processing, Cirenceister, UK, December 14–December 16, 2004, pp. 23–26.
[36] J. Via, D. Ramirez, I. Santamaria, Properness and widely linear processing of quaternion random vectors, IEEE Transactions on Information Theory 56 (7) (2010) 3502–3515.
[37] C. Cheong-Took, D.P. Mandic, A quaternion widely linear adaptive filter, IEEE Transactions on Signal Processing 58 (8) (2010) 4427–4431.
[38] C. Cheong-Took, D.P. Mandic, Augmented second-order statistics of quaternion random signals, Signal Processing 91 (2) (2011) 214–224.
[39] J. Navarro-Moreno, R.M. Fernandez-Alcala, J.C. Ruiz-Molina, A quaternion widely linear series expansion and its applications, IEEE Signal Processing Letters 19 (12) (2012) 868–871.

[40] Y. Xia, C. Jahanchahi, T. Nitta, D.P. Mandic, Performance bounds of quaternion estimators, IEEE Transactions on Neural Networks and Learning Systems 26 (12) (2015) 3287–3292.

[41] Y. Xia, B. Jelfs, M.M. Van Hulle, J.C. Principe, D.P. Mandic, An augmented echo state networks for nonlinear adaptive filtering of complex noncircular signals, IEEE Transactions on Neural Networks 22 (1) (2011) 74–83.

[42] Y. Xia, C. Jahanchahi, D.P. Mandic, Quaternion-valued echo state networks, IEEE Transactions on Neural Networks and Learning Systems 26 (4) (2015) 663–673.

[43] D.P. Mandic, C. Jahanchahi, C. Cheong-Took, A quaternion gradient operator and its applications, IEEE Signal Processing Letters 18 (1) (2011) 47–50.

[44] C.A. Deavours, The quaternion calculus, American Mathematical Monthly 80 (9) (1973) 995–1008.

[45] M. Buehner, P. Young, A tighter bound for the echo state property, IEEE Transactions on Neural Networks 17 (3) (2006) 820–824.

[46] M.C. Ozturk, D. Xu, J.C. Principe, Analysis and design of echo state networks, Neural Computations 19 (1) (2007) 111–138.

[47] B. Zhang, D.J. Miller, Y. Wang, Nonlinear system modeling with random matrices: echo state networks revisited, IEEE Transactions on Neural Networks and Learning Systems 23 (1) (2012) 175–182.

[48] E. Soria-Olivas, J. Calpe-Maravilla, J.F. Guerrero-Martinez, M. Martinez-Sober, J. Espi-Lopez, An easy demonstration of the optimum value of the adaptation constant in the LMS algorithm [FIR filter theory], IEEE Transactions on Education 41 (1) (1998) 81.

[49] S.H. Stogatz, Nonlinear Dynamics and Chaos: With Applications to Physics, Biology, Chemistry and Engineering, Studies in Nonlinearity, Westview, Boulder, CO, USA, 2001.

[50] J.F. Manwell, J.G. Mcgowan, A.L. Rogers, Wind Energy Explained: Theory, Design and Application, Wiley, West Sussex, UK, 2002.

[51] C. Cheong-Took, G. Strbac, A. Aihara, D.P. Mandic, Quaternion-valued short-term joint forecasting of three-dimensional wind and atmospheric parameters, Renewable Energy 36 (6) (2011) 1754–1760.

[52] M. Kobayashi, Fixed points of split quaternionic Hopfield neural networks, Signal Processing 136 (2017) 38–42.

CHAPTER

# 13

# IDENTIFICATION OF SHORT-TERM AND LONG-TERM FUNCTIONAL SYNAPTIC PLASTICITY FROM SPIKING ACTIVITIES

**Dong Song, Brian S. Robinson, Theodore W. Berger**
*University of Southern California, Los Angeles, CA, United States*

## CONTENTS



## CHAPTER POINTS

- This chapter introduces the challenging task of the identification of synaptic plasticity from spiking activities.
- A nonlinear dynamical modeling approach is proposed for identifying short-term and long-term synaptic plasticity using spiking activities.
- Synaptic learning rules are finally defined as functions that map input–output signals to the nonstationarity.

**Adaptive Learning Methods for Nonlinear System Modeling. DOI: 10.1016/B978-0-12-812976-0.00017-8**

## 13.1 INTRODUCTION

Synaptic plasticity refers to the phenomenon that strength of synaptic connections between neurons changes over time. Depending on its timescale, synaptic plasticity can be divided into short-term synaptic plasticity (STSP), which lasts for milliseconds to minutes [74,63,64], and long-term synaptic plasticity (LTSP), which lasts for at least tens of minutes to hours or longer, even permanently [12] (Fig. 13.1). Due to the operating timescale, STSP has been proposed to be a neural substrate for online processing of temporal information, while LTSP has long been postulated as the underlying mechanism of learning and memory formation [33,1,3,4,40,44–48,68,36].

Synaptic plasticity is typically studied by using electrical stimulations to mimic presynaptic action potentials and measuring corresponding postsynaptic responses, *e.g.*, postsynaptic potentials (PSPs) or postsynaptic currents (PSCs). Different presynaptic patterns lead to different forms of synaptic plasticity. With paired-pulse stimulation, STSP with increased response, *i.e.*, facilitation [34], or decreased response, *i.e.*, depression [70], can be revealed. With fixed-interval train stimulation, more complex patterns indicating mixtures of facilitation and depression can be observed [20]. With high-frequency stimulation (HFS) and low-frequency stimulation (LFS), two forms of LTSP, long-term potentiation (LTP) and long-term depression (LTD), can be elicited, respectively [12,14,35]. The theta burst pattern, which consists of brief high-frequency bursts delivered at a theta frequency, has been reported to generate a maximum level of LTP [38]. LTP and LTD can also be induced with repetitive paired-pulse stimulations to presynaptic and postsynaptic neurons with certain sequences and intervals. Due to its more naturalistic induction pattern compared with HFS/LFS-induced LTP/LTD and its closer connection to Hebbian learning, this so-called spike timing–dependent plasticity (STDP) has drawn intensive attention and investigation [11,39,41,17,21].

Despite the large body of data collected and general agreements on its functions at the theoretical level, how synaptic plasticity determines high-level cognitive functions such as learning and memory remains controversial [13,67] and the exact nature of synaptic plasticity during cognitive operations is still not well known. One of the main reasons is the lack of a computational tool for the identification of both STSP and LTSP from *natural* ensemble spiking activities, which is required for studying neural correlates of cognitive functions in behaving animals.

Identification of synaptic plasticity from spiking activities is a challenging task. Brain regions underlying cognition are large-scale, point-process, multiinput multioutput (MIMO) systems. In any given region, information is encoded in the ensemble firing of populations of neurons [15,16,24,27,28, 53,55]. The input–output signals are stimulus- or behavior-driven spike trains, as opposed to electrical stimuli and evoked responses typically used in *in vitro* studies of synaptic plasticity [2,5–7]. Inferring neuron-to-neuron connection strength such as PSP and further changes of such a strength with multiple timescales from ensemble spiking activities, which is a necessary step for studying synaptic plasticity, poses serious modeling challenges [56,65]. Furthermore, the effects of input spike patterns to postsynaptic responses are often mixtures of LTSP and STSP. Most LTSP induction patterns mentioned above can elicit STSP. For example, the interpulse intervals of HFS patterns fall in the timescale of STSP, and in fact they have also been used to study STSP [20,58,59]. Conversely, many STSP induction patterns can also induce LTSP. The interplay between STSP and LTSP becomes even more intertwined and complex when input and output patterns are irregular and continuous as in natural conditions.

To solve this problem, we formulate a nonlinear dynamical modeling approach for identifying STSP and LTSP using spiking activities. In this approach, we define STSP as a form of nonlinear dynamics

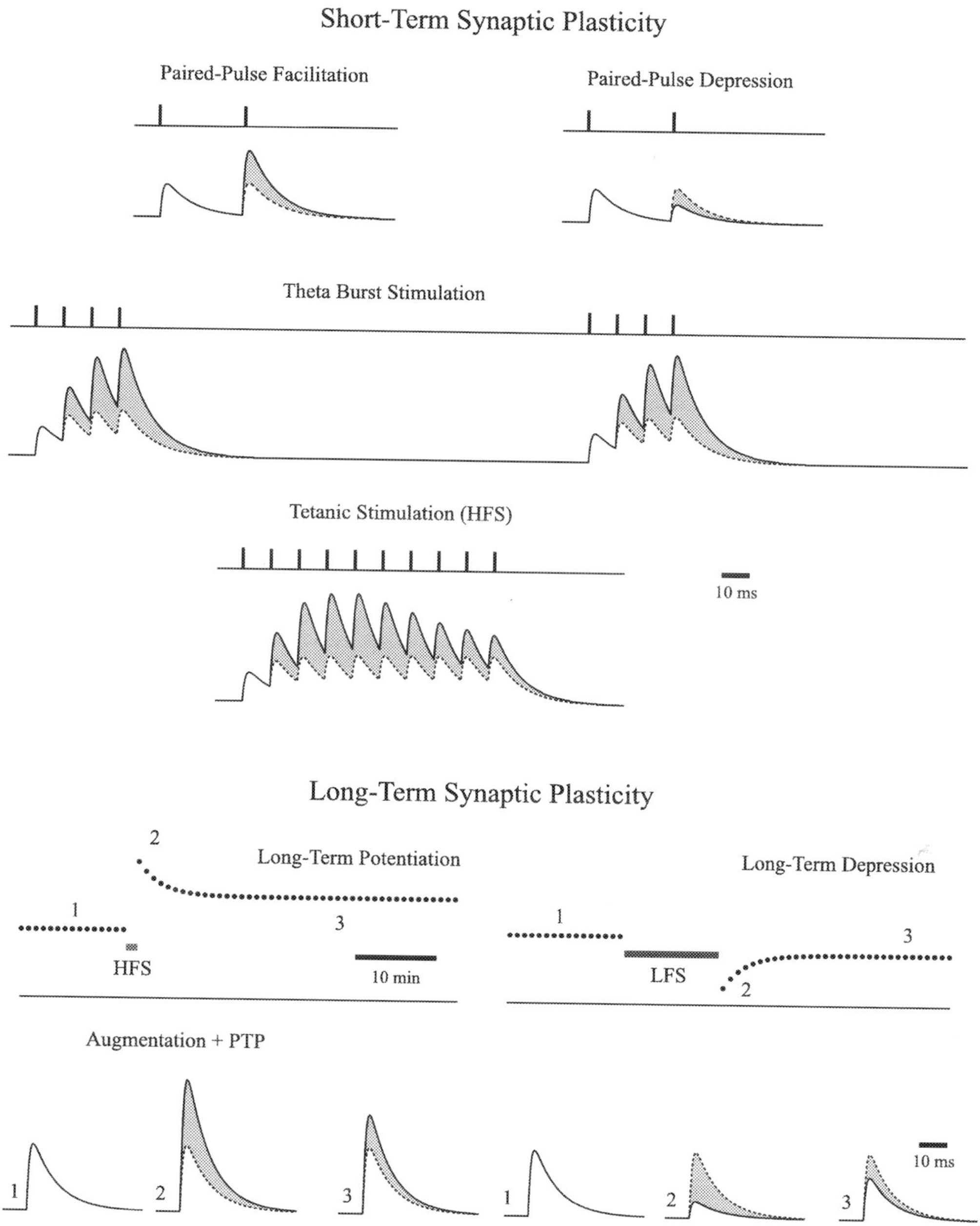


**FIGURE 13.1**

Different forms of synaptic plasticity. The shaded areas represent nonlinear dynamics and nonstationarity caused by short-term and long-term synaptic plasticity, respectively.

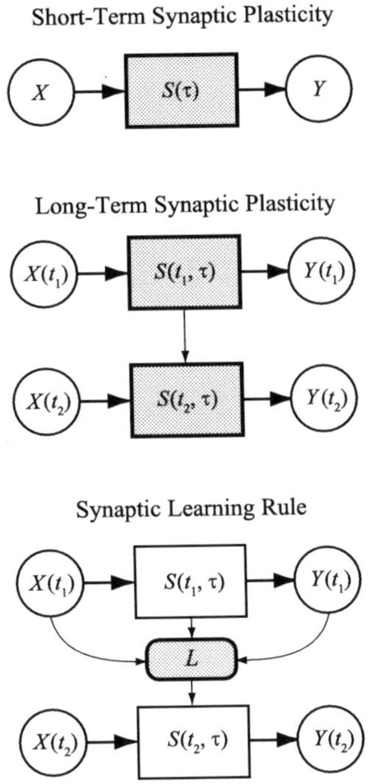


**FIGURE 13.2**

Schematic diagram of the three modeling steps. $X$: input sequences; $Y$: output sequences; $S$: spiking neuron model; $L$: learning rule for $S$. In Step 1, $S$ is not a function of time. In Steps 2 and 3, $S$ varies with time. During learning, $S$ evolves as the consequence of input and output activities following a learning rule. Colored boxes indicate the functions that need to be identified in each modeling step.

that can be represented and identified with a biologically plausible structure. LTSP is defined as a system's nonstationarity that can be expressed as time varying nonlinear dynamical models. Furthermore, synaptic learning rules are defined as functions that map input–output signals to the nonstationarity. Identification of synaptic plasticity becomes the estimation of the nonlinear dynamics, nonstationarity and learning rule function from point-process input–output signals.

## 13.2 IDENTIFICATION OF STSP WITH NONLINEAR DYNAMICAL MODEL

### 13.2.1 THEORY

Due to the short timescales of STSP, it can be studied as a form of nonlinear dynamics. In this context, dynamics refers to finite-memory history dependency of PSPs, *i.e.*, the magnitude of PSP is influenced not only by the current input spike or stimulation, but also preceding inputs within a memory window. Nonlinear dynamics happens when the dynamical interaction does not obey the additivity property. Such a system can be expressed by a time-invariant function (Fig. 13.2, top panel).

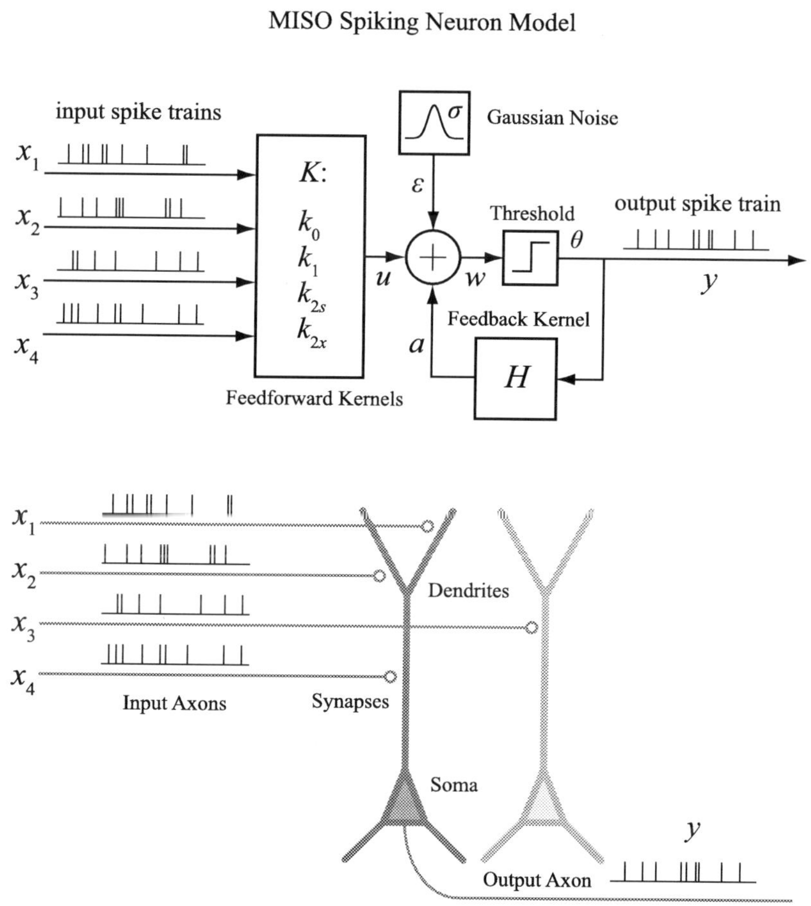


**FIGURE 13.3**

Multiinput, single-output (MISO) nonlinear dynamical model of spiking neuron. The MISO model takes a physiologically plausible structure. All model variables are simultaneously estimated from the input–output spiking activities.

We use a nonlinear dynamical and biologically plausible model structure to extract PSP and STSP from input–output spiking activities [9,56–59,62,65]. This is a nontrivial task since both inputs and output are point-process signals containing only spike timings; PSP and STSP need to be inferred instead of directly measured. The model takes the form of a multiinput, single-output (MISO) model that can be expressed as (Fig. 13.3)

$$w = u(K, x) + a(H, y) + \varepsilon(\sigma), \tag{13.1}$$

$$y = \begin{cases} 0 & \text{when } w < \theta, \\ 1 & \text{when } w \geq \theta. \end{cases} \tag{13.2}$$

Variable $x$ represents input spike trains; $y$ represents the output spike train; $w$ represents the prethreshold membrane potential of the output neurons, which is expressed as the summation of the postsynaptic potential $u$ caused by input spike trains, the output spike–triggered after-potential $a$ and a Gaussian white noise $\varepsilon$ with standard deviation $\sigma$. A threshold, $\theta$, determines the generation of the output spike and the associated feedback after-potential ($a$).

The feedforward transformation from $x$ to $u$ and the feedback transformation from $y$ to $a$ take the form of a second-order Volterra model $K$ and a first-order Volterra model $H$, respectively. We have

$$u(t) = k_0 + \sum_{n=1}^{N}\sum_{\tau=0}^{M_k} k_1^{(n)}(\tau)x_n(t-\tau) + \sum_{n=1}^{N}\sum_{\tau_1=0}^{M_k}\sum_{\tau_2=0}^{M_k} k_{2s}^{(n)}(\tau_1,\tau_2)x_n(t-\tau_1)x_n(t-\tau_2)$$
$$+ \sum_{n_1=1}^{N}\sum_{n_2=1}^{n_1-1}\sum_{\tau_1=0}^{M_k}\sum_{\tau_2=0}^{M_k} k_{2x}^{(n_1,n_2)}(\tau_1,\tau_2)x_{n_1}(t-\tau_1)x_{n_2}(t-\tau_2) + \ldots, \quad (13.3)$$

$$a(t) = \sum_{\tau=1}^{M_h} h(\tau)y(t-\tau). \quad (13.4)$$

The zero order kernel, $k_0$, is the value of $u$ when there is no input. It determines the spontaneous firing rate of the output neuron. First-order kernels $k_1^{(n)}$ describe the linear relation between the $n$th input $x_n$ and $u$, as functions of the time intervals ($\tau$) between the present time and the past time. Second-order self-kernels $k_{2s}^{(n)}$ describe the nonlinear interactions between pairs of spikes within the $n$th input $x_n$ as they affect $u$. Second-order cross kernels $k_{2x}^{(n_1,n_2)}$ describe the nonlinear interactions between pairs of spikes from different inputs ($x_{n_1}$ and $x_{n_2}$) as they affect $u$. $N$ is the number of inputs. Higher-order kernels can also be included to capture higher-order nonlinearities. The term $h$ represents a linear feedback kernel that transforms the output spike to the after-potential $a$. $M_k$ and $M_h$ are memory lengths of feedforward and feedback processes, respectively.

To reduce number of parameters and avoid overfitting, basis functions are utilized in model estimations [62]. Basis functions $b$ can take the forms of Laguerre basis [42,43,60] or B-spine basis [22]. With input and output spike trains $x$ and $y$ convolved with $b$, we have

$$v_j^{(n)}(t) = \sum_{\tau=0}^{M_k} b_j(\tau)x_n(t-\tau), \quad (13.5)$$

$$v_j^{(h)}(t) = \sum_{\tau=1}^{M_h} b_j(\tau)y(t-\tau). \quad (13.6)$$

Eqs. (13.3) and (13.4) are rewritten

$$u(t) = c_0 + \sum_{n=1}^{N}\sum_{j=1}^{L} c_1^{(n)}(j)v_j^{(n)}(t) + \sum_{n=1}^{N}\sum_{j_1=1}^{L}\sum_{j_2=1}^{j_1} c_{2s}^{(n)}(j_1,j_2)v_{j_1}^{(n)}(t)v_{j_2}^{(n)}(t)$$
$$+ \sum_{n_1=1}^{N}\sum_{n_2=1}^{n_1-1}\sum_{j_1=1}^{L}\sum_{j_2=1}^{L} c_{2x}^{(n_1,n_2)}(j_1,j_2)v_{j_1}^{(n_1)}(t)v_{j_2}^{(n_2)}(t) + \ldots, \quad (13.7)$$

$$a(t) = \sum_{j=1}^{L} c_h(j) v_j^{(h)}(t), \tag{13.8}$$

where $c_1^{(n)}$, $c_{2s}^{(n)}$, $c_{2x}^{(n_1,n_2)}$ and $c_h$ are the model coefficients of $k_1^{(n)}$, $k_{2s}^{(n)}$, $k_{2x}^{(n_1,n_2)}$ and $h$, respectively, and $c_0$ is equal to $k_0$. Given the kernels are smooth and continuous functions, the number of basis functions $L$ is much smaller than the memory length ($M_k$ and $M_h$).

In Eqs. (13.7) and (13.8), $v$ and $vv$ can be calculated from the known $x$, $y$ and $b$. The Volterra series essentially expresses the nonlinear relationship between $u$ and $x$ into a linear relationship between $u$ and $[v, vv]$. The joint effect of the threshold $\theta$ and the Gaussian noise $\varepsilon$ is equivalent to a *probit* link function that maps the value of $u + a$ into the probability of $y$ is equal to 1. The whole model thus can be expressed as a generalized linear model with the nonlinearity structured in the Volterra series. Therefore, this MISO model can be termed a generalized Volterra model (GVM) [58,59]. As a special case, the first-order GVM is equivalent to the commonly used generalized linear models [50,49,52,69, 26,19,73].

Due to the point-process nature of the input and output signals, all model variables are dimensionless. In addition, the values of $k$, $h$, $u$, $a$, $w$, $\varepsilon$, $\sigma$ and $\theta$ can be scaled; $\theta$ and $k_0$ can be translated, without influencing the probability of generating output spikes. So, in practice, $k_0$, $k_1$, $k_{2s}$, $k_{2x}$ and $h$ are first estimated with a unitary $\sigma$ and a zero-valued $\theta$ using the Matlab® *glmfit* function and then normalized with the absolute value of $k_0$. In the final format, $\sigma$ is equal to $1/|k_0|$; $\theta$ is equal to zero; the baseline value of $w$ (*i.e.*, $k_0$) is $-1$ [57,58]. The Laguerre parameter controlling the asymptotic decaying rate of the basis functions and the total number of basis functions $L$ are optimized with respect to the log-likelihood function [62].

The Volterra kernels quantitatively describe the input–output nonlinear dynamics of the neuron. A more intuitive representation is given by the single-pulse and paired-pulse response functions ($r_1$ and $r_2$) derived from the kernels. We have

$$r_1^{(n)}(\tau) = k_1^{(n)}(\tau) + k_{2s}^{(n)}(\tau, \tau), \tag{13.9}$$

$$r_{2s}^{(n)}(\tau_1, \tau_2) = 2k_{2s}^{(n)}(\tau_1, \tau_2), \tag{13.10}$$

$$r_{2x}^{(n_1,n_2)}(\tau_1, \tau_2) = k_{2x}^{(n_1,n_2)}(\tau_1, \tau_2), \tag{13.11}$$

where $r_1^{(n)}$ is essentially the PSP elicited by a single spike from the $n$th input neuron; $r_2^{(n)}$ describes the joint nonlinear effect of pairs of spikes from the $n$th input neuron in addition to the summation of their first-order responses, *i.e.*, $r_1^{(n)}(\tau_1) + r_1^{(n)}(\tau_2)$. The term $r_{2x}^{(n_1,n_2)}(\tau_1, \tau_2)$ represents the joint nonlinear effect of pairs of spikes with one spike from neuron $n_1$ and one spike from neuron $n_2$ (Fig. 13.4).

In this formulation, the ensemble neuronal input–output properties are represented by model coefficients $S = [k, h]$. All coefficients are simultaneously estimated from input and output spike trains ($x$ and $y$). Specifically, the first-order response function $r_1$ can be interpreted as the PSP; the second-order response function $r_2$ is equivalent to the paired-pulse STSP.

### 13.2.2 NONLINEAR DYNAMICAL MODEL OF THE HIPPOCAMPAL CA3-CA1

We have applied nonlinear dynamical modeling to the hippocampal CA3 to CA1 in rodents, nonhuman primates and human subjects [8,10,30–32,57,58,66] and the prefrontal cortex layer 2/3 to layer 5

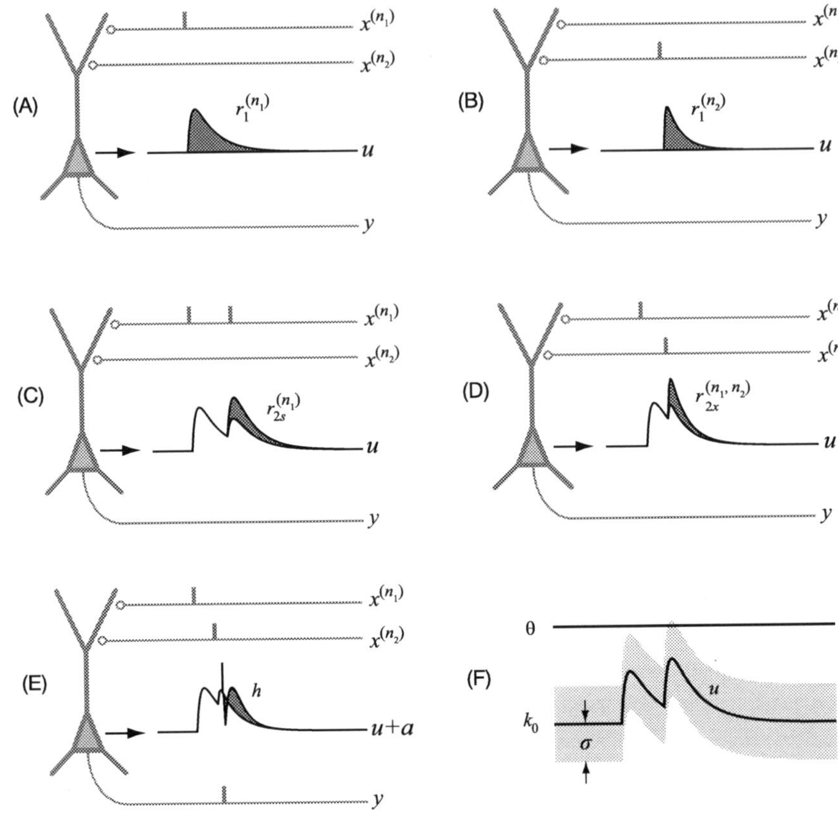


**FIGURE 13.4**

Physiological interpretation of the MISO spiking neuron model variables. Here, $k_0$ ($r_0$) is the resting membrane potential, $r_1$ is the postsynaptic potential, $r_2$ is the paired-pulse facilitation/depression function, $h$ is the output spike-triggered after-potential, $u$ is the synaptic potential and $\sigma$ is the standard deviation of the Gaussian noise.

in nonhuman primates [29,61]. In these applications, the input–output spike trains are recorded from well-trained animals, where the input and output signals as well as the input–output transformations are stabilized and thus sufficiently described by stationary GVMs. Fig. 13.5 shows a GVM of a hippocampal CA1 neuron estimated from a rodent performing a delayed-nonmatch-to-sample (DNMS) task. Results show that this neuron receives inputs from 6 out of 24 CA3 neurons. The single-pulse and paired-pulse response functions ($r_1$ and $r_2$) are the PSPs and paired-pulse facilitation/depression functions inferred from the CA3-CA1 spiking activities. The zero order kernel is $-1$. The threshold is 0. The standard deviation of the Gaussian noise is estimated to be 0.418. This MISO model can accurately predict the output spike train based on the input spike trains (Fig. 13.5, bottom-middle) as verified with the out-of-sample Kolmogorov–Smirnov test based on the time rescaling theorem (Fig. 13.5, bottom-right). In this stationary model, all model variables are time-invariant.

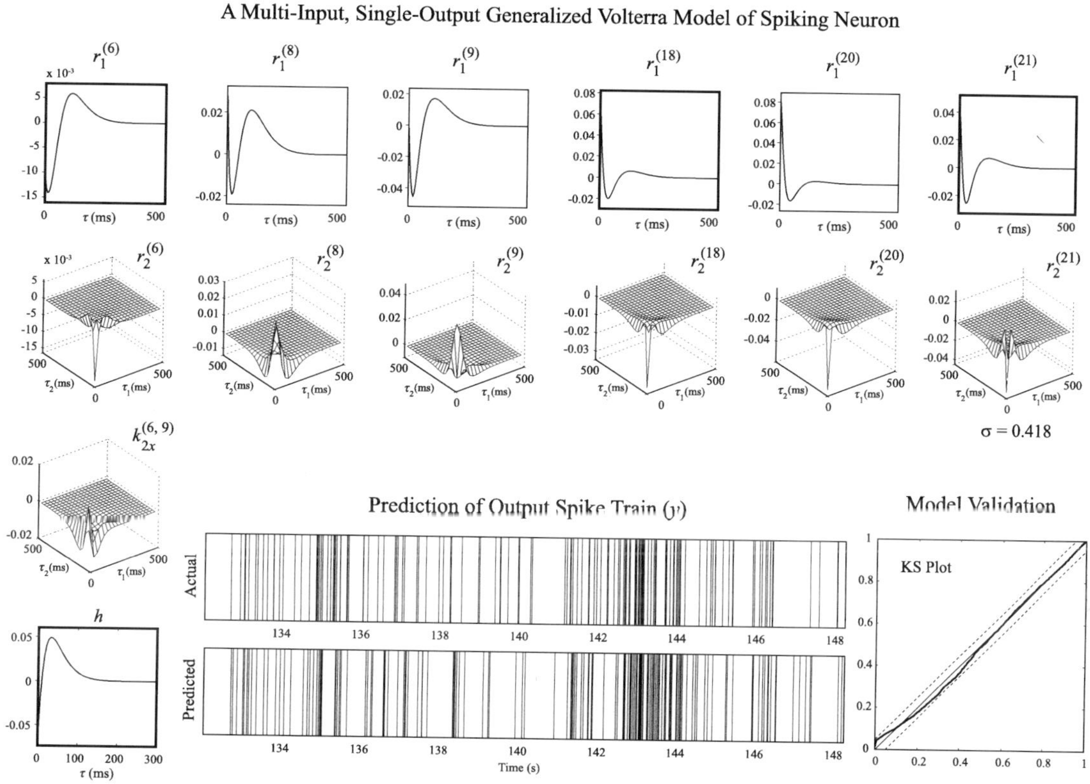


**FIGURE 13.5**

A stationary MISO GVM of a hippocampal CA1 neuron. Here, $r_1$ are the single-pulse response functions, $r_2$ are the paired-pulse response functions for the same input neuron, $k_{2x}$ are cross kernels for pairs of neurons and $h$ is the feedback kernel. The output spike train is predicted with the model and the input spike trains. The model is validated with an out-of-sample, Kolmogorov–Smirnov (KS) test based on the time rescaling theorem. In the KS plot, dashed lines are the 95% confidence bounds.

## 13.3 IDENTIFICATION OF LTSP WITH NONSTATIONARY MODEL

### 13.3.1 THEORY

The nonlinear dynamical spiking neuron model has provided a quantitative way to infer PSP and STSP from spiking input–output activities. To further infer LTSP, we extend the GVM to be time varying to track system nonstationarity. GVM coefficients are estimated recursively from the input–output spikes and the changes of the nonlinear dynamics are described by the temporal evolution of the kernel functions (Fig. 13.2, middle).

We develop the nonstationary model by combining the GVM and the point-process adaptive filtering method [18,25,65]. GVM coefficients ($c$) are taken as state variables while the input–output spikes are taken as observable variables. Using adaptive filtering methods, state variables are recursively updated as the observable variables unfold in time. The underlying change of system input–output properties is then represented by the time varying GVM ($S(t) = [k(t), h(t)]$) reconstructed with the time varying coefficients $c(t)$.

Specifically, the probability of observing an output spike at time $t$, *i.e.*, $P(t)$, is predicted by the GVM at time $t-1$ based on the inputs up to $t$ and output before $t$. Secondly, the difference between $P(t)$ and the new observation of output $y(t)$ is used to update the GVM coefficients. Using the stochastic state point process filtering algorithm, coefficient vector $C(t)$ and its covariance matrix $W(t)$ are both updated iteratively at each time step $t$ as follows:

$$P(t) = 0.5 - 0.5 \cdot \mathrm{erf}\left(\frac{\theta - u(t) - a(t)}{\sqrt{2}\sigma}\right), \tag{13.12}$$

$$c(t) = c(t-1) + W(t)\left[\left(\frac{\partial \log P(t)}{\partial C(t)}\right)^T (y(t) - P(t))\right], \tag{13.13}$$

$$\begin{aligned} W(t)^{-1} &= [W(t-1) + Q]^{-1} + \\ &\left[\left(\frac{\partial \log P(t)}{\partial c(t)}\right)^T P(t)\left(\frac{\partial \log P(t)}{\partial c(t)}\right) - (y(t) - P(t))\frac{\partial^2 \log P(t)}{\partial c(t)\partial c(t)'}\right], \end{aligned} \tag{13.14}$$

where $Q$ is the coefficient noise covariance matrix and erf is the Gauss error function. In this method, $W$ acts as the adaptive "learning rate" that allows reliable and rapid tracking of the model coefficients.

### 13.3.2 SIMULATION STUDIES ON NONSTATIONARY NONLINEAR DYNAMICAL MODEL

We have tested intensively the nonstationary nonlinear dynamical modeling algorithm with synthetic input–output spike train data obtained through simulations. In all simulations, the inputs are Poisson random spike trains with mean firing rates ranging from 2 Hz to 6 Hz.

#### 13.3.2.1 Step Changes

First, we simulate a second-order, two-input, single-output spiking neuron model with step changes. The total simulation length is 8000 s. At 4000 s, the first-order kernel and the second-order self-kernel of the first input (*i.e.*, $k_1^{(1)}$ and $k_{2s}^{(1)}$) decrease the amplitudes by half with the same waveforms; the first-order kernel and the second-order self kernel of the second input (*i.e.*, $k_1^{(2)}$ and $k_{2s}^{(2)}$) double the amplitudes with the same waveforms. Other kernels (*i.e.*, $k_0$, $k_{2x}^{(1,2)}$, and $h$) remain constant during the whole simulation. Using the simulated input–output spiking data, we apply the nonstationary nonlinear dynamical modeling algorithm to track the changes of the kernels. Results show that all kernels as well as the changes of the kernels can be recovered accurately from the simulated input–output spike trains (Fig. 13.6). In this nonstationary model, model variables become time varying.

#### 13.3.2.2 LTP and LTD-Like Changes

We further test the capability of the nonstationary modeling algorithm to track biologically plausible forms of synaptic plasticity. A first-order, two-input system with concurrent LTP- and LTD-like

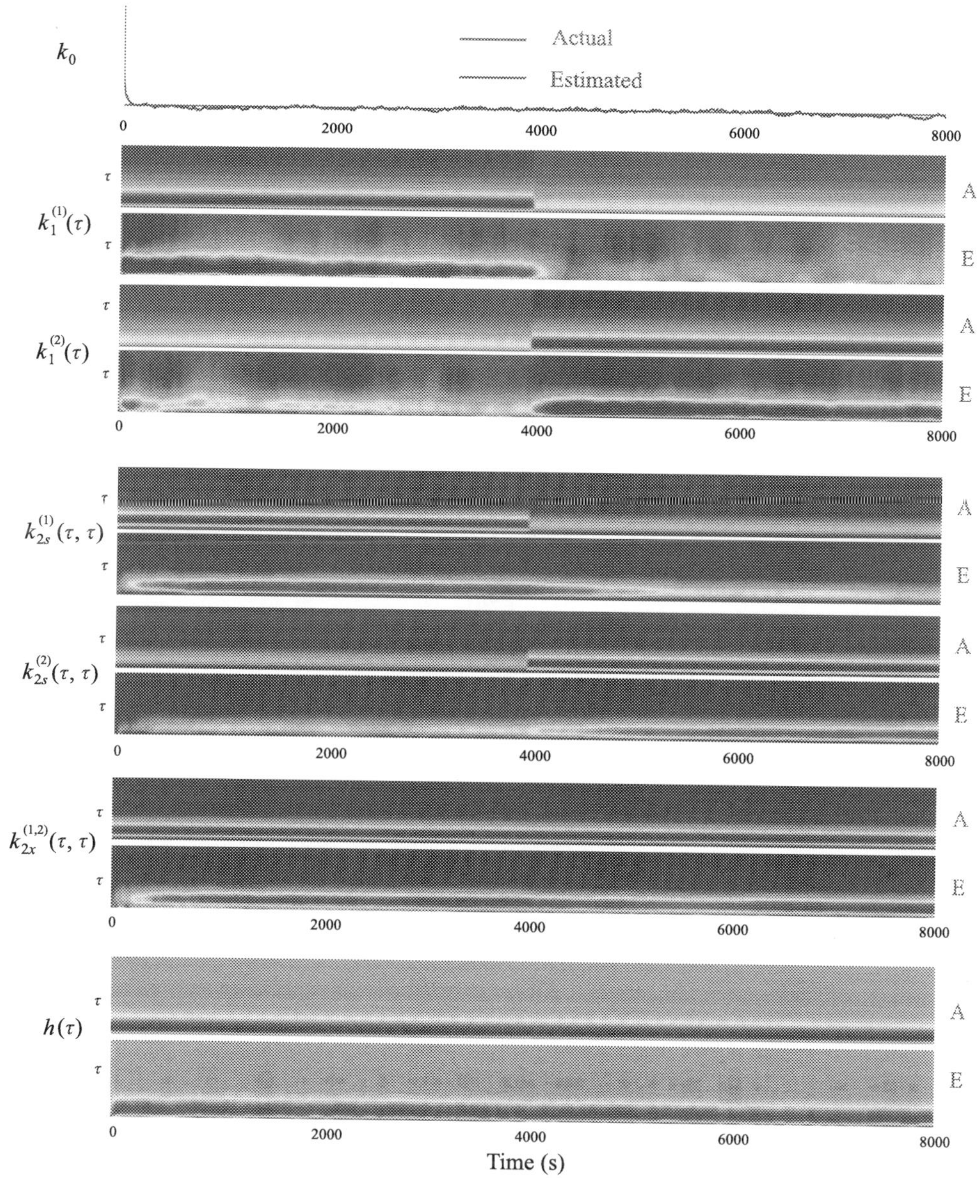


**FIGURE 13.6**

Tracking a second-order, two-input, single-output system with step changes. First-order kernels ($k_1$) and second-order self-kernels ($k_{2s}$) have step changes at 4000 s. Zero order kernels ($k_0$), second-order cross kernels ($k_{2x}$) and feedback kernels ($h$) remain constant. Memory length is 500 ms for $k_1$, $k_{2s}$ and $k_{2x}$, 1000 ms for $h$. The kernel amplitude is color-coded. Only diagonal values of second-order kernels are plotted for simplicity. (A): actual kernels; E: estimated kernels.

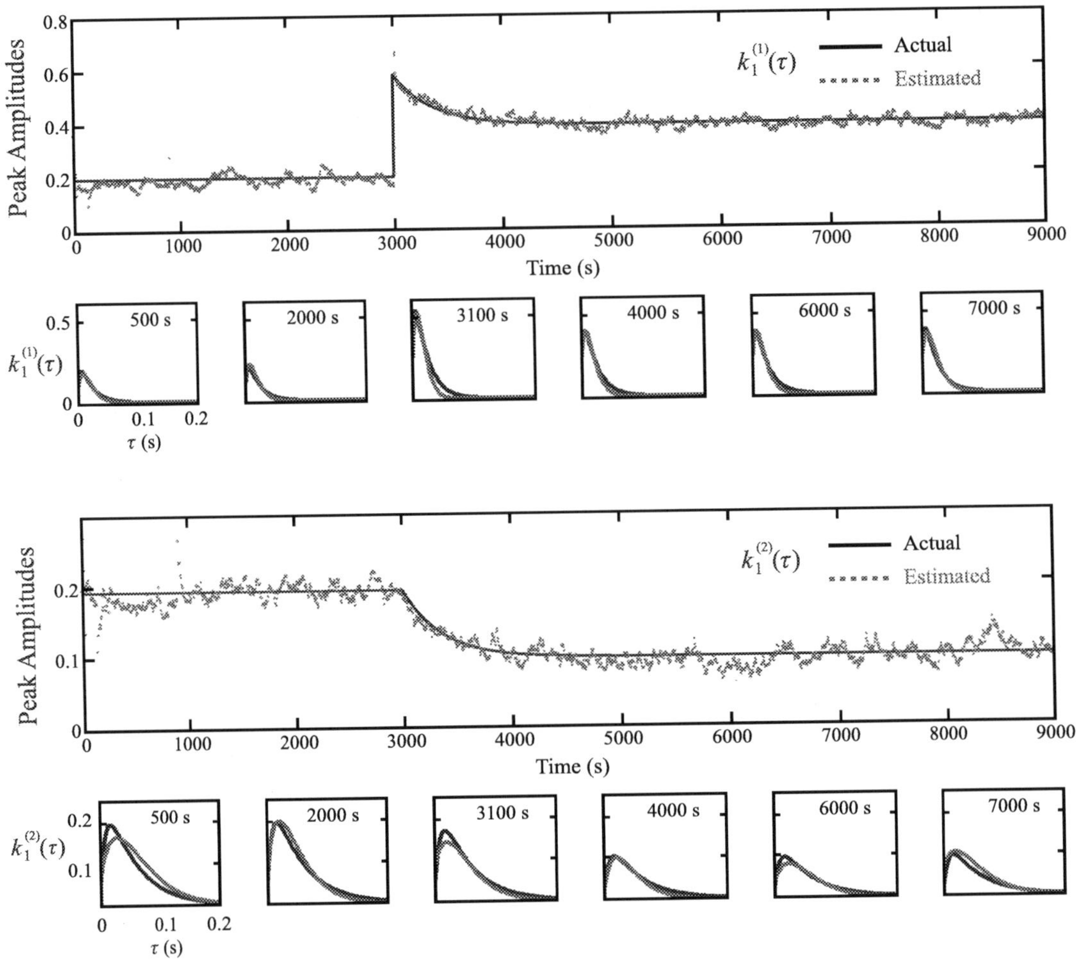


**FIGURE 13.7**

Tracking a first-order, two-input system with LTP- and LTD-like changes. Black line: peak amplitudes of the actual kernels; red (gray in print version) dots: peak amplitudes of the estimated kernels. Interval between successive estimation points plotted is 20 s.

changes at different inputs is simulated (Fig. 13.7). The total simulation length is 9000 s. At 3000 s, the kernels of the first input and the second input have LTP-like change (*i.e.*, a near instantaneous increase followed by an exponential decay to a potentiated level double the initial amplitude) and LTD-like change (*i.e.*, an exponential decay to half of the initial amplitude), respectively. Results show that the estimated kernels (red (gray in print version)) track faithfully the actual kernels (black) during the whole simulation, given the LTP-like change is much more challenging than the step change.

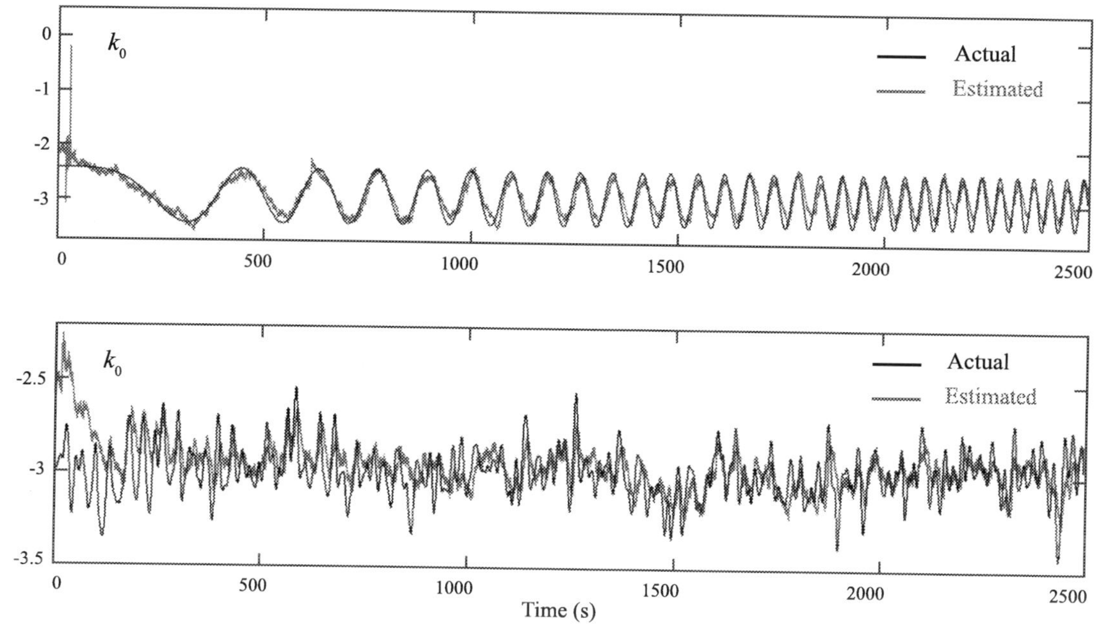


**FIGURE 13.8**

Tracking the input-independent baseline $k_0$ ($c_0$). Top panel: $k_0$ is a linear chirp from 0 Hz at 0 s to 0.0125 Hz at 2500 s. Bottom panel: $k_0$ is a random noise sequence with mean at −3 and low pass filtered at 0.05 Hz.

#### *13.3.2.3 Input-Independent Changes*

In our formulation of the nonstationary model, the zero order time varying kernel $k_0(t)$ (*i.e.*, $c_0(t)$) in Eq. (13.13) describes the input-independent nonstationarity, *i.e.*, changes of the output firing probability caused by latent factors other than the observed input spike trains. To test whether the algorithm can gracefully handle this type of nonstationarity, we modify the step change simulation in Section 13.3.2.1 by varying the input-independent baseline firing probability following a linear chirp (Fig. 13.8, top) and a random noise sequence (Fig. 13.8, bottom). Results show that the estimated baselines (red (gray in print version)) rapidly converge to the actual baselines (black) in both simulations without interfering with the estimation of higher-order kernels, which represent the input-dependent nonstationarities.

#### *13.3.2.4 Larger Number of Inputs*

To validate the capability of the algorithm to tackle simultaneous gradual changes with larger number of inputs, we design a first-order, eight-input system with sigmoidal changes. Among the eight inputs, three inputs increase amplitudes following a sigmoidal curve; three inputs decrease amplitudes following a sigmoidal curve; two inputs remain constant. The output mean firing rates before and after the sigmoidal changes are 3.677 Hz and 5.112 Hz, respectively. The total simulation length is 9000 s. Results show that all eight estimated kernels converge to the actual kernels during the simulation (Fig. 13.9).

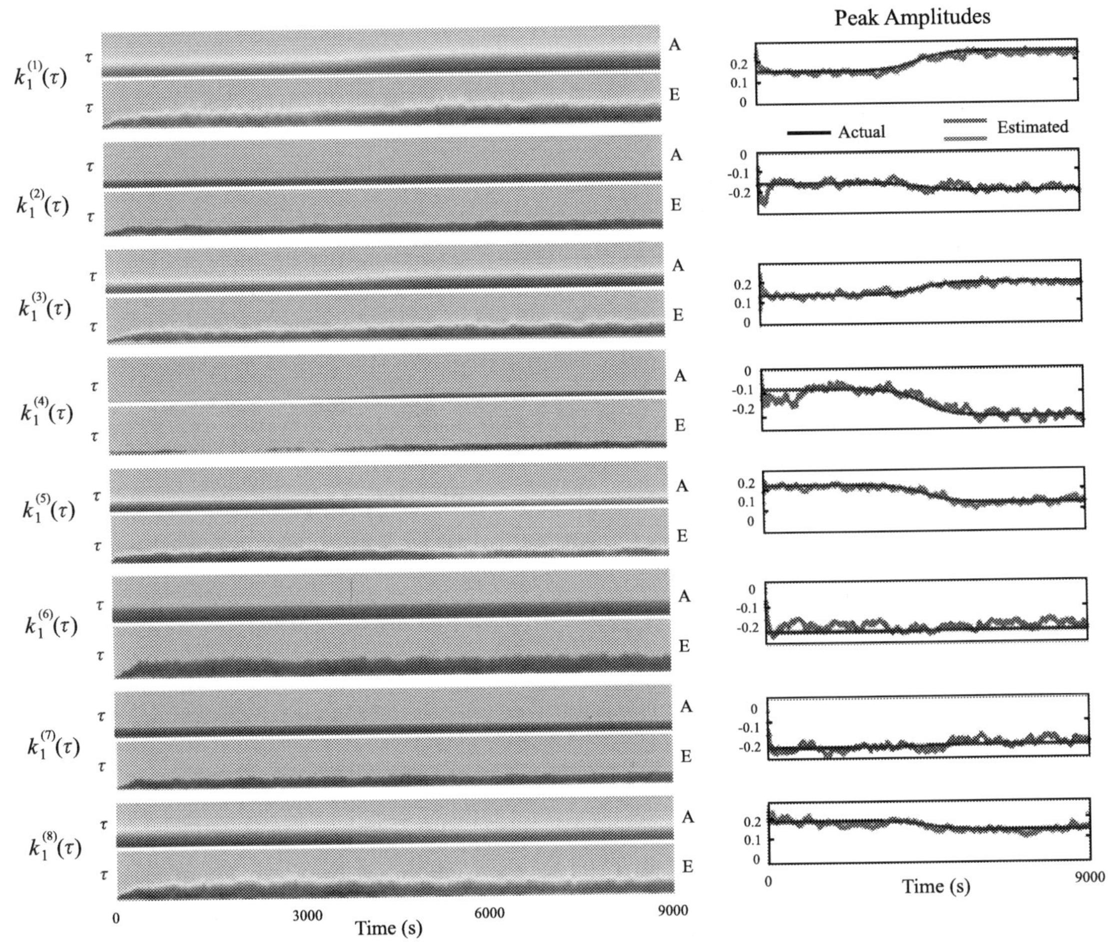


**FIGURE 13.9**

Tracking a first-order, eight-input system with sigmoidal changes. Left column: actual and estimated kernels across simulation time. The memory length is 200 ms. Right column: peak amplitudes of actual kernels (black) and estimated kernel (red (mid gray in print version) for positive kernels and blue (dark gray in print version) for negative kernels) across simulation time with a 1-s resolution.

## 13.4 IDENTIFICATION OF SYNAPTIC LEARNING RULE

### 13.4.1 THEORY

The nonstationary GVM above describes how LTSP evolves over time. In the last step, we seek to *explain* how LTSP is induced by a synaptic learning rule (Fig. 13.2, bottom).

Given the estimated nonstationary model $S(t)$, the change at time $t$ can be calculated as $\Delta S(t) = S(t) - S(t-1)$, *i.e.*, $\Delta k(t) = k(t) - k(t-1)$, $\Delta h(t) = h(t) - h(t-1)$ and $\Delta S(t) = [\Delta k(t), \Delta h(t)]$.

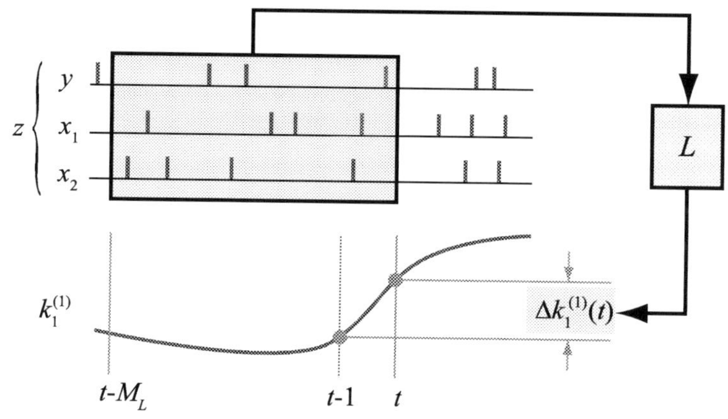


**FIGURE 13.10**

Schematic diagram showing the learning rule ($L$) from the ensemble spatiotemporal pattern of input/output spike trains ($z$: $[x, y]$) to the changes of the MISO GVM ($\Delta k_1$). For simplicity, only two inputs and one first-order kernel ($k_1$ between input $x_1$ and the output $y$) are shown.

Given that $\Delta S(t)$ is caused by the preceding input and output activities ($z = [x, y]$), the sought ensemble synaptic learning rule $L$ is essentially the causal relationship between the spatiotemporal pattern $z$ and $\Delta S(t)$ (Fig. 13.10). The $\Delta S(t)$ can be expressed as a Volterra functional power series of $z$ in a finite memory window ($M_L$) as

$$\begin{aligned}\Delta S(t) = L_0 &+ \sum_{n=1}^{N+1}\sum_{\tau=0}^{M_L} L_1^{(n)}(\tau) z_n(t-\tau)\\ &+ \sum_{n_1=1}^{N+1}\sum_{n_2=1}^{n_1}\sum_{\tau_1=0}^{M_L}\sum_{\tau_2=0}^{M_L} L_2^{(n_1,n_2)}(\tau_1,\tau_2) z_{n_1}(t-\tau_1) z_{n_2}(t-\tau_2)\\ &+ \sum_{n_1=1}^{N+1}\sum_{n_2=1}^{n_1}\sum_{n_3=1}^{n_2}\sum_{\tau_1=0}^{M_L}\sum_{\tau_2=0}^{M_L}\sum_{\tau_3=0}^{M_L} L_3^{(n_1,n_2,n_3)}(\tau_1,\tau_2,\tau_3) z_{n_1}(t-\tau_1) z_{n_2}(t-\tau_2) z_{n_3}(t-\tau_3) + \dots\end{aligned} \tag{13.15}$$

In this formulation, $L_0$ is the zero order learning rule describing the input/output activity–independent drift of the system; $L_1$ is the first-order learning rule describing the linear relation between the changes of the GVM and the input or output activities; $L_2$ is the second-order learning rule describing how the pair-wise nonlinear interactions between input/output spikes change the GVM; $L_3$ is the third-order learning rule describing how the triplet-wise nonlinear interactions between input/output spikes change the MIMO model (note that there are redundancies in $L_2$ and $L_3$ when $n_1$, $n_2$ and $n_3$ contain the same inputs. Eq. (13.15) is used nonetheless for its simplicity).

Since $\Delta S(t)$ has been estimated, $x$ and $y$ are known, Eq. (13.15) is essentially a linear model and the learning rule $L$ can thus be estimated with the standard least squares method. In practice, basis functions and penalized likelihood estimation methods can be utilized to reduce the total number of coefficients and select the significant terms in Eq. (13.15). Compared with existing learning rules, such as (a) the classical Bienenstock–Cooper–Munro (BCM) model of synaptic modification and (b) the STDP, this formulation provides a general form of ensemble learning rule defining how the changes of GVM are determined by the spatio-temporal patterns of the input–output spikes. Linear and nonlinear interactions between multiple input/output spikes with different interspike intervals are explicitly included in this formulation (Fig. 13.10).

### 13.4.2 RELATIONSHIP BETWEEN VOLTERRA KERNELS AND STDP

Eq. (13.15) provides a general representation of the ensemble learning rule underlying the changes of the neuronal input–output function. It utilizes a Volterra power series to describe how the GVM variables change as a consequence of the interactions between the input–output spikes. Before testing the identification method with simulation, it is instructive to elucidate the relation between the Volterra kernels and the most accepted synaptic learning rule, *i.e.*, STDP (Fig. 13.11).

According to the STDP rule, the change of synaptic weight between a presynaptic neuron and a postsynaptic neuron is determined by the timing of the presynaptic and postsynaptic spiking activities ($\tau_x$ and $\tau_y$). If a presynaptic (*i.e.*, input) spike precedes a postsynaptic spike within a short time window (*i.e.*, $\tau_x > \tau_y$), the synaptic weight will be enhanced; if the opposite happens (*i.e.*, $\tau_x < \tau_y$), the synaptic weight will be reduced. The relationship between the amount of synaptic weight change and the spike timings, *i.e.*, the STDP function, can be described with two exponential curves (Fig. 13.11, left). Since in this STDP expression the synaptic weight change depends only on the pair-wise interaction between a presynaptic spike and a postsynaptic spike, the Volterra expression of the learning rule contains only the second-order cross term between $x$ and $y$. In the case of a single-input system, Eq. (13.15) is reduced to

$$\Delta S(t) = \sum_{\tau_x=0}^{M_L} \sum_{\tau_y=0}^{M_L} L_2^{(x,y)}(\tau_x, \tau_y)x(t-\tau_x)y(t-\tau_y). \tag{13.16}$$

In $L_2^{(x,y)}$, the STDP function describes only the steady-state level of synaptic weight change. Another factor that contributes to $L_2^{(x,y)}$ is the induction dynamics of the STDP, *i.e.*, the synaptic weight does not change instantaneously following a delta function; instead, it relatively slowly builds up and then stabilizes following a smooth induction function $I$ (Fig. 13.11, right). The STDP function and the induction function $I$ jointly determine the shape of $L_2^{(x,y)}$ (Fig. 13.11, middle-top). In other words, this form of cross kernel between $x$ and $y$ is the Volterra representation of the STDP and induction functions.

### 13.4.3 SIMULATION STUDIES ON LEARNING RULE IDENTIFICATION

We further test the learning rule identification method with simulated input–output spiking data. A first-order, single-input, single-output spiking neuron model is built with the GVM structure (Fig. 13.12). In this model, the first-order feedforward kernel $k_1$ has a typical EPSP shape determined by two exponentials with the time constants at 30 ms and 150 ms, controlling the onset and decay rates, respectively. The peak amplitude is 0.3. The feedback kernel has a negative exponential shape to model the refrac-

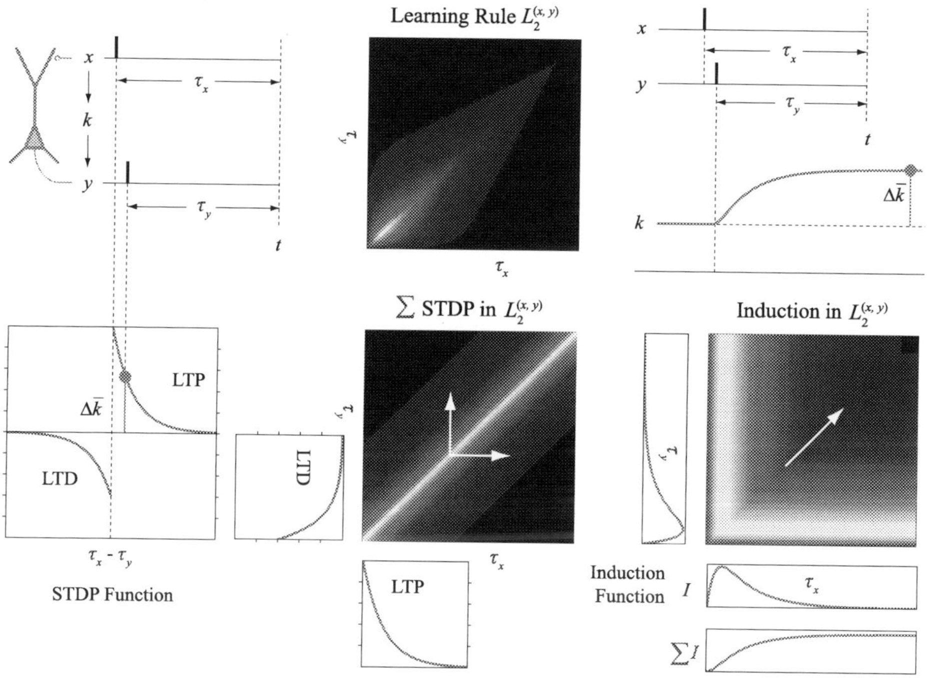


**FIGURE 13.11**

Relationship between the STDP function, the induction function ($I$) and the second-order cross kernel ($k_{2x}$) of input ($x$) and output ($y$). Left column: STDP function. Center middle: the integral of the STDP function in $k_{2x}$. Right column: the induction function $I$ and its representation in $k_{2x}$. The integral of $I$ describes the STDP dynamics. The STDP function determines the steady-state change ($\Delta\bar{k}$) of the synaptic weight $k$. Center top: the $k_{2x}$ representation of the STDP function and the induction function. This $k_{2x}$ is calculated as the element-wise product of the $k_{2x}$ of STDP function and the $k_{2x}$ of $I$. Note that $I$ is not plotted in scale for better visualization. White arrows indicate the directions of the STDP and induction functions in the cross kernel.

tory period of spike generation. The time constant and peak amplitude are 10 ms and $-1$. The noise standard deviation $\sigma$ is 0.4. During the simulation, the input spike train $x$ is fed into the model to generate the output spike train $y$. The synaptic strength between the input and output neurons, *i.e.*, $k_1$, is changed following the standard STDP rule and the induction function. In this simulation, the shape of $k_1$ remains the same; only the amplitude is changed. For convenience, we use $k_1$ to represent both the peak amplitude and the whole kernel function. The left (LTD) half of the STDP function is a single exponential with time constant and peak amplitude at 33.7 ms and $-0.018$. The right (LTP) half of the STDP function is a single exponential with time constant and peak amplitude at 16.8 ms and 0.032. Both sides share the same induction function determined by two exponentials with time constants at 900 ms and 4500 ms, for onset and decay rates, respectively. Without loss of generality, the integral of the induction function is set at 1. To prevent the intrinsic instability of the simulated system, we adjust the input patterns based on the level of $k_1$. When $k_1$ reaches a value below a certain threshold, the input $x$ changes from a 5-Hz Poisson random train to a 50-Hz burst to cause more potentiation and increase

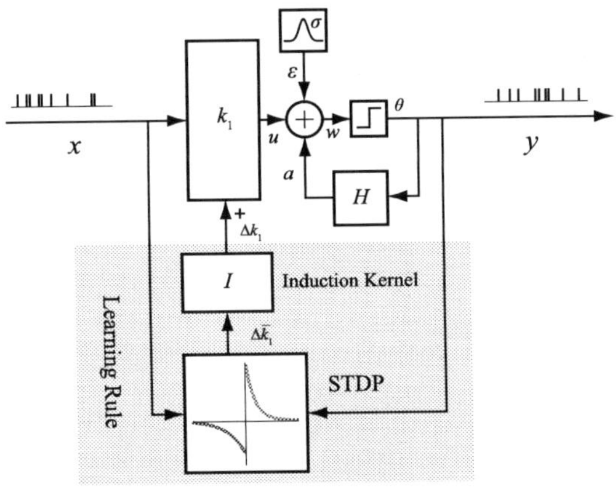


**FIGURE 13.12**

Simulation of a first-order, single-input, single-output spiking neuron model with STDP learning rule. During the simulation, presynaptic spikes $x$ and postsynaptic spikes $y$ change the feedforward kernel $k_1$ following the STDP and the induction functions. The STDP function determines the steady-state change $\Delta\bar{k}_1$; the induction function $I$ defines the STDP dynamics.

of $k_1$; when $k_1$ is above the threshold, $x$ is shifted to spike within 10 ms after output spikes to cause more depression and decrease of $k_1$. This simple method ensures $k_1$ to fluctuate in a stable manner. The total length of simulation is 200 s.

Using the nonstationary GVM method, we first track the changes of the synaptic strength, *i.e.*, $k_1(t)$, and compare them with the actual values. Results show that $k_1(t)$ can be accurately recovered from the simulated input–output spike trains (Fig. 13.13, top).

Given $k_1(t)$, we further estimate the learning rule, *i.e.*, the second-order cross kernel $L_2^{(x,y)}$. In order to reduce the number of open parameters, we expand the kernel with three sets of Laguerre basis functions as

$$\Delta k_1(t) = \sum_{j_A=1}^{L_A} \sum_{j_\psi=1}^{L_\psi} c_{xy}(j_A, j_\psi) v_{xy}^{j_A}(t) v_{xy}^{j_\psi}(t) + \sum_{j_A=1}^{L_A} \sum_{j_\psi=1}^{L_\psi} c_{yx}(j_A, j_\psi) v_{yx}^{j_A}(t) v_{yx}^{j_\psi}(t), \tag{13.17}$$

where

$$v_{xy}^{j_A}(t) = \sum_{\tau_x=1}^{M_\psi} b_{xy}^{j_A}(\tau_x - \tau_y) x(t - \tau_x), \tag{13.18}$$

$$v_{xy}^{j_\psi}(t) = \sum_{\tau_y=0}^{\tau_x - 1} b_{xy}^{j_\psi}(\tau_y) y(t - \tau_y), \tag{13.19}$$

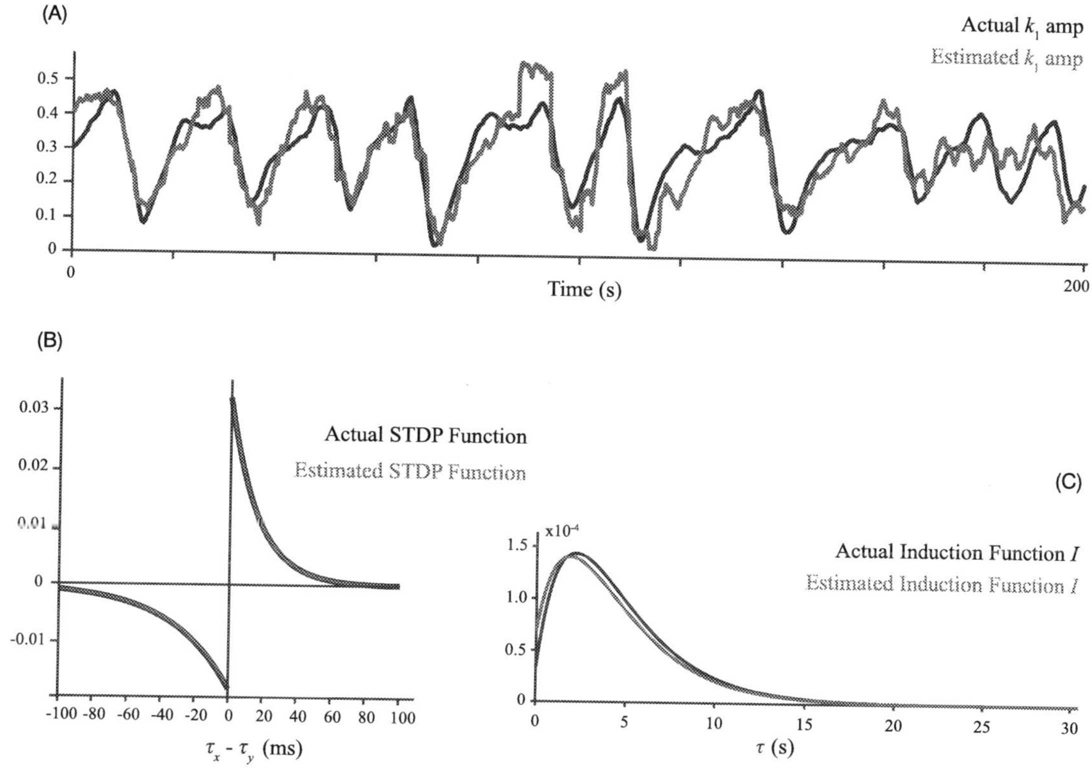


**FIGURE 13.13**

Identification of STDP and induction functions from spiking input–output data. (A): peak amplitudes of the first-order feedforward kernel $k_1$; (B): STDP functions; (C): induction functions. In all plots, back: actual; red (gray in print version): estimated.

$$v_{yx}^{jA}(t) = \sum_{\tau_y=1}^{M_\psi} b_{yx}^{jA}(\tau_y - \tau_x)y(t - \tau_y), \tag{13.20}$$

$$v_{yx}^{j\psi}(t) = \sum_{\tau_x=0}^{\tau_y-1} b_{yx}^{j\psi}(\tau_y)x(t - \tau_x). \tag{13.21}$$

In these equations, $c$ represents the sought-after learning rule coefficients. They are split into $c_{xy}$ and $c_{yx}$ to represent the two halves of the cross kernel for $x$ preceding $y$ and $y$ preceding $x$, respectively. The subscript $A$ represents the STDP amplitude. The subscript $\psi$ represents the STDP induction. Since all $v$ can be calculated with the predefined basis functions and the known $x$ and $y$, Eq. (13.17) is essentially linear and $c$ can be estimated with a least squares method. With the estimated coefficients, $L_2^{(x,y)}$ can be reconstructed as

$$L_2^{(x,y)}(\tau_x, \tau_y) = \sum_{j_A=1}^{L_A} \sum_{j_\psi=1}^{L_\psi} c_{xy}(j_A, j_\psi) b_{xy}^{j_A}(\tau_x - \tau_y) b_{xy}^{j_\psi}(\tau_y) + \sum_{j_A=1}^{L_A} \sum_{j_\psi=1}^{L_\psi} c_{yx}(j_A, j_\psi) b_{yx}^{j_A}(\tau_y - \tau_x) b_{yx}^{j_\psi}(\tau_x). \tag{13.22}$$

Here, $L_2^{(x,y)}$ constitutes a general representation of the synaptic learning since it does not require the independence of the STDP function and the induction function. In this study, however, since these two functions are assumed to be independent, $L_2^{(x,y)}$ is equal to the outer product of the two functions. The left (LTD) side of the STDP function can be obtained by integrating the vertical vectors in the upper half of $L_2^{(x,y)}$ along its diagonal line; similarly, the right (LTP) side of the STDP function can be obtained by integrating the horizontal vectors in the lower half of $L_2^{(x,y)}$ along its diagonal line. The induction function for LTD can be obtained by integrating the diagonal vectors in the upper half of $L_2^{(x,y)}$ along the vertical axis; similarly, the induction function for LTP can be obtained by integrating the diagonal vectors in the lower half of $L_2^{(x,y)}$ along the horizontal axis. The induction function is then normalized to have a unitary integral. The STDP functions are scaled accordingly.

Fig. 13.13 shows the final identification results using the same 200-s input–output spiking data. It is evident that both the STDP function and the induction function are faithfully recovered with this method (Fig. 13.13B, C).

## 13.5 SUMMARY AND DISCUSSION

In this chapter, we describe a systems identification approach for studying different forms of synaptic plasticity using spiking activities. In this formulation, synaptic strength is represented as input–output (linear or nonlinear) dynamics between neurons; STSP is defined as input–output nonlinear dynamics; LTSP is defined as the nonstationarity of such nonlinear dynamics; synaptic learning rule is essentially the function governing the formation of the LTSP based on the input–output spiking patterns. As a special case, STDP is equivalent to a second-order learning rule describing the pair-wise interactions between single-input spikes and single-output spikes. Using experimental and simulated input–output data, we have shown that STSP, LTSP and the learning rule can be accurately identified with a set of nonstationary nonlinear dynamical models.

The MISO model and the multiinput, multioutput (MIMO) model, which consists of a series of MISO models, have been used intensively as a tool to identify the neuronal functional connectivity [56,62], *e.g.*, hippocampal CA3-CA1, and build cortical prostheses for restoring cognitive functions, *e.g.*, hippocampal memory prosthesis [8,10,23,30–32,66]. Previous results have shown that the MIMO model can accurately predict the output (*e.g.*, hippocampal CA1) spike trains based on the ongoing input (*e.g.*, hippocampal CA3) spike trains. Electrical stimulation with the spatiotemporal patterns of the predicted output spike trains can restore or even enhance the memory function during a spatial delayed nonmatch-to-sample task (DNMS) in rodents. Since all these previous models are stationary, their success must be due to the fact that these applications only involve learned behaviors in well-trained animals, where the behavioral performance and the spatiotemporal pattern representations of behaviors (or memory events) are all stabilized, and as a consequence, the MIMO input–output transformations can be sufficiently described by stationary MIMO models. The resulting hippocampal prostheses essentially replicate the input–output properties of the hippocampal CA3-CA1 after learning and memory formation. To build hippocampal memory prostheses capable of learning and memory formation in a self-organizing manner, on the other hand, it is required to identify and further mimic the nonstationary behaviors and the underlying learning rules of the MIMO input–output properties. The identification methods described in this chapter provide a computational framework for studying such nonstationary properties and the identified functional learning rule may potentially be used as the computational basis for developing the next-generation, adaptive hippocampal memory prostheses.

We have tested the learning rule identification algorithm with a simple form of STDP. In future studies, we may use the full Volterra expression (Eq. (13.15)), as opposed to the reduced second-order cross kernel expression of STDP in this study, to identify the ensemble synaptic learning rule. Previous studies have shown the existence of a single-spike rule [37,71], triplet rule and quadruplet rule of STDP [51,72]. These learning rules, together with their induction functions, are equivalent to the first-order, third-order and fourth-order learning rule kernels of the Volterra expression, respectively. A formal approach of identifying the learning rule from experimental data would be including all possible terms (*e.g.*, zero order, first-order, second-order, third-order and even fourth-order terms) in the model and then using statistical model estimation methods (*e.g.*, regularized estimation) to characterize and select the significant terms [62]. The resulting ensemble learning rule is expected to provide a more thorough description of LTSP.

It should be noted that a spiking neuron with the standard paired spike STDP rule is intrinsically unstable since the STDP rule is unbounded, *i.e.*, there is no mechanism to saturate the synaptic weight or prevent it from being negative. In the example shown here, we solve this problem with a rather empirical method: the input patterns, rather than the neuronal input–output properties, are adjusted to keep the synaptic weight fluctuating within a certain range. The main reason behind this strategy is to use the standard STDP function in the simulation and show how it is related to the Volterra kernels. A more realistic approach is to modify the simple STDP rule shown here into a stable STDP rule, where the weight change in the LTD portion of the STDP rule is multiplicative instead of additive [54]. Consequently, a more sophisticated multistage generalized multilinear modeling procedure needs to be utilized to estimate the STDP functions [54].

## ACKNOWLEDGMENTS

This work is supported partially by the Defense Advanced Research Projects Agency (DARPA) Restoring Active Memory (RAM) program (N66001-14-C-4016), partially by the National Institute of Health (NIH) U01 GM104604 and partially by the NIH Biomedical Simulations Resource (BMSR) 5P41EB001978-32.

## REFERENCES

[1] D.L. Alkon, D.G. Amaral, M.F. Bear, J. Black, T.J. Carew, N.J. Cohen, J.F. Disterhoft, H. Eichenbaum, S. Golski, L.K. Gorman, G. Lynch, B.L. Mcnaughton, M. Mishkin, J.R. Moyer, J.L. Olds, D.S. Olton, T. Otto, L.R. Squire, U. Staubli, L.T. Thompson, C. Wible, Learning and memory, Brain Research Reviews 16 (1991) 193–220.

[2] P. Anderson, T.V. Bliss, K.K. Skrede, Lamellar organization of hippocampal pathways, Experimental Brain Research 13 (1971) 222–238.

[3] C.A. Barnes, Memory deficits associated with senescence – neurophysiological and behavioral-study in the rat, Journal of Comparative & Physiological Psychology 93 (1979) 74–104.
[4] T.W. Berger, Long-term potentiation of hippocampal synaptic transmission affects rate of behavioral learning, Science 224 (1984) 627–630.
[5] T.W. Berger, G. Chauvet, R.J. Sclabassi, A biological based model of functional properties of the hippocampus, Neural Networks 7 (1994) 1031–1064.
[6] T.W. Berger, J.L. Eriksson, D.A. Ciarolla, R.J. Sclabassi, Nonlinear systems analysis of the hippocampal perforant path-dentate projection. II. Effects of random impulse train stimulation, Journal of Neurophysiology 60 (1988) 1076–1094.
[7] T.W. Berger, J.L. Eriksson, D.A. Ciarolla, R.J. Sclabassi, Nonlinear systems analysis of the hippocampal perforant path-dentate projection. III. Comparison of random train and paired impulse stimulation, Journal of Neurophysiology 60 (1988) 1095–1109.
[8] T.W. Berger, R.E. Hampson, D. Song, A. Goonawardena, V.Z. Marmarelis, S.A. Deadwyler, A cortical neural prosthesis for restoring and enhancing memory, Journal of Neural Engineering 8 (2011) 046017.
[9] T.W. Berger, D. Song, R.H.M. Chan, V.Z. Marmarelis, The neurobiological basis of cognition: identification by multi-input, multioutput nonlinear dynamic modeling, Proceedings of the IEEE 98 (2010) 356–374.
[10] T.W. Berger, D. Song, R.H.M. Chan, V.Z. Marmarelis, J. LaCoss, J. Wills, R.E. Hampson, S.A. Deadwyler, J.J. Granacki, A hippocampal cognitive prosthesis: multi-input, multi-output nonlinear modeling and VLSI implementation, IEEE Transactions on Neural Systems and Rehabilitation Engineering 20 (2012) 198–211.
[11] G.Q. Bi, M.M. Poo, Synaptic modifications in cultured hippocampal neurons: dependence on spike timing, synaptic strength, and postsynaptic cell type, The Journal of Neuroscience 18 (1998) 10464–10472.
[12] T.V. Bliss, T. Lomo, Long-lasting potentiation of synaptic transmission in the dentate area of the anaesthetized rabbit following stimulation of the perforant path, Journal of Physiology 232 (1973) 331–356.
[13] C.R. Bramham, LTP not equal learning: lessons from short-term plasticity, Frontiers in Behavioral Neuroscience 4 (2010) 3.
[14] C.R. Bramham, B. Srebro, Induction of long-term depression and potentiation by low- and high-frequency stimulation in the dentate area of the anesthetized rat: magnitude, time course and EEG, Brain Research 405 (1) (1987 Mar 3) 100–107.
[15] E.N. Brown, R.E. Kass, P.P. Mitra, Multiple neural spike train data analysis: state-of-the-art and future challenges, Nature Neuroscience 7 (2004) 456–461.
[16] G. Buzsaki, Large-scale recording of neuronal ensembles, Nature Neuroscience 7 (2004) 446–451.
[17] N. Caporale, Y. Dan, Spike timing-dependent plasticity: a Hebbian learning rule, Annual Review of Neuroscience 31 (2008) 25–46.
[18] R.H.M. Chan, D. Song, T.W. Berger, Tracking temporal evolution of nonlinear dynamics in hippocampus using time-varying Volterra kernels, in: Proceedings of the IEEE EMBS Conferen., 2008, pp. 4996–4999.
[19] Z. Chen, D.F. Putrino, S. Ghosh, R. Barbieri, E.N. Brown, Statistical inference for assessing functional connectivity of neuronal ensembles with sparse spiking data, IEEE Transactions on Neural Systems and Rehabilitation Engineering 19 (2011) 121–135.
[20] J.S. Dittman, A.C. Kreitzer, W.G. Regehr, Interplay between facilitation, depression, and residual calcium at three presynaptic terminals, Journal of Neuroscience 20 (2000) 1374–1385.
[21] D.E. Feldman, The spike timing dependence of plasticity, Neuron 75 (2012) 556–571.
[22] C. de Boor, On calculating with B-splines, Journal of Approximation Theory 6 (1972) 50–62.
[23] S.A. Deadwyler, T.W. Berger, A.J. Sweatt, D. Song, R.H. Chan, I. Opris, G.A. Gerhardt, V.Z. Marmarelis, R.E. Hampson, Donor/recipient enhancement of memory in rat hippocampus, Frontiers in Systems Neuroscience 7 (2013) 120.
[24] S.A. Deadwyler, R.E. Hampson, Ensemble activity and behavior – whats the code?, Science 270 (1995) 1316–1318.
[25] U.T. Eden, L.M. Frank, R. Barbieri, V. Solo, E.N. Brown, Dynamic analysis of neural encoding by point process adaptive filtering, Neural Computation 16 (2004) 971–998.
[26] S. Eldawlatly, R. Jin, K.G. Oweiss, Identifying functional connectivity in large-scale neural ensemble recordings: a multi-scale data mining approach, Neural Computation 21 (2009) 450–477.
[27] W.J. Freeman, How Brains Make up Their Minds, Columbia University Press, 1999.
[28] A.P. Georgopoulos, A.B. Schwartz, R.E. Kettner, Neuronal population coding of movement direction, Science 233 (1986) 1416–1419.
[29] R.E. Hampson, G.A. Gerhardt, V. Marmarelis, D. Song, I. Opris, L. Santos, T.W. Berger, S.A. Deadwyler, Facilitation and restoration of cognitive function in primate prefrontal cortex by a neuroprosthesis that utilizes minicolumn-specific neural firing, Journal of Neural Engineering 9 (2012) 056012.

[30] R.E. Hampson, D. Song, R.H.M. Chan, A.J. Sweatt, M.R. Riley, G.A. Gerhardt, D.C. Shin, V.Z. Marmarelis, T.W. Berger, S.A. Deadwyler, A nonlinear model for hippocampal cognitive prosthesis: memory facilitation by hippocampal ensemble stimulation, IEEE Transactions on Neural Systems and Rehabilitation 20 (2012) 184–197.

[31] R.E. Hampson, D. Song, R.H.M. Chan, A.J. Sweatt, M.R. Riley, A.V. Goonawardena, V.Z. Marmarelis, G.A. Gerhardt, T.W. Berger, S.A. Deadwyler, Closing the loop for memory prosthesis: detecting the role of hippocampal neural ensembles using nonlinear models, IEEE Transactions on Neural Systems and Rehabilitation 20 (2012) 510–525.

[32] R.E. Hampson, D. Song, I. Opris, L.M. Santos, D.C. Shin, G.A. Gerhardt, V.Z. Marmarelis, T.W. Berger, S.A. Deadwyler, Facilitation of memory encoding in primate hippocampus by a neuroprosthesis that promotes task-specific neural firing, Journal of Neural Engineering 10 (2013) 066013.

[33] D.O. Hebb, The Organization of Behavior, Wiley & Sons, New York, 1949.

[34] G. Hess, U. Kuhnt, L.L. Voronin, Quantal analysis of paired-pulse facilitation in Guinea pig hippocampal slices, Neuroscience Letters 77 (2) (1987) 187–192.

[35] M. Ito, Long-term depression, Annual Review of Neuroscience 12 (1989) 85–102.

[36] D. Johnston, S.M. Wu, Foundations of Cellular Neurophysiology, MIT Press, Cambridge, Mass., 1995.

[37] R. Kempter, W. Gerstner, J.L. von Hemmen, Hebbian learning and spiking neurons, Physical Review E 59 (1999) 4498–4514.

[38] J. Larson, D. Wong, G. Lynch, Patterned stimulation at the theta frequency is optimal for the induction of hippocampal long-term potentiation, Brain Research 368 (1986) 347–350.

[39] W.B. Levy, O. Steward, Temporal contiguity requirements for long term associative potentiation depression in the hippocampus, Neuroscience 8 (1983) 791–797.

[40] G. Lynch, M. Baudry, The biochemistry of memory – a new and specific hypothesis, Science 224 (1984) 1057–1063.

[41] H. Markram, J. Lubke, M. Frotscher, B. Sakmann, Regulation of synaptic efficacy by coincidence of postsynaptic APs and EPSPs, Science 275 (1997) 213–215.

[42] V.Z. Marmarelis, Identification of nonlinear biological systems using Laguerre expansions of kernels, Annals of Biomedical Engineering 21 (1993) 573–589.

[43] V.Z. Marmarelis, Nonlinear Dynamic Modeling of Physiological Systems, Wiley-IEEE Press, Hoboken, 2004.

[44] K.C. Martin, D. Michael, J.C. Rose, M. Barad, A. Casadio, H.X. Zhu, E.R. Kandel, MAP kinase translocates into the nucleus of the presynaptic cell and is required for long-term facilitation in Aplysia, Neuron 18 (1997) 899–912.

[45] T.J. McHugh, K.I. Blum, J.Z. Tsien, S. Tonegawa, M.A. Wilson, Impaired hippocampal representation of space in CA1-specific NMDAR1 knockout mice, Cell 87 (1996) 1339–1349.

[46] B.L. McNaughton, C.A. Barnes, G. Rao, J. Baldwin, M. Rasmussen, Long-term enhancement of hippocampal synaptic transmission and the acquisition of spatial information, The Journal of Neuroscience 6 (1986) 563–571.

[47] R.G. Morris, Synaptic plasticity and learning: selective impairment of learning rats and blockade of long-term potentiation in vivo by the N-methyl-D-aspartate receptor antagonist AP5, The Journal of Neuroscience 9 (1989) 3040–3057.

[48] K. Nakazawa, M.C. Quirk, R.A. Chitwood, M. Watanabe, M.F. Yeckel, L.D. Sun, A. Kato, C.A. Carr, D. Johnston, M.A. Wilson, S. Tonegawa, Requirement for hippocampal CA3 NMDA receptors in associative memory recall, Science 297 (2002) 211–218.

[49] M. Okatan, M.A. Wilson, E.N. Brown, Analyzing functional connectivity using a network likelihood model of ensemble neural spiking activity, Neural Computation 17 (2005) 1927–1961.

[50] L. Paninski, J.W. Pillow, E.P. Simoncelli, Maximum likelihood estimation of a stochastic integrate-and-fire neural encoding model, Neural Computation 16 (2004) 2533–2561.

[51] J.P. Pfister, W. Gerstner, Triplets of spikes in a model of spike timing-dependent plasticity, The Journal of Neuroscience 26 (2006) 9673–9682.

[52] J.W. Pillow, L. Paninski, V.J. Uzzell, E.P. Simoncelli, E.J. Chichilnisky, Prediction and decoding of retinal ganglion cell responses with a probabilistic spiking model, Journal of Neuroscience 25 (2005) 11003–11013.

[53] A. Pouget, P. Dayan, R.S. Zemel, Inference and computation with population codes, Annual Review of Neuroscience 26 (2003) 381–410.

[54] B.S. Robinson, T.W. Berger, D. Song, Identification of stable spike-timing-dependent plasticity from spiking activity with generalized multilinear modeling, Neural Computation 28 (2016) 2320–2351.

[55] E. Salinas, L.F. Abbott, Vector reconstruction from firing rates, Journal of Computational Neuroscience 1 (1994) 89–107.

[56] D. Song, T.W. Berger, Identification of nonlinear dynamics in neural population activity, in: K.G. Oweiss (Ed.), Statistical Signal Processing for Neuroscience and Neurotechnology, McGraw-Hill/Irwin, Boston, 2009.

[57] D. Song, R.H. Chan, V.Z. Marmarelis, R.E. Hampson, S.A. Deadwyler, T.W. Berger, Nonlinear dynamic modeling of spike train transformations for hippocampal-cortical prostheses, IEEE Transactions on Biomedical Engineering 54 (2007) 1053–1066.

[58] D. Song, R.H. Chan, V.Z. Marmarelis, R.E. Hampson, S.A. Deadwyler, T.W. Berger, Nonlinear modeling of neural population dynamics for hippocampal prostheses, Neural Networks 22 (2009) 1340–1351.

[59] D. Song, R.H. Chan, V.Z. Marmarelis, R.E. Hampson, S.A. Deadwyler, T.W. Berger, Sparse generalized Laguerre–Volterra model of neural population dynamics, in: Proceedings of the IEEE EMBS Conference, 2009, pp. 4555–4558.

[60] D. Song, B.S. Robinson, R.H.M. Chan, V.Z. Marmarelis, R.E. Hampson, S.A. Deadwyler, T.W. Berger, Identification of functional synaptic plasticity from ensemble spiking activities: a nonlinear dynamical modeling approach, in: Proceedings of the IEEE EMBC Neural Engineering Conference, 2013, pp. 617–620.

[61] D. Song, I. Opris, R.H. Chan, V.Z. Marmarelis, R.E. Hampson, S.A. Deadwyler, T.W. Berger, Functional connectivity between layer 2/3 and layer 5 neurons in prefrontal cortex of nonhuman primates during a delayed match-to-sample task, in: Proceedings of the IEEE EMBS Conference, 2012, pp. 2555–2558.

[62] D. Song, H. Wang, C.Y. Tu, V.Z. Marmarelis, R.E. Hampson, S.A. Deadwyler, T.W. Berger, Identification of sparse neural functional connectivity using penalized likelihood estimation and basis functions, Journal of Computational Neuroscience 35 (2013) 335–357.

[63] D. Song, V.Z. Marmarelis, T.W. Berger, Parametric and non-parametric modeling of short-term synaptic plasticity. Part I: computational study, Journal of Computational Neuroscience 26 (2009) 1–19.

[64] D. Song, Z. Wang, V.Z. Marmarelis, T.W. Berger, Parametric and non-parametric modeling of short-term synaptic plasticity. Part II: experimental study, Journal of Computational Neuroscience 26 (2009) 21–37.

[65] D. Song, R.H.M. Chan, B.S. Robinson, V.Z. Marmarelis, I. Opris, R.E. Hampson, S.A. Deadwyler, T.W. Berger, Identification of functional synaptic plasticity from spiking activities using nonlinear dynamical modeling, Journal of Neuroscience Methods 244 (2015) 123–135, https://doi.org/10.1016/j.jneumeth.2014.09.023.

[66] D. Song, B.S. Robinson, R.E. Hampson, V.Z. Marmarelis, S.A. Deadwyler, T.W. Berger, Sparse large-scale nonlinear dynamical modeling of human hippocampus for memory prostheses, IEEE Transactions on Neural Systems and Rehabilitation Engineering 26 (2) (2018) 272–280, https://doi.org/10.1109/TNSRE.2016.2604423.

[67] C.F. Stevens, A million dollar question: does LTP = memory?, Neuron 20 (1998) 1–2.

[68] S. Tonegawa, J.Z. Tsien, T.J. McHugh, P. Huerta, K.I. Blum, M.A. Wilson, Hippocampal CA1-region-restricted knockout of NMDAR1 gene disrupts synaptic plasticity, place fields, and spatial learning, Cold Spring Harbor Symposia 61 (1996) 225–238.

[69] W. Truccolo, U.T. Eden, M.R. Fellows, J.P. Donoghue, E.N. Brown, A point process framework for relating neural spiking activity to spiking history, neural ensemble, and extrinsic covariate effects, Journal of Neurophysiology 93 (2005) 1074–1089.

[70] L.P. Tuff, R.J. Racine, R. Adamec, The effects of kindling on GABA-mediated inhibition in the dentate gyrus of the rat. I. Paired-pulse depression, Brain Research 277 (1) (1983) 79–90.

[71] M.C. van Rossum, G.Q. Bi, G.G. Turrigiano, Stable Hebbian learning from spike timing-dependent plasticity, The Journal of Neuroscience 20 (2000) 8812–8821.

[72] H.X. Wang, R.C. Gerkin, D.W. Nauen, G.Q. Bi, Coactivation and timing-dependent integration of synaptic potentiation and depression, Nature Neuroscience 8 (2005) 187–193.

[73] M.Y. Zhao, A. Batista, J.P. Cunningham, C. Chestek, Z. Rivera-Alvidrez, R. Kalmar, An L(1)-regularized logistic model for detecting short-term neuronal interactions, Journal of Computational Neuroscience 32 (2012) 479–497.

[74] R.S. Zucker, W.G. Regehr, Short-term synaptic plasticity, Annual Review of Physiology 64 (1) (2002) 355–405.

CHAPTER

# ADAPTIVE $H_\infty$ TRACKING CONTROL OF NONLINEAR SYSTEMS USING REINFORCEMENT LEARNING

# 14

**Hamidreza Modares*, Bahare Kiumarsi†, Kyriakos G. Vamvoudakis‡, Frank L. Lewis†,§**

**Missouri University of Science and Technology, Rolla, MO, United States* †*UTA Research Institute, University of Texas at Arlington, Fort Worth, TX, United States* ‡*Virginia Tech, Blacksburg, VA, United States* §*State Key Laboratory of Synthetical Automation for Process Industries, Northeastern University, Shenyang, China*

## CONTENTS



## CHAPTER POINTS

- The result of this approach is to design an online data-based solution to the $H_\infty$ tracking control problem.
- Reinforcement learning is employed to learn the solution to the $H_\infty$ tracking in real-time and without requiring the system dynamics.

Adaptive Learning Methods for Nonlinear System Modeling. DOI: 10.1016/B978-0-12-812976-0.00018-X

## 14.1 INTRODUCTION

Reinforcement learning (RL) [1–3], inspired by learning mechanisms observed in animals, is concerned with how an agent or decision maker takes actions so as to optimize a cost of its long-term interactions with the environment. The cost function is prescribed and captures some desired system behaviors such as minimizing the transient error and minimizing the control effort for achieving a specific goal. The agent learns an optimal policy so that, by taking actions produced based on this policy, the long-term cost function is optimized. Similar to RL, optimal control involves finding an optimal policy by optimizing a long-term performance criterion. Strong connections between RL and optimal control have prompted a major effort towards introducing and developing online and model-free RL algorithms to learn the solution to optimal control problems [4–6].

RL methods have been successfully used to solve the optimal regulation problems by learning the solution to the so-called Hamilton–Jacobi equations arising from both optimal $H_2$ [7–18] and $H_\infty$ [19–30] regulation problems. For continuous-time (CT) systems, [8,9] proposed a promising RL algorithm, called integral RL (IRL), to learn the solution to the Hamilton–Jacobi–Bellman (HJB) equations using only partial knowledge about the system dynamics. They used an iterative online policy iteration [31] procedure to implement their IRL algorithm. The original IRL algorithm and many of its extensions are on-policy algorithms. That is, the policy that is applied to the system to generate data for learning (behavior policy) is the same as the policy that is being updated and learned about (target policy). The work [15] presented an off-policy RL algorithms for CT systems in which the behavior policy could be different from the target policy. This algorithm does not require any knowledge of the system dynamics and is data efficient because it reuses the data generated by the behavior policy to learn as many target policies as required. Many variants and extensions of off-policy RL algorithms are presented in the literature. Other than the IRL-based PI algorithms and off-policy RL algorithms, efficient synchronous PI algorithms with guaranteed closed-loop stability were proposed for CT systems in [7,11,12] to learn the solution to the HJB equation. Synchronous IRL algorithms were also presented for solving the HJB equation in [23,32].

Although RL algorithms have been widely used to solve the optimal regulation problems, few results considered solving the optimal tracking control problem (OTCP) for both discrete-time [33–36] and continuous-time systems [6,37]. Moreover, existing methods for continuous-time systems require the exact knowledge of the system dynamics *a priori* while finding the feedforward part of the control input using either the dynamic inversion concept or the solution of output regulator equations [39–41]. While the importance of the RL algorithms is well understood for solving optimal regulation problems for uncertain systems, the requirement of the exact knowledge of the system dynamics for finding the steady-state part of the control input in the existing OTCP formulation does not allow for direct extending of the IRL algorithm for solving the OTCP.

In this chapter, we develop adaptive optimal controllers based on the RL techniques to learn the optimal $H_\infty$ tracking control solutions for nonlinear continuous-time systems without knowing the system dynamics or the command generator dynamics. An augmented system is first constructed from the tracking error dynamics and the command generator dynamics to introduce a new discounted performance function for the OTCP. The tracking Hamilton–Jacobi–Isaac (HJI) equations are then derived to solve OTCPs. Off-policy RL algorithms, implemented on an actor-critic structure, are used to find the solution to the tracking HJI equations online using only measured data along the augmented system trajectories. These algorithms are developed for both affine and nonaffine nonlinear systems. Therefore, they can be employed in control of many real-world applications, including robot manipulators, mobile robots, unmanned aerial vehicles (UAVs), power systems and human–robot interaction systems.

## 14.2 $H_\infty$ OPTIMAL TRACKING CONTROL FOR NONLINEAR AFFINE SYSTEMS

Existing solutions to the $H_\infty$ tracking problem are composed of two steps [38–41]. A feedforward control input is designed to guarantee perfect tracking using either dynamic inversion or by solving the so-called output regulator equations in the first step. A feedback control input is designed in the second step by solving an HJI equation to stabilize the tracking error dynamics. In these methods, procedures for computing the feedback and feedforward terms are based on offline solution methods which require complete knowledge of the system dynamics. In this section, a new formulation for the $H_\infty$ tracking is presented which allows developing model-free RL solutions.

Consider the nonlinear time-invariant system given as

$$\dot{\mathbf{x}}(t) = \mathbf{f}(\mathbf{x}(t)) + \mathbf{g}(\mathbf{x}(t))\,\mathbf{u}(t) + \mathbf{k}(\mathbf{x}(t))\,\mathbf{w}(t), \tag{14.1}$$

where $\mathbf{x}(t) \in \mathbb{R}^n$, $\mathbf{u}(t) \in \mathbb{R}^m$ and $\mathbf{w}(t) \in \mathbb{R}^p$ represent the state of the system, the control input and the external disturbance of the system, respectively. The drift dynamics is represented by $\mathbf{f}(\mathbf{x}(t)) \in \mathbb{R}^n$, $\mathbf{g}(\mathbf{x}(t)) \in \mathbb{R}^{n\times m}$ is the input dynamics and $\mathbf{k}(\mathbf{x}(t)) \in \mathbb{R}^p$ is the disturbance dynamics. It is assumed that $\mathbf{f}(\mathbf{0}) = \mathbf{0}$ and $\mathbf{f}(\mathbf{x}(t))$, $\mathbf{g}(\mathbf{x}(t)$ and $\mathbf{k}(\mathbf{x}(t))$ are unknown Lipschitz functions and the system is stabilizable.

**Assumption 1.** Let $\mathbf{r}(t)$ be the bounded reference trajectory and assume that there exists a Lipschitz continuous command generator function $\mathbf{h_d}(t) \in \mathbb{R}^n$ with $\mathbf{h_d}(\mathbf{0}) = \mathbf{0}$ such that

$$\dot{\mathbf{r}}(t) = \mathbf{h_d}(t)\,\mathbf{r}(t). \tag{14.2}$$

Define the tracking error

$$\mathbf{e_d}(t) \triangleq \mathbf{x}(t) - \mathbf{r}(t). \tag{14.3}$$

Using (14.1)–(14.3), the tracking error dynamics is given by

$$\dot{\mathbf{e}}_\mathbf{d}(t) = \mathbf{f}(\mathbf{x}(t)) + \mathbf{g}(\mathbf{x}(t))\,\mathbf{u}(t) + \mathbf{k}(\mathbf{x}(t))\,\mathbf{w}(t) - \mathbf{h_d}(\mathbf{r}(t)). \tag{14.4}$$

The performance output to be controlled is defined such that it satisfies

$$\|z(t)\|^2 = \mathbf{e_d}^T\mathbf{Q}\,\mathbf{e_d} + \mathbf{u}^T\mathbf{R}\,\mathbf{u}. \tag{14.5}$$

The goal of the $H_\infty$ tracking is to attenuate the effect of the disturbance input $\mathbf{w}$ on the performance output $\mathbf{z}$. Before defining the $H_\infty$ tracking control problem, we define the following general $L_2$-gain or disturbance attenuation condition.

**Definition 1** (Bounded $L_2$-gain or disturbance attenuation)**.** The nonlinear system (14.1) is said to have $L_2$-gain less than or equal to $\gamma$ if the following disturbance attenuation condition is satisfied for all $\mathbf{w} \in L_2[0, \infty)$:

$$\frac{\int_t^\infty e^{-\alpha(\tau-t)} \|\mathbf{z}(\tau)\|^2 d\tau}{\int_t^\infty e^{-\alpha(\tau-t)} \|\mathbf{w}(\tau)\|^2 d\tau} \leqslant \gamma^2, \tag{14.6}$$

where $\alpha > 0$ is the discount factor and $\gamma$ represents the amount of attenuation from the disturbance input $\mathbf{w}(t)$ to the defined performance output variable $\mathbf{z}(t)$.

The disturbance attenuation condition (14.6) implies that the effect of the disturbance input to the desired performance output is attenuated by a degree at least equal to $\gamma$. The desired performance output represents a meaningful cost in the sense that it includes a positive penalty on the tracking error and a positive penalty on the control effort. The use of the discount factor is essential. This is because the feedforward part of the control input does not converge to zero in general and thus penalizing the control input in the performance function without a discount factor makes the performance function unbounded.

Using (14.5) in (14.6) one has

$$\int_t^\infty e^{-\alpha(\tau-t)}(\mathbf{e_d}^T \mathbf{Q} \mathbf{e_d} + \mathbf{u}^T \mathbf{R} \mathbf{u}) d\tau \leqslant \gamma^2 \int_t^\infty e^{-\alpha(\tau-t)} (\mathbf{w}^T \mathbf{w})\, d\tau. \tag{14.7}$$

**Definition 2** ($H_\infty$ **optimal tracking).** The $H_\infty$ tracking control problem is to find a control policy $\mathbf{u} = \beta(\mathbf{e_d}, \mathbf{r})$ for some smooth function $\beta$ depending on the tracking error $\mathbf{e}$ and the reference trajectory $\mathbf{r}$, such that:
(i) The closed-loop system $\dot{\mathbf{x}} = \mathbf{f}(\mathbf{x}) + \mathbf{g}(\mathbf{x})\, \beta(\mathbf{e_d}, \mathbf{r}) + \mathbf{k}(\mathbf{x})\, \mathbf{w}$ satisfies the attenuation condition (14.7).
(ii) The tracking error dynamics (14.4) with $\mathbf{w} = 0$ is locally asymptotically stable.

The main difference between Definition 2 and the standard definition of the $H_\infty$ tracking control problem (see [38], Definition 5.2.1) is that a more general disturbance attenuation condition is defined here. Previous work on the $H_\infty$ optimal tracking divides the control input into feedback and feedforward parts. The feedforward part is first obtained separately without considering any optimality criterion. Then, the problem of optimal design of the feedback part is reduced to an $H_\infty$ optimal regulation problem. In contrast, in the new formulation, both feedback and feedforward parts of the control input are obtained simultaneously and optimally as a result of the defined $L_2$-gain with discount factor in (14.7).

### 14.2.1 HJI EQUATION FOR $H_\infty$ OPTIMAL TRACKING

In this section, it is first shown that the problem of solving the $H_\infty$ tracking problem can be transformed into a min–max optimization problem subject to an augmented system composed of the tracking error dynamics and the command generator dynamics. A tracking HJI equation is then developed which gives the solution to the min–max optimization problem. The stability and $L_2$-gain boundedness of the tracking HJI control solution are discussed.

Define the augmented system state

$$\mathbf{X}(t) = [\mathbf{e_d}(t)^T\ \mathbf{r}(t)^T]^T \in \mathbb{R}^{2n},$$

where $\mathbf{e_d}(t)$ is the tracking error defined in (14.3) and $\mathbf{r}(t)$ is the reference trajectory.

Using (14.2) and (14.4), define the augmented system

$$\dot{\mathbf{X}}(t) = \mathbf{F}(\mathbf{X}(t)) + \mathbf{G}(\mathbf{X}(t))\,\mathbf{u}(t) + \mathbf{K}(\mathbf{X}(t))\,\mathbf{w}(t), \tag{14.8}$$

where $\mathbf{u}(t) = \mathbf{u}(\mathbf{X}(t))$ and

$$\mathbf{F}(\mathbf{X}) = \begin{bmatrix} \mathbf{f}(\mathbf{e_d} + \mathbf{r}) - \mathbf{h_d}(\mathbf{r}) \\ \mathbf{h_d}(\mathbf{r}) \end{bmatrix}, \mathbf{G}(\mathbf{X}) = \begin{bmatrix} \mathbf{g}(\mathbf{e_d} + \mathbf{r}) \\ \mathbf{0} \end{bmatrix}, \mathbf{K}(\mathbf{X}) = \begin{bmatrix} \mathbf{k}(\mathbf{e_d} + \mathbf{r}) \\ \mathbf{0} \end{bmatrix}.$$

The disturbance attenuation condition (14.7) using the augmented state becomes

$$\int_t^\infty e^{-\alpha(\tau - t)} (\mathbf{X}^T \mathbf{Q_T} \mathbf{X} + \mathbf{u}^T \mathbf{R} \mathbf{u}) d\tau \leqslant \gamma^2 \int_t^\infty e^{-\alpha(\tau - t)} (\mathbf{w}^T \mathbf{w})\, d\tau, \tag{14.9}$$

where

$$\mathbf{Q_T} = \begin{bmatrix} \mathbf{Q} & \mathbf{0} \\ \mathbf{0} & \mathbf{0} \end{bmatrix}.$$

Based on (14.9), define the performance function

$$J(\mathbf{u}, \mathbf{w}) = \int_t^\infty e^{-\alpha(\tau - t)} (\mathbf{X}^T \mathbf{Q_T} \mathbf{X} + \mathbf{u}^T \mathbf{R}\, \mathbf{u} - \gamma^2 \mathbf{w}^T \mathbf{w})\, d\tau. \tag{14.10}$$

Solvability of the $H_\infty$ control problem is equivalent to solvability of the following zero-sum game [42]:

$$V^\star(\mathbf{X}(t)) = J(\mathbf{u}^\star, \mathbf{w}^\star) = \min_{\mathbf{u}} \max_{\mathbf{d}} \; J(\mathbf{u}, \mathbf{w}), \tag{14.11}$$

where $J$ is defined in (14.10) and $V^\star(\mathbf{X}(t))$ is defined as the optimal value function. This two-player zero-sum game control problem has a unique solution if a game theoretic saddle point exists, *i.e.*, if the following Nash condition holds:

$$V^\star(\mathbf{X}(t)) = \min_{\mathbf{u}} \max_{\mathbf{d}} \; J(\mathbf{u}, \mathbf{w}) = \max_{\mathbf{d}} \min_{\mathbf{u}} \; J(\mathbf{u}, \mathbf{w}).$$

Differentiating (14.10), note that $V(\mathbf{X}(t)) = J(\mathbf{u}(t), \mathbf{w}(t))$ gives the following Bellman equation:

$$H(V, \mathbf{u}, \mathbf{w}) \triangleq \mathbf{X}^T \mathbf{Q_T} \mathbf{X} + \mathbf{u}^T \mathbf{R}\, \mathbf{u} - \gamma^2 \mathbf{w}^T \mathbf{w} - \alpha V + {V_\mathbf{X}}^T (\mathbf{F} + \mathbf{G}\, \mathbf{u} + \mathbf{K}\, \mathbf{w}) = 0, \tag{14.12}$$

where $\mathbf{F} \triangleq \mathbf{F}(\mathbf{X})$, $\mathbf{G} \triangleq \mathbf{G}(\mathbf{X})$, $\mathbf{K} \triangleq \mathbf{K}(\mathbf{X})$ and $V_\mathbf{X} = \partial V / \partial \mathbf{X}$.

Applying stationarity conditions $\partial H(V^\star, \mathbf{u}, \mathbf{w})/\partial \mathbf{u} = 0$, $\partial H(V^\star, \mathbf{u}, \mathbf{w})/\partial \mathbf{w} = 0$ [43] gives the optimal control and disturbance inputs as

$$\mathbf{u}^\star = -\frac{1}{2} \mathbf{R}^{-1} \mathbf{G}^T\, {V_\mathbf{X}}^\star, \tag{14.13}$$

$$\mathbf{w}^\star = \frac{1}{2\gamma^2} \mathbf{K}^T\, {V_\mathbf{X}}^\star, \tag{14.14}$$

where $V^\star$ is the optimal value function defined in (14.11). Substituting the control input (14.13) and the disturbance (14.14) into (14.12), the following tracking HJI equation is obtained:

$$\begin{aligned} H(V^\star, \mathbf{u}^\star, \mathbf{w}^\star) &\triangleq \mathbf{X}^T \mathbf{Q}_\mathbf{T} \mathbf{X} + {V_\mathbf{X}^\star}^T \mathbf{F} - \alpha V_\mathbf{X} \\ &- \frac{1}{4} {V_\mathbf{X}^\star}^T \mathbf{G}^T \mathbf{R}^{-1} \mathbf{G} V_\mathbf{X}^\star + \frac{1}{4\gamma^2} {V_\mathbf{X}^\star}^T \mathbf{K}\mathbf{K}^\mathbf{T} V_\mathbf{X}^\star = 0. \end{aligned} \tag{14.15}$$

It is shown in [44] that the control solution (14.13)–(14.15) satisfies the disturbance attenuation condition (14.9) (part (i) of Definition 2) and that it guarantees the stability of the tracking error dynamics (14.4) without the disturbance (part (ii) of Definition 2), if the discount factor is less than an upper bound.

### 14.2.2 OFF-POLICY IRL FOR LEARNING THE TRACKING HJI EQUATION

In this section, an off-policy RL algorithm is first given to learn this control solution online and without requiring any knowledge of the system dynamics.

The Bellman equation (14.12) is linear in the cost function $V$, while the HJI equation (14.15) is nonlinear in the value function $V^\star$. Therefore, solving the Bellman equation for $V$ is easier than solving the HJI for $V^\star$. Instead of directly solving for $V^\star$, a policy iteration (PI) algorithm iterates on both control and disturbance players to break the HJI equation into a sequence of differential equations linear in the cost. An offline PI algorithm for solving the $H_\infty$ optimal tracking problem is given as follows.

---

**Algorithm 1** Offline RL algorithm.

1: **procedure**

2: Start with an admissible stabilizing control policy $\mathbf{u}^0$.

3: For a control input $\mathbf{u}^j$ and disturbance policy $\mathbf{w}^j$, find $V^j$ using the following Bellman equation:

$$H(V^j, \mathbf{u}^j, \mathbf{w}^j) = \mathbf{X}^T \mathbf{Q}_\mathbf{T} \mathbf{X} + (V_\mathbf{X}{}^j)^T (\mathbf{F} + \mathbf{G}\mathbf{u}^j + \mathbf{K}\mathbf{w}^j) - \alpha V^j + (\mathbf{u}^j)^T \mathbf{R}\mathbf{u}^j - \gamma^2 (\mathbf{w}^j)^T \mathbf{w}^j = 0. \tag{14.16}$$

4: Update the disturbance using

$$\mathbf{w}^{j+1} = \arg \max_{\mathbf{d}} \left[ H(V^j, \mathbf{u}^j, \mathbf{w}) \right] = \frac{1}{2\gamma^2} \mathbf{K}^T V_\mathbf{X}{}^j \tag{14.17}$$

and the control policy using

$$\mathbf{u}^{j+1} = \arg \min_{\mathbf{u}} \left[ H(V^j, \mathbf{u}, \mathbf{w}^{j+1}) \right] = -\frac{1}{2} \mathbf{R}^{-1} \mathbf{G}^T V_\mathbf{X}{}^j. \tag{14.18}$$

5: Go to 3.

6: **end procedure**

---

Algorithm 1 extends the results of the simultaneous RL algorithm in [27] to the tracking problem. The convergence of this algorithm to the minimal nonnegative solution of the HJI equation was shown in [27]. In fact, similar to [27], the convergence of Algorithm 1 can be established by proving that iteration on (14.16) is essentially a Newton iterative sequence which converges to the unique solution of the HJI equation (14.15).

Algorithm 1 requires complete knowledge of the system dynamics. In the following, the off-policy IRL algorithm, which was presented in [14,15] for solving the $H_2$ optimal regulation problem, is extended here to solve the $H_\infty$ optimal tracking for systems with completely unknown dynamics. To this end, the system dynamics (14.8) is first written as

$$\dot{\mathbf{X}} = \mathbf{F} + \mathbf{G}\,\mathbf{u}^j + \mathbf{K}\,\mathbf{w}^j + \mathbf{G}\,(\mathbf{u} - \mathbf{u}^j) + \mathbf{K}\,(\mathbf{w} - \mathbf{w}^j), \tag{14.19}$$

where $\mathbf{u}^j \in \mathbb{R}^m$ and $\mathbf{w}^j \in \mathbb{R}^q$ are policies to be updated. In this equation, the control input $\mathbf{u}$ is the behavior policy which is applied to the system to generate data for learning, while $\mathbf{u}^j$ is the target policy which is evaluated and updated using data generated by the behavior policy. The fixed control policy $\mathbf{u}$ should be a stable and exploring control policy. Moreover, the disturbance input $\mathbf{w}$ is the actual external disturbance that comes from an external source and is not under our control. However, the disturbance $\mathbf{w}^j$ is the disturbance that is evaluated and updated. One advantage of this off-policy IRL Bellman equation is that, in contrast to on-policy RL-based methods, the disturbance input that is applied to the system does not require to be adjustable.

Differentiating $V^j(\mathbf{X})$ along with the system dynamics (14.19) and using (14.16)–(14.18) gives

$$\begin{aligned}
&\dot{V}^j = (V_{\mathbf{X}}{}^j)^T(\mathbf{F} + \mathbf{G}\,\mathbf{u}^j + \mathbf{K}\,\mathbf{w}^j) + (V_{\mathbf{X}}{}^j)^T\mathbf{G}(\mathbf{u} - \mathbf{u}^j) + (V_{\mathbf{X}}{}^j)^T\mathbf{K}\,(\mathbf{w} - \mathbf{w}^j)\\
&= \alpha\, V^j - \mathbf{X}^T\mathbf{Q}_\mathbf{T}\,\mathbf{X} - (\mathbf{u}^j)^T\mathbf{R}\,\mathbf{u}^j + \gamma^2(\mathbf{w}^j)^T\mathbf{w}^j -\\
&2\,(\mathbf{u}^{j+1})^T\mathbf{R}\,(\mathbf{u} - \mathbf{u}^j) + 2\gamma^2(\mathbf{w}^{j+1})^T(\mathbf{w} - \mathbf{w}^j).
\end{aligned} \tag{14.20}$$

Multiplying both sides of (14.20) by $e^{-\alpha(\tau - t)}$ and integrating from both sides yields the following off-policy IRL Bellman equation:

$$\begin{aligned}
&e^{-\alpha T}\,V^j(\mathbf{X}(t+T)) - V^j(\mathbf{X}(t)) =\\
&\int_t^{t+T} e^{-\alpha(\tau - t)}(-\mathbf{X}^T\mathbf{Q}_\mathbf{T}\,\mathbf{X} - (\mathbf{u}^j)^T\mathbf{R}\,\mathbf{u}^j + \gamma^2(\mathbf{w}^j)^T\mathbf{w}^j)\,d\tau\\
&+ \int_t^{t+T} e^{-\alpha(\tau - t)}(-2\,(\mathbf{u}^{j+1})^T\mathbf{R}\,(\mathbf{u} - \mathbf{u}^j) + 2\gamma^2(\mathbf{w}^{j+1})^T(\mathbf{w} - \mathbf{w}^j))\,d\tau.
\end{aligned} \tag{14.21}$$

Note that, for a fixed control policy $\mathbf{u}$ (the policy that is applied to the system) and a given disturbance $\mathbf{w}$ (the actual disturbance that is applied to the system), Eq. (14.21) can be solved for both the value function $V^j$ and the updated policies $\mathbf{u}^{j+1}$ and $\mathbf{w}^{j+1}$ simultaneously.

**Lemma 1.** *The off-policy IRL equation* (14.21) *gives the same solution for the value function as the Bellman equation* (14.16) *and the same updated control and disturbance policies as* (14.18) *and* (14.17).

*Proof.* See [44]. □

The following algorithm uses the off-policy tracking Bellman equation (14.21) to iteratively solve the HJI equation (14.15) without requiring any knowledge of the system dynamics. The implementation of this algorithm is discussed in the next subsection. It is shown how the data collected from a fixed control policy $u$ are reused to evaluate many updated control policies $\mathbf{u}_i$ sequentially until convergence to the optimal solution is achieved.

---

**Algorithm 2** Online off-policy RL algorithm for solving the tracking HJI equation.

---

1: **procedure**

2: **Phase 1 (data collection using a fixed control policy):** Apply a fixed control policy $\mathbf{u}$ to the system and collect required system information about the state, control input and disturbance at $N$ different sampling intervals $T$.

3: For a control input $\mathbf{u}^j$ and disturbance policy $\mathbf{w}^j$, find $V^j$ using the following Bellman equation:

$$H(V^j, \mathbf{u}^j, \mathbf{w}^j) = \mathbf{X}^T\,\mathbf{Q_T}\,\mathbf{X} + (V_{\mathbf{X}}{}^j)^T\,(\mathbf{F} + \mathbf{G}\mathbf{u}^j + \mathbf{K}\,\mathbf{w}^j) - \alpha V^j + (\mathbf{u}^j)^T\mathbf{R}\mathbf{u}^j - \gamma^2(\mathbf{w}^j)^T\mathbf{w}^j = 0. \tag{14.22}$$

4: **Phase 2 (reuse of collected data sequentially to find an optimal policy iteratively):** Given $\mathbf{u}^j$ and $\mathbf{w}^j$, use collected information in phase 1 to Solve the following Bellman equation for $V^j$, $\mathbf{u}^{j+1}$ and $\mathbf{w}^{j+1}$ simultaneously:

$$\begin{aligned} &e^{-\alpha T}V^j(\mathbf{X}(t+T)) - V^j(\mathbf{X}(t)) = \\ &\int_t^{t+T} e^{-\alpha(\tau - t)}(-\mathbf{X}^T\mathbf{Q_T}\,\mathbf{X} - (\mathbf{u}^j)^T\mathbf{R}\mathbf{u}^j + \gamma^2(\mathbf{w}^j)^T\mathbf{w}^j)\,d\tau \\ &+ \int_t^{t+T} e^{-\alpha(\tau - t)}(-2(\mathbf{u}^{j+1})^T\mathbf{R}(\mathbf{u} - \mathbf{u}^j) + 2\gamma^2(\mathbf{w}^{j+1})^T(\mathbf{w} - \mathbf{w}^j))\,d\tau. \end{aligned} \tag{14.23}$$

5: Stop if a stopping criterion is met, otherwise set $j = j + 1$ and go to 3.

6: **end procedure**

---

Inspired by the off-policy algorithm in [14], Algorithm 2 has two separate phases. First, a fixed initial exploratory control policy $\mathbf{u}$ is applied and the system information is recorded over the time interval T. Second, without requiring any knowledge of the system dynamics, the information collected in phase 1 is repeatedly used to find a sequence of updated policies $\mathbf{u}^j$ and $\mathbf{w}^j$ converging to $\mathbf{u}^\star$ and $\mathbf{w}^\star$. Note that Eq. (14.23) is a scalar equation and can be solved in a least square sense after collecting enough data samples from the system. It is shown in the following section how to collect required information in phase 1 and reuse it in phase 2 in a least square sense to solve (14.23) for $V^j$, $\mathbf{u}^{j+1}$ and $\mathbf{w}^{j+1}$ simultaneously. After the learning is done and the optimal control policy $\mathbf{u}^\star$ is found, it can be applied to the system.

**Theorem 1** (Convergence of Algorithm 2)**.** *The off-policy Algorithm 2 converges to the optimal control and disturbance solutions given by* (14.13) *and* (14.14) *where the value function satisfies the tracking HJI equation* (14.15).

*Proof.* See [44]. □

### 14.2.3 IMPLEMENTING ALGORITHM 2 USING NEURAL NETWORKS

In order to implement the off-policy RL Algorithm 2, it is required to reuse the collected information found by applying a fixed control policy $\mathbf{u}$ to the system to solve Eq. (14.23) for $V^j$, $\mathbf{u}^{j+1}$ and $\mathbf{w}^{j+1}$

iteratively. Three neural networks (NNs), *i.e.*, the actor NN, the critic NN and the disturber NN, are used here to approximate the value function and the updated control and disturbance policies in the Bellman equation (14.23). That is, the solution $V^j$, $\mathbf{u}^{j+1}$ and $\mathbf{w}^{j+1}$ of the Bellman equation (14.23) is approximated by three NNs as

$$\hat{V}^j(\mathbf{X}) = \hat{\mathbf{W}}_1^T \sigma(\mathbf{X}), \tag{14.24}$$

$$\hat{\mathbf{u}}^{j+1}(\mathbf{X}) = \hat{\mathbf{W}}_2^T \phi(\mathbf{X}), \tag{14.25}$$

$$\hat{\mathbf{w}}^{j+1}(\mathbf{X}) = \hat{\mathbf{W}}_3^T \varphi(\mathbf{X}), \tag{14.26}$$

where $\sigma = [\sigma_1, ..., \sigma_{l_1}] \in \mathbb{R}^{l_1}$, $\phi = [\phi_1, ..., \phi_{l_2}] \in \mathbb{R}^{l_2}$ and $\varphi = [\varphi_1, ..., \varphi_{l_3}] \in \mathbb{R}^{l_3}$ provide suitable basis function vectors, $\hat{\mathbf{W}}_1 \in \mathbb{R}^{l_1}$, $\hat{\mathbf{W}}_2 \in \mathbb{R}^{m \times l_2}$ and $\hat{\mathbf{W}}_3 \in \mathbb{R}^{q \times l_3}$ are constant weight vectors and $l_1$, $l_2$ and $l_3$ are the number of neurons. Define $\mathbf{v}^1 = [v_1^1, ..., v_1^m]^T = \mathbf{u} - \mathbf{u}^j$, $\mathbf{v}^2 = [v_1^2, ..., v_q^2]^T = \mathbf{w} - \mathbf{w}^j$ and assume $\mathbf{R} = diag(r, ..., r_m)$. Then, substituting (14.24)–(14.26) in (14.23) yields

$$\begin{aligned} e(t) &= \hat{\mathbf{W}}_1^T (e^{-\alpha T} \sigma(\mathbf{X}(t+T)) - \sigma(\mathbf{X}(t))) \\ &- \int_t^{t+T} e^{-\alpha(\tau - t)} (-\mathbf{X}^T \mathbf{Q_T} \mathbf{X} - (\mathbf{u}^j)^T \mathbf{R} \mathbf{u}^j + \gamma^2 (\mathbf{w}^j)^T \mathbf{w}^j) d\tau \\ &+ 2 \sum_{l=1}^{m} r_l \int_t^{t+T} e^{-\alpha(\tau - t)} \hat{\mathbf{W}}_{2,l}^T \phi(\mathbf{X}(t)) v_l^1 \, d\tau \\ &- 2\gamma^2 \sum_{k=1}^{q} \int_t^{t+T} e^{-\alpha(\tau - t)} \hat{\mathbf{W}}_{3,k}^T \varphi(\mathbf{X}(t)) v_k^2 \, d\tau, \end{aligned} \tag{14.27}$$

where $e(t)$ is the Bellman approximation error, $\hat{\mathbf{W}}_{2,\mathbf{l}}$ is the $l$th column of $\hat{\mathbf{W}}_2$ and $\hat{\mathbf{W}}_{3,\mathbf{k}}$ is the $k$th column of $\hat{\mathbf{W}}_3$. The Bellman approximation error is the continuous-time counterpart of the temporal difference (TD) [10]. In order to bring the TD error to its minimum value, the least squares method is used. To this end, rewrite Eq. (14.27) as

$$y(t) + e(t) = \hat{\mathbf{W}}^\mathbf{T} \mathbf{h(t)}, \tag{14.28}$$

where

$$\hat{\mathbf{W}} = [\hat{\mathbf{W}}_1^T, \hat{\mathbf{W}}_{2,1}^T, ..., \hat{\mathbf{W}}_{2,\mathbf{m}}^T, \hat{\mathbf{W}}_{3,1}^T, ..., \hat{\mathbf{W}}_{3,\mathbf{q}}^T]^T \in \mathbb{R}^{l_1 + m \times l_2 + q \times l_3},$$

$$\mathbf{h(t)} = \begin{bmatrix} e^{-\alpha T} \sigma(\mathbf{X}(t+T)) - \sigma(\mathbf{X}(t))) \\ 2r_1 \int_t^{t+T} e^{-\alpha(\tau - t)} \, \phi(\mathbf{X}(t)) \, v_1^1 \, d\tau \\ \vdots \\ 2r_m \int_t^{t+T} e^{-\alpha(\tau - t)} \, \phi(\mathbf{X}(t)) \, v_m^1 \, d\tau \\ -2\gamma^2 \int_t^{t+T} e^{-\alpha(\tau - t)} \, \varphi(\mathbf{X}(t)) \, v_1^2 \, d\tau \\ \vdots \\ -2\gamma^2 \int_t^{t+T} e^{-\alpha(\tau - t)} \, \varphi(\mathbf{X}(t)) \, v_q^2 \, d\tau \end{bmatrix}, \tag{14.29}$$

$$y(t) = \int_t^{t+T} e^{-\alpha(\tau-t)}(-\mathbf{X}^T \mathbf{Q_T X} - (\mathbf{u}^j)^T \mathbf{R u}^j + \gamma^2 (\mathbf{w}^j)^T \mathbf{w}^j) d\tau. \quad (14.30)$$

The parameter vector $\hat{\mathbf{W}}$, which gives the approximated value function, actor and disturbance (14.24)–(14.26), is found by minimizing, in the least squares sense, the Bellman error. Assume that the systems state, input and disturbance information are collected at $N \geqslant l_1 + m \times l_2 + q \times l_3$ (the number of independent elements in $\hat{\mathbf{W}}$) points $t_1$ to $t_N$ in the state space, over the same time interval T in phase 1. Then, for a given $\mathbf{u}^j$ and $\mathbf{w}^j$, one can use this information to evaluate (14.29) and (14.30) at $N$ points to form

$$\mathbf{H} = [\mathbf{h}(t_1), \ldots., \mathbf{h}(t_N)],$$
$$\mathbf{Y} = [y(t_1), \ldots., y(t_N)]^T.$$

The least squares solution to (14.28) is then equal to

$$\hat{\mathbf{W}} = (\mathbf{H}\mathbf{H}^T)^{-1}\mathbf{H}\mathbf{Y},$$

which gives $V^j$, $\mathbf{u}^{j+1}$ and $\mathbf{w}^{j+1}$. Note that although $\mathbf{X}(t+T)$ appears in Eq. (14.27), this equation is solved in a least square sense after observing N samples $\mathbf{X}(t), \mathbf{X}(t+T), \ldots, \mathbf{X}(t+NT)$. Therefore, the knowledge of the system is not required to predict the future state $\mathbf{X}(t+T)$ at time t to solve (14.27).

## 14.3 $H_\infty$ OPTIMAL TRACKING CONTROL FOR A CLASS OF NONLINEAR NONAFFINE SYSTEMS

This section considers the design of an RL-based optimal tracking control solution for a class of non-affine systems.

### 14.3.1 A CLASS OF NONAFFINE DYNAMICAL SYSTEMS

A special class of nonaffine systems can be described as

$$\dot{\mathbf{X}}(t) = \mathbf{f}(\mathbf{X}(t)) + \mathbf{g}(\mathbf{X}(t))\mathbf{L}(\mathbf{u}) + \mathbf{D}\mathbf{w}(t), \quad (14.31)$$

where $\mathbf{X}(t) \in \mathbb{R}^n$, $\mathbf{u}(t) \in \mathbb{R}^m$ and $\mathbf{w}(t) \in \mathbb{R}^p$ are the state of the system, the control input and the external disturbance input, respectively. The functions $\mathbf{f}(\mathbf{X}(t))$ and $\mathbf{g}(\mathbf{X}(t))$ are Lipschitz functions. This system is affine in a nonlinear function $\mathbf{L}(.)$ of the control input $\mathbf{u}(t)$. This class of nonaffine systems allows the definition of a new performance function for the optimal $H_\infty$ problem such that the existence of the constrained optimal control is assured (if any exists).

The following example shows that the UAV as a real-world application can be presented in the form of (14.31).

**Example 1.** A general class of nonlinear nonaffine UAV systems has the following well-known form:

$$\dot{x}_1 = V \cos\gamma \, \cos\psi + d_1 w_1,$$

$$
\begin{aligned}
\dot{x}_2 &= V \cos\gamma \, \sin\psi + d_2 w_2, \\
\dot{x}_3 &= -V \sin\gamma \, + d_3 w_3, \\
\dot{V} &= -\alpha_2 V^2 - g \sin\gamma + \alpha_1 \bar{T} - \alpha_3 n_z - \alpha_4 \frac{n_z^2}{V^2}, \\
\dot{\gamma} &= \frac{g}{V}(n_z \cos\phi - \cos\gamma), \\
\dot{\psi} &= \frac{g}{V\cos\gamma} n_z \sin\phi,
\end{aligned}
\tag{14.32}
$$

with

$$
\begin{aligned}
n_x &= \frac{\bar{T}\bar{T}_{\max}\cos\alpha - D}{mg}, \\
n_x &= \frac{\bar{T}\bar{T}_{\max}\sin\alpha + K}{mg},
\end{aligned}
$$

where $x_1$, $x_2$, $x_3$ are the UAV location coordinates, $\gamma$ is the pitch angle, $\psi$ is the heading angle, $\phi$ is bank angle, $V$ is the UAV velocity and $m$ is the mass of the UAV. The terms $n_x$ and $n_z$ denote longitudinal and normal components of the load factor, depending on the current thrust $\bar{T}$, drag force $D$ and lift force $K$ ($g$ is the acceleration due to gravity) [45].

Define the state of the UAV as

$$
\mathbf{X} = \{x_1, x_2, x_3, V, \gamma, \psi\}^{\mathrm{T}} \tag{14.33}
$$

and the control input and disturbance inputs (wind velocity) as $\mathbf{u}(t) = [\bar{T}, n_z, \phi]^{\mathrm{T}} = [\begin{array}{ccc} u_1 & u_2 & u_3 \end{array}]^{\mathrm{T}}$ and $\mathbf{w}(t)$, respectively. The constraints on the control input are as follows:

$$
\begin{aligned}
|u_1| &\leqslant \bar{u}_1, \\
|u_2| &\leqslant \bar{u}_2.
\end{aligned}
\tag{14.34}
$$

Using (14.32) and (14.33), the UAV dynamics can be written as a nonlinear nonaffine CT system as

$$
\dot{\mathbf{X}}(t) = M(\mathbf{X}(t), \mathbf{u}(t)) + \mathbf{D}\mathbf{w}(t), \tag{14.35}
$$

with

$$
\mathbf{D} = [\begin{array}{cccccc} d_1 & d_2 & d_3 & 0 & 0 & 0 \end{array}]^{\mathrm{T}},
$$

$$
\mathbf{M}(\mathbf{X}, \mathbf{u}) = \begin{bmatrix}
x_4 \cos(x_5)\cos(x_6) \\
x_4 \cos(x_5)\sin(x_6) \\
-x_4 \sin(x_5) \\
-\alpha_2 x_4^2 - g\sin(x_5) + \alpha_1 u_1 - \alpha_3 u_2 - \alpha_4 \frac{u_2^2}{x_4^2} \\
\frac{g}{x_4}(-\cos(x_5) + u_2 \cos(u_3)) \\
\frac{g}{x_4 \cos(x_5)} u_2 \sin(u_3)
\end{bmatrix}.
$$

The UAV dynamics (14.35) can be written in the form of (14.31) with

$$\mathbf{f}(\mathbf{X}(t)) = \begin{bmatrix} x_4\cos(x_5)\cos(x_6) \\ x_4\cos(x_5)\sin(x_6) \\ -x_4\sin(x_5) \\ -\alpha_2 x_4^2 - g\sin(x_5) \\ \frac{g}{x_4}(-\cos(x_5)) \\ 0 \end{bmatrix}, \quad \mathbf{g}(\mathbf{X}(t)) = \begin{bmatrix} 0 & 0 & 0 & 0 & 0 \\ 0 & 0 & 0 & 0 & 0 \\ 0 & 0 & 0 & 0 & 0 \\ \alpha_1 & -\alpha_3 & -\frac{\alpha_4}{x_4^2} & 0 & 0 \\ 0 & 0 & 0 & 1 & 0 \\ 0 & 0 & 0 & 0 & \frac{g}{x_4\cos(x_5)} \end{bmatrix},$$

$$\mathbf{L}(\mathbf{u}(t)) = \begin{bmatrix} L_1 \\ L_2 \\ L_3 \\ L_4 \\ L_5 \end{bmatrix} = \begin{bmatrix} u_1 \\ u_2 \\ u_2^2 \\ u_2\cos(u_3) \\ u_2\sin(u_3) \end{bmatrix}.$$

Eq. (14.31) represents a large class of nonaffine systems far larger than the systems that are affine in the control itself. In fact, most aircraft dynamics can be expressed in the form of (14.31) if the lift equation satisfies certain assumptions [45].

### 14.3.2 PERFORMANCE FUNCTION AND $H_\infty$ CONTROL TRACKING FOR NONAFFINE SYSTEMS

It is shown in [46] that the existence of an admissible optimal control solution for nonaffine systems depends on how the utility function $r(\mathbf{X}, \mathbf{u})$ is defined. Moreover, to deal with the input constraints, a nonquadratic performance index needs to be defined as follows.

Let the reference trajectory be generated by the command generator dynamics (14.2). The performance or control output $\mathbf{z}(t)$ is defined such that it satisfies

$$\|\mathbf{z}(t)\|^2 = (\mathbf{X} - \mathbf{r})^{\mathrm{T}}\mathbf{Q}(\mathbf{X} - \mathbf{r}) + W(\mathbf{L}(\mathbf{u})), \tag{14.36}$$

where $\mathbf{Q} \succeq 0$ and $W(\mathbf{L}(\mathbf{u}))$ is a positive definite nonquadratic function of $L(\mathbf{u})$ which penalizes the control effort and is chosen as follows to assure the constrained control effort:

$$W(\mathbf{L}(\mathbf{u})) = \int_0^{\mathbf{L}(\mathbf{u})} \mathbf{w}(s)\,ds = \sum_{j=1}^{l} \Big(\int_0^{\mathbf{L}_j(\mathbf{u})} w_j(s_j)\,ds_j\Big),$$

where $\mathbf{w}(s) = \tanh^{-1}(\bar{\mathbf{L}}^{-1}s) = \begin{bmatrix} w_1(s_1) & \cdots & w_l(s_l) \end{bmatrix}^{\mathrm{T}}$ and $\bar{\mathbf{L}}$ is the constant diagonal matrix given by $\bar{\mathbf{L}} = diag(\bar{L}_1, \ldots, \bar{L}_l)$, which determines the bounds on $\mathbf{L}(\mathbf{u})$. Note that the bounds are originally given for the control input $\mathbf{u}(t)$ itself. However, one can transform these bounds to bounds on $L(\mathbf{u})$.

The $H_\infty$ control is to develop a control input such that (1) the system (1) with $\mathbf{w} = \mathbf{0}$ is asymptotically stable and (2) the $L_2$ gain condition (14.6) with $\mathbf{z}(t)$ defined in (14.36) is satisfied in the presence of $\mathbf{w} \in L_2[0, \infty)$.

The disturbance attenuation condition is satisfied if the following cost function is nonpositive:

$$J(\mathbf{X}) = \int_t^{\infty} e^{-\alpha(\tau - t)} \left[ (\mathbf{X} - \mathbf{r})^{\mathrm{T}} \mathbf{Q} (\mathbf{X} - \mathbf{r}) + W(\mathbf{L}(\mathbf{u})) - \gamma^2 w^{\mathrm{T}} w \right] d\tau. \tag{14.37}$$

### 14.3.3 SOLUTION OF THE $H_\infty$ CONTROL TRACKING PROBLEM OF NONAFFINE SYSTEMS

Define the tracking error as (14.3). Then, using (14.2) and (14.31), the tracking error dynamics becomes

$$\dot{\mathbf{e}}_{\mathbf{d}}(t) = \dot{\mathbf{X}}(t) - \dot{\mathbf{r}}(t) = \mathbf{f}(\mathbf{X}(t)) + \mathbf{g}(\mathbf{X}(t))\mathbf{L}(\mathbf{u}) + \mathbf{D}\,\mathbf{w}(t) - \mathbf{h}_{\mathbf{d}}(\mathbf{r})(t). \tag{14.38}$$

Based on (14.2) and (14.38), an augmented system can be constructed in terms of the tracking error $\mathbf{e}(t)$ and the reference trajectory $\mathbf{r}(t)$ as

$$\begin{aligned} \dot{\mathbf{Z}}(t) &= \begin{bmatrix} \dot{\mathbf{e}}(\mathbf{t}) \\ \dot{\mathbf{r}}(\mathbf{t}) \end{bmatrix} = \begin{bmatrix} \mathbf{f}(\mathbf{e}(t) + \mathbf{r}(t)) - \mathbf{h}_{\mathbf{d}}(\mathbf{r}(t)) \\ \mathbf{h}_{\mathbf{d}}(\mathbf{r}(t)) \end{bmatrix} + \begin{bmatrix} \mathbf{g}(\mathbf{e}(t) + \mathbf{r}(t)) \\ \mathbf{0} \end{bmatrix} \mathbf{L}(\mathbf{u}) + \begin{bmatrix} \mathbf{D} \\ \mathbf{0} \end{bmatrix} \mathbf{w}(t) \\ &\equiv \mathbf{F}(\mathbf{Z}(t)) + \mathbf{G}(\mathbf{Z}(t))\mathbf{L}(\mathbf{u}) + \mathbf{K}\mathbf{w}(t), \end{aligned} \tag{14.39}$$

where the augmented state is

$$\mathbf{Z}(t) = \begin{bmatrix} \mathbf{e}(t) \\ \mathbf{r}(t) \end{bmatrix}.$$

The performance index (14.37) can be rewritten as

$$J(\mathbf{L}(\mathbf{u}), \mathbf{w}) = \int_t^{\infty} e^{-\alpha(\tau - t)} \left( \mathbf{Z}^T(\tau) \mathbf{Q}_1 \mathbf{Z}(\tau) + W(\mathbf{L}(\mathbf{u})) - \gamma^2 \mathbf{w}^{\mathrm{T}} \mathbf{w} \right) d\tau, \tag{14.40}$$

with $\mathbf{Q}_1 = \begin{bmatrix} \mathbf{Q} & \mathbf{0} \\ \mathbf{0} & \mathbf{0} \end{bmatrix}$.

The $H_\infty$ control problem can be expressed as a two-player zero-sum differential game in which the control effort policy player $\mathbf{L}(\mathbf{u})$ seeks to minimize the value function, while the disturbance policy player $\mathbf{w}(t)$ desires to maximize it. The goal is to find the feedback saddle point $(\mathbf{L}^\star(\mathbf{u}), \mathbf{w}^\star)$ such that [42]

$$V^\star(\mathbf{Z}(t)) = \min_{\mathbf{L}(\mathbf{u})} \max_{\mathbf{w}} J(\mathbf{L}(\mathbf{u}), \mathbf{w}). \tag{14.41}$$

On the basis of (14.40) and noting that $V(\mathbf{Z}(t)) = J(\mathbf{L}(\mathbf{u}), \mathbf{w})$, the $H_\infty$ tracking Bellman equation is

$$\mathbf{Z}^{\mathrm{T}} \mathbf{Q}_1 \mathbf{Z} + W(\mathbf{L}(\mathbf{u})) - \gamma^2 \mathbf{w}^{\mathrm{T}} \mathbf{w} - \alpha V(\mathbf{Z}) + \dot{V}(\mathbf{Z}) = 0 \tag{14.42}$$

and the Hamiltonian is given by

$$H(\mathbf{Z}, L(\mathbf{u}), \mathbf{w}, V_{\mathbf{Z}}) = \mathbf{Z}^{\mathrm{T}} \mathbf{Q}_1 \mathbf{Z} + W(\mathbf{L}(\mathbf{u})) - \gamma^2 \mathbf{w}^{\mathrm{T}} \mathbf{w} - \alpha V(\mathbf{Z}) + V_{\mathbf{Z}}^{\mathrm{T}} (\mathbf{F}(\mathbf{Z}) + \mathbf{G}(\mathbf{Z})\mathbf{L}(\mathbf{u}) + \mathbf{K}\mathbf{w}).$$

Then the optimal control effort $L(\mathbf{u})$ and disturbance input $w(t)$ for the given problem are obtained by employing the stationarity condition

$$\mathbf{L}^\star(\mathbf{u}) = \arg\min_{\mathbf{L}(\mathbf{u})} H(\mathbf{Z}, \mathbf{L}(\mathbf{u}), \mathbf{w}, V^*) \triangleq \frac{d\left[\mathbf{Z}^{\mathrm{T}}\mathbf{Q}_1\mathbf{Z} + W(\mathbf{L}(\mathbf{u})) - \gamma^2\mathbf{w}^{\mathrm{T}}\mathbf{w} - \alpha V^\star + (V_{\mathbf{Z}}^\star)^{\mathrm{T}}\dot{\mathbf{Z}}\right]}{d\,\mathbf{L}(\mathbf{u})},$$

$$\mathbf{w}^\star = \arg\max_{\mathbf{w}} H(\mathbf{Z}, \mathbf{L}(\mathbf{u}), \mathbf{w}, V^\star) \triangleq \frac{d\left[\mathbf{Z}^{\mathrm{T}}\mathbf{Q}_1\mathbf{Z} + W(\mathbf{L}(\mathbf{u})) - \gamma^2\mathbf{w}^{\mathrm{T}}\mathbf{w} - \alpha V^\star + (V_{\mathbf{Z}}^\star)^{\mathrm{T}}\dot{\mathbf{Z}}\right]}{d\,\mathbf{w}},$$

which give

$$\mathbf{L}^\star(\mathbf{u}) = -\bar{\mathbf{L}}\tanh^{\mathrm{T}}(\mathbf{v}^\star), \tag{14.43}$$

$$\mathbf{w}^\star = \frac{1}{2}\gamma^{-2}(V_{\mathbf{Z}}^*)^{\mathrm{T}}\mathbf{K}, \tag{14.44}$$

where

$$\mathbf{v}^\star = (V_{\mathbf{Z}}^\star)^{\mathrm{T}}\mathbf{G}. \tag{14.45}$$

Substituting (14.43) and (14.44) in Bellman equation (14.42) yields the HJI equation

$$\mathbf{Z}^{\mathrm{T}}\mathbf{Q}_1\mathbf{Z} + W(\mathbf{L}^\star(\mathbf{u})) - \gamma^2(\mathbf{w}^\star)^{\mathrm{T}}\mathbf{w}^\star - \alpha V^\star(\mathbf{Z}) + \dot{V}^\star(Z) = 0. \tag{14.46}$$

To find the optimal control solution, the tracking HJI equation (14.46) could first be solved and then the control effort $\mathbf{L}^\star(\mathbf{u})$ given by (14.43).

Note that the minimization problem (14.41) is defined in terms of $\mathbf{L}(\mathbf{u})$. Under certain conditions, this is equivalent to minimization in terms of $\mathbf{u}(t)$.

**Lemma 2.** *We have* $\min_{\mathbf{u}} H(\mathbf{Z}, \mathbf{L}(\mathbf{u}), \mathbf{w}, V_{\mathbf{Z}}) = \min_{\mathbf{L}(\mathbf{u})} H(\mathbf{Z}, \mathbf{L}(\mathbf{u}), \mathbf{w}, V_{\mathbf{Z}})$ *if the elements of* $\mathbf{L}(\mathbf{u})$ *are independent.*

*Proof.* The minimum of $H(\mathbf{Z}, \mathbf{L}(\mathbf{u}), \mathbf{w}, V_{\mathbf{Z}})$ with respect to $u$ is equal to

$$\min_{u} H(\mathbf{Z}, \mathbf{L}(\mathbf{u}), \mathbf{w}, V_{\mathbf{Z}}) = \left(\frac{\partial \mathbf{L}(\mathbf{u})}{\partial\,\mathbf{u}}\right)^T \frac{\partial H(\mathbf{Z}, \mathbf{L}(\mathbf{u}), \mathbf{w}, V_{\mathbf{Z}})}{\partial \mathbf{L}(\mathbf{u})} = 0 \tag{14.47}$$

and the minimum of $H(\mathbf{Z}, \mathbf{L}(\mathbf{u}), \mathbf{w}, V_{\mathbf{Z}})$ with respect to $\mathbf{L}(\mathbf{u})$ is equal to

$$\min_{\mathbf{L}(\mathbf{u})} H(\mathbf{Z}, \mathbf{L}(\mathbf{u}), \mathbf{w}, V_{\mathbf{Z}}) = \frac{d\,H(\mathbf{Z}, \mathbf{L}(\mathbf{u}), \mathbf{w}, V_{\mathbf{Z}})}{d\,\mathbf{L}(\mathbf{u})} = 0. \tag{14.48}$$

Eqs. (14.47) and (14.48) are equivalent if and only if $J = d\mathbf{L}(\mathbf{u})/d\,\mathbf{u}$ is a nonsingular matrix which guarantees the elements of $\mathbf{L}(\mathbf{u})$ are independent [46]. □

Note that if the elements of $\mathbf{L}^\star(\mathbf{u})$ are independent, then the optimal control is given by

$$\mathbf{u}^\star = -\mathbf{L}^{-1}(\bar{L}\tanh^T(\mathbf{v}^\star)), \tag{14.49}$$

thus $\mathbf{L}(\mathbf{u}^\star) = \mathbf{L}^\star(\mathbf{u})$. Otherwise, it is shown in the subsequent sections how to use (14.43) to find $\mathbf{v}^\star$ and $\mathbf{u}^\star$ consequently to assure $\mathbf{L}(\mathbf{u}^\star) = \mathbf{L}^\star(\mathbf{u})$. The next result holds for both independent and dependent $\mathbf{L}(\mathbf{u})$.

**Theorem 2** (Solution to bounded $L_2$ gain problem). *Assume that there exists a continuous-time positive semidefinite solution $V^\star(\mathbf{Z})$ to the tracking HJI equation* (14.46). *Let $\mathbf{L}^\star(\mathbf{u})$ be given by* (14.43). *Then $\mathbf{L}^\star(\mathbf{u})$ in* (14.31) *makes the $L_2$ gain from the disturbance to the performance output less than or equal to $\gamma$.*

*Proof.* See [46]. □

If the elements of $\mathbf{L}(\mathbf{u})$ are independent, then there exists a $\mathbf{u}^\star$ such that $\mathbf{L}(\mathbf{u}^\star) = \mathbf{L}^\star(\mathbf{u})$ and this $\mathbf{u}^\star$ makes the $L_2$ gain less than or equal to $\gamma$. On the other hand, if the elements of $\mathbf{L}^\star(\mathbf{u})$ are dependent, a method of solution is suggested in subsequent sections.

### 14.3.4 OFF-POLICY REINFORCEMENT LEARNING FOR NONAFFINE SYSTEMS

In this section, the off-policy RL is presented to solve the optimal $H_\infty$ control of nonaffine nonlinear systems. In the proposed method, no knowledge about the system dynamics and the reference trajectory dynamics is needed. Moreover, it does not require an adjustable disturbance input and it avoids bias in finding the value function. Two algorithms are developed for two different cases: (1) for nonaffine systems with independent elements in $\mathbf{L}(\mathbf{u})$ and (2) for nonaffine systems with dependent elements in $\mathbf{L}(\mathbf{u})$. Then the implementation of these two algorithms is given.

The system dynamics (14.39) can be rewritten as

$$\dot{\mathbf{Z}}(t) = \mathbf{F}(\mathbf{Z}(t)) + \mathbf{G}(\mathbf{Z}(t))\mathbf{L}^j(\mathbf{u}) + \mathbf{K}\mathbf{w}^j + \mathbf{G}(\mathbf{Z}(t))(\mathbf{L}(\mathbf{u}) - \mathbf{L}^j(\mathbf{u})) + \mathbf{K}(\mathbf{w} - \mathbf{w}^j), \tag{14.50}$$

where $\mathbf{L}^j(\mathbf{u})$ and $\mathbf{w}^j(t)$ are the policies that are updated. By contrast, $\mathbf{L}(\mathbf{u})$ and $\mathbf{w}(t)$ are the policies that are applied to the system to collect the data.

By the definition, it is easy to see that

$$e^{-\alpha(t_k - t_{k-1})} V^{j+1}(\mathbf{Z}(t_k)) - V^{j+1}(\mathbf{Z}(t_{k-1})) = \int_{t_{k-1}}^{t_k} e^{-\alpha(\tau - t_{k-1})} (V_{\mathbf{Z}}^{j+1})^{\mathrm{T}} \dot{\mathbf{Z}}(t) - \alpha V^{j+1}\, dt. \tag{14.51}$$

Substituting (14.50) into (14.51) yields

$$\begin{aligned} e^{-\alpha(t_k - t_{k-1})} V^{j+1}(\mathbf{Z}(t_k)) - V^{j+1}(\mathbf{Z}(t_{k-1})) = \int_{t_{k-1}}^{t_k} e^{-\alpha(\tau - t_{k-1})} (V_{\mathbf{Z}}^{j+1})^{\mathrm{T}} \Big[\mathbf{F}(\mathbf{Z}(t)) + \mathbf{G}(\mathbf{Z}(t))\mathbf{L}^j(\mathbf{u}) \\ + \mathbf{K}\mathbf{w}^j + \mathbf{G}(\mathbf{Z}(t)(\mathbf{L}(\mathbf{u}) - \mathbf{L}^j(\mathbf{u})) + \mathbf{K}(\mathbf{w} - \mathbf{w}^j)\Big]\, dt. \end{aligned} \tag{14.52}$$

On the other hand, one has

$$(V_{\mathbf{Z}}^{j+1})^{\mathrm{T}} \Big[\mathbf{F}(\mathbf{Z}) + \mathbf{G}(\mathbf{Z})\mathbf{L}^j(\mathbf{u}) + \mathbf{K}\mathbf{w}^j\Big] = \alpha V^{j+1} - r_a(\mathbf{Z}(t), L^j(\mathbf{u}), \mathbf{w}^j), \tag{14.53}$$

where

$$r_a(\mathbf{Z}(t), \mathbf{L}^j(\mathbf{u}), \mathbf{w}^j) = \mathbf{Z}^{\mathrm{T}}\mathbf{Q}_1\mathbf{Z} + W(\mathbf{L}(\mathbf{u}^j)) - \gamma^2 (\mathbf{w}^j)^{\mathrm{T}}\mathbf{w}^j.$$

Substituting (14.53) into (14.52) yields

$$e^{-\alpha(t_k-t_{k-1})}V^{j+1}(\mathbf{Z}(t_k)) - V^{j+1}(\mathbf{Z}(t_{k-1})) = \int_{t_{k-1}}^{t_k} e^{-\alpha(\tau-t_{k-1})}((V_{\mathbf{Z}}^{j+1})^{\mathrm{T}}$$
$$\times\left[\mathbf{G}(\mathbf{Z}(t)(\mathbf{L}(\mathbf{u}) - \mathbf{L}^j(\mathbf{u})) + \mathbf{K}(\mathbf{w} - \mathbf{w}^j)\right] - r_a(\mathbf{Z}(t), \mathbf{L}^j(\mathbf{u}), \mathbf{w}^j))dt. \tag{14.54}$$

Using (14.43)–(14.45) in (14.54) yields the following off-policy $H_\infty$ Bellman equation:

$$e^{-\alpha(t_k-t_{k-1})}V^{j+1}(\mathbf{Z}(t_k)) - V^{j+1}(\mathbf{Z}(t_{k-1})) = \int_{t_{k-1}}^{t_k} e^{-\alpha(\tau-t_{k-1})}((\mathbf{v}^{j+1}\bar{\mathbf{L}}(\tanh^{\mathrm{T}}(\mathbf{v}^j) - \tanh^{\mathrm{T}}(v))$$
$$+ 2\gamma^2\mathbf{w}^{j+1}(\mathbf{w} - \mathbf{w}^j) - r_a(\mathbf{Z}(t), \mathbf{v}^j, \mathbf{w}^j))dt. \tag{14.55}$$

Note that if $\mathbf{v}^j$ and $\mathbf{w}^j$ are given, the unknown functions $V^{j+1}(\mathbf{Z})$, $\mathbf{v}^{j+1}$ and $\mathbf{w}^{j+1}$ can be approximated using (14.55). Then $\mathbf{L}^{j+1}(\mathbf{u})$ is found from $\mathbf{v}^{j+1}$.

The elements of $\mathbf{L}^{j+1}(\mathbf{u})$ can be either dependent or independent. If elements in $\mathbf{L}^{j+1}(\mathbf{u})$ are independent, then the Bellman equation (14.55) can be solved iteratively using stored data to find $\mathbf{L}^{\star}(\mathbf{u})$ and the optimal control policy is $\mathbf{u}^{\star}$. The following algorithm shows how to iterate on (14.55) to find the optimal control policy in this case.

**Algorithm 3** Online off-policy RL algorithm for nonaffine system with independent elements in $\mathbf{L}(\mathbf{u})$.

1: **procedure**

2: Start with the control effort $\mathbf{L}(\mathbf{u})$ and disturbance input $\mathbf{w}$ and collect required system information at $N$ different sampling intervals $T$.

3: Given $\mathbf{v}^j$ and $\mathbf{w}^j$, use collected information in step 2 to solve the following Bellman equation for $V^{j+1}$, $\mathbf{v}^{j+1}$ and $\mathbf{w}^{j+1}$ simultaneously:

$$e^{-\alpha(t_k-t_{k-1})}V^{j+1}(\mathbf{Z}(t_k)) - V^{j+1}(\mathbf{Z}(t_{k-1})) = \int_{t_{k-1}}^{t_k} e^{-\alpha(\tau-t_{k-1})}((\mathbf{v}^{j+1}\bar{\mathbf{L}}(\tanh^{\mathrm{T}}(\mathbf{v}^j) - \tanh^{\mathrm{T}}(\mathbf{v}))$$
$$+ 2\gamma^2\mathbf{w}^{j+1}(\mathbf{w} - \mathbf{w}^j) - r_a(\mathbf{Z}(t), \mathbf{v}^j, \mathbf{w}^j))dt. \tag{14.56}$$

4: Stop if

$$\left|\mathbf{v}^{j+1} - \mathbf{v}^j\right| \leqslant \varepsilon \text{ and } \left|\mathbf{w}^{j+1} - \mathbf{w}^j\right| \leqslant \varepsilon.$$

5: Otherwise set $j = j + 1$ and go to 3.

6: **end procedure**

Algorithm 3 gives $\mathbf{L}^{j+1}(\mathbf{u})$ and, if the condition of Lemma 2 is satisfied, then the elements of the control input are $\mathbf{u}^{j+1} = -\mathbf{L}^{-1}(\bar{\mathbf{L}}\tanh^{\mathrm{T}}(\mathbf{v}^{j+1}))$. However, if elements in $\mathbf{L}^{j+1}(\mathbf{u})$ are dependent, then the dependency of its elements must be taken into account by encoding equality constraints while solve Eq. (14.55) for $\mathbf{v}^{j+1}$.

To find a form for solution constraints $\mathbf{L}(\mathbf{u})$ if it has dependent elements, consider the UAV system in Example 1 with

$$\mathbf{L}(\mathbf{u}(t)) = \begin{bmatrix} L_1 \\ L_2 \\ L_3 \\ L_4 \\ L_5 \end{bmatrix} = \begin{bmatrix} u_1 \\ u_2 \\ u_2^2 \\ u_2\cos(u_3) \\ u_2\sin(u_3) \end{bmatrix}.$$

Then, the dependency of the elements of $\mathbf{L}(\mathbf{u})$ becomes

$$L_3 = {L_2}^2 = {L_4}^2 + {L_5}^2.$$

This gives the following equality constraints:

$$\bar{L}_3\tanh(v_3) = (\bar{L}_2\tanh(v_2))^2 = (\bar{L}_4\tanh(v_4))^2 + (\bar{L}_5\tanh(v_5))^2.$$

In general, it is seen that one has a vector of equality functions

$$\mathbf{f}(\mathbf{L}) = [f_1(\mathbf{L}), ..., f_p(\mathbf{L})]^{\mathrm{T}} = \mathbf{0}, \tag{14.57}$$

with $p$ being the number of dependent elements in $\mathbf{L}(\mathbf{u})$. For example for the UAV system, one has $f_1 = \bar{L}_3\tanh(v_3) - (\bar{L}_2\tanh(v_2))^2$, $f_2 = (\bar{L}_2\tanh(v_2))^2 - (\bar{L}_4\tanh(v_4))^2 - (\bar{L}_5\tanh(v_5))^2$ and $f_3 = (\bar{L}_3\tanh(v_3)) - (\bar{L}_4\tanh(v_4))^2 - (\bar{L}_5\tanh(v_5))^2$. This constraint must be taken into account when solving (14.55) for $\mathbf{v}$ using NNs.

The following algorithm shows how to find the optimal control solution for the cases where $\mathbf{L}(\mathbf{u})$ has dependent elements. The details of implementation of solving (14.55) for $\mathbf{v}$ while considering the constraint imposed by the independency of elements of $\mathbf{v}$ are presented in the next subsection.

Before proceeding, $\bar{\mathbf{H}}$ is defined as

$$\bar{\mathrm{H}} = \int_{t_{k-1}}^{t_k} e^{-\alpha(\tau - t_{k-1})}\big((\mathbf{v}^{j+1}\bar{\mathbf{L}}(\tanh^{\mathrm{T}}(\mathbf{v}^j) - \tanh^{\mathrm{T}}(\mathbf{v})) + 2\gamma^2\mathbf{w}^{j+1}(\mathbf{w} - \mathbf{w}^j) - r_a(\mathbf{Z}(t), \mathbf{v}^j, \mathbf{w}^j)\big)dt$$
$$- e^{-\alpha(t_k - t_{k-1})}V^{j+1}(\mathbf{Z}(t_k)) + V^{j+1}(\mathbf{Z}(t_{k-1})).$$

**Algorithm 4** Online off-policy RL algorithm for nonaffine system with dependent elements in $\mathbf{L}(\mathbf{u})$.

1: **procedure**

2: Start with the control effort $\mathbf{L}(\mathbf{u})$ and disturbance input $\mathbf{w}$ and collect required system information at $N$ different sampling intervals $T$.

3: Given $\mathbf{v}^j$ and $\mathbf{w}^j$, use collected information in step 2 to solve the following Bellman equation for $V^{j+1}$, $\mathbf{v}^{j+1}$ and $\mathbf{w}^{j+1}$ simultaneously:

$$\min \bar{\mathbf{H}}^2 \ s.t.\ \mathbf{f}(\mathbf{v}) = \mathbf{0}. \tag{14.58}$$

4: Stop if

$$\left|\mathbf{v}^{j+1} - \mathbf{v}^j\right| \leqslant \varepsilon \ \text{ and } \ \left|\mathbf{w}^{j+1} - \mathbf{w}^j\right| \leqslant \varepsilon.$$

5: Otherwise set $j = j + 1$ and go to 3.

6: **end procedure**

The minimum value of $\bar{\mathbf{H}}$ in Algorithm 4 not considering the constraint (14.57) is zero. If this algorithm terminates, so that $\bar{\mathbf{H}} = 0$, then by Theorem 2 the $L_2$ gain problem is solved and there exists a $\mathbf{u}^\star$ such that $\mathbf{L}(\mathbf{u}^\star) = \mathbf{L}^\star(\mathbf{u})$.

The following subsection shows how to use NNs along with linear and nonlinear LS, respectively, to implement Algorithms 3 and 4.

### 14.3.5 NEURAL NETWORKS FOR IMPLEMENTATION OF OFF-POLICY RL ALGORITHMS

In this subsection, the solution of the off-policy $H_\infty$ Bellman equations (14.56) and Eq. (14.58) in Algorithms 3 and 4 using three NNs is presented. The unknown functions $V^{j+1}(Z)$, $\mathbf{v}^{j+1}$ and $\mathbf{w}^{j+1}$ can be approximated by three NNs as

$$\hat{V}^{j+1}(\mathbf{Z}) = \sum_{i=1}^{N_1} \hat{c}_i^{j+1} \phi_i(\mathbf{Z}) = \hat{\mathbf{C}}^{j+1} \phi(\mathbf{Z}), \tag{14.59}$$

$$\hat{v}_i^{j+1} = \sum_{k=1}^{N_2} \hat{p}_{i,k}^{j+1} \sigma_{i,k}(\mathbf{Z}) = \hat{\mathbf{P}}_{\mathbf{i}}^{j+1} \sigma_{\mathbf{i}}(\mathbf{Z}), \tag{14.60}$$

$$\hat{w}_i^{j+1} = \sum_{k=1}^{N_3} \hat{q}_{i,k}^{j+1} \rho_{i,k}(\mathbf{Z}) = \hat{\mathbf{Q}}_{\mathbf{i}}^{j+1} \rho_{\mathbf{i}}(\mathbf{Z}), \tag{14.61}$$

where $\hat{\mathbf{v}}^{j+1} = [\hat{v}_1^{j+1}, ..., \hat{v}_l^{j+1}]$, $\hat{\mathbf{w}}^{j+1} = [\hat{w}_1^{j+1}, ..., \hat{w}_q^{j+1}]$. The terms $\phi_{\mathbf{i}}(\mathbf{Z}) = [\phi_{i1}, ..., \phi_{iN_{i1}}]$, $\sigma_{\mathbf{i}}(\mathbf{Z}) = [\sigma_{i1}, ..., \sigma_{iN_{i2}}]$ and $\rho_{\mathbf{i}}(\mathbf{Z}) = [\rho_{i1}, ..., \rho_{iN_3}]$ are basis function vectors, $\hat{\mathbf{C}}^{j+1}$, $\hat{\mathbf{P}}_{\mathbf{i}}^{j+1}$ and $\hat{\mathbf{Q}}_{\mathbf{i}}^{j+1}$ are constant weight vectors and $N_1$, $N_2$ and $N_3$ are the number of neurons. Substituting (14.59)–(14.61) into the off-policy $H_\infty$ Bellman equation (14.55) yields

$$\begin{aligned} & e^{-\alpha(t_k - t_{k-1})} \hat{\mathbf{C}}^{j+1} \left[\phi(\mathbf{Z}(t_k)) - \phi(\mathbf{Z}(t_{k-1}))\right] = \\ & \int_{t_{k-1}}^{t_k} e^{-\alpha(\tau - t_{k-1})} \Big( \sum_{i=1}^{l} \hat{\mathbf{P}}_i^{j+1} \sigma_{\mathbf{i}}(\mathbf{Z}) \bar{\mathbf{L}}_i (\tanh^{\mathrm{T}}(\hat{\mathbf{v}}_i^j) - \tanh^{\mathrm{T}}(v_i)) \\ & + 2\gamma^2 \sum_{i=1}^{q} \hat{\mathbf{Q}}_i^{j+1} \rho_{\mathbf{i}}(\mathbf{Z})(w_i - w_i{}^j) - r_a(\mathbf{Z}(t), \hat{\mathbf{v}}^j, \hat{\mathbf{w}}^j)\Big) dt. \end{aligned} \tag{14.62}$$

By defining $\hat{\mathbf{P}} = \begin{bmatrix} \hat{P}_1 & \ldots & \hat{P}_l \end{bmatrix}$ and $\hat{\mathbf{Q}} = \begin{bmatrix} \hat{Q}_1 & \ldots & \hat{Q}_q \end{bmatrix}$, Eq. (14.62) can be rewritten as

$$\hat{\mathbf{W}}^{\mathrm{T}} \mathbf{h}(t_k) = y(t_k), \tag{14.63}$$

where

$$\hat{\mathbf{W}} = \begin{bmatrix} (\hat{\mathbf{C}}^{j+1})^{\mathrm{T}} & (\hat{\mathbf{P}}^{j+1})^{\mathrm{T}} & (\hat{\mathbf{Q}}^{j+1})^{\mathrm{T}} \end{bmatrix}^{\mathrm{T}},$$

$$\mathbf{h}(t_k) = \begin{bmatrix} e^{-\alpha(t_k - t_{k-1})}\phi(\mathbf{Z}(t_k)) - \phi(\mathbf{Z}(t_{k-1})) \\ \int_{t_{k-1}}^{t_k} e^{-\alpha(\tau - t_{k-1})}\sigma_1(\mathbf{Z})\bar{\mathbf{L}}_1(\tanh^{\mathrm{T}}(\hat{v}_1^j) - \tanh^{\mathrm{T}}(v_1))\, d\tau \\ \vdots \\ \int_{t_{k-1}}^{t_k} e^{-\alpha(\tau - t_{k-1})}\sigma_l(\mathbf{Z})\bar{\mathbf{L}}_l(\tanh^{\mathrm{T}}(\hat{v}_l^j) - \tanh^{\mathrm{T}}(v_l))\, d\tau \\ 2\gamma^2 \int_{t_{k-1}}^{t_k} e^{-\alpha(\tau - t_{k-1})}\rho_1(\mathbf{Z})(w_1 - {w_1}^j)\, d\tau \\ \vdots \\ 2\gamma^2 \int_{t_{k-1}}^{t_k} e^{-\alpha(\tau - t_{k-1})}\rho_q(\mathbf{Z})(w_q - {w_q}^j)\, d\tau \end{bmatrix},$$

$$y(t_k) = \int_{t_{k-1}}^{t_k} e^{-\alpha(\tau - t_{k-1})} r_a(\mathbf{Z}(t), \hat{\mathbf{v}}^j, \hat{\mathbf{w}}^j))\, dt.$$

### Case 1: Independency in Elements of $\mathbf{L}(\mathbf{u})$

Eq. (14.63) can be solved using the least square method for parameter vector $\hat{\mathbf{W}}$. Then the approximated value function and disturbance input are (14.59) and (14.61), respectively. The control input $\hat{\mathbf{u}}^{j+1}$ is found by determining $\mathbf{L}(\hat{\mathbf{u}}^{j+1})$ based on (14.45) from (14.60). The number of unknown parameters $\hat{\mathbf{W}}$ is $N_1 + N_2 + N_3$. Then, at least $N \geqslant N_1 + N_2 + N_3$ data sampled $t_1$ to $t_N$ should be collect before solving (14.63) in the least square sense,

$$\mathbf{Y} = \begin{bmatrix} y(t_1) & \dots & y(t_N) \end{bmatrix}^T,$$
$$\mathbf{H} = \begin{bmatrix} h(t_1) & \dots & h(t_N) \end{bmatrix}.$$

The least square solution is obtained as

$$\hat{\mathbf{W}} = (\mathbf{H}\mathbf{H}^{\mathrm{T}})^{-1}\mathbf{H}\mathbf{Y}.$$

### Case 2: Dependency in Elements of $\mathbf{L}(\mathbf{u})$

If the elements of $\mathbf{L}(\mathbf{u})$ are dependent, one has to solve a constrained nonlinear least square problem to take into account the equality constraints imposed by the dependency of the elements of $\mathbf{L}(\mathbf{u})$. To show this, consider the case of the UAV in Example 1. The following constraints are considered when finding the weights of NNs:

$$\bar{L}_3 \tanh(\hat{P}_3^{j+1}\sigma_3(\mathbf{Z})) = (\bar{L}_2 \tanh(\hat{P}_2^{j+1}\sigma_2(\mathbf{Z})))^2 =$$
$$(\bar{L}_4 \tanh(\hat{P}_4^{j+1}\sigma_4(\mathbf{Z})))^2 + (\bar{L}_5 \tanh(\hat{P}_5^{j+1}\sigma_5(\mathbf{Z})))^2.$$

This constraint is nonlinear in NN weights and thus requires using the nonlinear least square method. In general, (14.58) becomes

$$arg \min_{\hat{\mathbf{W}}} \left\| \hat{\mathbf{W}}\mathbf{H} - \mathbf{Y} \right\|^2 \quad s.t.\ \mathbf{f}(\hat{\mathbf{P}}^{j+1}, \sigma_1, \ldots, \sigma_l) = 0,$$

where the function $\mathbf{f}$ is defined in (14.57) and depends on how the elements of $\mathbf{L}(\mathbf{u})$ and consequently NN weights are related.

## REFERENCES


[1] R.S. Sutton, A.G. Barto, Reinforcement Learning: An Introduction, vol. 1, MIT Press, Cambridge, 2017.

[2] D.P. Bertsekas, J.N. Tsitsikli, Neuro-Dynamic Programming, Athena Scientific, MA, 1996.

[3] W.B. Powell, Approximate Dynamic Programming, Wiley, Hoboken, NJ, 2007.

[4] F.L. Lewis, D. Liu, Reinforcement Learning and Approximate Dynamic Programming for Feedback Control, Wiley, 2013.

[5] D. Vrabie, K.G. Vamvoudakis, F.L. Lewis, Optimal Adaptive Control and Differential Games by Reinforcement Learning Principles, Institution of Engineering and Technology, 2013.

[6] H. Zhang, L. Cui, X. Zhang, X. Luo, Data-driven robust approximate optimal tracking control for unknown general nonlinear systems using adaptive dynamic programming method, IEEE Transactions on Neural Networks 22 (2011) 2226–2236.

[7] K.G. Vamvoudakis, F.L. Lewis, Online actor–critic algorithm to solve the continuous-time infinite horizon optimal control problem, Automatica 46 (5) (2010) 878–888.

[8] D. Vrabie, F.L. Lewis, Neural network approach to continuous-time direct adaptive optimal control for partially unknown nonlinear systems, Neural Networks 22 (3) (2009) 237–246.

[9] D. Vrabie, O. Pastravanu, M. Abu-Khalaf, F.L. Lewis, Adaptive optimal control for continuous-time linear systems based on policy iteration, Automatica 45 (2) (2009) 447–484.

[10] R. Song, W. Xiao, H. Zhang, C. Sun, Adaptive dynamic programming for a class of complex-valued nonlinear systems, IEEE Transactions on Neural Networks and Learning Systems 25 (9) (2014) 1733–1739.

[11] H. Modares, F.L. Lewis, M.B. Naghibi-Sistani, Adaptive optimal control of unknown constrained-input systems using policy iteration and neural networks, IEEE Transactions on Neural Networks and Learning Systems 24 (10) (2013) 1513–1525.

[12] S. Bhasin, R. Kamalapurkar, M. Johnson, K.G. Vamvoudakis, F.L. Lewis, W.E. Dixon, A novel actor-critic-identifier architecture for approximate optimal control of uncertain nonlinear systems, Automatica 49 (1) (2013) 82–92.

[13] T. Bian, Y. Jiang, Z.-P. Jiang, Adaptive dynamic programming and optimal control of nonlinear nonaffine systems, Automatica 50 (10) (2014) 2624–2632.

[14] Y. Jiang, Z.-P. Jiang, Robust adaptive dynamic programming and feedback stabilization of nonlinear systems, IEEE Transactions on Neural Networks and Learning Systems 25 (5) (2014) 882–893.

[15] Y. Jiang, Z.-P. Jiang, Computational adaptive optimal control for continuous-time linear systems with completely unknown dynamics, Automatica 48 (10) (2012) 2699–2704.

[16] D. Liu, H. Li, D. Wang, Error bounds of adaptive dynamic programming algorithms for solving undiscounted optimal control problems, IEEE Transactions on Neural Networks and Learning Systems 26 (6) (2015) 1323–1334.

[17] B. Luo, H.-N. Wu, T. Huang, D. Liu, Data-based approximate policy iteration for affine nonlinear continuous-time optimal control design, Automatica 50 (12) (2014) 3281–3290.

[18] B. Luo, H.-N. Wu, H.-X. Li, Adaptive optimal control of highly dissipative nonlinear spatially distributed processes with neuro-dynamic programming, IEEE Transactions on Neural Networks and Learning Systems 26 (4) (2015) 684–696.

[19] M. Abu-Khalaf, F.L. Lewis, J. Huang, Neurodynamic programming and zero-sum games for constrained control systems, IEEE Transactions on Neural Networks and Learning Systems 19 (7) (2008) 1243–1252.

[20] H. Zhang, Q. Wei, D. Liu, An iterative adaptive dynamic programming method for solving a class of nonlinear zero-sum differential games, Automatica 47 (1) (2011) 207–214.

[21] K.G. Vamvoudakis, F.L. Lewis, Online solution of nonlinear two-player zero-sum games using synchronous policy iteration, International Journal of Robust and Nonlinear Control 22 (13) (2012) 1460–1483.

[22] K.G. Vamvoudakis, F.L. Lewis, Online Gaming: Real Time Solution of Nonlinear Two-Player Zero-Sum Games Using Synchronous Policy Iteration, INTECH Open Access Publisher, 2011.

[23] H. Modares, F.L. Lewis, M.-B. Naghibi-Sistani, Online solution of nonquadratic two-player zero-sum games arising in the $H_\infty$ control of constrained input systems, International Journal of Adaptive Control and Signal Processing 28 (3–5) (2014) 232–254.

[24] H. Zhang, C. Qin, B. Jiang, Y. Luo, Online adaptive policy learning algorithm for $H_\infty$ state feedback control of unknown affine nonlinear discrete-time systems, IEEE Transactions on Cybernetics 44 (12) (2014) 2706–2718.

[25] H.-N. Wu, B. Luo, Simultaneous policy update algorithms for learning the solution of linear continuous-time $H_\infty$ state feedback control, Information Sciences 222 (2013) 472–485.

[26] H.-N. Wu, B. Luo, Neural network based online simultaneous policy update algorithm for solving the HJI equation in nonlinear $H_\infty$ control, IEEE Transactions on Neural Networks and Learning Systems 23 (12) (2012) 1884–1895.

[27] B. Luo, H.-N. Wu, Computationally efficient simultaneous policy update algorithm for nonlinear $H_\infty$ state feedback control with Galerkin's method, International Journal of Robust and Nonlinear Control 23 (9) (2013) 991–1012.

[28] D. Vrabie, F.L. Lewis, Adaptive dynamic programming for online solution of a zero-sum differential game, Journal of Control Theory and Applications 9 (3) (2011) 353–360.

[29] H. Li, D. Liu, D. Wang, Integral reinforcement learning for linear continuous-time zero-sum games with completely unknown dynamics, IEEE Transactions on Automation Science and Engineering 11 (3) (2014) 706–714.

[30] B. Luo, H.-N. Wu, T. Huang, Off-policy reinforcement learning for $H_\infty$ control design, IEEE Transactions on Cybernetics 45 (1) (2015) 65–76.

[31] R.A. Howard, Dynamic Programming and Markov Processes, MIT Press, Cambridge, MA, 1960.

[32] K.G. Vamvoudakis, D. Vrabie, F.L. Lewis, Online adaptive algorithm for optimal control with integral reinforcement learning, International Journal of Robust and Nonlinear Control 24 (17) (2014) 2686–2710.

[33] B. Kiumarsi, F.L. Lewis, H. Modares, A. Karimpour, M.-B. Naghibi-Sistani, Reinforcement Q-learning for optimal tracking control of linear discrete-time systems with unknown dynamics, Automatica 50 (4) (2014) 1167–1175.

[34] H. Zhang, Q. Wei, Y. Luo, A novel infinite-time optimal tracking control scheme for a class of discrete-time nonlinear systems via the greedy HDP iteration algorithm, IEEE Transactions on Systems, Man, and Cybernetics, Part B (Cybernetics) 38 (2008) 937–942.

[35] D. Wang, D. Liu, Q. Wei, Finite-horizon neuro-optimal tracking control for a class of discrete-time nonlinear systems using adaptive dynamic programming approach, Neurocomputing 78 (2012) 14–22.

[36] T. Dierks, S. Jagannathan, Optimal tracking control of affine nonlinear discrete-time systems with unknown internal dynamics, in: Decision and Control, 2009 Held Jointly with the 2009 28th Chinese Control Conference, CDC/CCC 2009, Proceedings of the 48th IEEE Conference on, 2009, pp. 6750–6755.

[37] T. Dierks, S. Jagannathan, Optimal control of affine nonlinear continuous-time systems, in: American Control Conference, ACC, 2009, pp. 6750–6755.

[38] M.D.S. Aliyu, Nonlinear $H_\infty$ Control, Hamiltonian Systems and Hamilton–Jacobi Equations, CRC Press, 2017.

[39] S. Devasia, D. Chen, B. Paden, Nonlinear inversion-based output tracking, IEEE Transactions on Automatic Control 41 (7) (1996) 930–942.

[40] G.J. Toussaint, T. Basar, F. Bullo, $H_\infty$-optimal tracking control techniques for nonlinear underactuated systems, in: Decision and Control, 2000, Proceedings of the 39th IEEE Conference on, vol. 3, 2000, pp. 2078–2083.

[41] J.A. Ball, P. Kachroo, A.J. Krener, $H_\infty$ tracking control for a class of nonlinear systems, IEEE Transaction on Automatic Control 44 (6) (1999) 1202–1206.

[42] T. Basar, P. Bernard, $H_\infty$ Optimal Control and Related Minimax Design Problems, Birkhäuser, Boston, MA, 1995.

[43] F.L. Lewis, D. Vrabie, V. Syrmos, Optimal Control, 3rd edition, Wiley, 2012.

[44] H. Modares, F.L. Lewis, Z.-P. Jiang, $H_\infty$ tracking control of completely unknown continuous-time systems via off-policy reinforcement learning, IEEE Transactions on Neural Networks and Learning Systems 26 (2015) 2550–2562.

[45] B.L. Stevens, F.L. Lewis, E.N. Johnson, Aircraft Control and Simulation, third edition, Wiley-Blackwell, 2015.

[46] B. Kiumarsi, W. Kang, F.L. Lewis, $H_\infty$ control of non-affine aerial systems using off-policy reinforcement learning, Unmanned Systems 4 (1) (2016) 51–60.

CHAPTER

# ADAPTIVE DYNAMIC PROGRAMMING FOR OPTIMAL CONTROL OF NONLINEAR DISTRIBUTED PARAMETER SYSTEMS

15

**Biao Luo*, Derong Liu†, Huai-Ning Wu‡**

*State Key Laboratory of Management and Control for Complex Systems, Institute of Automation, Chinese Academy of Sciences, Beijing, China †School of Automation, Guangdong University of Technology, Guangzhou, China ‡School of Automation Science and Electrical Engineering, Beihang University, Beijing, China

## CONTENTS



## CHAPTER POINTS

- Through model reduction, the distributed parameter system is converted to a finite-dimensional slow subsystem of ordinary differential equations that accurately describes its dominant dynamics.

Adaptive Learning Methods for Nonlinear System Modeling. DOI: 10.1016/B978-0-12-812976-0.00019-1

- An adaptive optimal control method based on neurodynamic programming is developed for distributed parameter systems with full system model.
- An adaptive optimal control method based on policy iteration is developed for partially unknown distributed parameter systems.

## 15.1 INTRODUCTION

Over the past few decades, the research on distributed parameter systems (DPSs) [1–13] has received extensive attention. Many industrial processes are essentially DPSs due to the spatial variations of the underlying physical phenomena, such as diffusion, convection, phase-dispersion, vibration and flow. Typical examples of industrial DPSs include the catalytic packed-bed reactors [3] used to convert methanol to formaldehyde, the Czochralski crystallization [14] of high-purity crystals, the temperature profile [15] of high-speed aerospace vehicles, the chemical vapor deposition [2] of thin films for microelectronics manufacturing, the aerosol-based production of nanoparticles [16] used in medical applications, the Tokamak device [17] for nuclear fusion that is an intangible doughnut-shaped bottle created by magnetic lines used for confining the high-temperature plasma and air traffic flow in the National Airspace System [18], as well as wavy behavior [19] of excitable media in biology, etc. These DPSs are essentially described by a set of nonlinear partial differential equations (PDEs) with homogeneous or mixed boundary conditions. Thus, the DPSs are also called PDE systems.

It is noted that DPSs are infinite-dimensional systems, where their system state is a spatiotemporal function. So far, many results [2,20,3,21,22,5–13,23–25] have been reported to investigate its control problem. Optimal control of PDE systems refers to a class of methods that can be used to synthesize a control policy which results in the best possible behavior with respect to the prescribed performance criterion. In [22], based on the empirical eigenfunctions (EEFs) computed via Karhunen–Loève decomposition (KLD), the infinite-dimensional optimization problem of transport–reaction processes was solved with successive quadratic programming (SQP). In [26], the PDE system was formulated as a finite-dimensional system of ordinary differential equations (ODEs) via spatial discretization and then an optimal feedback control approach was proposed. In [15], a dual heuristic programming approach was introduced to solve the temperature profile control problem of high-speed aerospace vehicles. In [27], Xu *et al.* considered a class of bilinear parabolic PDE systems and suggested a sequential linear quadratic regulator approach based on an iterative scheme, and in [17] this approach was successfully applied to the optimal tracking control of current profile in Tokamaks. The proposed control schemes in [23–25] can be utilized to design the controllers for the DPSs with constraints and could be a new route for developing advanced control techniques in practice.

Reinforcement learning is a powerful tool for solving the optimal control problem of Markov decision processes, which has been widely investigated as a machine learning technique in the artificial intelligence community [28,29]. In recent years, the thought of reinforcement learning has been introduced to develop adaptive dynamic programming (ADP) methods for solving optimization control problems of finite-dimensional systems, and many meaningful results have been reported [30–36,6,37,10–12,38–41,13,42–49]. For optimal control problems, it is usually required to solve the Hamilton–Jacobi–Bellman equation (HJBE) [50,51], which is extremely difficult to solve analytically. To overcome this difficulty, ADP has appeared as an efficient tool to solve the HJBE by using the system model or real system data. For example, Murray *et al.* [52] suggested two policy iteration (PI) algorithms that avoid the necessity of knowing the internal system dynamics. Vrabie *et al.* [53] extended their result and proposed a new PI algorithm to solve the linear quadratic regulation problem. An integral reinforcement learning method [54] was introduced to solve the linear quadratic tracking problem of partially unknown continuous-time systems. An off-policy reinforcement learning approach was proposed to solve the data-based optimal control problem [37]. However, ADP methods have rarely been studied for control design of PDE systems. In this chapter, the optimal control problem of nonlinear PDE systems is considered. Two kinds of adaptive control methods based on ADP have been developed, *i.e.*, neurodynamic programming (NDP) and PI.

*Notation*: $\mathbb{R}$, $\mathbb{R}^n$ and $\mathbb{R}^{n\times m}$ are the set of real numbers, the $n$-dimensional Euclidean space and the set of all real matrices, respectively. The symbols $|\cdot|$, $\|\cdot\|$ and $\langle\cdot\rangle_{\mathbb{R}^n}$ denote the absolute value for scalars, the Euclidean norm and the inner product for vectors, respectively. Let $\mathbb{R}^\infty$ be the vector space of infinite sequences $a \triangleq [a_1 \ldots a_\infty]^\mathsf{T}$ of real numbers equipped with the norm $\|a\|_{\mathbb{R}^\infty} \triangleq \sqrt{\sum_{i=1}^{\infty} a_i^2}$, which is a natural generalization of space $\mathbb{R}^n$. The superscript $\mathsf{T}$ is used for the transpose and $\nabla \triangleq \partial/\partial x$ denotes a gradient operator notation. The terms $\bar{\sigma}(\cdot)$ and $\underline{\sigma}(\cdot)$ denote the maximum singular value and the minimum singular value. For a symmetric matrix $M$, $M > (\geq)\ 0$ means that it is a positive (semipositive) definite matrix. We have $\|v\|_M^2 \triangleq v^\mathsf{T} M v$ for some real vector $v$ and symmetric matrix $M > (\geq)\ 0$ with appropriate dimensions. The term $C^1(\Omega)$ represents the function space on compact set $\Omega$ with first derivatives being continuous. Also, $L_2([z_1, z_2], \mathbb{R}^n)$ is an infinite-dimensional Hilbert space of $n$-dimensional square integrable vector functions $\omega(z) \in L_2([z_1, z_2], \mathbb{R}^n)$, $z \in [z_1, z_2] \subset \mathbb{R}$ equipped with the inner product and norm: $\langle \omega_1(\cdot), \omega_2(\cdot)\rangle \triangleq \int_{z_1}^{z_2} \langle \omega_1(z), \omega_2(z)\rangle_{\mathbb{R}^n} \mathrm{d}z$ and $\|\omega_1(\cdot)\|_2 \triangleq \langle \omega_1(\cdot), \omega_1(\cdot)\rangle^{1/2}$ for $\forall \omega_1(z), \omega_2(z) \in L_2([z_1, z_2], \mathbb{R}^n)$.

## 15.2 PROBLEM DESCRIPTION

We consider a general class of DPSs which are described by highly dissipative nonlinear PDEs with the following state space representation:

$$\begin{cases} \dfrac{\partial y(z,t)}{\partial t} = \mathcal{L}\left(y, \dfrac{\partial y}{\partial z}, \dfrac{\partial^2 y}{\partial z^2}, \ldots, \dfrac{\partial^{n_0} y}{\partial z^{n_0}}\right) + \bar{B}(z)u(t), \\ y_h(t) = \displaystyle\int_{z_1}^{z_2} H(z)y(z,t)\mathrm{d}z, \end{cases} \tag{15.1}$$

subject to the mixed-type boundary conditions

$$q\left(y, \frac{\partial y}{\partial z}, \frac{\partial^2 y}{\partial z^2}, \ldots, \frac{\partial^{n_0-1} y}{\partial z^{n_0-1}}\right)\Bigg|_{z=z_1 \ or \ z_2} = 0 \tag{15.2}$$

and the initial condition

$$y(z,0) = y_0(z), \tag{15.3}$$

where $z \in [z_1, z_2] \subset \mathbb{R}$ is the spatial coordinate, $t \in [0, \infty)$ is the temporal coordinate, $y(z,t) = [y_1(z,t) \ \dots \ y_n(z,t)]^\mathsf{T} \in \mathbb{R}^n$ is the state and $u(t) \in \mathbb{R}^p$ is the manipulated input. The term $\mathcal{L} \in \mathbb{R}^n$ represents a sufficiently smooth nonlinear vector function that involves a highly dissipative, possibly nonlinear, spatial differential operator of order $n_0$ (an even number), $q$ is a sufficiently smooth nonlinear vector function, $\bar{B}(z)$ and $H(z)$ are known sufficiently smooth matrix functions of appropriate dimension and $y_0(z)$ is a smooth vector function representing the initial state profile.

In this chapter, we consider the optimal control problem of PDE system (15.1)–(15.3) with the following infinite-horizon linear quadratic performance index:

$$V(y_0(\cdot)) \triangleq \int_0^{+\infty} \left( \|y_h(t)\| + \|u(t)\|_R^2 \right) \mathrm{d}t, \tag{15.4}$$

where $R > 0$. The goal is to find a control policy $u^*(t)$ such that the above performance index is minimized, *i.e.*,

$$u(t) \triangleq u^*(t) \triangleq \arg\min_u V(y_0(\cdot, t)). \tag{15.5}$$

## 15.3 MODEL REDUCTION BASED ON KLD AND SINGULAR PERTURBATION TECHNIQUE

Karhunen–Loève decomposition (KLD) is a popular statistical pattern analysis method for seeking the dominant structures in an ensemble of a high-dimensional process and obtaining low-dimensional approximate descriptions in many engineering fields. We omit the derivation of KLD and directly give the implementation procedure as follows.

**Algorithm 3.1** KLD Implementation Procedure.

- *Step 1:* For the real DPS, collect an ensemble of states $\{y_i(z)\}, i = 1, \dots, M$.
- *Step 2:* Compute matrix $C = (c_{kj})_{M\times M}$ with $c_{kj} = \frac{1}{M}\int_{z_1}^{z_2} y_k^\mathsf{T}(\xi) y_j(\xi) \mathrm{d}\xi$.
- *Step 3:* Solve the eigenvalue problem $C\alpha_i = \lambda_i \alpha_i$, and assume that $\lambda_1 \geqslant \lambda_2 \geqslant \cdots \geqslant \lambda_M$.
- *Step 4:* Normalize the eigenvector $\alpha_i$ to satisfy $\langle \alpha_i, \alpha_i \rangle_{\mathbb{R}^n} = \frac{1}{M\alpha_i}$. Then EEFs can be computed with $\phi_i = Y(z)\alpha_i$, where $Y(z) = [y_1(z) \ \dots \ y_M(z)]$. □

For convenience, we denote $y(\cdot, t) \triangleq y(z,t) \in L_2([z_1, z_2], \mathbb{R}^n)$ and $M(\cdot) \triangleq M(z)$, $z \in [z_1, z_2]$ for some space varying matrix function $M(z)$. To simplify the notation, we consider the PDE system (15.1)–(15.3) with $n = 1$ without loss of generality. Assume that the PDE state $y(z,t)$ can be represented as an infinite weighted sum of a complete set of orthogonal basis functions $\{\phi_i(z)\}$, *i.e.*,

$$y(z,t) = \sum_{i=1}^{\infty} x_i(t)\phi_i(z), \tag{15.6}$$

where $x_i(t)$ is a time varying coefficient named the mode of the PDE system. By taking an inner product with $\phi_i(z)(i = 1, \dots)$ on both sides of the PDE system (15.1)–(15.3), we obtain the following infinite-dimensional ODE system:

$$\begin{cases} \dot{x}(t) = f_s(x(t), x_f(t)) + Bu(t), \\ \dot{x}_f(t) = \mathcal{L}_f(x(t), x_f(t)) + B_f u(t), \\ y_h(t) = H_s x(t) + H_f x_f(t), \end{cases} \tag{15.7}$$

where

$$x(t) = \langle y(\cdot, t), \Phi_s(\cdot)\rangle \triangleq [x_1(t) \ldots x_N(t)]^{\mathsf{T}} \tag{15.8}$$

and $x_f(t) = \langle y(\cdot, t), \Phi_f(\cdot)\rangle \triangleq [x_{N+1}(t) \ \ldots \ x_\infty(t)]^{\mathsf{T}} \in \mathbb{R}^\infty$, $f_s(x, x_f) \triangleq \langle \mathcal{L}, \Phi_s(\cdot)\rangle$, $\mathcal{L}_f(x, x_f) \triangleq \langle \mathcal{L}, \Phi_f(\cdot)\rangle$, $B \triangleq \langle \bar{B}(\cdot), \Phi_s(\cdot)\rangle$, $B_f \triangleq \langle \bar{B}(\cdot), \Phi_f(\cdot)\rangle$, $H_s \triangleq \int_{z1}^{z2} H(z)\Phi_s^{\mathsf{T}}(z)dz$, $H_f \triangleq \int_{z1}^{z2} H(z)\ \Phi_f^{\mathsf{T}}(z)dz$, with $\Phi_s(z) \triangleq [\phi_1(z) \ldots \phi_N(z)]^{\mathsf{T}}$ and $\Phi_f(z) \triangleq [\phi_{N+1}(z) \ldots \phi_\infty(z)]^{\mathsf{T}}$.

Owing to the highly dissipative nature of the PDE system (15.1)–(15.3), it is reasonable [55,56] to assume that $\mathcal{L}_f(x, x_f) = \frac{1}{\varepsilon} A_{f\varepsilon} x_f + f_f(x, x_f)$, where $\varepsilon > 0$ is a small positive parameter quantifying the separation between the slow (dominant) and fast (negligible) modes, $A_{f\varepsilon}$ is a matrix that is stable (in the sense that the state of the system $\dot{x}_f = A_{f\varepsilon} x_f$ tends exponentially to zero) and $f_f(x, x_f)$ satisfies

$$\|f_f(x, x_f)\| \leqslant k_1\|x\| + k_2\|x_f\|_{\mathbb{R}^\infty} \tag{15.9}$$

for $\|x\| \leqslant \beta_1$ and $\|x_f\| \leqslant \beta_2$ with $k_1, k_2 > 0$. Then system (15.7) can be rewritten as the following standard singularly perturbed form:

$$\begin{cases} \dot{x}(t) = f_s(x(t), x_f(t)) + Bu(t), \\ \varepsilon\dot{x}_f(t) = A_{f\varepsilon} x_f(t) + \varepsilon f_f(x(t), x_f(t)) + \varepsilon B_f u(t), \\ y_h(t) = H_s x(t) + H_f x_f(t). \end{cases} \tag{15.10}$$

Introducing the fast time scale $\tau = t/\varepsilon$ and setting $\varepsilon = 0$, we obtain the following infinite-dimensional fast subsystem form:

$$\frac{dx_f}{d\tau} = A_{f\varepsilon} x_f, \tag{15.11}$$

which is exponentially stable [55,56]. Then, setting $\varepsilon = 0$ in the system (15.10), we have $x_f = 0$. Thus, the following slow subsystem is obtained:

$$\begin{cases} \dot{x}(t) = f(x(t)) + Bu(t), \\ x(0) = x_0, \end{cases} \tag{15.12}$$

where $f(x) = f_s(x, x_f)$.

Observe that the performance index (15.4) can be rewritten as

$$V(x_0, x_{f0}) = \int_0^{+\infty} \left(\|x(t)\|_{Q_s}^2 + \|u(t)\|_R^2\right) dt + \Re(x, x_f), \tag{15.13}$$

where $\mathfrak{R}(x, x_f) \triangleq \int_0^{+\infty}(2x^{\mathsf{T}} Q_s f x_f + x_f^{\mathsf{T}} Q_f x_f)\mathrm{d}t$, $Q_s \triangleq H_s^{\mathsf{T}} H_s$, $Q_s f \triangleq H_s^{\mathsf{T}} H_f$, $Q_f \triangleq H_f^{\mathsf{T}} H_f$. By using $x_f = 0$, we have $\mathfrak{R}(x, x_f) = 0$. Then the expression (15.13) is written as

$$V(x_0) = \int_0^{+\infty} \left(\|x(\tau)\|_{Q_s}^2 + \|u(\tau)\|_R^2\right) \mathrm{d}\tau. \tag{15.14}$$

Now, the optimal control problem of the PDE system (15.1)–(15.3) can be reformulated as follows. Find a state feedback control policy to minimize the performance index (15.14) based on the finite-dimensional slow subsystem (15.12). It is known [50] that this optimal control problem relies on the solution of the following HJBE:

$$(\nabla V^*)^{\mathsf{T}} f + \|x\|_{Q_s}^2 - \frac{1}{4}(\nabla V^*)^{\mathsf{T}} B R^{-1} B^{\mathsf{T}} \nabla V^* = 0, \tag{15.15}$$

where $V^*(x) \in C^1(\Omega)$. Then the optimal control policy $u^*$ is given by

$$u^*(x) = -\frac{1}{2} R^{-1} B^{\mathsf{T}} \nabla V^*(x). \tag{15.16}$$

In this chapter, we use the EEFs for model reduction, *i.e.*, the basis function $\phi_i(z)$ is computed with KLD. The dimension of $\Phi_s$ (*i.e.*, $N$) is chosen such that it satisfies $\sum_{i=1}^N \lambda_i / \sum_{j=1}^M \lambda_j \geqslant 1 - \zeta$ for a small positive real number $\zeta > 0$.

## 15.4 ADAPTIVE OPTIMAL CONTROL DESIGN WITH NDP

In this section, we design an NDP-based adaptive optimal controller for the PDE system (15.1)–(15.3) on the basis of the finite-dimensional ODE model (15.12).

### 15.4.1 NEURODYNAMIC PROGRAMMING

To overcome the difficulty for solving the HJBE (15.15), we propose an adaptive optimal control method based on NDP, which can learn its solution online. NDP is implemented by using the critic neural network (NN) to approximate the value function. From the well-known high-order Weierstrass approximation theorem [57], a continuous function can be represented by an infinite-dimensional linearly independent basis function set $\{\psi_k(x)\}_{k=1}^{\infty}$. Thus, we consider the critic NN for approximating the solution $V^*(x)$ of the HJBE (15.15) on a compact set $\Omega$. Let $\Psi(x) \triangleq [\psi_1(x) \ \ \psi_2(x) \ \ \cdots \ \ \psi_L(x)]^{\mathsf{T}}$ be activation functions, where $L$ is the number of neurons in the NN hidden layer. Then $V^*(x)$ can be expressed with $\Psi(x)$ as follows:

$$V^*(x) = \theta^{*\mathsf{T}} \Psi(x) + \delta_1(x), \tag{15.17}$$

where $\theta^* \triangleq [\theta_1^* \ \ \theta_2^* \ \ \cdots \ \ \theta_L^*]^{\mathsf{T}}$ is the unknown ideal constant weight vector and $\delta_1(x)$ is the NN approximation error. The optimal control law (15.16) is rewritten as

$$u^*(x) = -\frac{1}{2} R^{-1} B^{\mathsf{T}} [\nabla \Psi(x)]^{\mathsf{T}} \theta^* - \frac{1}{2} R^{-1} B^{\mathsf{T}} \nabla \delta_1(x). \tag{15.18}$$

For convenience, define a matrix $D = BR^{-1}B^{\mathsf{T}}$. It follows from (15.17) that the HJBE (15.15) is represented as

$$\theta^{*\mathsf{T}}[\nabla\Psi]f + \|x\|_{Q_s}^2 - \frac{1}{4}\|\theta^{*\mathsf{T}}[\nabla\Psi]\|_D + \delta_2(x) = 0, \tag{15.19}$$

where $\delta_2(x)$ is the residual error of the HJBE due to the critic NN error that is given by

$$\delta_2(x) \triangleq [\nabla\delta_1(x)]^{\mathsf{T}}(f(x) + Bu^*(x)) - \frac{1}{4}\|\nabla\delta_1(x)\|_D. \tag{15.20}$$

Although the ideal critic NN weight vector $\theta^*$ provides the best approximate solution of the HJBE (15.15), it is unknown. Actually, the output of the critic NN is

$$\hat{V}(x) = \hat{\theta}^{\mathsf{T}}\Psi(x), \tag{15.21}$$

where $\hat{\theta} \triangleq [\hat{\theta}_1 \ \hat{\theta}_2 \ \cdots \ \hat{\theta}_L]^{\mathsf{T}}$ is the estimate of $\theta^*$. It follows from (15.18) that the NDP based adaptive optimal control is given by

$$\hat{u}(x) = -\frac{1}{2}R^{-1}B^{\mathsf{T}}[\nabla\Psi(x)]^{\mathsf{T}}\hat{\theta}. \tag{15.22}$$

Due to the NN approximation error, replacing $V^*(x)$ in the HJBE (15.15) with $\hat{V}(x)$ of (15.21) yields the following residual error: $(\nabla\hat{V}(x))^{\mathsf{T}}f(x) + \|x\|_{Q_s}^2 - \frac{1}{4}\|\nabla\hat{V}(x)\|_D = \delta_3(x)$, where

$$\delta_3(x) \triangleq \hat{\theta}^{\mathsf{T}}[\nabla\Psi(x)]f(x) + \|x\|_{Q_s}^2 - \frac{1}{4}\|\hat{\theta}^{\mathsf{T}}[\nabla\Psi(x)]\|_D. \tag{15.23}$$

It is observed from (15.19) and (15.23) that when $\hat{\theta} \to \theta^*$, $\delta_3 \to -\delta_2$. Therefore, it is desired to select the tuning law for $\hat{\theta}$ to minimize the squared residual error $E(\hat{\theta}) = \frac{1}{2}\delta_3^2(x)$. Besides, the tuning law is also expected to ensure stability of the closed-loop PDE system (under the adaptive optimal control law (15.22)) during the learning process of the critic NN weight vector. To attain the above two goals, we select the tuning law of $\hat{\theta}$ as follows:

$$\dot{\hat{\theta}} = -\alpha\mu\rho(x)\delta_3(x) + \alpha\kappa\nabla\Psi(x)D\nabla\bar{V}(x), \tag{15.24}$$

where $\alpha > 0$ is the adaptive gain of the critic NN, $\rho(x) \triangleq \nabla\Psi(x)(f(x) + B\hat{u}(x))$, $\mu \triangleq 1/(1+\rho^{\mathsf{T}}\rho)^2$,

$$\kappa = \begin{cases} 0, & if\ (\nabla\bar{V})^{\mathsf{T}}(f + B\hat{u}), \\ 0.5, & else \end{cases} \tag{15.25}$$

and $\bar{V}(x) \in C^1(\Omega)$ is a radially unbounded Lyapunov function satisfying

$$(\nabla\bar{V})^{\mathsf{T}}f - \frac{1}{2}(\nabla\bar{V})^{\mathsf{T}}D(\nabla\Psi)^{\mathsf{T}}\theta^* \leqslant 0. \tag{15.26}$$

### 15.4.2 STABILITY ANALYSIS

Notice that the adaptive optimal control is designed based on slow subsystems. Thus, we should analyze the stability of the original closed-loop PDE system. Before starting, we introduce a definition and some reasonable assumptions.

**Definition 1.** [58] Consider the system (15.12); the solution $x(t)$ is semiglobally uniformly ultimately bounded (SGUUB), if for any compact set $\Omega \subset \mathbb{R}^N$ and $x(t_0) = x_0 \in \Omega$, there exist positive constants $\mu_1$ and $T(\mu_1, x_0)$ such that $\|x(t)\| < \mu_1$ for all $t > t_0 + T$.

**Assumption 1.**

- The ideal weight vector of the critic NN is bound, *i.e.*, $\|\theta^*\| \leqslant \theta_M$, where $\theta_M$ is a positive number.
- On the compact set containing $\Omega$, $\|\Psi\| \leqslant \psi_M$, $\|\nabla\Psi\| \leqslant \psi_{DM}$, $|\delta_1| \leqslant \delta_{1M}$ and $\|\nabla\delta_1\| \leqslant \delta_{1DM}$, where $\psi_M$, $\psi_{DM}$, $\delta_{1M}$ and $\delta_{1DM}$ are positive numbers.
- The function $f(x)$ is bounded on a compact set containing $\Omega$, *i.e.*, $\|f\| \leqslant \upsilon_{fM}$, where $\upsilon_{fM}$ is a positive number.

Define the critic NN weight estimation error as $\tilde{\theta}(t) \triangleq \hat{\theta}(t) - \theta^*$, the value function estimation error as $\tilde{V}(x) \triangleq \hat{V}(x) - V^*(x)$ and the control estimation error as $\tilde{u}(x) \triangleq \hat{u}(x) - u^*(x)$, respectively. Then we have the following result.

**Theorem 1.** *Consider the optimal control problem of the PDE system* (15.1)–(15.3) *and the adaptive optimal control policy* (15.22) *with the tuning law* (15.24) *for critic NN weight vector. Let Assumption 1 hold. Then there exist positive real constants $\sigma_1, \sigma_2$ and $\varepsilon^*$, such that if $\|x_0\| \leqslant \sigma_1$, $\|x_{f0}\|_{\mathbb{R}^\infty} \leqslant \sigma_2$ and $\varepsilon \in (0, \varepsilon^*)$:*
*(1) The critic NN weight estimation error $\tilde{\theta}$ is SGUUB and the state $y(z,t)$ of the original closed-loop PDE system* (15.1)–(15.3) *is SGUUB in $L_2$-norm,* i.e., *there exists a positive number $\sigma_3$ such that $\|y(\cdot, t)\|_2 \leqslant \sigma_3$.*
*(2) The value function estimation error $\tilde{V}(x)$ and the control estimation error $\tilde{u}(x)$ are also SGUUB.*

*Proof.* (1) Under the adaptive optimal control policy (15.22) with the tuning law (15.24) of $\hat{\theta}$, the closed-loop PDE system is written as

$$\begin{cases} \dfrac{\partial y}{\partial t} = \mathcal{L}\left(y, \dfrac{\partial y}{\partial z}, \ldots, \dfrac{\partial^{n_0} y}{\partial z^{n_0}}\right) - \dfrac{1}{2}\bar{B}(z) R^{-1} B^T [\nabla\Psi]^T \hat{\theta}, \\ \dot{\hat{\theta}} = -\alpha\mu\rho(x)\delta_3(x) + \alpha\kappa\nabla\Psi(x) D \nabla\bar{V}(x). \end{cases} \tag{15.27}$$

Using the model reduction procedure presented in Section 15.3, the system (15.27) is equivalently rewritten as

$$\begin{cases} \dot{x} = f_s(x,0) - \dfrac{1}{2} B R^{-1} B^T [\nabla\Psi]^T \hat{\theta} + f_s(x, x_f) - f_s(x,0), \\ \varepsilon\dot{x}_f = A_{f\varepsilon} x_f + \varepsilon f_f(x, x_f) - \dfrac{1}{2}\varepsilon B_f R^{-1} B^T [\nabla\Psi]^T \hat{\theta}, \\ \dot{\hat{\theta}} = -\alpha\mu\rho\delta_3 + \alpha\kappa\nabla\Psi D \nabla\bar{V}. \end{cases} \tag{15.28}$$

Since $\mathcal{L}$ is a sufficiently smooth nonlinear vector function, it is obvious that there exists a constant $k_3 > 0$ such that

$$\|f_s(x, x_f) - f_s(x, 0)\| \leqslant k_3 \|x_f\|_{\mathbb{R}^\infty} \tag{15.29}$$

for $\|x\| \leqslant \sigma_1$.

According to the exponential stability property of the $x_f$-subsystem (15.11) and the converse Lyapunov theorem [59], we know that there exist a Lyapunov function candidate $V_f(x_f)$, and positive real numbers $l_1, l_2, l_3$ and $l_4$ such that the following conditions hold:

$$\begin{cases} l_1 \|x_f\|^2_{\mathbb{R}^\infty} \leqslant V_f(x_f) \leqslant l_2 \|x_f\|^2_{\mathbb{R}^\infty}, \\ \dot{V}_f(x_f) \leqslant \dfrac{1}{\varepsilon} \nabla V_f^T A_{f\varepsilon} x_f \leqslant -\dfrac{l_3}{\varepsilon} \|x_f\|^2_{\mathbb{R}^\infty}, \\ \|\nabla V_f\|_{\mathbb{R}^\infty} \leqslant l_4 \|x_f\|_{\mathbb{R}^\infty}, \end{cases} \tag{15.30}$$

where $\nabla V_f \triangleq \partial V_f / \partial x_f$.

Before proving the stability of the closed-loop PDE system, we derive some useful conditions as follows:

$$\|\nabla V^*\| = \|(\nabla \Psi)^T \theta^* + \nabla \delta_1\| \leqslant \upsilon_{DVM}, \tag{15.31}$$

$$\|u^*\| = \left\| -\frac{1}{2} R^{-1} B^T \nabla V^* \right\| \leqslant \upsilon_{uM}, \tag{15.32}$$

$$\|f + Bu^*\| \leqslant \upsilon_{fBM}, \tag{15.33}$$

$$\upsilon_{Dm} \leqslant \|D\| = \|BR^{-1}B^T\| \leqslant \upsilon_{DM}, \tag{15.34}$$

where $\upsilon_{Rm} \triangleq \underline{\sigma}(R)$, $\upsilon_{RM} \triangleq \overline{\sigma}(R)$, $\upsilon_{Bm} \triangleq \underline{\sigma}(B)$, $\upsilon_{BM} \triangleq \overline{\sigma}(B)$, $\upsilon_{DVM} \triangleq \psi_{DM}\theta_M + \delta_{1DM}$, $\upsilon_{uM} \triangleq \frac{1}{2}\upsilon_{Rm}^{-1}\upsilon_{RM}\upsilon_{DVM}$, $\upsilon_{fBM} \triangleq \upsilon_{fM} + \upsilon_{BM}\upsilon_{uM}$, $\upsilon_{Dm} \triangleq \upsilon_{RM}^{-1}\upsilon_{Bm}^2$ and $\upsilon_{DM} \triangleq \upsilon_{Rm}^{-1}\upsilon_{BM}^2$. Moreover, considering the fact that $\int_{z_1}^{z_2} \overline{B}^T(z)\overline{B}(z)dz = B^T B + B_f^T B_f$, we have

$$\|B_f^T \nabla V_f\| = ((\nabla V_f)^T B_f B_f^T \nabla V_f)^{\frac{1}{2}} \leqslant l_4 l_5 \|x_f\|_{\mathbb{R}^\infty}, \tag{15.35}$$

where $l_5 \triangleq \left(\overline{\sigma}\left(\int_{z_1}^{z_2} \overline{B}^T(z)\overline{B}(z)dz - B^T B\right)\right)^{\frac{1}{2}}$.

Let us consider the following Lyapunov function candidate:

$$L(t) = V^*(x) + V_f(x_f) + \bar{V}(x) + V_o(\tilde{\theta}), \tag{15.36}$$

where $V^*(x)$ is the solution of the HJBE (15.15) and $V_o(\tilde{\theta}) = \frac{1}{2\alpha}\tilde{\theta}^T\tilde{\theta}$. Differentiating $V^*(x)$, $V_f(x_f)$, $\bar{V}(x)$ and $V_o(\tilde{\theta})$ with respect to time along the trajectory of the system (15.28) yields

$$\begin{aligned} \dot{V}^*(x) =& (\nabla V^*)^T \left[f_s(x, 0) + B\hat{u} + f_s(x, x_f) - f_s(x, 0)\right] \\ =& -\|x\|^2_{Q_s} - -\|u^*\|^2_R + (u^*)^T B^T (\nabla \Psi)^T \tilde{\theta} - (u^*)^T B^T \nabla \delta_1 + (\nabla V^*)^T [f_s(x, x_f) - f_s(x, 0)] \\ \leqslant& -\varpi_1 \|x\|^2 + \varpi_2 \|x_f\|_{\mathbb{R}^\infty} + \varpi_3 \|\tilde{\theta}\| + \varpi_4, \end{aligned} \tag{15.37}$$

where $\varpi_1 \triangleq \upsilon_{Qm}$, $\varpi_2 \triangleq k_3\upsilon_{DVM}$, $\varpi_3 \triangleq \upsilon_{uM}\upsilon_{BM}\upsilon_{DM}$ and $\varpi_4 \triangleq \upsilon_{uM}\upsilon_{BM}\delta_{1DM}$ with $\upsilon_{Qm} \triangleq \underline{\sigma}(Q_s)$ and $\upsilon_{QM} \triangleq \bar{\sigma}(Q_s)$. We have

$$
\begin{aligned}
\dot{V}_f(x_f) =& \frac{1}{\varepsilon}(\nabla V_f)^T A_{f\varepsilon} x_f + (\nabla V_f)^T f_f(x, x_f) + (\nabla V_f)^T B_f u^* - \frac{1}{2}(\nabla V_f)^T B_f R^{-1} B^T (\nabla \Psi)^T \tilde{\theta} \\
&+ \frac{1}{2}(\nabla V_f)^T B_f R^{-1} B^T \nabla \delta_1 \\
\leqslant& (-\frac{l_3}{\varepsilon} + \varpi_5)\|x_f\|^2_{\mathbb{R}^\infty} + \varpi_6 \|x\| \|x_f\|_{\mathbb{R}^\infty} + \varpi_7 \|x_f\|_{\mathbb{R}^\infty} \|\tilde{\theta}\| + \varpi_8 \|x_f\|_{\mathbb{R}^\infty},
\end{aligned}
\tag{15.38}
$$

where $\varpi_5 \triangleq l_4 k_2$, $\varpi_6 \triangleq l_4 k_1$, $\varpi_7 \triangleq \frac{1}{2} l_4 l_5 \upsilon_{Rm}^{-1} \upsilon_{BM} \psi_{DM}$ and $\varpi_8 \triangleq l_4 l_5 (\upsilon_{uM} + \frac{1}{2}\upsilon_{Rm}^{-1}\upsilon_{BM}\delta_{1DM})$,

$$
\dot{\bar{V}}(x) = (\nabla \bar{V})^T (f + B\hat{u}) = (\nabla \bar{V})^T f - \frac{1}{2}(\nabla \bar{V})^T D (\nabla \Psi)^T (\tilde{\theta} + \theta^*),
\tag{15.39}
$$

$$
\begin{aligned}
\dot{V}_o(\tilde{\theta}) =& -\mu \tilde{\theta}^T \rho \delta_3 + \kappa \tilde{\theta}^T \nabla \Psi D \nabla \bar{V} \\
=& \Big[ -\mu \tilde{\theta}^T \nabla \Psi (f + Bu^*) + \frac{1}{2}\mu \|\tilde{\theta}^T \nabla \Psi\|^2_D - \frac{1}{2}\mu \tilde{\theta}^T \nabla \Psi D \nabla \delta_1 \Big] \Big[ (\tilde{\theta}^T \nabla \Psi - (\nabla \delta_1)^T)(f + Bu^*) \\
&- \frac{1}{4}\|\tilde{\theta}^T \nabla \Psi\|^2_D + \frac{1}{4}\|\nabla \delta_1\|^2_D \Big] + \kappa \tilde{\theta}^T \nabla \Psi D \nabla \bar{V} \\
=& -\varpi_9 \|\tilde{\theta}\|^4 + \varpi_{10}\|\tilde{\theta}\|^3 + \varpi_{11}\|\tilde{\theta}\|^2 + \varpi_{12}\|\tilde{\theta}\| + \kappa \tilde{\theta}^T \nabla \Psi D \nabla \bar{V},
\end{aligned}
\tag{15.40}
$$

where $\varpi_9 \triangleq \frac{1}{8}\mu \psi_{Dm}^2 \upsilon_{Dm}^2$, $\varpi_{10} \triangleq \frac{3}{4}\mu \psi_{DM}^3 \upsilon_{DM} \upsilon_{fBM} + \frac{1}{8}\mu \psi_{DM}^3 \delta_{1DM} \upsilon_{DM}^2$, $\varpi_{11} \triangleq \frac{1}{2}\mu \psi_{DM}^2 \delta_{1DM} \upsilon_{DM} \upsilon_{fBM}$ and $\varpi_{12} \triangleq \mu \psi_{DM} \delta_{1DM} \upsilon_{fBM}^2 + \frac{3}{4}\mu \psi_{DM} \delta_{1DM}^2 \upsilon_{DM} \upsilon_{fBM} + \frac{1}{2}\mu \psi_{DM}^2 \delta_{1DM} \upsilon_{DM} \upsilon_{fBM} + \frac{1}{8}\mu \psi_{DM} \delta_{1DM}^3 \upsilon_{DM}^2$.

Moreover, it is easy to verify that with the choice of $\kappa$ in (15.25), the following condition holds:

$$
\dot{\bar{V}}(x) + \kappa \tilde{\theta}^T \nabla \Psi D \nabla \bar{V} \leqslant 0.
\tag{15.41}
$$

This is because, according to the definition (15.25) of $\kappa$ and using (15.39) and (15.26), we have when $\kappa = 0$, $\dot{\bar{V}}(x) + \kappa \tilde{\theta}^T \nabla \Psi D \nabla \bar{V} = \tilde{\theta}^T (f + B\hat{u}) \leqslant 0$ and when $\kappa = 0.5$,

$$
\begin{aligned}
\dot{\bar{V}}(x) + \kappa \tilde{\theta}^T \nabla \Psi D \nabla \bar{V} =& \tilde{\theta}^T (f + B\hat{u}) + 0.5 \tilde{\theta}^T \nabla \Psi D \nabla \bar{V} \\
=& (\nabla \bar{V})^T f - 0.5 (\nabla \bar{V})^T D (\nabla \Psi)^T \theta^* \leqslant 0.
\end{aligned}
$$

Thus, using (15.37)–(15.41), the time derivative of $L(t)$ in (15.36) satisfies

$$
\begin{aligned}
\dot{L}(t) =& \dot{V}^*(x) + \dot{V}_f(x_f) + \dot{\bar{V}}(x) + \dot{V}_o(\tilde{\theta}) \\
\leqslant& L_\theta(\|\tilde{\theta}\|) - v_{x\theta}^T \Pi(\varepsilon) v_{x\theta} + b_{\varpi}^T v_{x\theta} + \dot{\bar{V}}(x) + \kappa \tilde{\theta}^T \nabla \Psi D \nabla \bar{V},
\end{aligned}
\tag{15.42}
$$

where $\nu_{x\theta} \triangleq [\|x\| \;\; \|x_f\|_{\mathbb{R}^\infty} \;\; \|\tilde{\theta}\|]^T$, $L_\theta(\|\tilde{\theta}\|) \triangleq -\varpi_9\|\tilde{\theta}\|^4 + \varpi_{10}\|\tilde{\theta}\|^3 + (\varpi_{11}+\varpi_{13})\|\tilde{\theta}\|^2 + (\varpi_3 + \varpi_{12})\|\tilde{\theta}\| + \varpi_4$, $\varpi_13 > 0$, $b_\varpi \triangleq [0 \;\; \varpi_2+\varpi_8 \;\; 0]^T$ and

$$\Pi(\varepsilon) \triangleq \begin{bmatrix} \varpi_1 & -\varpi_6/2 & 0 \\ -\varpi_6/2 & l_3/\varepsilon - \varpi_5 & -\varpi_7/2 \\ 0 & -\varpi_7/2 & \varpi_{13} \end{bmatrix}.$$

It is noted that function $L_\theta(\|\tilde{\theta}\|)$ can be rewritten as

$$L_\theta(\|\tilde{\theta}\|) = \alpha_1(\|\tilde{\theta}\|) + \alpha_2(\|\tilde{\theta}\|) + \alpha_3(\|\tilde{\theta}\|) + \alpha_4(\|\tilde{\theta}\|), \tag{15.43}$$

where $\alpha_1(\|\tilde{\theta}\|) \triangleq -\|\tilde{\theta}\|^3(\frac{1}{4}\varpi_9\|\tilde{\theta}\| - \varpi_{10})$, $\alpha_2(\|\tilde{\theta}\|) \triangleq -\|\tilde{\theta}\|^2\,(\frac{1}{4}\varpi_9\|\tilde{\theta}\|^2 - (\varpi_{11}+\varpi_{13}))$, $\alpha_3(\|\tilde{\theta}\|) \triangleq -\|\tilde{\theta}\|(\frac{1}{4}\varpi_9\|\tilde{\theta}\|^3 - (\varpi_3+\varpi_{12}))$ and $\alpha_4(\|\tilde{\theta}\|) \triangleq -(\frac{1}{4}\varpi_9\|\tilde{\theta}\|^4 - \varpi_4)$.

Letting $\alpha_1(\|\tilde{\theta}\|) \leqslant 0$, $\alpha_2(\|\tilde{\theta}\| \leqslant 0$, $\alpha_3(\|\tilde{\theta}\| \leqslant 0$ and $\alpha_4(\|\tilde{\theta}\| \leqslant 0$, we get respectively $\|\tilde{\theta}\| \geqslant \vartheta_1$ with $\vartheta_1 \triangleq \sqrt{4\varpi_{10}/\varpi_9}$, $\|\tilde{\theta}\|) \geqslant \vartheta_2$ with $\vartheta_2 \triangleq \sqrt{4(\varpi_{11}+\varpi_{13}))/\varpi_9}$, $\|\tilde{\theta}\|) \geqslant \vartheta_3$ with $\vartheta_3 \triangleq \sqrt[3]{4(\varpi_3+\varpi_{12}))/\varpi_9}$ and $\|\tilde{\theta}\|) \geqslant \vartheta_4$ with $\vartheta_4 \triangleq \sqrt[4]{4\varpi_4/\varpi_9}$. Defining $\vartheta_{\max} \triangleq \max\{\vartheta_1, \vartheta_2, \vartheta_3, \vartheta_4\}$, it is clear from (15.43) that if $\|\tilde{\theta}\| \geqslant \vartheta_{\max}$, $L_\theta(\|\tilde{\theta}\|) \leqslant 0$. Letting $\varepsilon^* \triangleq \min\{\varepsilon_1, \varepsilon_2\}$, where $\varepsilon_1 \triangleq 4\varpi_1 l_3/(4\varpi_1\varpi_5 + \varpi_6^2)$ and $\varepsilon_2 \triangleq 4\varpi_1\varpi_{13}l_3/(\varpi_7^2\varpi_1 + \varpi_6^2\varpi_{13} + 4\varpi_1\varpi_5\varpi_{13})$, we have, if $\varepsilon \in (0, \varepsilon^*)$, $\Pi(\varepsilon) > 0$. Thus, we get $-\nu_{x\theta}^T\Pi(\varepsilon)\nu_{x\theta} + b_\varpi^T\nu_{x\theta} \leqslant -\varpi_{14}\|\nu_{x\theta}\|^2 + (\varpi_2+\varpi_8)\|\nu_{x\theta} \leqslant 0$ with $\varpi_{14} \triangleq \underline{\sigma}(\Pi)$. If $\|\nu_{x\theta}\| \geqslant ((\varpi_2+\varpi_8)/\varpi_{14}$, with these inequalities, it follows from (15.42) that $\dot{L}(t) \leqslant 0$. That is to say, for the closed-loop system (15.28), we have $\|\nu_{x\theta}\| < ((\varpi_2+\varpi_8)/\varpi_{14}$. This means that $x, x_f$ and $\tilde{\theta}$ are SGUUB. Due to the fact that $\|y(\cdot,t)\|_2^2 = \|x(t)\|^2 + \|x_f(t)\|_{\mathbb{R}^\infty}^2$, the state $y(z,t)$ of the original closed-loop PDE system (15.1)–(15.3) is SGUUB in $L_2$ norm. This completes the proof of part (1) of Theorem 1.

(2) It follows from part (1) of Theorem 1 that $\tilde{\theta}$ is SGUUB, thus there exists a $\upsilon_{\tilde{\theta}M} > 0$ such that $\|\tilde{\theta}\| \leqslant \upsilon_{\tilde{\theta}M}$. Considering $\tilde{V}(x) = \tilde{\theta}^T\Psi(x) - \delta_1$ and $\tilde{u}(x) = -\frac{1}{2}R^{-1}B^T(\nabla\Psi)^T\tilde{\theta} + \frac{1}{2}R^{-1}B^T\nabla\delta_1$, we have $\|\tilde{V}\| \leqslant \|\tilde{\theta}^T\Psi\| + \|\delta_1\| \leqslant \upsilon_{\tilde{\theta}M}\psi_M + \delta_{1M}$ and $\|\tilde{u}\| \leqslant \frac{1}{2}\|R^{-1}B^T(\nabla\Psi)^T\tilde{\theta}\| + \frac{1}{2}\|R^{-1}B^T\nabla\delta_1\| \leqslant \frac{1}{2}\upsilon_{Rm}^{-1}\upsilon_{BM}(\psi_{DM}\upsilon_{\tilde{\theta}M} + \delta_{1DM})$. This implies that $\tilde{V}$ and $\tilde{u}$ are SGUUB. □

### 15.4.3 SIMULATION STUDIES

Consider the following nonlinear diffusion–convection–reaction process [60,61,4]:

$$\begin{cases} \dfrac{\partial y(z,t)}{\partial t} &= \dfrac{\partial}{\partial z}\left(k(y)\dfrac{\partial y(z,t)}{\partial z}\right) + \beta_T(z)\left(e^{-\rho/(1+y)} - e^{-\rho}\right) + \beta_U(b(z)u(t) - y), \\ y_h(t) &= \displaystyle\int_{z_1}^{z_2} y(z,t)\mathrm{d}z, \end{cases} \tag{15.44}$$

subject to the Dirichlet boundary conditions

$$y(0,t) = y(\pi,t) = 0 \tag{15.45}$$

and the initial condition

$$y_0(t) = 0.3\sin(z), \tag{15.46}$$

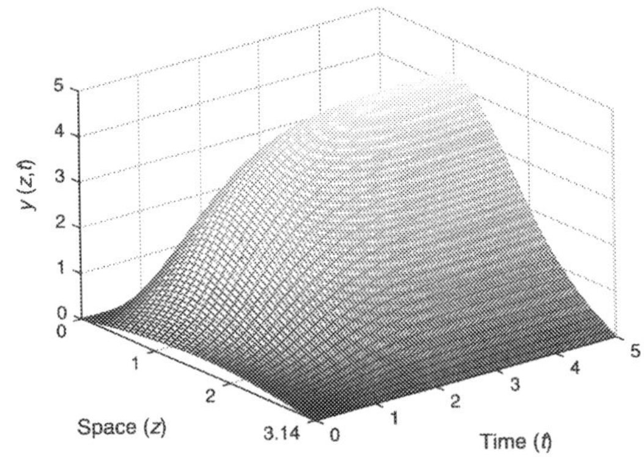


**FIGURE 15.1**

State profile $y(z,t)$ of open-loop PDE system.

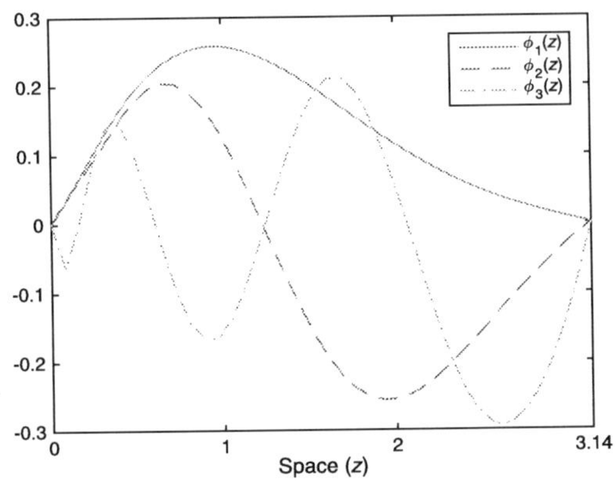


**FIGURE 15.2**

First three EEFs.

where $y(z,t)$ is the state, $z \in [0, \pi]$, $k(y)$ is the diffusion coefficient that may be constant or dependent on the state, $\beta_T(z)$ is the heat of reaction which is spatially varying, $\beta_U$ is the heat transfer coefficient, $\gamma$ is activation energy and $b(z)$ is the actuator distribution function. These parameters are given as $k(y) = 0.5 + 0.7/(1+y)$, $\beta_T(z) = 16(\cos(z)+1)$, $\beta_U = 2$, $\gamma = 2$, $b(z) = H(z-0.4\pi) - H(z-0.5\pi)$, where $H(\cdot)$ is the standard Heaviside function. The state profile of the open PDE system is given in Fig. 15.1, which shows that the steady-state profile $y(z) = 0$ is an unstable one. The weighting matrix $R$ in the performance index (15.4) is given as $R = 1$.

By using KLD to compute the EEFs, the first three EEFs are shown in Fig. 15.2. The first three EEFs are employed to compute the state of the slow subsystem. To use the developed NDP-based adaptive optimal control method, we select a critic NN activation function vector $\Psi(x)$ of size 18 (*i.e.*, $L = 18$) as

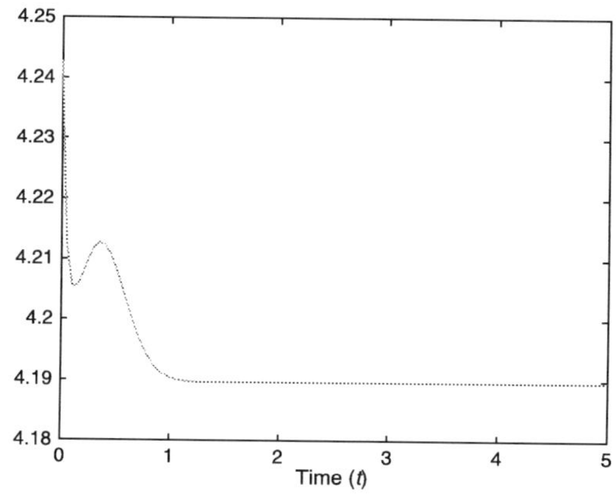


**FIGURE 15.3**

The trajectory of the critic NN weight norm $\|\hat{\theta}(t)\|$ with the NDP-based adaptive optimal control.

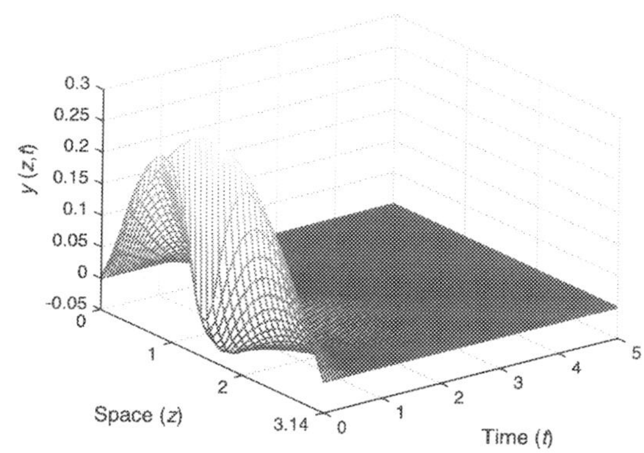


**FIGURE 15.4**

The state profile $y(z,t)$ of closed-loop PDE system with the NDP-based adaptive optimal control.

$$\Psi(x) = [x_1^2\ x_1x_2\ x_1x_3\ x_2^2\ x_2x_3\ x_3^2\ x_1^2x_2^2\ x_1^2x_3^2\ x_2^2x_3^2\ x_1x_2^3\ x_1x_3^3\ x_2x_3^3\ x_1^3x_2\ x_1^3x_3$$
$$x_2^3x_3\ x_1^4\ x_2^4\ x_3^4]^{\mathrm{T}}. \tag{15.47}$$

Each element of initial NN weight vector $\hat{\theta}(0)$ is given as 1. By using the adaptive optimal control policy (15.22) and the NN weights tuning law (15.24) with $\bar{V} = 0.5x^{\mathrm{T}}x$ and $\alpha = 10$, closed-loop simulations are conducted. Simulation results are shown in Figs. 15.3–15.6. Fig. 15.3 gives the norm of the critic NN weight vector $\|\hat{\theta}(t)\|$ and Fig. 15.4 shows the state profile $y(z,t)$ of the closed-loop PDE system. The state trajectories $x(t)$ of the closed-loop ODE system and the control $u(t)$ are demonstrated in Figs. 15.5 and 15.6, respectively.

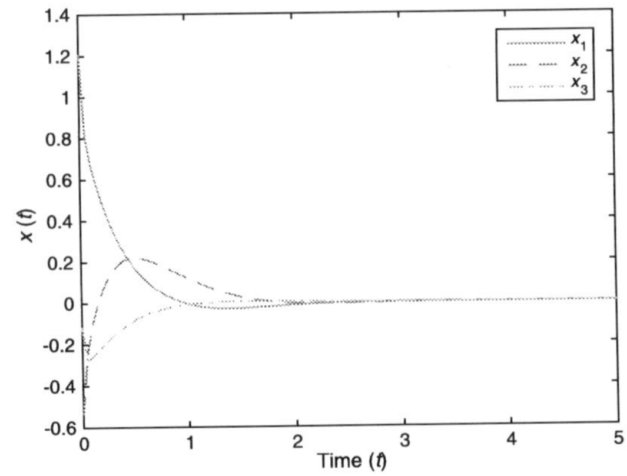


**FIGURE 15.5**

The trajectories $x(t)$ of the closed-loop ODE system with the NDP-based adaptive optimal control.

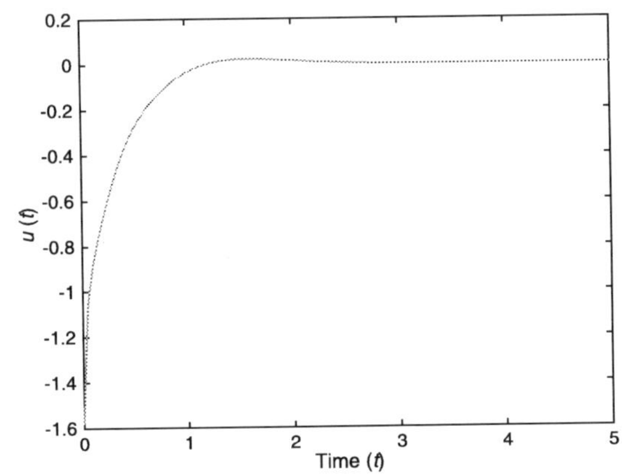


**FIGURE 15.6**

The trajectory of the NDP-based adaptive optimal control $u(t)$.

## 15.5 ADAPTIVE OPTIMAL CONTROL BASED ON POLICY ITERATION FOR PARTIALLY UNKNOWN DPSS

In this section, the optimal control problem of partially unknown DPS (15.1)–(15.3) is considered, *i.e.*, the internal dynamic model $\mathcal{L}$ is unknown and the model of $\bar{B}$ is known. It follows from Section 15.3 that the internal dynamic model $f(x)$ in the reduced-order model (15.12) is also unknown. This implies that explicit formulation of the HJBE (15.15) is unavailable, which further complicates the optimal control design problem. Thus, an adaptive optimal control based on PI is developed by using the reduced-order model (15.12) and the performance index (15.14) without requiring the internal system model.

### 15.5.1 POLICY ITERATION

Before starting, we introduce the following definition of admissible control.

**Definition 2.** (Admissible control) For the given system (15.12), $x \in \Omega \subset \mathbb{R}^N$, a control $u(x) : \Omega \to \mathbb{R}^p$ is defined to be admissible with respect to (15.14) on $\Omega$, denoted by $u(x) \in \mathfrak{U}(\Omega)$, if (1) $u$ is continuous on $\Omega$, (2) $u(0) = 0$, (3) $u(x)$ stabilizes the system and (4) $V_u(x) < \infty$, $\forall x \in \Omega$.

For convenience, define the notation $\mathcal{R}(x, u) \triangleq \|x\|^2_{Q_s} + \|u\|^2_R$. To solve the HJBE (15.15) without the internal system dynamics, the following PI is introduced.

**Algorithm 5.1** Policy Iteration.

- *Step 1:* Give an initial control policy $u^{(0)} \in \mathfrak{U}(\Omega)$.
- *Step 2:* (**Policy evaluation**) By using the control policy $u^{(i)}$, solve

$$V^{(i+1)}(x(t)) = \int_t^{t+\Delta t} \mathcal{R}(x(\tau), u^{(i)}(\tau))\mathrm{d}\tau + V^{(i+1)}(x(t + \Delta t)), \tag{15.48}$$

  where $V^{(i+1)}(x)$ is the unknown value function.
- *Step 3:* (**Policy improvement**) Update the control policy by

$$u^{(i+1)}(x) = -\frac{1}{2} R^{-1} B^{\mathsf{T}} \nabla V^{(i+1)}(x). \tag{15.49}$$

- *Step 4:* Set $i = i + 1$. Go back to Step 2 and continue. □

The following lemmas are given to demonstrate the convergence of the PI algorithm and its relationship with the HJBE (15.15).

**Lemma 1.** *[10] Consider the system* (15.12)*. Let $u^{(0)}(x) \in \mathfrak{U}(\Omega)$. The sequences $\{V^{(i)}(x)\}$ and $\{u^{(i)}(x)\}$ are generated by Algorithm 5.1. Then (1) $V^{(i+1)}(x)$ is the solution of Eq.* (15.48) *if and only if it is a solution of the Lyapunov function equation*

$$(\nabla V^{(i+1)})^{\mathsf{T}}(f + Bu^{(i+1)}) + \mathcal{R}(x, u^{(i)}) = 0 \tag{15.50}$$

*and (2) $u^{(i)} \in \mathfrak{U}(\Omega)$.*

**Lemma 2.** *[10] Consider the system* (15.12)*. For any $u(x) \in \mathfrak{U}(\Omega)$, the value function $V(x)$ associated with $u(x)$ satisfies the Lyapunov function equation*

$$(\nabla V)^{\mathsf{T}}(f + Bu) + \mathcal{R}(x, u) = 0. \tag{15.51}$$

From Lemmas 1 and 2, it follows that the solution $V^{(i+1)}(x)$ obtained by Eq. (15.48) in the PI algorithm is the value function associated with the control policy $u^{(i)}(x)$.

**Lemma 3.** *[10] Consider the system* (15.12)*. Given* $u^{(0)}(x) \in \mathfrak{U}(\Omega)$*, let* $V^*(x)$ *be the solution of the HJBE* (15.15) *and* $u^*(x)$ *given by* (15.16)*. The sequences* $\{V^{(i)}(x)\}$ *and* $\{u^{(i)}(x)\}$ *are generated by Algorithm 5.1. Then (1)* $V^{(i)}(x) \geqslant V^{(i+1)}(x)$ *and (2)* $\lim_{i\to\infty} V^{(i)}(x) = V^*(x)$ *and* $\lim_{i\to\infty} u^{(i)}(x) = u^*(x)$.

Lemma 1 shows that Eq. (15.48) in the PI algorithm is equivalent to the Lyapunov function equation (15.50). Moreover, Lemma 3 indicates that the solution of the Lyapunov function equation (15.50) converges the solution of the HJBE (15.15) when $i \to \infty$. Thus, it is concluded that the PI algorithm converges to the solution of the HJBE (15.15).

### 15.5.2 ADAPTIVE OPTIMAL CONTROL BASED ON POLICY ITERATION

From Lemmas 1 and 2, we prove that $V^{(i+1)}(x)$ of Eq. (15.48) in the PI algorithm is the value function of the control policy $u^{(i)}(x)$. Thus, the problem of solving Eq. (15.48) can be generally formulated as follows. Solve the equation

$$V(x(t)) = \int_t^{t+\Delta t} \mathcal{R}(x(\tau), u(\tau))d\tau + V(x(t+\Delta t)) \tag{15.52}$$

for the value function $V(x)$ of the control policy $u(x) \in \mathfrak{U}(\Omega)$. Similar to (15.17), $V(x)$ can be expressed with $\Upsilon(x)$ as follows:

$$V(x) = w^{\mathsf{T}}\Upsilon(x) + \varepsilon_1, \tag{15.53}$$

where $\Upsilon(x) = [\gamma_1(x)\ \gamma_2(x)\ \cdots\ \gamma_L(x)]^{\mathsf{T}}$ represents the activation functions of the critic NN, $w = [w_1\ w_2\ \cdots\ w_L]^{\mathsf{T}}$ is the unknown ideal constant weight vector and $\varepsilon_1$ is the NN approximation error. The substitution of (15.53) into (15.52) yields

$$\int_t^{t+\Delta t} \mathcal{R}(x(\tau), u(\tau))d\tau + w^{\mathsf{T}}\Delta\Upsilon(t) = \varepsilon_2, \tag{15.54}$$

where $\varepsilon_2 \triangleq \varepsilon_1(x(t)) - \varepsilon_1(x(t+\Delta t))$ and $\Delta\Upsilon(t) \triangleq \Upsilon(x(t+\Delta t)) - \Upsilon(x(t))$. Since $w$ is usually unknown, the output of the critic NN is

$$\hat{V}(x) = \hat{w}^{\mathsf{T}}\Upsilon(x), \tag{15.55}$$

where $\hat{w} = [\hat{w}_1\ \hat{w}_2\ \cdots\ \hat{w}_L]^{\mathsf{T}}$ is the estimated vector of the ideal weight vector $w$.

With (15.55), the Eq. (15.52) is rewritten as

$$\int_t^{t+\Delta t} \mathcal{R}(x(\tau), u(\tau))d\tau + \hat{w}^{\mathsf{T}}\Delta\Upsilon(t) = e_1. \tag{15.56}$$

It follows from (15.54) and (15.56) that when $\hat{w} \to w$, $e_1 \to \varepsilon_2$. Thus, by choosing the objective function as the squared residual error,

$$E_1(\hat{w}) = \frac{1}{2}e_1^{\mathsf{T}}e_1. \tag{15.57}$$

It is desired to select $\hat{w}$ such that $E_1(\hat{w})$ is minimized. Select the tuning law of the weight vector for the critic NN as a gradient descent scheme, which is given by

$$\dot{\hat{w}} = -\alpha_1 \frac{\partial E_1}{\partial \hat{w}} = -\alpha_1 \left( \int_t^{t+\Delta t} \mathcal{R}(x(\tau), u(\tau)) d\tau + \hat{w}^{\mathsf{T}} \Delta \Upsilon(t) \right) \Delta \Upsilon(t), \tag{15.58}$$

where $\alpha_1 > 0$ is the adaptive gain of the critic NN. Defining the critic NN weight estimation error as $\tilde{w} \triangleq \hat{w} - w$. Then it follows from (15.54) and (15.58) that

$$\begin{aligned} \dot{\tilde{w}} &= -\alpha_1 \left( \int_t^{t+\Delta t} \mathcal{R}(x(\tau), u(\tau)) d\tau + (\tilde{w} + w)^{\mathsf{T}} \Delta \Upsilon(t) \right) \Delta \Upsilon(t) \\ &= -\alpha_1 (\tilde{w}^{\mathsf{T}} \Delta \Upsilon(t) + \varepsilon_2) \Delta \Upsilon(t). \end{aligned} \tag{15.59}$$

After the critic NN weight is convergent, it follows from (15.55) and (15.49) that the PI-based adaptive optimal control is given by

$$\hat{u}(x) = -\frac{1}{2} R^{-1} B^{\mathsf{T}} \nabla \Upsilon^{\mathsf{T}}(x) \hat{w}, \tag{15.60}$$

where the parameters $\hat{w}$ are updated with (15.58).

### 15.5.3 IMPLEMENTATION PROCEDURE

In the Subsection 15.5.2, the adaptive method is designed for solving one iterative equation (15.48) in the PI algorithm. Based on (15.59) and (15.60), the following complete adaptive optimal control design procedure is given for the original PDE system (15.1)–(15.3).

---

**Algorithm 5.2** Adaptive Optimal Control Based on PI.

---

- *Step 1:* Use KLD (Algorithm 3.1) to compute EEFs.
- *Step 2:* Select $\Upsilon(x)$ for citric NN and its initial $w^{(0)}$ such that $\hat{u}^{(0)} \in \mathfrak{U}(\Omega)$. Let $i = 0$.
- *Step 3:* Apply the following PI-based adaptive optimal control $\hat{u}^{(i)}$ to the real DPS:

$$\hat{u}(x) = \hat{u}^{(i)}(x) = -\frac{1}{2} R^{-1} B^{\mathsf{T}} \nabla \Upsilon^{\mathsf{T}}(x) w^{(i)}. \tag{15.61}$$

  Tune the critic NN weight vector $\hat{w}$ with

$$\dot{\hat{w}} = -\alpha_1 \left( \int_t^{t+\Delta t} \mathcal{R}(x(\tau), \hat{u}^{(i)}(\tau)) d\tau + \hat{w}^{\mathsf{T}} \Delta \Upsilon(t) \right) \Delta \Upsilon(t) \tag{15.62}$$

  until its convergence. Then take $w^{(i+1)}$ as the convergent $\hat{w}$.
- *Step 4:* If $\|w^{(i+1)} - w^{(i)}\| \leqslant \epsilon$ for a small $\epsilon > 0$, stop tuning $w^{(i)}$ and keep it invariable from that instant on; else, $i = i + 1$, go back to Step 3 and continue.

---

### 15.5.4 STABILITY ANALYSIS

Similar to Assumption 1, Assumption 2 is required for the stability analysis of the PI-based adaptive optimal control.

**Assumption 2.**

- For $\forall u \in \mathfrak{U}(\Omega)$, the residual error $\varepsilon_2$ in (15.54) is bounded on a compact set containing $\Omega$ such that $|\varepsilon_2| \leqslant \varepsilon_{2M}$ and the ideal weight vector $w$ is bound, *i.e.*, $\|w\| \leqslant \upsilon_{wM}$, where $\varepsilon_{2M}$ and $\upsilon_{wM}$ are positive numbers.
- The critic NN activation functions $\Upsilon$ and their gradients $\nabla\Upsilon$ are bounded on a compact set containing $\Omega$, *i.e.*, $\|\Upsilon\| \leqslant \upsilon_{\gamma M}$ and $\|\nabla\Upsilon\| \leqslant \upsilon_{D\gamma M}$, where $\upsilon_{\gamma M}$ and $\upsilon_{D\gamma M}$ are positive numbers.
- There exist $\eta_1, \eta_2, T_0 > 0$ such that $\eta_1 I \leqslant \int_t^{t+T_0} \Delta\Upsilon(\tau)\Delta\Upsilon^T(\tau)d\tau \leqslant \eta_2 I$.

The following theorem shows that the critic NN weight vector $\hat{w}$ converges to the ideal unknown weight vector $w$ in (15.53) by using the NN weight tuning rule (15.58). This means that $\hat{V}(x)$ given in (15.55) converges to the actual value function $V(x)$ of the control policy $u$.

**Theorem 2.** *Consider the system* (15.12) *and the performance index* (15.14). *Let Assumption 2 hold; the tuning law for $\hat{w}$ is given by* (15.58). *Then the critic NN weight estimation error $\tilde{w}$ is SGUUB.*

*Proof.* Select the following Lyapunov function candidate:

$$L(t) = \frac{1}{2\alpha_1}\tilde{w}^T\tilde{w}. \tag{15.63}$$

By taking the derivative along the solution of (15.59), we have $\dot{L}(t) = \alpha_1^{-1}\tilde{w}^T\dot{\tilde{w}} = -\tilde{w}^T(\tilde{w}^T\Delta\Upsilon + \varepsilon_2)\Delta\Upsilon = -|\tilde{w}^T\Delta\Upsilon|(|\tilde{w}^T\Delta\Upsilon| - |\varepsilon_2|)$. Thus, $\dot{L}(t) \leqslant 0$ if $|\tilde{w}^T\Delta\Upsilon| \geqslant |\varepsilon_{2M}|$, which means that it is SGUUB, *i.e.*, there exists a $\mu_1 > 0$ such that $|\tilde{w}^T\Delta\Upsilon| < \mu_1$. According to Assumption 2 on $\Delta\Upsilon$ and the proof of the technical Lemma 2(b) in [62],

$$\begin{aligned}
\|\tilde{w}\| &< \frac{\sqrt{\eta_2 T_0}}{\eta_1}\mu_1 + \frac{\eta_2}{\eta_1}\int_t^{t+T_0} \|\alpha_1\Delta\Upsilon\| \cdot \|\tilde{w}^T\Delta\Upsilon + \varepsilon_2\| d\tau \\
&< \frac{\sqrt{\eta_2 T_0}}{\eta_1}\mu_1 + \frac{\alpha_1\eta_2}{\eta_1}(\mu_1 + \varepsilon_{2M})\left(\int_t^{t+T_0} \|\Delta\Upsilon(\tau)\|^2 d\tau\right)^{\frac{1}{2}}\left(\int_t^{t+T_0} d\tau\right)^{\frac{1}{2}} \\
&= \frac{\sqrt{\eta_2 T_0}}{\eta_1}(\mu_1 + \alpha_1\eta_2(\mu_1 + \varepsilon_{2M})).
\end{aligned}$$

This means that $\tilde{w}$ is SGUUB. □

Note that Theorem 2 gives the convergence of the critic NN weights. However, this method is just used for solving one iterative equation (15.48) in the PI algorithm. Thus, we also need to show that the tuning law (15.58) for a critic NN weight vector can guarantee the convergence of the PI algorithm, the stabilities of the closed-loop reduced-order ODE system (15.12) and the original PDE system (15.1)–(15.3).

With the NN activation function vector $\Upsilon(x)$, the solution of the HJBE (15.15) is represented as

$$V^*(x) = (w^*)^{\mathsf{T}}\Upsilon(x) + \varepsilon_3, \tag{15.64}$$

where $w^* = [w_1^* \; w_2^* \; \cdots \; w_L^*]^{\mathsf{T}}$ is the unknown ideal constant weight vector and $\varepsilon_3$ is the approximation error. It is known that the HJBE (15.15) can also be written as

$$V^*(x(t+\Delta t)) - V^*(x(t)) + \int_t^{t+\Delta t} \mathcal{R}(x(\tau), u^*(\tau))\mathrm{d}\tau = 0. \tag{15.65}$$

The substitution of (15.64) into the HJBE (15.65) yields

$$\int_t^{t+\Delta t} \mathcal{R}(x(\tau), u^*(\tau))\mathrm{d}\tau + w^{*\mathsf{T}}\Delta\Upsilon(t) = \varepsilon_4, \tag{15.66}$$

where $\varepsilon_4 \triangleq \varepsilon_3(x(t)) - \varepsilon_3(x(t+\Delta t))$.

Define the weight estimation error $\tilde{w}_1 \triangleq \hat{w} - w^*$. By using the control policy $u^{(i)}$ in one iterative step of the PI algorithm, it follows from (15.62) that

$$\begin{aligned}\dot{\tilde{w}}_1 &= -\alpha_1\left(\int_t^{t+\Delta t} \mathcal{R}(x(\tau), \hat{u}^{(i)}(\tau))\mathrm{d}\tau + (\hat{w}_1 + w^*)^{\mathsf{T}}\Delta\Upsilon(t)\right)\Delta\Upsilon(t)\\ &= -\alpha_1(\hat{w}_1^{\mathsf{T}}\Delta\Upsilon(t) + \varepsilon_2^{(i)})\Delta\Upsilon(t),\end{aligned} \tag{15.67}$$

where $\varepsilon_2^{(i)} \triangleq \int_t^{t+\Delta t} \mathcal{R}(x(\tau), \hat{u}^{(i)}(\tau))\mathrm{d}\tau + w^{*\mathsf{T}}\Delta\Upsilon(t)$. It is noted form Lemma 3 that $u^{(i)} \to u^*$ when $i \to \infty$; this implies that $\varepsilon_2^{(i)} \to \varepsilon_4$, *i.e.*, $\hat{w} \to w^*$.

To show the convergence of $\tilde{w}_1$, the following assumption is required.

**Assumption 3.** The NN approximation error $\varepsilon_3$ of $V^*$ and its derivative $\nabla\varepsilon_3$ are bounded on a compact set containing $\Omega$, *i.e.*, $\|\varepsilon_3\| \leqslant \varepsilon_{3M}$ and $\|\nabla\varepsilon_3\| \leqslant \varepsilon_{3DM}$, where $\varepsilon_{3M}$ and $\varepsilon_{3DM}$ are positive numbers.

**Theorem 3.** *Consider the PDE system* (15.1)–(15.3), *use* (15.62) *as the tuning law for critic NN weight vector, let the control policy be given by* (15.61), *select an initial critic NN weight vector* $w^{(0)}$ *such that* $\hat{u}^{(0)} \in \mathfrak{U}(\Omega)$ *and let Assumptions 2 and 3 hold. Then there exist positive real constants* $\sigma_1$, $\sigma_2$ *and* $\varepsilon_1^*$, *such that if* $\|x_0\| \leqslant \sigma_1$, $\|x_{f0}\| \leqslant \sigma_2$ *and* $\varepsilon \in (0, \varepsilon_1^*)$, *the estimation error* $\tilde{w}_1$ *of* (15.67) *and the solution* $x$ *of the closed-loop reduced-order ODE system* (15.14) *are SGUUB. Moreover, the state* $y(z,t)$ *of the original closed-loop PDE system* (15.1)–(15.3) *is SGUUB in* $L_2$ *norm,* i.e., *there exists a positive number* $\sigma_3$ *such that* $\|y(\cdot,t)\| < \sigma_3$.

*Proof.* Under the control policy (15.61) and the critic NN weight vector tuning law (15.62), the closed-loop PDE system is represented as

$$\begin{cases}\dfrac{\partial y}{\partial t} = \mathcal{L}\left(y, \dfrac{\partial y}{\partial z}, \dfrac{\partial^2 y}{\partial z^2}, \ldots, \dfrac{\partial^{n_0} y}{\partial z^{n_0}}\right) - \dfrac{1}{2}\bar{B}(z)R^{-1}B^T\nabla\Upsilon^T w^{(i)},\\ \dot{\hat{w}} = -\alpha_1\left(\displaystyle\int_t^{t+\Delta t} \mathcal{R}(x(\tau), \hat{u}^{(i)}(\tau))d\tau + \hat{w}^T\Delta\Upsilon\right)\Delta\Upsilon.\end{cases} \tag{15.68}$$

Using the singular perturbation technique in Subsection 15.3 for model reduction, the system (15.68) is equivalently rewritten as

$$\begin{cases} \dot{x} = f_s(x,0) + -\frac{1}{2}BR^{-1}B^T\nabla\Upsilon^T w^{(i)} + f_s(x,x_f) - f_s(x,0), \\ \varepsilon\dot{x}_f = A_{f\varepsilon}x_f + \varepsilon f_f(x,x_f) - \frac{1}{2}\varepsilon B_f R^{-1}B^T\nabla\Upsilon^T w^{(i)}, \\ \dot{\hat{w}} = -\alpha_1\left(\int_t^{t+\Delta t}\mathcal{R}(x(\tau),\hat{u}^{(i)}(\tau))d\tau + \hat{w}^T\Delta\Upsilon\right)\Delta\Upsilon. \end{cases} \tag{15.69}$$

Similar to the proof of Theorem 1, the conditions (15.29)–(15.32) and (15.35) also hold and are required in this proof.

Let us consider the following Lyapunov function candidate:

$$L(t) = V^*(x) + V_f(x_f) + L_1(t), \tag{15.70}$$

where $L_1(t) \triangleq \frac{1}{2\alpha_1}\tilde{w}_1^T\tilde{w}_1$ and $V^*$ is the solution of the HJBE (15.15). Differentiating $L(t)$ with respect to the closed-loop system (15.69) yields

$$\dot{L}(t) = \dot{V}^*(x) + \dot{V}_f(x_f) + \dot{L}_1(t). \tag{15.71}$$

It is required to compute each part of (15.71). According to the HJBE (15.15), we get

$$(\nabla V^*)^T f = -\|x\|^2_{Q_s} + \frac{1}{4}(\nabla V^*)^T BR^{-1}B^T\nabla V^* = -\|x\|^2_{Q_s} + \|u^*\|^2_R. \tag{15.72}$$

Taking the derivatives of $V^*(x)$, $V_f(x_f)$ and $L_1(t)$ along the trajectories of (15.69) yields

$$\begin{aligned} \dot{V}^*(x) &= -\|x\|^2_{Q_s} + \|u^*\|^2_R + (u^*)^T B^T\nabla\Upsilon^T w^{(i)} + (\nabla V^*)^T[f_s(x,x_f) - f_s(x,0)] \\ &\leqslant -\upsilon_{Qm}\|x\|^2 + \upsilon_{RM}\upsilon^2_{uM} + \upsilon_{uM}\upsilon_{BM}\upsilon_{D\gamma M}\upsilon_{wM} + k_3\upsilon_{DVM}\|x_f\|_{\mathbb{R}^\infty} \\ &= m_1\|x\|^2 + m_2\|x_f\|_{\mathbb{R}^\infty} + m_3, \end{aligned} \tag{15.73}$$

where $\upsilon_{Qm} \triangleq \underline{\sigma}(Q_s)$, $\upsilon_{RM} \triangleq \bar{\sigma}(R)$, $m_1 \triangleq \upsilon_{Qm}$, $m_2 \triangleq k_3\upsilon_{DVM}$ and $m_3 \triangleq \upsilon_{uM}\upsilon_{BM}\upsilon_{D\gamma M}\upsilon_{wM}$. We have

$$\begin{aligned} \dot{V}_f(x_f) &\leqslant -\frac{l_3}{\varepsilon}\|x_f\|^2_{\mathbb{R}^\infty} + l_4\|x_f\|_{\mathbb{R}^\infty}(k_1\|x\| + k_2\|x_f\|_{\mathbb{R}^\infty}) + \frac{1}{2}l_4l_5\upsilon^{-1}_{Rm}\upsilon_{BM}\upsilon_{wM}\|x_f\|_{\mathbb{R}^\infty} \\ &= (-\frac{l_3}{\varepsilon} + m_4)\|x_f\|^2_{\mathbb{R}^\infty} + m_5\|x\|\|x_f\|_{\mathbb{R}^\infty} + m_6\|x_f\|_{\mathbb{R}^\infty}, \end{aligned} \tag{15.74}$$

where $m_4 \triangleq l_4k_2$, $m_5 \triangleq l_4k_1$ and $m_6 \triangleq \frac{1}{2}l_4l_5\upsilon^{-1}_{Rm}\upsilon_{BM}\upsilon_{wM}$. Also,

$$\dot{L}_1(t) = -\tilde{w}_1^T(\tilde{w}_1^T\Delta\Upsilon + \varepsilon_2^{(i)})\Delta\Upsilon \leqslant -m_6\|\tilde{w}_1\|^2 + m_7, \tag{15.75}$$

where $m_6 \triangleq \underline{\sigma}_{x\in\Omega}\{\Delta\Upsilon(x)\Delta\Upsilon^T(x)\}$ and $m_7 \triangleq \bar{\sigma}_{x\in\Omega}\{\varepsilon_{2M}\Delta\Upsilon(x)\}$.

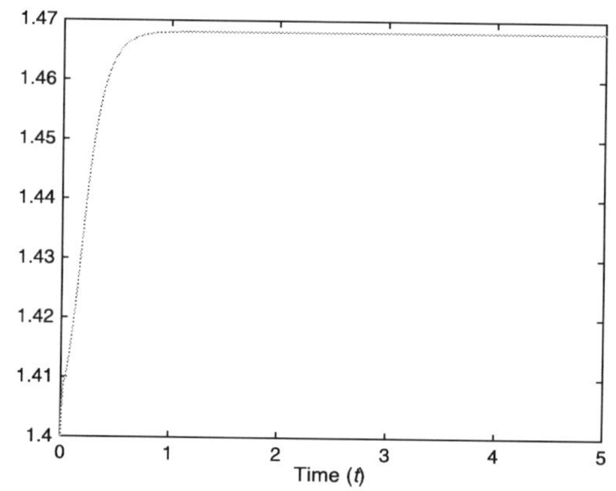


**FIGURE 15.7**

The trajectory of the critic NN weight norm $\|\hat{w}(t)\|$ with the PI-based adaptive optimal control.

Define $v_{xw} \triangleq [\|x\| \;\; \|x_f\|_{\mathbb{R}^\infty} \;\; \|\tilde{w}_1\|]^T$. Then it follows from (15.71), (15.73)–(15.75) that

$$\dot{L}(t) \leqslant -v_{xw}^T \Pi(\varepsilon) v_{xw} + b_m^T v_{xw} + m_9, \tag{15.76}$$

where

$$\Pi(\varepsilon) \triangleq \begin{bmatrix} m_1 & -m_5/2 & 0 \\ -m_5/2 & l_3/\varepsilon - m_4 & 0 \\ 0 & 0 & m_6 \end{bmatrix},$$

$b_m \triangleq [0 \;\; m_8 \;\; 0]^T$, $m_8 \triangleq m_2 + m_6$ and $m_9 \triangleq m_3 + m_7$. Let $\varepsilon^* \triangleq (4m_1 l_3)/(m_5^2 + 4m_1 m_4)$. Then we have, if $\varepsilon \in (0, \varepsilon^*)$, $\Pi(\varepsilon) > 0$. Letting $m_{10} \triangleq \underline{\sigma}(\Pi)$, it follows from (15.76) that $\dot{L}(t) \leqslant -m_{10}\|v_{xw}\|^2 + b_m^T\|v_{xw}\| + m_9$. Thus, $\dot{L}(t) \leqslant 0$ if $\|v_{xw}\| \geqslant m_8/2m_{10} + \sqrt{m_8^2 + 4m_{10}m_9/2m_{10}}$. This means that $x, x_f$ and $\tilde{w}_1$ are SGUUB. Considering $\|y(\cdot,t)\|_2^2 = \|x(t)\|^2 + \|x_f(t)\|_{\mathbb{R}^\infty}^2$, this means that the state $y(z,t)$ of the original closed-loop PDE system (15.1)–(15.3) is SGUUB in $L_2$ norm. □

### 15.5.5 SIMULATION STUDIES

To test the efficiency of the PI-based adaptive optimal control method in this section, the diffusion–convection–reaction process (15.44)–(15.46) in Section 15.4.3 is employed for simulation studies. The critic NN activation function vector $\Upsilon(x)$ is selected the same as in (15.47). Let the initial NN weight vector be given by $\hat{w}_1(0) = 1.4$ and $\hat{w}_i(0) = 0$ for $i = 2, ..., 18$. Choose $\alpha_1 = 5$, and the adaptive optimal control policy (15.61) with the NN weights tuning law (15.62) has been employed for closed-loop simulations. Figs. 15.7–15.10 demonstrate the norm of the critic NN weight vector $\|\hat{w}(t)\|$, the state profile $y(z,t)$ of the closed-loop PDE system, the state trajectories $x(t)$ of the closed-loop ODE system and the control $u(t)$, respectively.

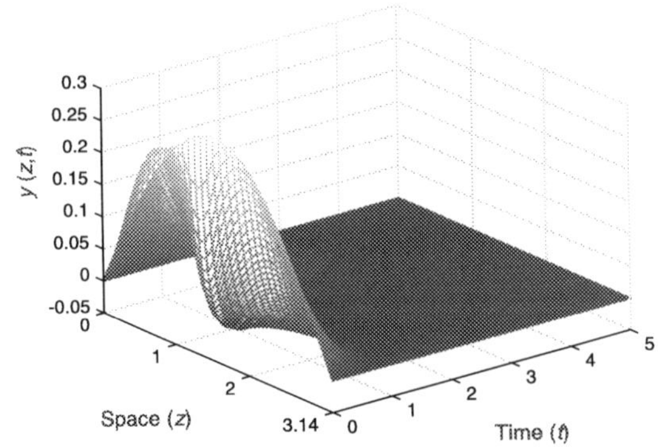


**FIGURE 15.8**

The state profile $y(z,t)$ of the closed-loop PDE system with the PI-based adaptive optimal control.

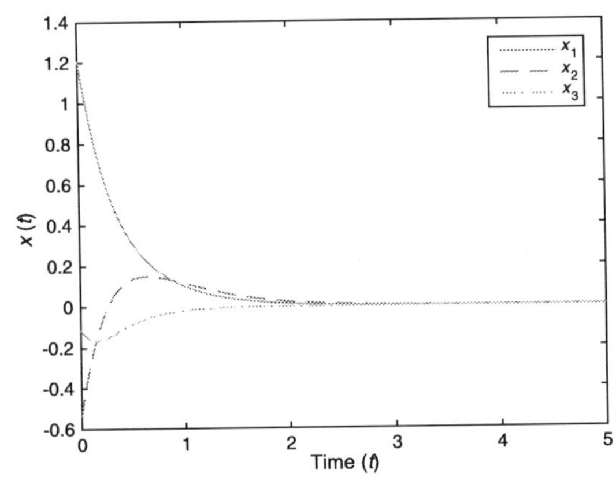


**FIGURE 15.9**

The trajectories $x(t)$ of the closed-loop ODE system with the PI-based adaptive optimal control.

## 15.6 CONCLUSIONS

Two ADP-based adaptive optimal control methods, *i.e.*, NDP and PI, have been developed for the nonlinear DPSs. First, KLD and the singular perturbation technique have been employed for model reduction and a finite-dimensional ODE system is derived. Based on the ODE system, the optimal control problem has been converted to solve the HJBE. To solve the HJBE, the NDP-based adaptive optimal control method has been developed by using the critic NN to estimate the value function of the HJBE. Furthermore, the optimal control problem of partially unknown DPS has been investigated and the PI-based adaptive optimal control method has been developed. The convergence of both control methods has been analyzed. Through the simulation studies, the results demonstrate the efficiency of the two developed ADP-based adaptive optimal control methods.

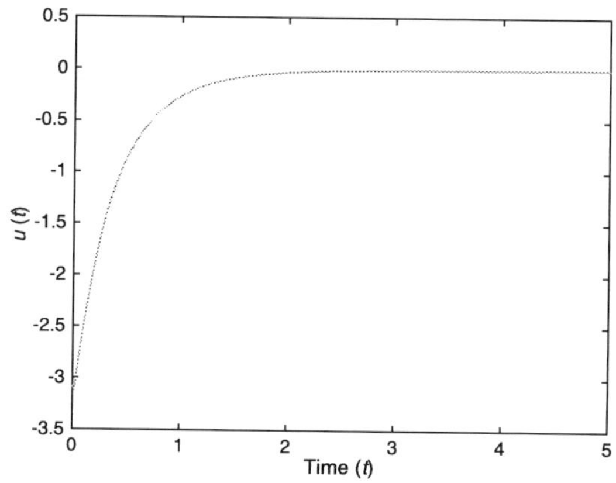


**FIGURE 15.10**

The trajectory of PI-based adaptive optimal control $u(t)$.

## ACKNOWLEDGMENTS

T This work was supported in part by the National Natural Science Foundation of China under Grants 61503377, 61533017, 61625302, 61473011 and U1501251.

## REFERENCES

[1] M. Krstic, A. Smyshlyaev, Boundary Control of PDEs: A Course on Backstepping Designs, vol. 16, SIAM, 2008.
[2] W.H. Ray, Advanced Process Control, McGraw-Hill, New York, 1981.
[3] P.D. Christofides, Nonlinear and Robust Control of PDE Systems: Methods and Applications to Transport-Reaction Processes, Birkhäuser, Boston, MA, 2001.
[4] H.X. Li, C. Qi, Spatio-Temporal Modeling of Nonlinear Distributed Parameter Systems, Springer, 2011.
[5] B. Luo, H.N. Wu, Approximate optimal control design for nonlinear one-dimensional parabolic PDE systems using empirical eigenfunctions and neural network, IEEE Transactions on Systems, Man, and Cybernetics, Part B: Cybernetics 42 (6) (2012) 1538–1549.
[6] B. Luo, H.N. Wu, Online policy iteration algorithm for optimal control of linear hyperbolic PDE systems, Journal of Process Control 22 (7) (2012) 1161–1170.
[7] H.N. Wu, B. Luo, $L_2$ disturbance attenuation for highly dissipative nonlinear spatially distributed processes via HJI approach, Journal of Process Control 24 (5) (2014) 550–567.
[8] W. He, S.S. Ge, Vibration control of a flexible beam with output constraint, IEEE Transactions on Industrial Electronics 62 (8) (2015) 5023–5030.
[9] W. He, S.S. Ge, Cooperative control of a nonuniform gantry crane with constrained tension, Automatica 66 (4) (2016) 146–154.
[10] B. Luo, H.N. Wu, H.X. Li, Data-based suboptimal neuro-control design with reinforcement learning for dissipative spatially distributed processes, Industrial & Engineering Chemistry Research 53 (29) (2014) 8106–8119.
[11] B. Luo, H.N. Wu, H.X. Li, Adaptive optimal control of highly dissipative nonlinear spatially distributed processes with neuro-dynamic programming, IEEE Transactions on Neural Networks and Learning Systems 26 (4) (2015) 684–696.
[12] B. Luo, T. Huang, H.N. Wu, X. Yang, Data-driven $H_\infty$ control for nonlinear distributed parameter systems, IEEE Transactions on Neural Networks and Learning Systems 26 (11) (2015) 2949–2961.
[13] H.N. Wu, B. Luo, Heuristic dynamic programming algorithm for optimal control design of linear continuous-time hyperbolic PDE systems, Industrial & Engineering Chemistry Research 51 (27) (2012) 9310–9319.

[14] A. Armaou, P.D. Christofides, Crystal temperature control in the Czochralski crystal growth process, AIChE Journal 47 (1) (2001) 79–106.

[15] V. Yadav, R. Padhi, S. Balakrishnan, Robust/optimal temperature profile control of a high-speed aerospace vehicle using neural networks, IEEE Transactions on Neural Networks 18 (4) (2007) 1115–1128.

[16] A. Kalani, P.D. Christofides, Simulation, estimation and control of size distribution in aerosol processes with simultaneous reaction, nucleation, condensation and coagulation, Computers & Chemical Engineering 26 (7) (2002) 1153–1169.

[17] Y. Ou, C. Xu, E. Schuster, T.C. Luce, J.R. Ferron, M.L. Walker, et al., Optimal tracking control of current profile in Tokamaks, IEEE Transactions on Control Systems Technology 19 (2) (2011) 432–441.

[18] A.M. Bayen, R.L. Raffard, C.J. Tomlin, Adjoint-based control of a new Eulerian network model of air traffic flow, IEEE Transactions on Control Systems Technology 14 (5) (2006) 804–818.

[19] J.W. Wang, H.N. Wu, H.X. Li, Distributed proportional–spatial derivative control of nonlinear parabolic systems via fuzzy PDE modeling approach, IEEE Transactions on Systems, Man, and Cybernetics, Part B: Cybernetics 42 (3) (2012) 927–938.

[20] M.J. Balas, The Galerkin method and feedback control of linear distributed parameter systems, Journal of Mathematical Analysis and Applications 91 (2) (1983) 527–546.

[21] R.F. Curtain, H. Zwart, An Introduction to Infinite-Dimensional Linear Systems Theory, vol. 21, Springer, 1995.

[22] E. Bendersky, P.D. Christofides, Optimization of transport-reaction processes using nonlinear model reduction, Chemical Engineering Science 55 (19) (2000) 4349–4366.

[23] J.-W. Wang, H.-N. Wu, Exponential pointwise stabilization of semi-linear parabolic distributed parameter systems via the Takagi–Sugeno fuzzy PDE model, IEEE Transactions on Fuzzy Systems 26 (1) (2018) 155–173.

[24] W. He, S. Zhang, Control design for nonlinear flexible wings of a robotic aircraft, IEEE Transactions on Control Systems Technology 25 (1) (2017) 351–357.

[25] W. He, C. Sun, S.S. Ge, Top tension control of a flexible marine riser by using integral-barrier Lyapunov function, IEEE/ASME Transactions on Mechatronics 20 (2) (2015) 497–505.

[26] D.V. Prokhorov, Optimal neurocontrollers for discretized distributed parameter systems, in: Proceedings of the 2003 American Control Conference, vol. 1, IEEE, 2003, pp. 549–554.

[27] C. Xu, Y. Ou, E. Schuster, Sequential linear quadratic control of bilinear parabolic PDEs based on POD model reduction, Automatica 47 (2) (2011) 418–426.

[28] R.S. Sutton, A.G. Barto, Reinforcement Learning: An Introduction, The MIT Press, Cambridge, 1998.

[29] D.P. Bertsekas, J.N. Tsitsiklis, Neuro-Dynamic Programming, Athena Scientific, Belmont, Mass., 1996.

[30] Q. Wei, F. Wang, D. Liu, X. Yang, Finite-approximation-error-based discrete-time iterative adaptive dynamic programming, IEEE Transactions on Cybernetics 44 (12) (2014) 2820–2833.

[31] Q. Wei, D. Liu, A novel iterative $\theta$-adaptive dynamic programming for discrete-time nonlinear systems, IEEE Transactions on Automation Science and Engineering 11 (4) (2014) 1176–1190.

[32] D. Liu, H. Li, D. Wang, Online synchronous approximate optimal learning algorithm for multi-player non-zero-sum games with unknown dynamics, IEEE Transactions on Systems, Man, and Cybernetics: Systems 44 (8) (2014) 1015–1027.

[33] D. Wang, D. Liu, H. Li, Policy iteration algorithm for online design of robust control for a class of continuous-time nonlinear systems, IEEE Transactions on Automation Science and Engineering 11 (2) (2014) 627–632.

[34] Q. Wei, D. Liu, Numerical adaptive learning control scheme for discrete-time non-linear systems, IET Control Theory & Applications 7 (11) (2013) 1472–1486.

[35] D. Liu, H. Liang, D. Wang, Neural-network-based zero-sum game for discrete-time nonlinear systems via iterative adaptive dynamic programming algorithm, Neurocomputing 110 (2013) 92–100.

[36] H. Li, D. Liu, Optimal control for discrete-time affine non-linear systems using general value iteration, IET Control Theory & Applications 6 (18) (2012) 2725–2736.

[37] B. Luo, H.N. Wu, T. Huang, D. Liu, Data-based approximate policy iteration for affine nonlinear continuous-time optimal control design, Automatica 50 (12) (2014) 3281–3290.

[38] B. Luo, H.N. Wu, T. Huang, Off-policy reinforcement learning for $H_\infty$ control design, IEEE Transactions on Cybernetics 45 (1) (2015) 65–76.

[39] B. Luo, H.N. Wu, T. Huang, D. Liu, Reinforcement learning solution for HJB equation arising in constrained optimal control problem, Neural Networks 71 (2015) 150–158.

[40] B. Luo, D. Liu, T. Huang, D. Wang, Model-free optimal tracking control via critic-only Q-learning, IEEE Transactions on Neural Networks and Learning Systems 27 (10) (2016) 2134–2144.

[41] B. Luo, D. Liu, H.N. Wu, D. Wang, F.L. Lewis, Policy gradient adaptive dynamic programming for data-based optimal control, IEEE Transactions on Cybernetics 47 (10) (2017) 3341–3354.

[42] H.N. Wu, B. Luo, Neural network based online simultaneous policy update algorithm for solving the HJI equation in nonlinear $H_\infty$ control, IEEE Transactions on Neural Networks and Learning Systems 23 (12) (2012) 1884–1895.
[43] H.N. Wu, B. Luo, Simultaneous policy update algorithms for learning the solution of linear continuous-time $H_\infty$ state feedback control, Information Sciences 222 (2013) 472–485.
[44] B. Xu, C. Yang, Z. Shi, Reinforcement learning output feedback NN control using deterministic learning technique, IEEE Transactions on Neural Networks and Learning Systems 25 (3) (2014) 635–641.
[45] D. Zhao, Y. Zhu, MEC-a near-optimal online reinforcement learning algorithm for continuous deterministic systems, IEEE Transactions on Neural Networks and Learning Systems 26 (2) (2015) 346–356.
[46] Y. Zhu, D. Zhao, D. Liu, Convergence analysis and application of fuzzy-HDP for nonlinear discrete-time HJB systems, Neurocomputing 149, Part A (2015) 124–131.
[47] F.L. Lewis, D. Liu, Reinforcement Learning and Approximate Dynamic Programming for Feedback Control, vol. 17, John Wiley & Sons, Hoboken, NJ, 2013.
[48] B. Luo, D. Liu, H.N. Wu, Adaptive constrained optimal control design for data-driven nonlinear discrete-time systems with critic-only structure, IEEE Transactions on Neural Networks and Learning Systems (2018), https://doi.org/10.1109/TNNLS.2017.2751018, in press.
[49] B. Luo, D. Liu, T. Huang, X. Yang, H. Ma, Multi-step heuristic dynamic programming for optimal control of nonlinear discrete-time systems, Information Sciences 411 (2017) 66–83.
[50] F.L. Lewis, D. Vrabie, V.L. Syrmos, Optimal Control, John Wiley & Sons, Hoboken, NJ, 2013.
[51] D.P. Bertsekas, Dynamic Programming and Optimal Control, vol. 1, Athena Scientific, Nashua, 2005.
[52] J.J. Murray, C.J. Cox, G.G. Lendaris, R. Saeks, Adaptive dynamic programming, IEEE Transactions on Systems, Man, and Cybernetics, Part C: Applications and Reviews 32 (2) (2002) 140–153.
[53] D. Vrabie, O. Pastravanu, M. Abu-Khalaf, F.L. Lewis, Adaptive optimal control for continuous-time linear systems based on policy iteration, Automatica 45 (2) (2009) 477–484.
[54] H. Modares, F.L. Lewis, Linear quadratic tracking control of partially-unknown continuous-time systems using reinforcement learning, IEEE Transactions on Automatic Control 59 (11) (2014) 3051–3056.
[55] P.D. Christofides, Robust control of parabolic PDE systems, Chemical Engineering Science 53 (16) (1998) 2949–2965.
[56] A. Armaou, P.D. Christofides, Finite-dimensional control of nonlinear parabolic PDE systems with time-dependent spatial domains using empirical eigenfunctions, Applied Mathematics and Computer Science 11 (2) (2001) 287–318.
[57] R. Courant, D. Hilbert, Methods of Mathematical Physics, vol. 1, Wiley, 2004.
[58] S. Ge, C.C. Hang, T.H. Lee, T. Zhang, Stable Adaptive Neural Network Control, Kluwer Academic Publishers, Norwell, USA, 2001.
[59] H.K. Khalil, J. Grizzle, Nonlinear Systems, vol. 3, Prentice Hall, Upper Saddle River, 2002.
[60] A. Armaou, P.D. Christofides, Dynamic optimization of dissipative PDE systems using nonlinear order reduction, Chemical Engineering Science 57 (24) (2002) 5083–5114.
[61] A. Varshney, S. Pitchaiah, A. Armaou, Feedback control of dissipative PDE systems using adaptive model reduction, AIChE Journal 55 (4) (2009) 906–918.
[62] K.G. Vamvoudakis, F.L. Lewis, Online actor–critic algorithm to solve the continuous-time infinite horizon optimal control problem, Automatica 46 (5) (2010) 878–888.

# Index

## D



## E

Printed in the United States
By Bookmasters